PARTICLES AND FIELDS IN THE MAGNETOSPHERE

ASTROPHYSICS AND SPACE SCIENCE LIBRARY

A SERIES OF BOOKS ON THE RECENT DEVELOPMENTS

OF SPACE SCIENCE AND OF GENERAL GEOPHYSICS AND ASTROPHYSICS

PUBLISHED IN CONNECTION WITH THE JOURNAL

SPACE SCIENCE REVIEWS

VOLUME 17

PARTICLES AND FIELDS IN THE MAGNETOSPHERE

PROCEEDINGS OF A SYMPOSIUM ORGANIZED BY THE SUMMER ADVANCED STUDY INSTITUTE, HELD AT THE UNIVERSITY OF CALIFORNIA, SANTA BARBARA, CALIF., AUGUST 4–15, 1969

Edited by

B. M. McCORMAC

Lockheed Palo Alto Research Laboratory,

Palo Alto, Calif., U.S.A.

D. REIDEL PUBLISHING COMPANY

DORDRECHT-HOLLAND

The symposium was sponsored by the Army Research Office;
Defense Atomic Support Agency; Lockheed Palo Alto Research Laboratory;
Office of Naval Research; University of California, Berkeley, Space Science Laboratory;
University of California, San Diego, Department of Applied Electrophysics

Library of Congress Catalog Card Number 78-115884

ISBN-13: 978-94-010-3286-5 e-ISBN-13: 978-94-010-3284-1
DOI: 10.1007/978-94-010-3284-1

Softcover reprint of the hardcover 1st edition 1970

PREFACE

This book contains the lectures presented at the Advanced Study Institute, 'Earth's Particles and Fields, 1969', which was held at the University of California, Santa Barbara, during the period August 4 through 15, 1969. One hundred seventy persons from twelve different countries attended the Institute.

The authors and the publisher have made a special effort for rapid publication of an up-to-date status of the particles and fields in the earth's magnetosphere, which is an ever changing research area. Special thanks are due to the lecturers for their diligent preparation and excellent presentations. The individual lectures and the published papers were deliberately limited; the author's cooperation in conforming to these specifications is greatly appreciated. The contents of the book are organized by subject area rather than in the order in which papers were presented during the Institute. Many thanks are due to Drs. Kinsey Anderson, Sam Bame, Leverett Davis, Gilbert Mead, Harry Elliot, Kenneth Behannon, Reimar Lüst, A. W. Schardt, Carl-Gunne Fälthammar, and Martin Walt who served as session chairmen during the Institute and contributed greatly to its success by skillfully directing the discussion period in a stimulating manner after each lecture. Dr. Martin Walt and the Summary Panel worked hard to prepare an excellent summary of various aspects of particles and fields in the magnetosphere at the end of the Institute.

Many persons contributed to the success of the Institute. Drs. Kinsey Anderson, Harry Elliot, Carl-Gunne Fälthammar, Reimar Lüst, A. W. Schardt, James A. Van Allen, and Martin Walt were especially helpful in preparing the technical program. Mrs. Lillian McClellan and Miss Virginia James, from the Department of Conferences, and Mr. Jeri Mock, Resident Halls, University of California, Santa Barbara, were most cooperative in providing very excellent living and conference facilities. Mr. Götz Paschmann helped with many of the arrangements. The assistant editor, Mrs. Diana McCormac, checked the manuscripts and proofs and worked hard to achieve a uniform style in this book.

Direct financial support was provided the Institute by: the Army Research Office, the Defense Atomic Support Agency, Lockheed Palo Alto Research Laboratory, and the Office of Naval Research. The University of California, Berkeley, Space Science Laboratory, and the University of California, San Diego, Department of Applied Electrophysics also sponsored the Institute.

BILLY M. MCCORMAC

Palo Alto,
March 1970

TABLE OF CONTENTS

PREFACE V

PART I: MAGNETOSPHERIC MODELS

K. A. ANDERSON / Entry of Solar Cosmic Rays Into the Earth's Magnetosphere 3
A. J. DESSLER / Formation and Geometry of Geomagnetic Tail 18
EDWARD W. HONES, JR. / Magnetotail Plasma and Magnetospheric Substorms 24
S.-I. AKASOFU / A Model Current System for the Magnetospheric Substorm 34
W. I. AXFORD / On the Origin of Radiation Belt and Auroral Primary Ions 46
VYTENIS M. VASYLIUNAS / Mathematical Models of Magnetospheric Convection and Its Coupling to the Ionosphere 60

PART II: THE BOW SHOCK

S. J. BAME / Solar Wind Stimulation of the Magnetosphere 75
A. J. HUNDHAUSEN / Shock Waves in the Solar Wind 79
K. W. OGILVIE and L. F. BURLAGA / Hydromagnetic Observations in the Solar Wind 82
MICHAEL D. MONTGOMERY / Plasma Measurements Near the Earth's Bow Shock: Vela 4 95
F. L. SCARF, R. W. FREDRICKS, and C. F. KENNEL / AC Electric and Magnetic Fields and Collisionless Shock Structures 102

PART III: MAGNETOSPHERIC PARTICLES

JAMES A. VAN ALLEN / Energetic Particle Phenomena in the Earth's Magnetospheric Tail 111
S. SINGER and S. J. BAME / Anisotropic Distributions of Energetic Electrons in the Earth's Magnetotail and Magnetosheath 122
P. L. ROTHWELL, V. H. WEBB, and L. KATZ / Trapped and Polar Particles During the June 9, 1968 Magnetic Storm 132
G. A. PAULIKAS, J. B. BLAKE and A. L. VAMPOLA / Solar Particle Observations Over the Polar Caps 141
C. R. CHAPPELL, K. K. HARRIS, and G. W. SHARP / The Reaction of the Plasmapause to Varying Magnetic Activity 148

PART IV: MAGNETIC AND ELECTRIC FIELDS IN THE MAGNETOSPHERE

KENNETH W. BEHANNON / Magnetic Fields in the Earth's Tail 157
M. SUGIURA, T. L. SKILLMAN, B. G. LEDLEY, and J. P. HEPPNER / Magnetic Field Observations in High β Regions of the Magnetosphere 165
PAUL J. COLEMAN, JR. and ROBERT L. McPHERRON / Fluctuations in the Distant Geomagnetic Field During Substorms: ATS 1 171
CHRISTOPHER T. RUSSELL and ROBERT E. HOLZER / AC Magnetic Fields 195
G. HAERENDEL and R. LÜST / Electric Fields in the Ionosphere and Magnetosphere 213
EUGENE M. WESCOTT, JOHN D. STOLARIK, and JAMES P. HEPPNER / Auroral and Polar Cap Electric Fields from Barium Releases 229
DONALD A. GURNETT / Satellite Measurements of DC Electric Fields in the Ionosphere 239
N. C. MAYNARD and J. P. HEPPNER / Variations in Electric Fields from Polar Orbiting Satellites 247

PART V: WAVE-PARTICLE INTERACTIONS

C. F. KENNEL, R. W. FREDRICKS, and F. L. SCARF / High Frequency Electrostatic Waves in the Magnetosphere 257
JOHN M. CORNWALL / Mutually Interacting Instabilities in the Magnetosphere 266
F. L. SCARF, C. F. KENNEL, R. W. FREDRICKS, I. M. GREEN, and G. M. CROOK / AC Fields and Wave Particle Interactions 275
A. HASEGAWA / Hydromagnetic Waves and Instabilities in the Magnetosphere 284
ROBERT A. HELLIWELL / Intensity of Discrete VLF Emissions 292

PART VI: RADIATION BELT OBSERVATIONS

JAMES I. VETTE / Summary of Particle Populations in the Magnetosphere 305
L. A. FRANK / Further Comments Concerning Low Energy Charged Particle Distributions Within the Earth's Magnetosphere and Its Environs 319
J. R. WINCKLER / The Origin and Distribution of Energetic Electrons in the Van Allen Radiation Belts 332
H. ELLIOT and R. J. HYNDS / Recent Measurements of Inner Belt Protons 353
S. M. KRIMIGIS / Alpha Particles Trapped in the Earth's Magnetic Field 364
J. B. BLAKE and G. A. PAULIKAS / Measurements of Trapped α-Particles: $2 \leqslant L \leqslant 4.5$ 380

PART VII: ACCELERATION AND MOTION OF PARTICLES

CARL-GUNNE FÄLTHAMMAR / Introductory Survey of Radiation Belt Diffusion 387
D. J. WILLIAMS / Trapped Protons $\geqslant$100 keV and Possible Sources 396
MARTIN WALT / Radial Diffusion of Trapped Particles 410
GERHARD HAERENDEL / On the Balance Between Radial and Pitch Angle Diffusion 416

PART VIII: SUMMARY

MARTIN WALT / Panel Report 431
BILLY M. McCORMAC / Conclusions 442

APPENDIX I: GLOSSARY 450

INDEX OF SUBJECTS 451

Participants

1. L. L. Newkirk
2. Jan Trulsen
3. George J. Lewak
4. Egil Leer
5. Michael P. Gough
6. Bruce Tsurtani
7. Ching-I. Meng
8. Frank R. Harris
9. Richard Chase
10. Robert Spiger
11. Willard P. Olson
12. Godehard A. Guenther
13. Horacio A. Cazeneuve
14. J. B. Reagan
15. William L. Imhof
16. K. W. Ogilvie
17. Akira Hasegawa
18. Christopher T. Russell
19. John Olson
20. William P. Jones
21. Bertram M. Shuman
22. A. L. Vampola
23. James Waggoner
24. J. R. Burrows
25. George L. Mason
26. E. Scott Claflin
27. William G. Stanley
28. Paul Perrault
29. John Boyd
30. Alan Johnstone
31. D. E. Page
32. B. A. Whalen
33. D. H. Fairfield
34. Joseph M. Grebowsky
35. J. D. Winningham
36. Pompiliu Verzariu
37. Richard P. Vondrak
38. Warner L. Ecklund
39. Jose Lourenco
40. Jean-Francois Crifo
41. Michael Montgomery
42. Sidney Singer
43. James H. Coon
44. Sidney L. Ossakow
45. G. Davidson
46. Giovanni Perona
47. W. Ian Axford
48. Donald L. Carpenter
49. Abel J.-J. Chen
50. Leverett Davis, Jr.
51. Lawrence D. Kavanaugh, Jr.
52. Edward C. Stone
53. Christine M. Davies
54. Harald Trefall
55. Eugene M. Wescott
56. Ernest G. Holeman
57. Alois W. Schardt
58. Kenneth W. Behannon
59. John D. Craven
60. James I. Vette
61. S. J. Bame
62. George A. Paulikas
63. Stanley C. Freden
64. Carlos S. Warren
65. James E. McCoy
66. Thomas A. Parnell
67. James G. Moore
68. Syun-I. Akasofu
69. Franklin D. Barish
70. Chuan Sheng Liu
71. Emile Aerts
72. Heinz F. Pellkofer
73. Eberhard Achtermann
74. Donald L. Evans
75. Walter E. Smith
76. Gustav J. Freyer
77. M. Sugiura
78. Joseph G. Kelley
79. Paul Rothwell
80. Ludwig Katz
81. Anne B. Kahle
82. M. B. Baker
83. Paul R. Satterblom
84. Richard H. Ball
85. Theodore A. Fritz
86. Klaus Wilhelm
87. Kristian E. Schlegel
88. Robert G. Willis
89. Sidney L. Russak
90. Harry Elliot
91. R. J. Hynds
92. Dennis S. Peacock
93. G. Stevens
94. Aharon Eviatar
95. Patrick Cassen
96. Fred F. Tomblin
97. Leonard L. DeVries
98. Henri Reme
99. John B. Cladis
100. Kinsey A. Anderson
101. Robert A. Helliwell
102. Alex Dessler
103. Thomas A. Farley
104. Raymond G. D'Arcy
105. Harry I. West, Jr.
106. Gilbert D. Mead
107. Richard G. Johnson
108. Frederic Sage
109. Renee Prange
110. Richard D. Sharp
111. E. G. Shelley
112. Charles R. Chappell
113. Götz Paschmann
114. G. H. Nakano
115. David T. Young
116. Bernd Häusler
117. Donald D. Wallis
118. Carl E. Nielsen
119. Chung G. Park
120. George A. Kuck
121. Jacqueline Etcheto
122. Martha Mangan
123. Daniel Hale
124. Gerhard Haerendel
125. Martin Walt
126. Billy M. McCormac
127. Diana McCormac
128. E. W. Hones, Jr.
129. Carl-Gunne Fälthammar
130. Theodore Northrop
131. Charles F. Kennel
132. Lothar Rossberg
133. Vytenis M. Vasyliunas

PART I

MAGNETOSPHERIC MODELS

ENTRY OF SOLAR COSMIC RAYS INTO THE EARTH'S MAGNETOSPHERE

K. A. ANDERSON

Physics Dept. and Space Sciences Laboratory, University of California, Berkeley, Calif., U.S.A.

If the solar wind flowed past the magnetosphere smoothly with no instabilities and no tension force the pressure of the wind would close off the geomagnetic field at some distance behind the earth. The angle of closing would be the Mach angle which is equal to the reciprocal of the Mach number. Since $M \sim 6$ the length of the tail would be ~ 6 times the radius of the magnetosphere or about 100 R_E. A well developed neutral sheet would not be found in this magnetosphere since the normal component is not very much different from the horizontal component (see Figure 1a from Johnson, 1960). But the magnetosphere is observed to have a well developed neutral sheet extending behind earth at least 80 R_E. It is thus clear that the solar wind interacts

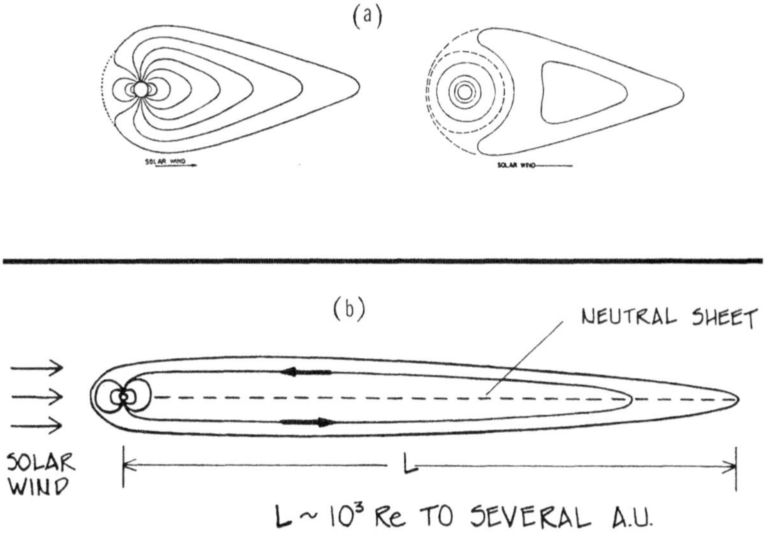

Fig. 1. Two versions of a closed magnetosphere. There is now general theoretical and experimental agreement that the earth's magnetosphere is like neither of these. (a) No viscous drag. (b) With viscous drag and very slow field to line merging.

strongly with the magnetosphere. If a viscous interaction were to draw the tail out to distances of many thousand R_E or even several astronomical units (1 AU $\cong 1.5 \times 10^8$ km) as shown in Figure 1b, there would be a neutral sheet. However, for a tail of great length to develop, the rate at which field lines merge across the neutral sheet would have to be exceedingly slow.

The detailed nature of the viscous drag that might arise as the solar wind passes over the magnetosphere is not known. One possibility is that the magnetopause is unstable to the flow of solar wind past it. This instability results in long length hydromagnetic waves being generated and the solar wind pushing on the wave crests can exert the required tension to extend the tail. Disturbances of amplitude up to 2 R_E are observed on the magnetopause in a direction transverse to the field lines. However, it appears that the amplitude of the waves varies greatly from day to day, yet the tail is evidently always very long although little is known of how the tail length varies. Another mechanism that provides for an extended magnetotail is the connection of interplanetary field lines to geomagnetic field lines. Then, due to the frozen-in condition on magnetic field lines in highly conducting plasmas the magnetotail is drawn back.

Models of the latter type are called *open* magnetospheres. Figures 2 and 3 illustrate

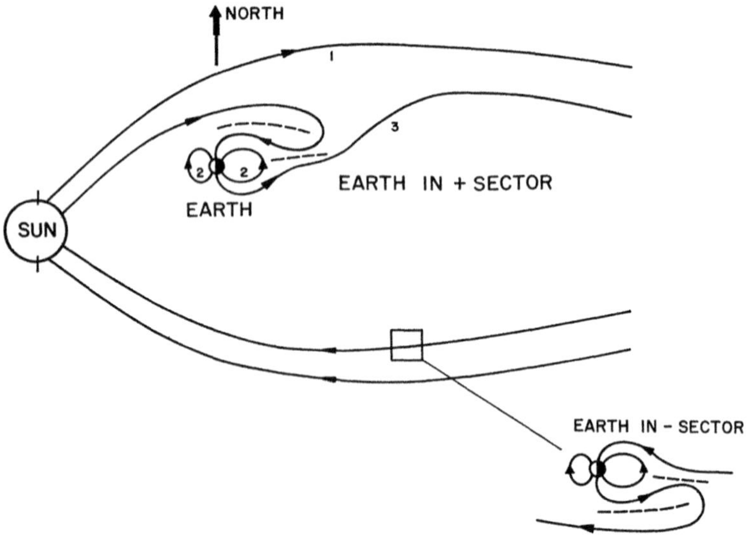

Fig. 2. The field line topology of the open magnetosphere in the interplanetary field. One polar cap but not the other is connected to interplanetary field lines near earth. Two neutral surfaces result, the neutral sheet in the center of the magnetotail, and the magnetopause.

the essential field line topology. Those figures must be understood as only a snapshot since the loops are dragged by the solar wind. Open models of the magnetosphere with field line merging are quite attractive because they afford a means of releasing energy from the solar wind inside the magnetosphere. The energy appears as the magnetic field lines annihilate in the neutral sheet. Also, the inward motion of the field lines provides a way of moving plasma on a large scale through the magnetosphere.

Dungey (1958) suggested that interplanetary magnetic field lines might connect to the geomagnetic field. This field line topology, in a form convenient for the discussion to follow, is shown in Figure 2. This figure shows three kinds of magnetic field lines in the solar system:

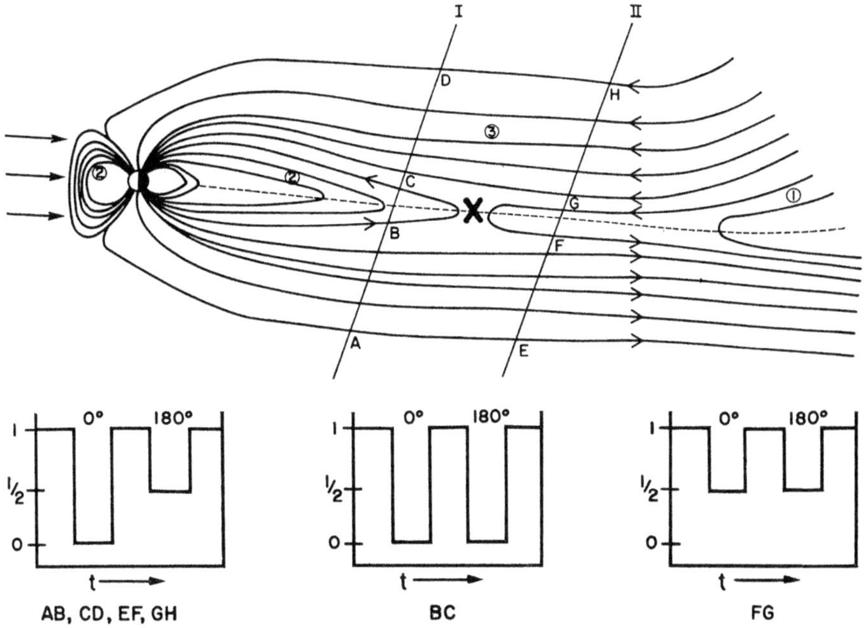

Fig. 3. This figure shows how the magnetotail might be built up mostly of type 3 field lines. Other regions implied by the open magnetosphere model are shown also. The moon, or another large absorber, will shadow energetic solar particles in differing ways depending on the type of field line. Thus, study of the shadow patterns around the moon when it is in the tail allows the topology of the field lines to be found. The shadow pattern for region FG is constructed for the special case of a pitch angle distribution symmetrical about 90° and fairly strong scattering across field lines in interplanetary space.

(1) Ordinary interplanetary field lines extending from the sun to several astronomical units perhaps tens of AU;

(2) Field lines that close in the neighborhood of the earth and that originate from currents inside the earth; and

(3) Field lines resulting from a connection of the first two types.

Due to the frozen-in field condition in interplanetary space the loop in the type 3 field line must move with the solar wind. When this loop contacts the portion of the field line emerging from the southern hemisphere it reconnects, forming a type 1 and a type 2 line. Type 3 lines are then reformed near the front of the earth in the manner suggested by Dungey. Disordering of the field takes place across the bow shock and the presence of waves (Petschek and Thorne, 1967) will facilitate this connection in the absence of a southward component in the interplanetary field. Two surfaces of magnetic field reversal in Figure 2 are indicated by dashed lines. Physically these can be identified as the magnetopause and the neutral sheet in the magnetotail.

The question of open vs. closed magnetosphere is of fundamental importance. Since no detailed theory of either model can be given we must turn to experiments to decide the issue. To date the best experimental approach involves use of high speed 'test charges' supplied by the sun to trace out field lines. This general idea was suggested by Michel and Dessler (1965). They attempted to compare how the arrival of particles in the polar cap would occur under open and closed magnetospheric models. They used PCA records that were available at the time. It is fair to say that no conclusive test could result due to uncertainties in the observations. Direct particle measurements have now become available and are superior to PCA observations for the question at hand.

At the present time the best evidence that the magnetosphere consists mostly of connected field lines comes from solar flare electrons of about 50 keV kinetic energy. These particles have a cyclotron radius of only about 100 km and move with a speed of about 10^{10} cm/sec. They appear with little delay in the geomagnetic tail and this led Lin and Anderson (1966) to the conclusion that most of the field lines in the tail were connected. They estimated the length of the tail, defined as the distance from the earth to the place where most of the electrons have entered the tail, to be less than 5000 R_E.

The topology and large scale motion of the geomagnetic field are of such basic significance to space physics that additional experimental information about this problem should be sought. In principle this can be done by a network of simultaneous, precise vector magnetic field measurements over the magnetosphere and in the surrounding interplanetary space. In practice it is unlikely this can be carried out due to the vast number of simultaneous measurements required. The second possibility is to study the angular distribution of energetic particles on the field lines. From their symmetry characteristics and the loss cones, it is possible in principle to find the topology of the field lines. The practical difficulty here is that the precision to which the very small loss cones must be measured is on the order of 10^{-3} radians, far beyond present and contemplated experimental and spacecraft capabilities. In addition, there is the possibility that pitch angle diffusion would quickly obscure the loss cones. The third possible method is to make use of large absorbers on the field lines whose topology is to be investigated. By large absorber is meant: (a) An object which is thick enough to absorb the particles in question. Since the most interesting particles have low energy this condition is easily met. (b) The size of the absorber must be appreciably larger than the gyroradius of the particles. (c) An object whose dimension normal to the field lines is large compared to $V\tau$ where τ is the bounce time of the particles and V the flow velocity of the field lines relative to the object. Estimates for the geomagnetic field place V at about 10 km/sec and the moon-earth travel time for these particles is about 10 sec. The moon has a diameter of 3476 km and thus fulfills the condition of a large absorber which is demonstrated directly by the observed sharpness of the shadows (Figure 5).

The reversal of the interplanetary field direction from one sector to another (Wilcox and Ness, 1965) leads to a N–S asymmetry (Anderson and Lin, 1969) in the topology of Figure 2. When the interplanetary field is directed in the sense outward from the sun the northern half of the tail will be partly or completely connected directly to the sun. The field lines in the southern half of the tail go far beyond the earth. When the earth is in a negative sector (sense of field toward the sun) the situation is reversed.

Both situations are shown in Figure 2. If the solar source of particles were localized, solar particles might appear in only one-half of the magnetotail. Evans and Stone (1969) have found a large N–S asymmetry in 1 to 40 MeV protons. After persisting for many hours the asymmetry disappeared after a sector boundary was crossed. While it persisted the asymmetry was consistent with the direction of the interplanetary field in terms of the topology shown in Figure 2.

While a N–S asymmetry is clearly shown by this experiment, on many occasions both halves of the magnetotail have approximately equal particle intensity. It is also not clear why the intensities become equal after the reversal of the interplanetary field direction and remained that way. On the other hand, the Evans and Stone result does show that diffusive entry into the magnetosphere requires $\gtrsim$20 hr at least at certain times. Access to one hemisphere and not the other seems totally incomprehensible in terms of diffusive entry. The geomagnetic activity in the two hemispheres is closely the same at all times, so that this could not account for a large N–S difference. In any case there is no theory connecting geomagnetic activity indices and diffusion coefficients.

It should be noted here that it is not possible to test the asymmetric connection of the magnetosphere to the interplanetary field by correlating differences between the geomagnetic indices K_N and K_S to the direction of the interplanetary field. It is

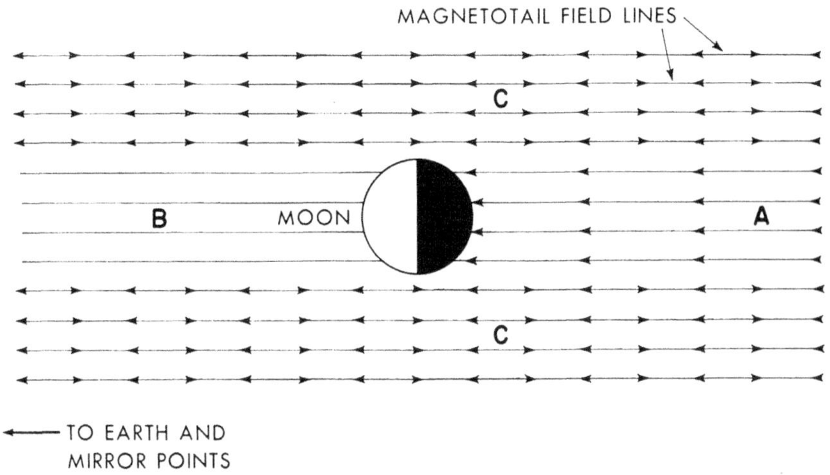

Fig. 4. Motion of solar electrons in the magnetotail assuming small diffusion coefficients. The source of particles is to the right and the earth is to the left. The arrows indicate the motion of solar electrons on type 3 field lines. Region A contains only the incoming solar particle distribution. Region C also contains this distribution which then mirrors near the earth so that a symmetrical bidirectional distribution (except for a small loss cone) appears on these field lines. In absence of drifts and diffusion region B contains no solar particles.

easy to imagine how, in a closed magnetosphere, the viscous drag on a particular hemisphere could depend on the direction of the interplanetary field. To be more specific, if tangential drag results due to wave disturbances on an unstable magnetopause, it is entirely possible that the wave amplitude (and hence the hemisphere disturbance index) could depend on the direction of the field just outside the magnetopause.

The use of electrons to examine the question of the open magnetosphere has been pursued further by making use of the shadows formed by absorption of these particles when they strike the moon. Figure 4 shows the nature of these shadows when the source of particles lies further down the tail than the moon and if there is no diffusion of particles across lines of force in the magnetotail after they leave the source. It is also assumed that the particles leaving the source mirror near the earth and return. Then if the omnidirectional particle intensity on these lines is J, it will be zero between the moon and earth and one-half J between the moon and the source.

This pattern of zero intensity in front of the moon and half intensity behind it

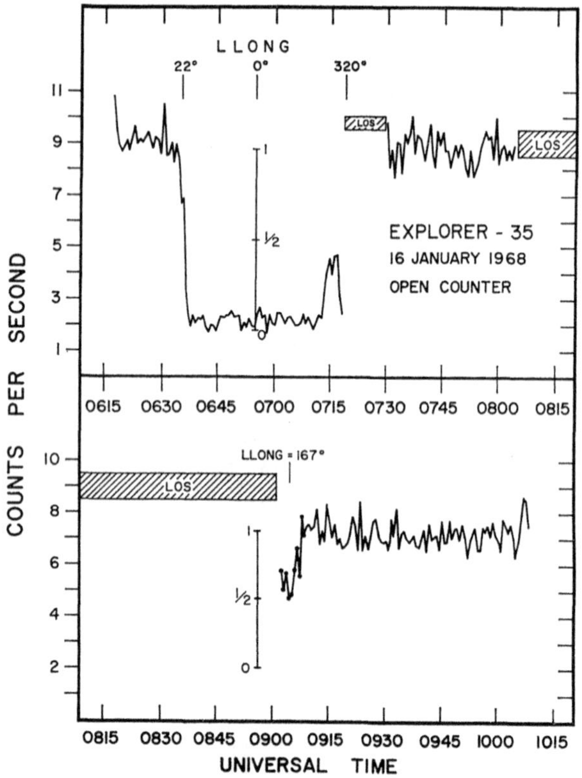

Fig. 5. Full shadows are almost always observed in front of the moon when it is in the magnetotail. At times it has been possible to measure the shadow behind the moon as in this example; they have been found to be half shadows.

has been observed by Anderson and Lin (1969). They found twelve such examples from various places within the tail. One of these cases is shown here as Figure 5. This shows that the source of solar electrons for the magnetotail lay behind the moon ($\gtrsim 60\ R_E$ geocentric distance) on these occasions. This important result can be restated as follows: solar electrons of energy 20 keV to several hundred keV do not enter the magnetosphere by penetration of the magnetopause on the front side of the earth, nor do they enter through the flanks of the magnetotail back to geocentric distances of 60 R_E. Thus diffusion through these boundaries and across magnetotail lines of force does not occur in the range $-60\ R_E < X_{SE} < +10\ R_E$. As far as the author knows this is the only rigorously proven statement that can be made about particle diffusion in the magnetotail at the present time. It is certainly the case, as will be shown later that no conclusive experimental demonstration that diffusion occurs in the magnetotail has ever been performed.

We next turn to the question of the source in Figure 3. Ultimately the source of the particles is the sun. However, the direct source feeding the tail conceivably might be a diffusion region which brings particles through the magnetopause and into the interior of the magnetotail. Before we can conclude that the direct source is indeed the sun and therefore that the tail is open we must eliminate the possibility of diffusion. The presence of the half-shadow is crucial in eliminating diffusion as a physically plausible mechanism in this case.

The existence of half shadows behind the moon in a closed magnetotail places inordinately severe requirements on the diffusion. The argument briefly is that since the unidirectional intensity of solar electrons in the particle shadow behind the moon is the same as the unidirectional particle intensity on interplanetary field lines, the diffusion would have to bring the particles deep into the magnetotail ($\gtrsim 10\ R_E$ from the magnetopause and $\gtrsim 10\ R_E$ from the neutral sheet) at speeds comparable to their full speed of $\sim 10^{10}$ cm/sec. This means that the diffusion must be so fast that there is free access to the inside of the magnetosphere; thus the particle must move inward about 10 R_E in 1 sec. The diffusion coefficient is $D \approx L^2/T$ where $L = 10\ R_E$ and $T \approx 1$ sec. Also $D = (\Delta L)^2/\Delta t$ where ΔL = displacement per scattering and Δt = time per scattering $= 1/\Omega = mc/eB$. From equation of motion ($F = ma$) we have $e\delta E = m(\Delta L/\Delta t^2)$. Solving for δE and substituting $B = 20\gamma$, we obtain $\delta E \approx 70$ V/m for the amplitude of the ac electric field. These intense fields would have to extend over regions $\sim 10\ R_E$ in at least one dimension.

The dc electric fields can be eliminated easily as a direct source feeding solar particles into a closed magnetosphere from the outside. The potential drop across the tail would have to be $\sim 10^8$ V ($E \sim 10$ V/m). Currently, potentials of 10^4 V are considered plausible.

There seems to be no remaining physical possibilities for transporting particles deep into the magnetotail and it has been concluded that the sun is the direct source and hence most of the field lines in the magnetotail are open.

There is further evidence from the study of lunar shadowing effects on solar particles that the earth's magnetosphere is open. From Figure 3 it is seen that in an open magnetosphere a tongue of interplanetary field lines surround the neutral sheet behind the line of reconnection (marked $\times$ in that Figure). Such field lines do not pass through the earth and hence there would not necessarily be a full shadow in front of the moon. The details of the shadow pattern will now depend on the angular distribution and the parallel and perpendicular scattering coefficients in the interplanetary medium. If, for example, $k_\perp$ is sufficiently large and the distribution of pitch angles is bidirectionally symmetric there would be half shadows in front of and behind the moon.

Half shadows in front of the moon near the neutral sheet have been observed on one crossing of the plasma sheet out of a total of seventeen crossings (Anderson and Lin, 1969). The details of this observation should be discussed in more detail. The spacecraft entered the magnetotail late on November 14. The two shadow patterns that occurred on November 15, 1967, are characteristic of the type 3 field lines; the same was true on November 16. Then on November 17 the shadow patterns were completely different. Half shadows appear in front of the moon. The shadow ratio is close to 2 for both these shadows. From the analysis above we are dealing with fluxes having isotropic pitch angle distribution on interplanetary lines of force. Van Allen and Ness (1969) have reported that the pitch angle distribution is indeed fairly uniform for this solar event. After the appearance of the interplanetary field lines near the center of the tail the neutral sheet is crossed. Referring to Figure 3 it can be seen that is exactly what is expected from the open model of the magnetosphere. Both shadows on November 18 are deep and there is a half shadow behind the moon, the pattern characteristic of the type 3 lines. The return of the full shadows is exactly what is expected from the open model. The appearance of interplanetary field in the center of the magnetotail implies that the reconnection region, marked by an $\times$ in Figure 3, lay at a distance less than 60 R_E from the earth on this occasion. We note that the days November 17 to 20 have QQ geomagnetic index.

Figure 6 plots the shadow ratio as a function of time from November 14 to 20. In the magnetosheath region (interplanetary, or type 1 topology), the ratio is closely 2 just as it was on November 17 near the center of the magnetotail. This result must be interpreted to mean that interplanetary field lines were present near the neutral sheet at a geocentric distance of $X_{SE} \cong -60\ R_E$. The absence of half shadows on the other sixteen crossings evidently means that interplanetary field lines are not usually present at this distance. The interpretation is that the neutral line moves. It usually, but not always, lies beyond $X_{SE} \cong -60\ R_E$. When the half shadows were observed the magnetosphere was unusually free from disturbance.

Evidence using low rigidity solar electrons as field line tracers thus favors a mostly open magnetosphere with tail length $<5000\ R_E$. The questions of exactly how many field lines are connected, and their location in the magnetotail remain to be answered in detail. It may turn out to be the case that (a) The field lines in the plasma sheet all close across the neutral sheet; and (b) All the other field lines to the north and to the south of the plasma sheet enter the interplanetary medium. If this is the case, about 25% of the magnetotail lines close on themselves and the other 75% are 'open'. This estimate is based on measured properties of the plasma sheet at $\sim 20\ R_E$.

Another item for speculation is that the theta structure persists until the plasma sheet field lines have all closed, and then the remaining field lines break up into the strands containing thin neutral sheets apparently observed on the Pioneer spacecraft.

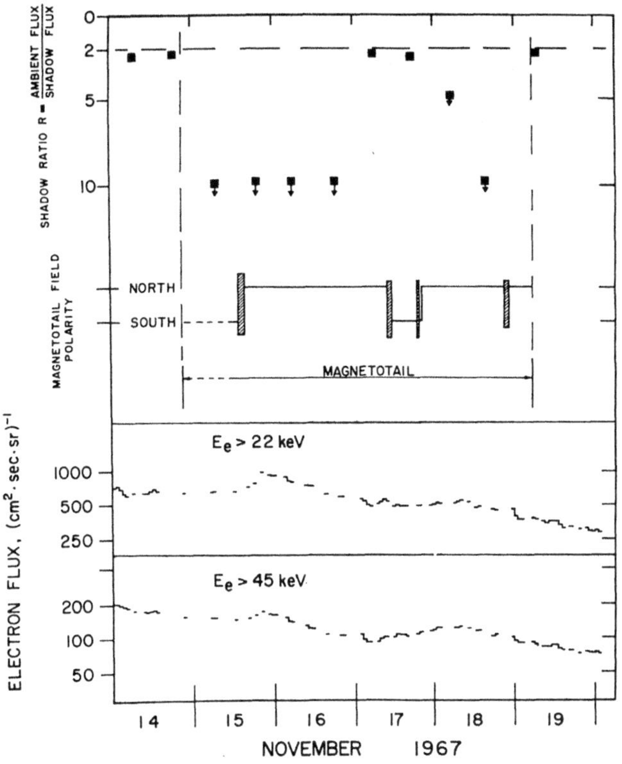

Fig. 6. The shadow ratio as a function of time during the tail crossing. The two anomalous half shadows occur near the neutral sheet and full shadows occur on either side of it. Note that the electron intensity does not change noticeably as the spacecraft crosses first the dusk and then the dawn magnetopause.

Thus the 'length' of the tail may not be a simple quantity but may depend on what method is used to measure it.

The experimental evidence for field line closure across the neutral sheet is somewhat indirect. There are three arguments in favor of such field lines:

(1) It would be difficult (although not impossible) to imagine how the hot plasma in the plasma sheet could be confined.

(2) Energetic electron fluxes deep in the magnetotail sometimes show characteristics similar to those of trapped fluxes closer to the earth, specifically in the skirt.

(3) The radial dependence of the magnetotail field magnitude is best explained according to Behannon (1969) if some closure across the neutral sheet is assumed.

Next, solar proton measurements bearing on the question of the magnetosphere's openness will be reviewed. This rather extensive set of observations has recently been reviewed by Montgomery and Singer (1969). Their summary is repeated here:

(1) The 1 to 10 MeV solar proton fluxes are often somewhat lower in the magnetotail than in interplanetary space. When differences appear they are typically 10 to 50%.

(2) Rapid changes in the proton intensity appear in the magnetotail but are delayed from 15 min up to 1 or 2 hr. Approximately 8 or 10 cases exist in which the delay can be readily observed. There are no known cases of 'immediate access' in which the delay into the tail can be shown to be seconds or even a few minutes.

(3) Over a wide energy range, the delays show only a small energy dependence or even none at all.

(4) The rapid variations that appear in the proton fluxes are often smoothed out somewhat when observed in the magnetotail. But more significant is the fact that rapid changes are largely preserved.

(5) Inside the magnetotail the particle angular distributions are nearly isotropic even when outside the tail they are quite anisotropic.

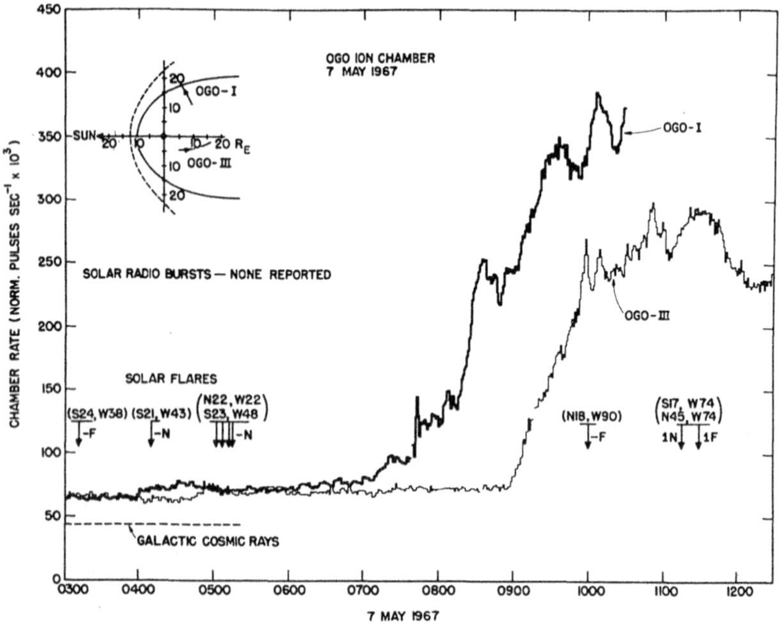

Fig. 7. An example of a large delay in proton arrival inside the magnetosphere (Kane *et al.*, 1968)

Figure 7 is an example from Kane *et al.* (1968) of a large delay in proton arrival in the magnetosphere. Proton measurements will probably prove to be the most sensitive method for gathering information on the length of the magnetotail. Solar electrons move much more rapidly, hence the dispersion times are much less and are harder to measure at present. Furthermore it is known that the electron fluxes do not show much filamentary structure (Lin and Anderson, 1967). Thus the only way to time electron events is to use flare events with abrupt onsets.

Various interpretations have been given the proton observations. Lanzerotti (1968) postulates very rapid diffusion across the magnetopause in order to account for the rapid appearance of solar protons $\sim 7\ R_E$ from the earth. However, there is nothing in the measurements to uniquely demonstrate the need for such a process and the fact that the energy spectra are closely the same inside and outside the tail argues against strong diffusion. Furthermore, Lin (1968) has conclusively shown that solar flare electrons of energy 20 to 200 keV do not enter the magnetosphere through the magnetopause within 60 R_E of the earth. It seems unacceptable to further postulate that the diffusion selects out the high rigidity protons and rapidly diffuses them without diffusing the electrons at all.

Williams and Bostrom (1969) also postulate diffusion into the long cylindrical magnetotail to account for the imperfect tracking of proton fluxes over the polar cap and in the magnetotail. However, they do not present direct physical evidence to show that diffusion is the only way to account for the observations. In fact, in this case it may be that small differences in detector counting rates inside and outside the magnetosphere are not due to real differences in proton intensity. The observed counting rate differences may be due to changes in the latitude angle of the interplanetary magnetic field. This could have varied the counting rate of the detector situated in interplanetary space. Such changes may be several tens of degrees and large anisotropies in the proton fluxes are frequently observed. Effects of this kind, however, would not change the flux observed in the magnetotail by very much. Montgomery and Singer (1969) conclude that the proton observations can be best accounted for by a combination of openness and diffusion.

Here we develop the idea that all the observed features of the simultaneous proton measurement can be understood in terms of an open magnetosphere. In this picture neither pitch angle nor perpendicular diffusion will play an important role in this first order description.

Kane (1969) has suggested that the proton observations can be accounted for simply in terms of an open magnetosphere. He points out that the filamentary structure frequently observed in the proton fluxes requires that a spatial variation seen in interplanetary space near the earth will be seen in the magnetotail near the earth after a delay equal to $\sim L/V$ where L is the distance to where connection of the magnetotail and interplanetary field takes place, and V is the speed of the solar wind (Figure 8a).

For the long decay of about 1 hr seen in the event shown in Figure 7, L would be on the order of 300 R_E. For shorter delay times of 15 min the length of the tail becomes less than 100 R_E. This seems to be shorter than any theoretical estimates for a tail length. Very short delays (<15 min) might be accounted for if the filament is very large (Figure 8b). Then the delay time would be

$$T_d \sim \frac{L(1 + \sin \alpha)}{V_p},$$

where V_p is the speed of the solar proton parallel to the field lines. Figure 8b represents the situation when a large bundle of magnetic field lines is uniformly filled with solar

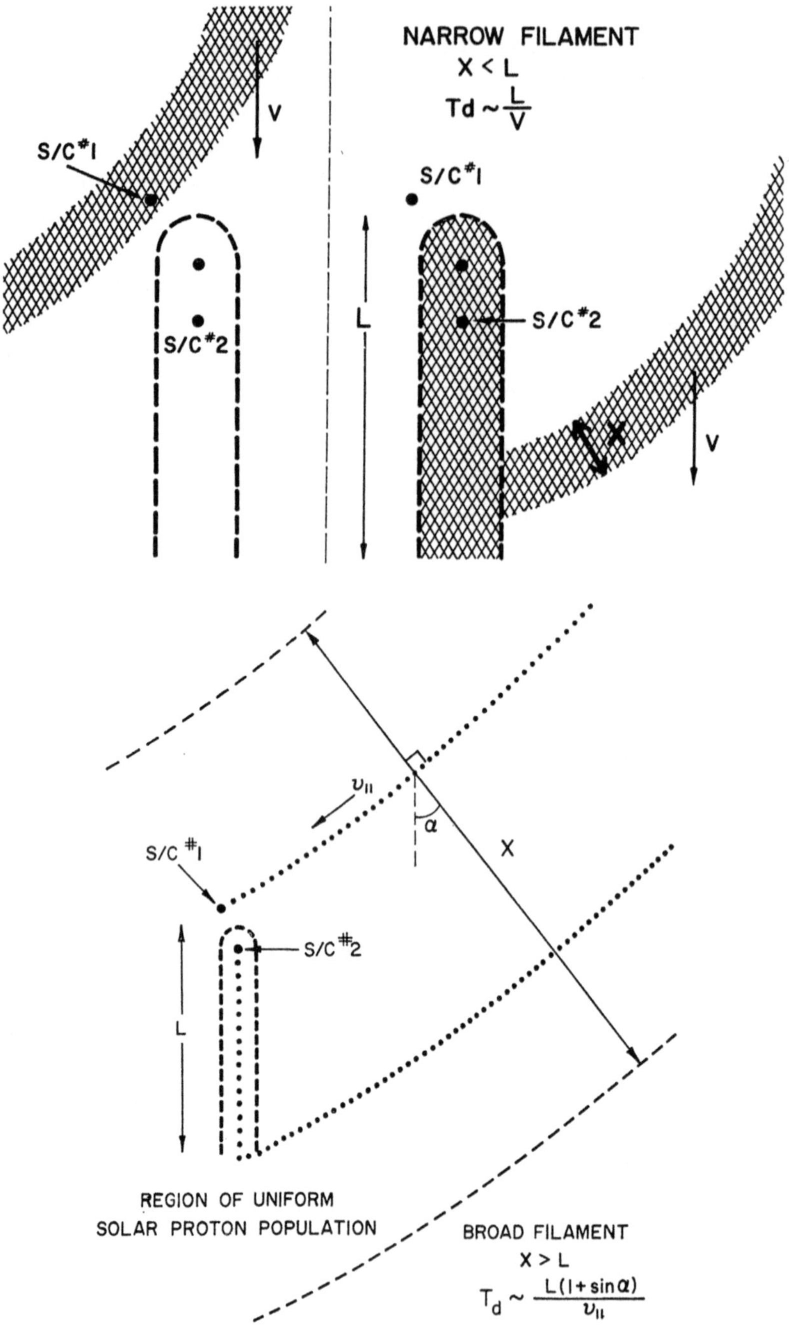

Fig. 8. Model to explain delayed arrival of protons in the magnetotail. It is based on filamentary character of proton fluxes in the interplanetary field and connection of these field lines to the earth's magnetic field.

particles. Such a situation might be expected to occur at the onset of a prompt flare event, particularly when the flare is not too close to 60°W solar longitude.

The filament explanation can rather naturally account for the great variability of the delay times. In the case of the small filament ($X<L$) several factors contribute to the variable delay:

(1) Both V and L may vary. How V does so is quite well known but we have little quantitative knowledge of L.

(2) The delay times may depend on where in the magnetotail the observation is made. Presumably the field line connection proceeds from the outside inward. This process need not be entirely systematic in that there may be a dawn-dusk asymmetry and variable connection rates as a function of distance along the magnetotail. These effects could account for the patchiness of the proton flux over the polar caps during early phases of proton events.

Little information is yet available for the relative arrival times of particles of different energy following solar proton intensity changes. Montgomery and Singer (1969) report that the differences are rather small by which they mean that the arrival

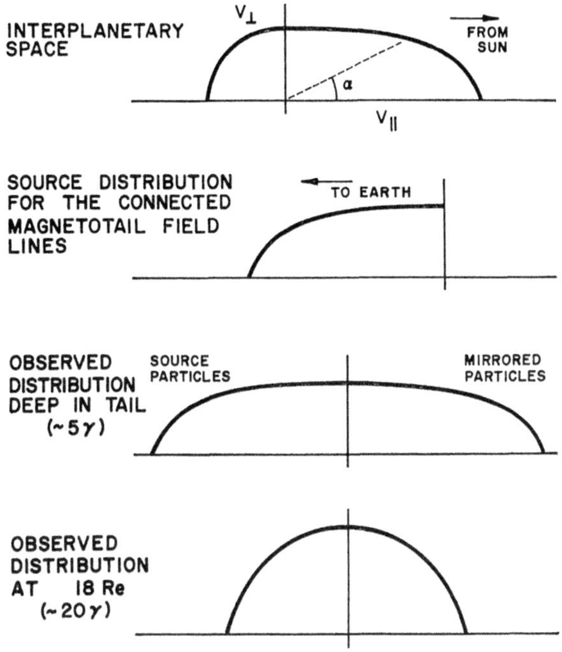

Fig. 9. Explanation of the approach to isotropy at 18 R_E.

times of protons ~1 and ~20 MeV differ by no more than a few minutes. The timing will need to be more precise, however, if tail lengths of a few hundred to $\sim 10^3$ R_E occur (the difference in arrival times for 1 and 20 MeV protons in a 300 R_E tail is only ~2 min).

Another observation of great importance to be explained is the apparent isotropy of the proton fluxes in the magnetotail at distances of $\sim 18\ R_E$ (the Vela orbit). This can be explained by noting that:

(1) In the tail, the incoming protons mirror near the earth; hence the pitch angle distribution in the tail becomes bidirectional and symmetrical (see Figure 9).

(2) The pitch angle distribution is altered at $18\ R_E$ since the magnetic field is about 20γ, four times as strong as in the interplanetary field on the average. This causes the distribution to open up and appear less anisotropic (Figure 9).

Thus, diffusion need not be postulated to obtain near-isotropic pitch angle distributions at $18\ R_E$ geocentric distance in the magnetotail when the distribution is anisotropic outside.

Lanzerotti (1968) has reported that the energy spectra of solar protons when measured simultaneously inside and outside the magnetosphere are closely the same. Van Allen (1969) has reported a similar result for solar electrons. These observations provide important and independent evidence that both species of particles have entered the magnetosphere on smoothly connected field lines. It seems highly unlikely that the energy of these particles could remain constant if it were hypothesized that they were diffused into a closed magnetosphere. The required diffusion rates and hence ac electric fields are so large that energy changes would almost surely result.

It has been assumed tacitly that if an interplanetary field line has connected to a magnetotail line, solar particles will follow this new configuration into the magnetosphere. The initial radius of curvature of the field line in the merging region of the magnetopause is of the order of the boundary thickness, presumably tens or hundreds of kilometers and the angle at which the field lines meet is unusually close to 45° (the normal garden hose angle). The solar electrons, with cyclotron radius of a few tens of kilometers, can therefore readily follow the field lines with nearly adiabatic motion. This would not be the case for protons. They may or may not be reflected depending on their pitch angle. Since the magnetopause is subject to large amplitude oscillations the radius of curvature of the segment may enlarge after a wave period (~ 5 min). Actually, particle reflection at the magnetopause provides a way of accounting for the intensity being reduced in the magnetosphere by 10 to 50% compared to outside.

References

Anderson, K. A. and Lin, R. P.: 1969, *J. Geophys. Res.* **74**, 3953.
Behannon, K. W.: 1969, *Trans. Amer. Geophys. Union* **50**, 295.
Dungey, J. W.: 1958, *Cosmic Electrodynamics*, Cambridge University Press, New York.
Evans, L. C. and Stone, E. C.: 1969, California Institute of Technology Preprint, **SRL-69-6**.
Johnson, F. S.: 1960, *J. Geophys. Res.* **65**, 3049.
Kane, S. R.: 1969, private communication.
Kane, S. R., Winckler, J. R., and Hofmann, D. J.: 1968, *Planetary Space Sci.* **16**, 1381.
Lanzerotti, A. J.: 1968, *Phys. Rev. Letters* **21**, 929.
Lin, R. P.: 1968, *J. Geophys. Res.* **73**, 3066.
Lin, R. P. and Anderson, K. A.: 1966, *J. Geophys. Res.* **71**, 4213.
Lin, R. P. and Anderson, K. A.: 1967, *Solar Phys.* **1**, 446.
McCracken, K. G. and Ness, N. F.: 1966, *J. Geophys. Res.* **71**, 3315.

Michel, F. C. and Dessler, A. J.: 1965, *J. Geophys. Res.* **70**, 4305.
Montgomery, M. D. and Singer, S.: 1969, *J. Geophys. Res.* **74**, 2869.
Ness, N. F., Scearce, C. S., and Cantarano, S. C.: 1967, *J. Geophys. Res.* **72**, 3769.
Petschek, H. E. and Thorne, R. M.: 1967, *Astrophys. J.* **147**, 1157.
Van Allen, J. A.: 1969, Univ. of Iowa Report **69-29**.
Van Allen, J. A. and Ness, N. F.: 1969, *J. Geophys. Res.* **74**, 71.
Wilcox, J. M. and Ness, N. F.: 1965, *Science* **148**, 1592.
Williams, D. J. and Bostrom, C. O.: 1969, *J. Geophys. Res.* **74**, 3019.

FORMATION AND GEOMETRY OF GEOMAGNETIC TAIL

A. J. DESSLER
Dept. of Space Science, Rice University, Houston, Tex., U.S.A.

1. Introduction

The idea of a geomagnetic field confined by solar plasma is an old one (e.g., Chapman and Ferraro, 1931; Dungey, 1955; Hoyle, 1956; Dessler, 1958; Parker, 1958; Piddington, 1959). By 1960, with the publication by Johnson (1960) of his teardrop model, the position of theorists was unified to the point that they all were in agreement that the outer geomagnetic field was seriously distorted by solar plasma and that beyond 6 to 10 R_E the geomagnetic field was confined asymmetrically by this plasma. However, as is often the case in space physics, what theorists believe has perhaps less influence on scientific opinion than even the more modest of them realize. A figure used in several recent reviews shows the 'magnetosphere of the 1950's' as a dipole field extending presumably to infinity while the 'magnetosphere of the 1960's' is the negligible merging model put forth by Dessler (1964).

It is commonly accepted that magnetic merging at the magnetopause and in the neutral sheet of the geomagnetic tail has an effect on magnetospheric phenomenon. Discussion now centers on how the magnetic merging process works, precisely how and to what extent it influences the magnetosphere, and what other important processes, such as field aligned currents (Birkeland currents), might be influencing magnetospheric phenomena.

2. Formation of the Tail

Although Parker (1958) and Piddington (1959) suggested an elongated or stretched out tail, the teardrop model of Johnson (1960) was the preferred model until the publication of the IMP 1 results by Ness (1965). For example, the perspicacious magnetospheric convection theory of Axford and Hines (1961) started with Johnson's magnetospheric model as the basic configuration.

The early magnetospheric models that employed magnetic merging derived a magnetospheric configuration that resembled Johnson's teardrop model in many respects (cf., Dungey, 1961; Axford *et al.*, 1965). In contrast, the magnetospheric configuration derived by considering merging as a negligible factor closely anticipated the experimental results (cf., Figure 1 of Dessler, 1964; Figure 14 of Ness, 1965). This perhaps fortuitous agreement should not be taken to indicate that merging does not take place, but simply that it is not sufficiently fast to grossly modify the magnetospheric model that one would expect in the absence of merging.

To account for the observed magnetospheric tail, we need to identify the force that

B. M. McCormac (ed.), Particles and Fields in the Magnetosphere.

balances the magnetic tension. Without this force the tail would close (and thus lead us back to the teardrop model). This force could be a tangential stress on the magnetosphere such as would be provided by either a viscous-like frictional interaction or the 'dragging' of field lines into the tail as would follow merging between interplanetary and geomagnetic field lines. These possibilities have been discussed by Axford (1969).

The other forces that could account for the tail are simple pressure stresses. It has been pointed out that the teardrop model of the magnetosphere is unstable owing to two possible sources of internal stress: (a) plasma pressure, and (b) hydromagnetic wave radiation pressure (Dessler, 1964). It was shown that hydromagnetic wave pressure of an occasional large amplitude wave could blow open the tail, and it was pointed out that "if solar plasma is brought into the magnetospheric tail, the conclusion that the tail is open will be enforced because the momentum of the flowing solar wind would be adequate by itself to force the tail to open".

Once the tail is blown open it will remain open because of the pressure of the solar wind blowing against the flanks of the magnetosphere and blowing into the neutral sheet. This point of view has been examined quantitatively by Wentworth (1967) who found a stable magnetospheric tail which has dimensions that agree well with recent experimental results. Beyond about 300 R_E, the tension must be balanced by plasma in the tail flowing away from the earth.

However, to explain the dynamical behavior of the magnetosphere, the flow of plasma within the magnetosphere, and the correlation of geomagnetic activity with a southward interplanetary field some magnetic merging at the magnetopause and in the neutral sheet is required. These phenomena and their relationship to magnetic merging have been reviewed recently by Axford (1969).

There is little agreement on the merging mechanism, although several quantitative models have been put forth. Several merging mechanisms of the type that involve an instability are cited in the reviews by Axford (1969) and Hultqvist (1969). In this approach, it is tacitly assumed that the merging mechanism in the tail is the same as the mechanism causing merging to occur at the magnetopause. However, this may not be the case. A 'universal' merging mechanism that works both in the tail and at the magnetopause (and in solar flares) may well also operate at the sector boundaries in the interplanetary magnetic field. Such a mechanism would act to either annihilate the interplanetary magnetic field or lead to the formation of magnetic bubbles in the interplanetary medium. Magnetic merging at sector boundaries has not been reported. (For a fuller discussion of this point see Dessler and Michel, 1966, and Dessler, 1967.)

The 'vacuum' merging mechanism described by Dessler (1968) is not subject to the above criticism. This mechanism proceeds in regions where there is not quite enough plasma pressure to hold apart oppositely directed magnetic fields. There should be a tendency to form a plasma void in the region immediately behind the earth (such as is formed behind the moon). Magnetic merging can take place in the neutral sheet region out to about 30 R_E where the solar wind is unable to enter directly. The merging process brings plasma into the neutral sheet, and it is lost out the sides by gradient drift motion. The rate of merging is determined by the balance between the rate at which plasma is brought into the near portion of the neutral sheet and the rate at which it is lost out the sides.

The region of the tail in which merging occurs has not yet been identified experimentally. The measurements are difficult because the neutral sheet is quite thin so the satellite passes through it quickly. Also, the magnetic fields in the neutral sheet are very weak; they are expected to be smaller than the acknowledged spacecraft fields.

The vacuum merging process can proceed at a time varying rate, and at a rate different from that that would occur at either the magnetopause or at sector boundaries. It is not necessary that the merging rate at the front of the magnetosphere and in the tail be equal (on the average). As merging occurs in the tail, and the tail field is pushed into the inner magnetosphere, the front of the magnetosphere will distend to the point that the magnetospheric field in front is too weak to hold back the solar wind. Magnetic field at the front of the magnetosphere weaker than about 50γ will be blown back to form a tail as suggested by Parker in 1958.

3. Tail Cross Section

The expected shape of the magnetospheric tail at large distances from the earth has not been investigated in any detail, although Dessler and Juday (1965) suggested that the tail field should take the form of two cylinders with a figure 8 cross sectional

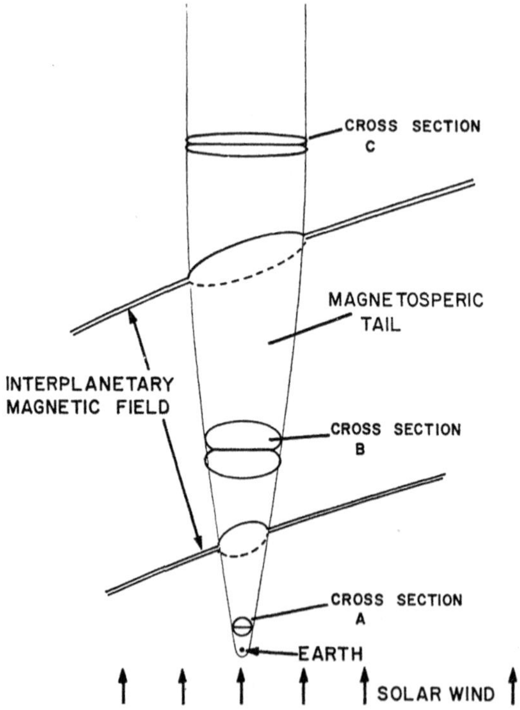

Fig. 1. Schematic sketch showing progressive flattening of geomagnetic tail with increasing geocentric distance.

pattern at distances of several hundred earth radii. The suggestion of the Figure 8 configuration was made without taking into account the anisotropy of the interplanetary magnetic field pressure.

Davis (1969) has pointed out that the interplanetary magnetic field exerts an anisotropic pressure on the magnetospheric tail so that the tail should tend to flatten in the ecliptic plane. This flattening should become effective for any tail length longer than a few hundred earth radii, whether or not magnetic merging is physically significant.

We should expect the interplanetary magnetic field to loop over the magnetospheric tail as shown in Figure 1. This is shown in more detail in Figure 2. The interplanetary field exerts a pressure on the top and bottom of the tail but not on the sides. The pressure arises from an isotropic magnetic pressure $\mathbf{B}^2/2\mu_0$ and a magnetic tension $\mathbf{B}^2/\mu_0$ that tends to straighten out the magnetic lines of force.

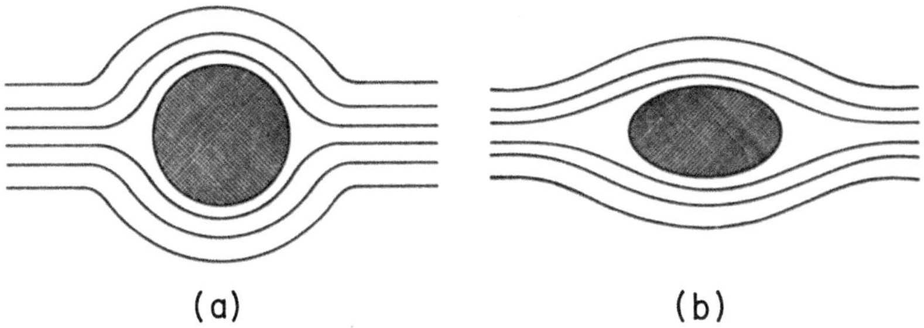

Fig. 2. Projection on a plane perpendicular to the geomagnetic tail axis showing the interplanetary magnetic field lines looping over the geomagnetic tail (shown cross hatched). Note that the pressure of the interplanetary magnetic field acts only on the top and bottom of the tail. The two cross sections (a) and (b) shown here correspond respectively to cross sections *A* and *B* in Figure 1.

Thus, near the earth the tail cross section should be nearly circular, or slightly elliptical with the major axis parallel to the earth's dipole axis. Beyond about 300 R_E where, according to Wentworth (1967), the total pressure inside the geomagnetic tail is in equilibrium with the pressure of the interplanetary medium, the cross section of the tail should become elliptical with the major axis of the ellipse perpendicular to the earth's dipole axis.

4. Tail Length

The choice between the short tailed teardrop model and the open tail has been settled by direct spacecraft measurements. The length of the tail is still a subject of uncertainty however. Direct measurement of the magnetospheric tail at large distances is difficult because the tail shape may be modified, as discussed above, and the tail position may change with time due to variations in the nonradial components of the solar wind flow.

It is my opinion that the entry of energetic solar flare particles into the magnetosphere will eventually prove the most sensitive indicator of tail length and of whether the tail field is connected directly to the interplanetary magnetic field. The first tests that were made following the suggestion of Michel and Dessler (1965) that tail length could be measured in this way proved disappointing to advocates of slow merging. Lin and Anderson (1966) and Krimigis *et al.* (1967) reported tail lengths of the order of 0.1 AU or less. However, recent measurements utilizing roughly the same techniques have indicated strikingly different results. Montgomery and Singer (1969), Williams and Bostrom (1969), and Evans and Stone (1969) interpret their spacecraft data as (usually) indicating a longer tail with energetic protons entering by some mode other than direct connection.

The early predictions of diffusion of energetic protons into an unmerged tail (Michel, 1965; Michel and Dessler, 1965), which were made before any of the relevant properties of the tail were known, requires some iteration in light of these recent findings. However, the results are consistent with negligible merging between interplanetary and geomagnetic field lines out to distances as great as 1 AU. Beyond some reasonably large distance, which appears to vary with time, the energetic protons gain rapid access to the distant tail either through the flattening of the tail discussed above, or through the tail becoming fluted and striated (Michel and Dessler, 1965), or, of course, through direct magnetic connection.

5. Auroral Bombardment

It had been expected by some that auroral particle acceleration would occur on neutral sheet and magnetopause field lines (cf., Dungey, 1961; Dessler and Juday, 1965). It is now rather certain that these ideas are incorrect. The magnetic field lines that come into the equatorward portion of the auroral zone on the dayside do not reach the dayside magnetopause (Mead, 1964), and the field lines from the equatorward portion of the nightside auroral zone do not go into the neutral sheet (cf., Schield *et al.*, 1969). It appears, rather, that the principal auroral bombardment occurs on dipole like field lines that carry Birkeland currents (field aligned currents) into the ionosphere.

Schield *et al.* (1969) have suggested the Birkeland currents originate at the inner edge of the plasma sheet (the Alfvén layer) where there is a reasonably abrupt termination of the inward flow of plasma from the geomagnetic tail. Since the drift paths of protons and electrons in crossed **E** and **B** fields in the vicinity of a magnetic dipole are not coincident, a current must flow to maintain charge neutrality. This current is the Birkeland current.

The Birkeland current produced by the above mechanism is calculated to cause magnetic effects that are in agreement with the satellite observations of Zmuda *et al.* (1966, 1967). Furthermore, the Alfvén layer is located on magnetic field lines that pass down into the nightside auroral zone. Additional study is needed to learn what connections exist between the Birkeland currents and auroral bombardment.

Acknowledgment

This work was supported in part by the U.S. Air Force Cambridge Research Laboratories under Contract F19628-67-c-0166.

References

Axford, W. I.: 1969, *Rev. Geophys.* **7**, 421.
Axford, W. I. and Hines, C. O.: 1961, *Can. J. Phys.* **39**, 1433.
Axford, W. I., Petschek, H. E., and Siscoe, G. L.: 1965, *J. Geophys. Res.* **70**, 1231.
Chapman, S. and Ferraro, V. C. A.: 1931, *Terr. Mag.* **36**, 77.
Davis, L.: 1969, private communication.
Dessler, A. J.: 1958, *J. Geophys. Res.* **63**, 405.
Dessler, A. J.: 1964, *J. Geophys. Res.* **69**, 3913.
Dessler, A. J.: 1967, *Rev. G ophys.* **5**, 1.
Dessler, A. J.: 1968, *J. Geophys. Res.* **73**, 209.
Dessler, A. J. and Juday, R. D.: 1965, *Planetary Space Sci.* **13**, 63.
Dessler, A. J. and Michel, F. C.: 1966, in *adɛiation Trapped in the Earth's Magnetic Field* (ed. by B. M. McCormac), D. Reidel Publishing Company, Dordrecht-Holland, p. 447.
Dungey, J. W.: 1955, *The Physics of the Ionosphere*, The Physical Society, London, pp. 229–236 (also, Penn. State Univ. Sci. Rep. No. 69).
Dungey, J. W.: 1961, *Phys. Rev. Letters* **6**, 47.
Evans, L. C. and Stone, E. C.: 1969, *J. Geophys Res.* **74**, 5127.
Hoyle, F.: 1956, *Phys. Rev.* **104**, 269.
Hultqvist, B.: 1969, *Rev. Geophys.* **7**, 129.
Johnson, F. S.: 1960, *J. Geophys. Res.* **65**, 3049.
Krimigis, S. M., Van Allen, J. A. and Armstrong, T. P.: 1967, *Phys. Rev. Letters* **18**, 1204.
Lin, R. P. and Anderson, K. A.: 1966, *J. Geophys. Res.* **71**, 4213.
Mead, G. D.: 1964, *J. Geophys. Res.* **69**, 1181.
Michel, F. C.: 1965, *Planetary Space Sci.* **13**, 753.
Michel, F. C. and Dessler, A. J.: 1965, *J. Geophys. Res.* **70**, 4305.
Montgomery, M. D. and Singer, S.: 1969, *J. Geophys. Res.* **74**, 2869.
Ness, N. F.: 1965, *J. Geophys. Res.* **70**, 2989.
Parker, E. N.: 1958, *Phys. Fluids* **1**, 171.
Piddington, J. H.: 1959, *Geophys. J. Roy. Astron. Soc.* **2**, 173.
Schield, M. A., Freeman, J. W. and Dessler, A. J.: 1969, *J. Geophys. Res.* **74**, 247.
Wentworth, R. C.: 1967, *J. Geophys. Res.* **72**, 4582.
Williams, D. J. and Bostrom, C. O.: 1969, *J. Geophys. Res.* **74**, 3019.
Zmuda, A. J., Martin, J. H. and Heuring, F. T.: 1966, *J. Geophys. Res.* **71**, 5033.
Zmuda, A. J., Heuring, F. T. and Martin, J. H.: 1967, *J. Geophys. Res.* **72**, 1115.

MAGNETOTAIL PLASMA AND MAGNETOSPHERIC SUBSTORMS

EDWARD W. HONES, JR.
University of California, Los Alamos Scientific Laboratory, Los Alamos, N.M., U.S.A.

1. Introduction

One of the most discussed questions in connection with magnetospheric models in the past few years has been that of merging of magnetic field lines, both at the magnetopause and in the magnetotail, and of what role merging may play in driving magnetospheric convection. The plasma sheet in the magnetotail enters strongly in all such discussions and investigations of its properties and variations are of utmost importance in resolving the differences among the various models. This paper deals with two features of the plasma sheet as observed with the Vela satellites: (a) its time variations during magnetospheric substorms, and (b) the pressure of the magnetotail plasma as related to the dynamic pressure of the solar wind.

The instrumentation on the Vela satellites is not described here. The reader is simply reminded that the satellites are in nearly circular orbits of radius $\sim$ 18 R_E which are inclined at $\sim 60°$ relative to the ecliptic plane. They provide energy spectral measurements of electrons and protons between $\sim$100 eV and $\sim$20 keV and measurements of electrons with energies above several higher thresholds from $\sim$40 keV to several hundred keV.

2. Temporal Variations of the Plasma Sheet Associated with Magnetospheric Substorms

About 2 yr ago it was reported that the intensity of plasma in the plasma sheet at $\sim$18 R_E decreased during the development of negative magnetic bays in the nightside auroral zone, then suddenly increased again generally following the peak of the bays (Hones *et al.*, 1967; Hones, 1968). Our interpretation of this observation was that the plasma sheet grows thinner as plasma is squeezed earthward from the tail, then thickens again as a part of the plasma is forced back into the tail. Another possible interpretation – that the plasma sheet flapped back and forth over the satellite during substorms – was discounted because such a behavior might often have produced an initial *increase* of plasma intensity followed by a decrease.

Recently, studies based upon data from *two* Vela satellites passing through the plasma sheet close together have substantiated the conclusion of the earlier study, i.e., that the plasma sheet becomes thin and then expands in association with a substorm. Figure 1 illustrates this behavior rather well. Figure 1a traces the paths of Vela 3A and 4A in solar magnetospheric coordinates as they passed through the

B. M. McCormac (ed.), Particles and Fields in the Magnetosphere.

magnetotail between 1400 UT September 8 and 1400 UT September 9, 1967. Vela 4A crossed the neutral sheet slightly after 1400 UT September 8 and 3A crossed it ~20 hr later at 1000 UT September 9. Figure 1b shows properties of the plasma sheet electrons during this pass. Notable immediately in Figure 1b is the general *decrease* of the electrons' average energy and energy density at 4A with increasing time (i.e., as the satellite moves *away from* the neutral sheet) and the opposite general trend of the quantities at 3A as the latter moves *toward* the neutral sheet. Notice that the curves cross at ~2200 UT September 8 at which time the satellites are each ~3 R_E from the neutral sheet with 3A above it and 4A below it. The large excursions of the curves in Figure 1b are associated with negative bays at nighttime auroral zone stations as shown in Figure 1c. The decrease between ~1900 and 2000 and that between ~0200 and 0400 are particularly informative. The first of these, clearly associated with the

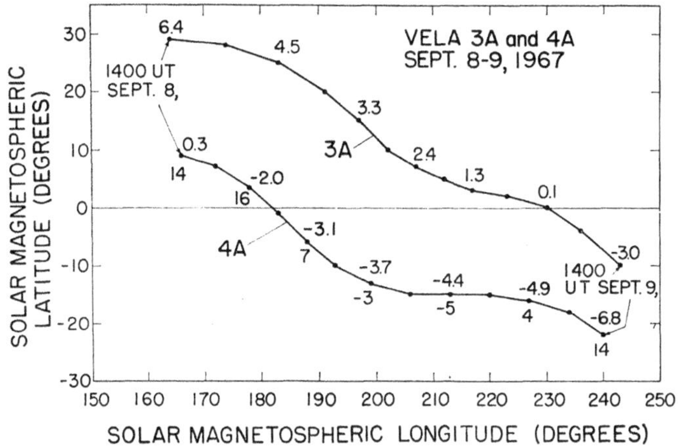

Fig. 1a. Passage of Vela satellites 3A and 4A through the magnetotail, September 8–9, 1967, mapped in solar magnetosphere coordinates. The dots are positions at 2 hr intervals. Above every second dot is the perpendicular distance (in R_E) of the satellite from the neutral sheet whose position was estimated by the formula of Russell and Brody (1967). The numbers below the curve for 4A are the tilt (in degrees) of the earth's dipole toward or away from the sun.

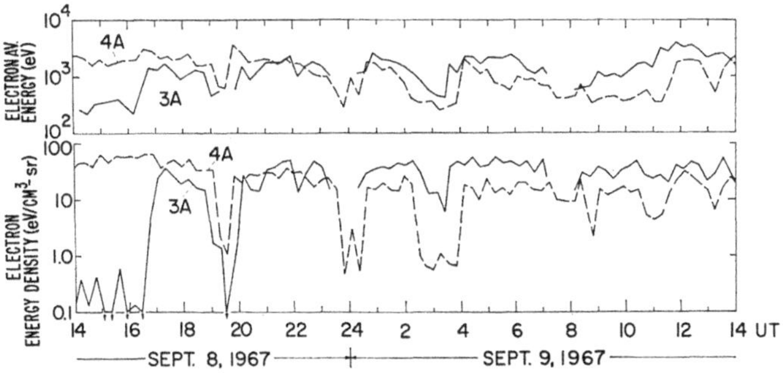

Fig. 1b. Average energy in eV (top curves) and energy density in eV/cm^3-sr (bottom curves) of plasma sheet electrons as measured with electrostatic analyzers on Vela 3A and 4A.

bay at Sodankyla (MLT≈2300), is stronger, starts earlier, and ends later at 3A than at 4A. Note that 3A is ~1.5 to 2 R_E further from the neutral sheet than is 4A. The decrease between 0200 and 0400 is evidently associated with the bays at Baker Lake (~1900 MLT), Fort Churchill (~2000 MLT), Great Whale River (~2100 MLT), Leirvogur (~0300 MLT), and Sodankyla (~0600 MLT). It is deeper and lasts longer at 4A than at 3A. At this time 3A is ~2 R_E closer to the neutral sheet than is 4A.

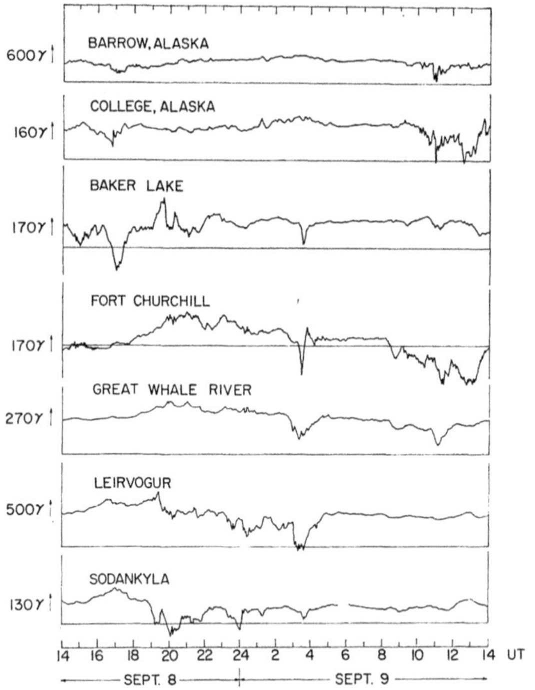

Fig. 1c. Traces of the *H* or *X*-component of the magnetic field at some auroral zone stations (September 8–9, 1967).

To summarize the September 8 to 9 pass, the general trends of the curves support the belief that the two satellites are on opposite sides of the neutral sheet during most of the pass. Given this, the excursions associated with bays can be interpreted only as contractions and re-expansions of the plasma sheet. The relatively greater excursion at the satellite estimated to be furthest from the neutral sheet lends credence to the accuracy of the formula of Russell and Brody (1967) whereby these estimates were made.

Some additional features of plasma sheet variations during substorms are illustrated in Figure 2. The *H*-traces of the nightside auroral zone stations, Meanook, Barrow, College and Sitka (Figure 2a) show a negative bay starting at ~0930 UT. Prior to this time the energetic electron flux at Vela, very near the neutral sheet, had been decreasing for about 1 hr. (Figure 2b). Riometer and rapid-run magnetometer records show that the bay at College ($\lambda_m = 64.6°$) started between 0930 and 0931. An increase of the X-rays at the balloon and a 'spike' of energetic electrons at Vela started at this same time within ± 30 sec. (The time resolution of the Vela and balloon measurements was ~ 2 sec.) Riometer and rapid-run magnetograms also show that the event started about 5 min later at Barrow ($\lambda_m = 68.5°$) and at Fort Yukon ($\lambda_m = 66.6°$). Interestingly, the energetic electron flux at Vela diminished rapidly after the peak at 09:34:30, reaching background levels at 0952, while electron precipitation at College and other nearby stations continued. Electron precipitation at College diminished sharply starting at 1015; at 1016 energetic electron fluxes returned suddenly at Vela.

The precise correlation of the onset of energetic electron flux at College and the onset of the 'spike' at Vela is a unique feature of this event and suggests that the two regions of the magnetosphere may have been coupled magnetically at $\sim$0930 to 0931. Models of the magnetosphere would hardly predict such a coupling, however, and it seems even less likely when one recalls that the electron flux at Vela drops away as the event begins at the higher latitude stations (Fort Yukon and Barrow) to which, it might be expected, Vela would be more nearly coupled. In short, the 0930 'spike' at Vela remains a mystery.

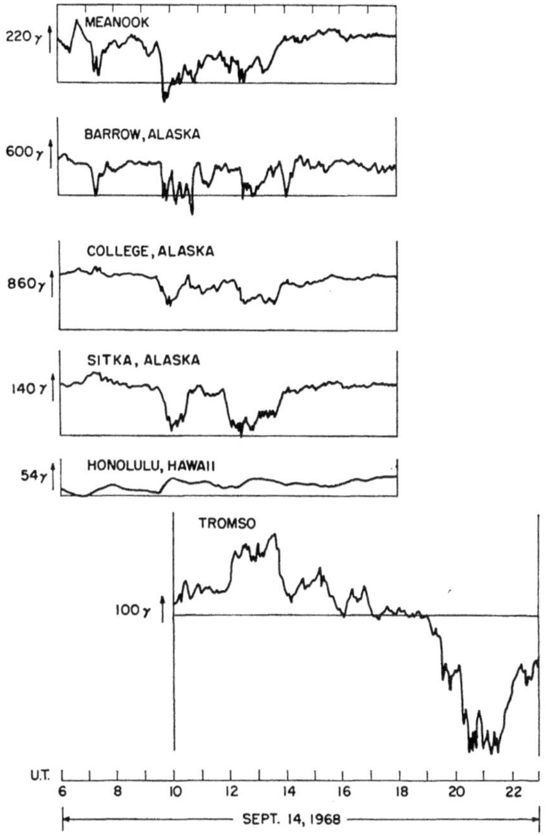

Fig. 2a. Traces of the *H* or *X*-component of the magnetic field at several auroral zone stations and at Honolulu (September 14, 1968).

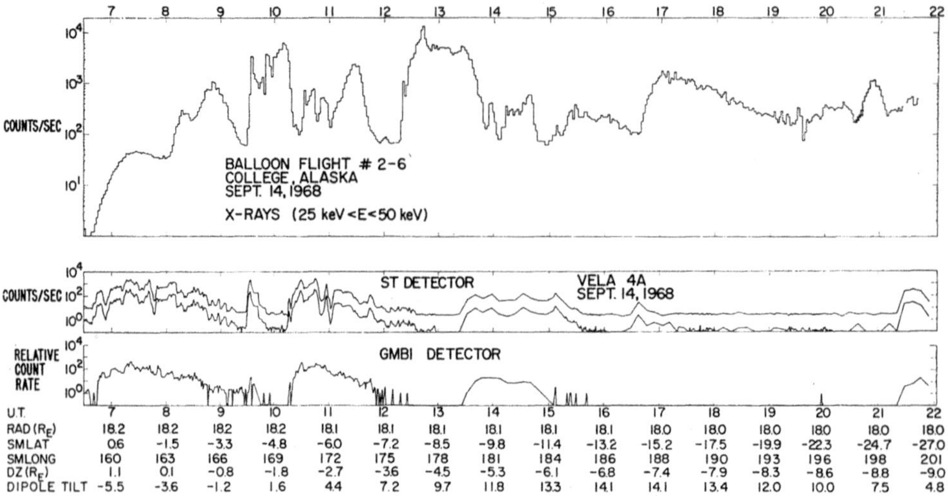

Fig. 2b. Responses of X-ray detector on a balloon near College, Alaska (top), and energetic electron detectors on Vela 4A (bottom) on September 14, 1968. The ST detector on Vela 4A measured electrons > 110 keV and > 460 keV. The GMBI detector measured electrons > 40 keV. Orbital parameters for Vela 4A are given at the bottom. DZ is the perpendicular distance of 4A from the estimated position of the neutral sheet. The balloon data were obtained by R. R. Brown and R. Karas, University of California, Berkeley.

The near-coincidence, at ~1015, of loss of electrons at College and their return at Vela is *not* an uncommon feature. On the contrary, the frequent occurrence of such anti-correlative associations has led us to contend (Hones *et al.*, 1967, 1968a, 1968b) that re-inflation of the plasma sheet is associated in a causative way with *relaxation* of auroral zone negative bays.

This anti-correlative behavior is illustrated again shortly after 1200 UT. Electron precipitation begins again sharply at College as the energetic electrons at Vela disappear. This time no 'spike' is seen at Vela, possibly because the satellite was much farther from the neutral sheet than it was at ~0930. Electrons reappear at Vela at ~1330 as precipitation at College decreases sharply.

3. Energy Density of Plasma in the Plasma Sheet

The energy density or pressure of plasma in the magnetotail at ~18 R_E is greatest near the neutral sheet and falls to very small values (relative to the magnetic field energy density) within a few earth radii perpendicular distance from the neutral sheet. The plasma consists of equal numbers of electrons of average energy ~1 keV and protons of average energy ~6 keV. Thus, most of the plasma kinetic energy is carried by the protons. The pressure is usually nearly isotropic. Figure 3a relates the magnetotail proton energy density with the *flow* energy density of the solar wind protons (i.e., the kinetic energy density associated with their radial velocity outward from the sun). The solar wind data were analyzed in the manner described by Hundhausen *et al.*

(1967). The magnetotail proton energy density is seen to range up to ~100% of the solar wind proton energy density and to average ~50% of it. The same kind of comparison for the magnetotail electrons (Figure 3b) shows that their energy density ranges up to about 20% and averages ~10% of that of the solar wind protons. Most of the very low values of proton and electron energy density in the plasma sheet were

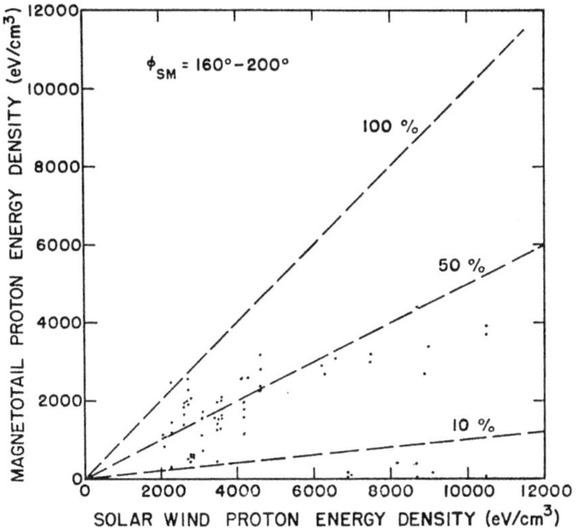

Fig. 3a. Energy density of magnetotail protons (eV/cm³) versus the *flow* energy density of the solar wind protons. Dashed lines indicate ratios of 1.0, 0.5, and 0.1, respectively. Each point represents nearly simultaneous measurements by a Vela satellite in the solar wind and another in the plasma sheet within one earth radius of the estimated neutral sheet position and between 160 and 200° solar magnetosphere longitude.

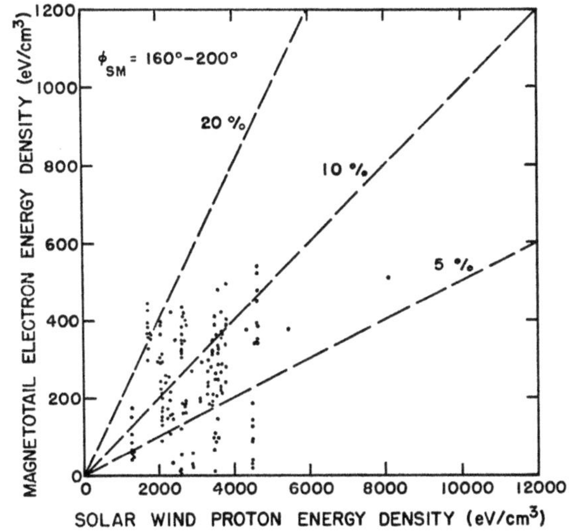

Fig. 3b. Energy density of magnetotail electrons (eV/cm³) vs. the *flow* energy density of the solar wind protons.

measured during substorms when the plasma density, as illustrated above, can become quite low even near the estimated neutral sheet position.

It is quite interesting to compare the (isotropic) pressure of the plasma sheet with the component of the magnetosheath plasma pressure which is normal to the magnetopause at $\sim$10 to 20 R_E downstream from the earth where the Vela measurements in the plasma sheet were taken. One suspects, of course, that in an equilibrium situation these two pressures will be equal; otherwise, the magnetotail would blow up or collapse laterally. Spreiter *et al.* (1968) have provided theoretical values of the magnetosheath plasma pressure. Detailed gasdynamic calculations indicate that $p_\perp$, the external pressure normal to the magnetotail about 10 to 12 R_E downstream is $\sim$5% of the stagnation pressure; the simple Newtonian pressure formula gives $p_\perp \approx 3\%$ of the stagnation pressure (see their Figure 34). Taking the stagnation pressure p_s to be $\varrho_{sw} V_{sw}^2 = 2U_{sw}$ (where the subscript *sw* stands for solar wind) and the plasma sheet pressure p_p to be $\varrho_p V_p^2 = \frac{2}{3} U_p$ we find that $p_p/p_s = \frac{1}{3} U_p/U_{sw} \approx \frac{1}{6}$. This value is significantly greater than the theoretical values and suggests that the interplanetary magnetic field, compressed in the magnetosheath plasma, adds importantly to the pressure there or

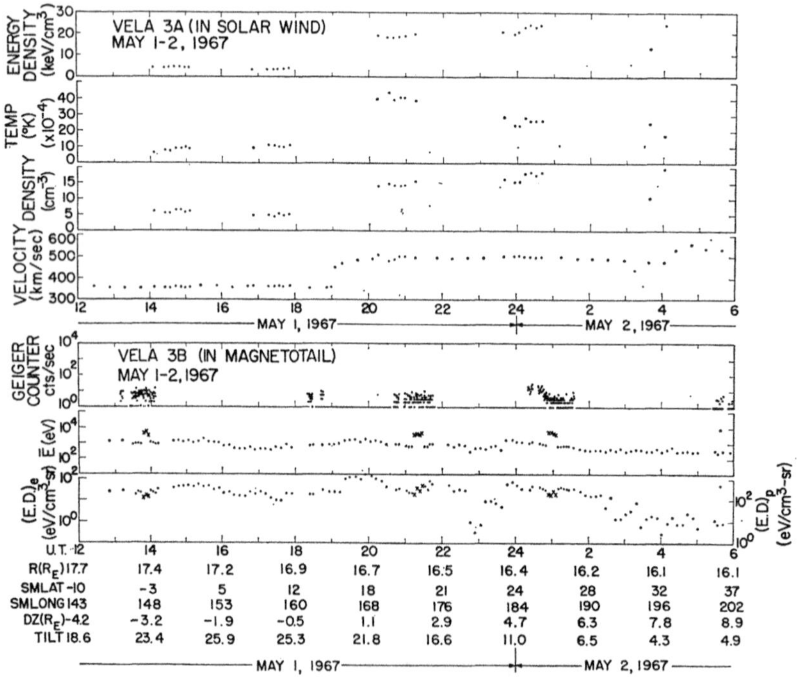

Fig. 4a. Solar wind and magnetotail measurements obtained with Vela 3A and Vela 3B, respectively, on May 1–2, 1967. Vela 3A was in the solar wind for the entire period shown and at the time of the sudden commencement, 1906 UT May 1, was very near the subsolar point at a geocentric distance of 17.5 R_E. Orbital parameters for 3B are given at the bottom of the Figure. DZ is the perpendicular distance (R_E) of the satellite from the estimated position of the neutral sheet. The *x*'s in the bottom two graphs for 3B are values for plasma sheet protons. The scale for the proton energy density $(ED)_p$ is shown at the right side of the bottom graph.

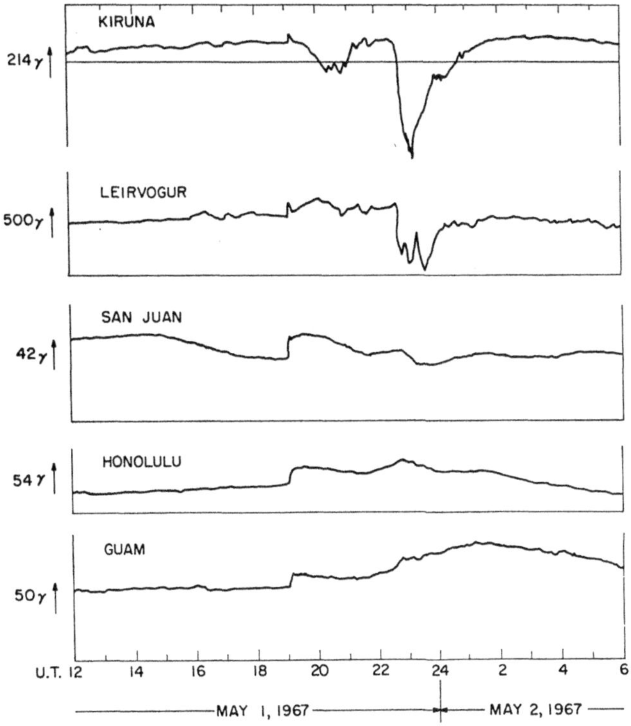

Fig. 4b. Traces of the *H* and *X* components of the magnetic field from auroral zone and low latitude stations on May 1–2, 1967.

that the real magnetotail flares out more rapidly downstream than does that of the model used by Spreiter *et al.*

How quickly and to what degree are changes in the solar wind dynamic pressure transferred to the plasma sheet? Figure 4a shows solar wind parameters measured by Vela 3A and plasma sheet electron and proton parameters measured by Vela 3B on May 1 to 2, 1967. A sudden commencement (SC) occurred at low latitude stations at 1906 UT May 1 (Figure 4b). Both Vela satellites were in store mode at this time, the analyzers providing particle energy spectra once every 512 sec. The spectra from 3A show that the solar wind velocity jumped from 355 km/sec to 460 km/sec sometime between 1900 and 1907, the result of an interplanetary shock wave passing the satellite. Measurements in real time mode about 1 hr before the SC and about 1 hr after it showed particle densities of 5/cm^3 and $\sim$15/cm^3, respectively. The proton flow energy density before the shock was $\sim$3700 eV/cm^3. After the shock, the flow energy density was $\sim$19000 eV/cm^3, up a factor of 5.1.

At Vela 3B, within $\sim$0.3 R_E of the neutral sheet and $\sim$1 hr from local midnight, the electron density jumped from $\sim$300 eV/cm^3 to 1200 eV/cm^3 sometime between 1914 and 1921 UT. Proton spectral measurements taken in real time mode at $\sim$1400 UT and $\sim$2130 UT showed that the proton energy density in the plasma sheet was $\sim$8 times that of the electrons. Thus, we estimate that the particle energy density in the plasma sheet jumped from ~2700 eV/cm^3 to ~11000 eV/cm^3, up a factor of 4. The change occurred between 8 and 15 min after the SC at the earth. The time delay (>8 min) is much longer than the 2 to 3 min that one would expect for transmission of the signal by Alfvén waves. As Figure 4a indicates, the change of energy density in the plasma sheet was caused mostly by a change in particle density. The electron average energy increased by a factor of ~1.5, while the particle number density increased by a factor of 2.7. From the pre-shock and post-shock solar wind proton density measurements the shock speed was estimated as ~500 km/sec (~4.7 R_E/min). Thus, the shock in the solar wind travelled 40 to 70 R_E along the magnetosphere's length in the time taken for its effect to be felt in the plasma sheet ~17 R_E behind the earth.

Before the SC the particle energy density in the plasma sheet was 2700/3700≈70% of the solar wind flow energy density; after the effect of the shock was felt in the plasma sheet this ratio was 11000/19000≈60%.

From ~2000 UT onward for many hours the velocity of the solar wind remained about constant. Its temperature decreased somewhat but its particle density and thus its energy density increased. The auroral zone stations (Leirvogur and Kiruna) near local midnight became more disturbed right after the SC (see Figure 4b). At ~2240 a negative bay started suddenly at Leirvogur and somewhat more gradually at Kiruna. The bays reached peaks at ~2330 and 2315 at the two stations and recovered by ~0100. At Vela 3B, ~3.5 R_E from the neutral sheet, the electron energy density dropped to a very low value at ~2240, recovered partially at ~2311 and then recovered fully at ~2340. The fact that the thinning of the plasma sheet and the associated bays occurred as long as 4 hr after the SC, when the solar wind plasma properties do not seem to have been changing drastically, suggests that this process is an internal one of readjustment of the magnetosphere to the environmental change imposed at the time of the SC. We are not able, however, to rule out the possibility that a change in the solar wind magnetic field occurred shortly before 2240 and promptly triggered the plasma sheet variation.

4. Conclusions

The data described above seem to imply the following things regarding the interaction of the solar wind with the magnetosphere and about the internal functioning of the magnetosphere itself:

(a) The magnetotail is blown out by the solar wind and inflated with plasma whose pressure at 18 R_E is often 15 to 20% of the solar wind's dynamic pressure. Such high values suggest that the magnetotail flares out faster than some present models predict.

(b) The plasma sheet pressure, even near the center of the tail, responds within minutes to changes in the dynamic pressure of the solar wind, and in approximate proportion to such changes.

(c) The plasma sheet becomes thinner with the development of auroral zone negative bays and then thickens again often as the bays recover. This process may be simply an internal adjustment of the magnetosphere to conditions of the solar wind existing for the preceding several hours. However, the possibility cannot be ruled out that the process occurs in quick response (within a few minutes) to changes of solar wind parameters (e.g., the magnetic field) which are not measured by the Vela satellites.

(d) The energetic electrons that precipitate in the auroral zone and populate the outer radiation belt during substorms are energized in the auroral zone and outer belt tubes of force. These tubes of force do not usually extend as far as 17 R_E into the tail during the energization process. Plasma may be ejected from these tubes of force into the tail during bay recovery and re-inflate the plasma sheet.

Acknowledgments

I am grateful to Mr. R. H. Karas and Professor R. R. Brown of the University of California, Berkeley, for permission to use data from their balloon flight #2–6 prior to its publication. The Vela nuclear test detection satellites have been designed, developed, and flown as a part of a joint program of the Advanced Research Projects Agency of the U.S. Department of Defense and the U.S. Atomic Energy Commission. The program is managed by the U.S. Air Force.

References

Hones, E. W., Jr.: 1968, in *Physics of the Magnetosphere* (ed. by R. L. Carovillano, J. F. McClay, and H. R. Radoski), D. Reidel Publishing Company, Dordrecht, Holland, p. 392.

Hones, E. W., Jr., Asbridge, J. R., Bame, S. J., and Strong, I. B.: 1967, *J. Geophys. Res.* **72**, 5879.

Hones, E. W., Jr., Bame, S. J., Singer, S., and Brown, R. R.: 1968a, *J. Geophys. Res.* **73**, 6189.

Hones, E. W., Jr., Singer, S., and Rao, C. S. R.: 1968b, *J. Geophys. Res.* **73**, 7339.

Hundhausen, A. J., Asbridge, J. R., Bame, S. J., Gilbert, H. E., and Strong, I. B.: 1967, *J. Geophys. Res.* **72**, 87.

Russell, C. T. and Brody, K. I.: 1967, *J. Geophys. Res.* **72**, 6104.

Spreiter, J. R., Alksne, A. Y., and Summers, A. L.: 1968, in *Physics of the Magnetosphere* (ed. by R. L. Carovillano, J. F. McClay, and H. R. Radoski), D. Reidel Publishing Company, Dordrecht, Holland, p. 301.

A MODEL CURRENT SYSTEM FOR THE MAGNETOSPHERIC SUBSTORM

S.-I. AKASOFU

Geophysical Institute, University of Alaska, College, Alaska, U.S.A.

Abstract. A model three dimensional current system for the magnetospheric substorm is proposed to account for the distribution of geomagnetic disturbance field vectors on the earth's surface and for the distribution of auroral electrons and protons at the ionospheric level and in the equatorial plane. It is suggested that the asymmetric ring current completes its circuit by accelerating auroral particles and by driving the auroral electrojet.

The development of a magnetospheric *storm* is determined by the intensity and occurrence frequency of magnetospheric *substorms* (cf., Akasofu, 1968). Thus, it is necessary to understand the basic processes related to this elementary phenomenon, the magnetospheric substorm. Further, many of the polar upper atmospheric disturbance phenomena (such as the auroral substorm and polar magnetic substorm) can be considered to be various manifestations of the magnetospheric substorm. These manifestations tell us some aspects of the processes associated with the magnetospheric substorm. For example, the polar magnetic substorm is a magnetic manifestation of the magnetospheric substorm. One of the objectives of studies of geomagnetic disturbance fields is to infer electric current systems associated with the magnetospheric substorm and also electric fields that generate the current systems.

In the past, the electric current system associated with magnetospheric substorms has been discussed in terms of the two dimensional 'equivalent current system'. It is a means to express the distribution of geomagnetic disturbance vectors by a current system on a spherical shell concentric with the earth (say, the ionosphere) and is not necessarily the true current system. One of the classical examples of such a current system, constructed by Silsbee and Vestine (1942), is shown in Figure 1. It consists of two concentrated currents along the auroral zone (one eastward and the other westward directed) and their 'return currents' in both the polar cap and mid-low latitudes. In mid-latitudes, the return currents are directed equatorward in the evening sector and poleward in the morning sector; in low latitudes, they are directed westward in the day sector and eastward in the dark sector.

In general, it is possible to reduce any three dimensional current system to an equivalent current distributed on a spherical shell, provided that a part of the current does not cross the earth's surface. In other words, there are a great number of possible three dimensional current systems which can give rise to the same distribution of equivalent currents and the same distribution of geomagnetic disturbance vectors. Now, with the valuable aid of satellite observations, such possibilities have been greatly reduced. Figure 2 shows a model current system proposed by Akasofu and Meng (1969) on the basis of a detailed analysis of the distribution of geomagnetic disturbance vectors on the ground for a number of polar magnetic substorms. A

B. M. McCormac (ed.), Particles and Fields in the Magnetosphere.

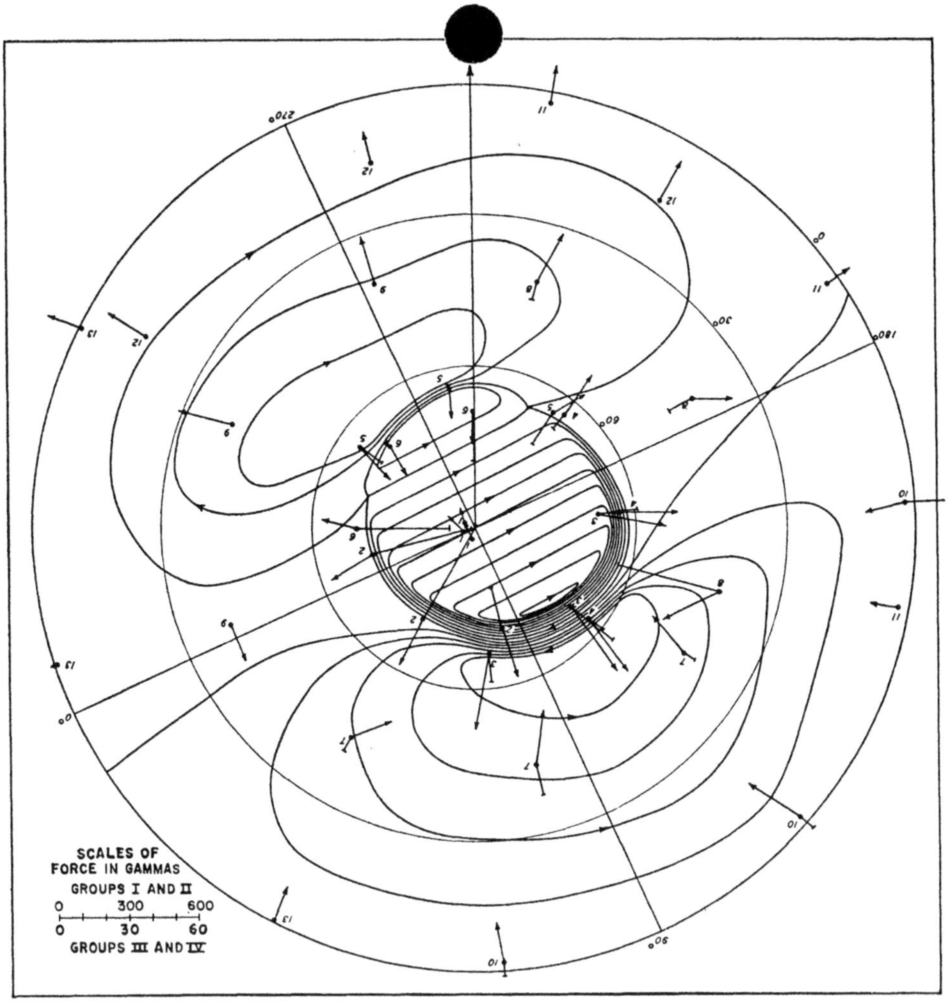

Fig. 1. Distribution of geomagnetic disturbance vectors and the corresponding equivalent current system for an average polar magnetic substorm constructed by Silsbee and Vestine (1942).

similar current system was discussed theoretically by Fejer (1961) under the assumption that the ring current particles are fed into the trapped region from the sunlit sector. Another type of asymmetric ring current belt was proposed by Cummings (1966), but it has a very limited longitudinal extent and generates an eastward ionospheric current to complete its circuit.

The proposed current system consists of an asymmetric ring current belt in the day-evening sector, current along the field lines from the morning end of the ring current to the northern and southern ionospheres, ionospheric current along the auroral oval in the dark sector (the auroral electrojet) and current along the field lines from the western end of the auroral electrojet to the ring current.

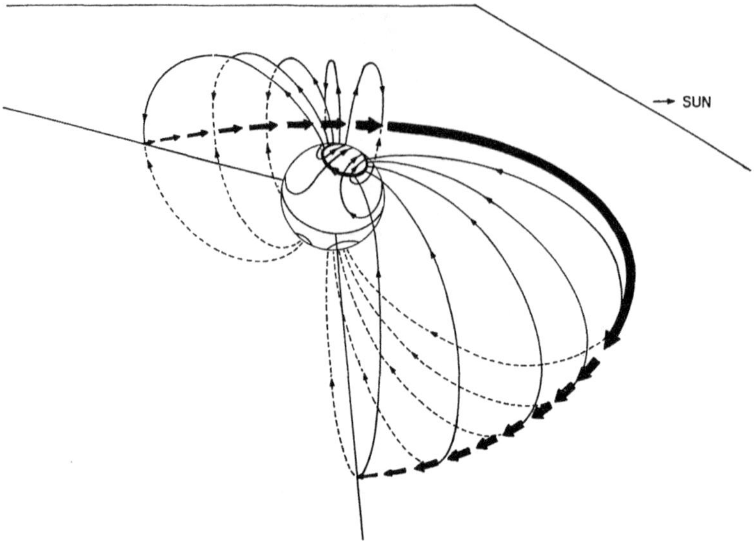

Fig. 2. Model three dimensional current system for a magnetospheric substorm proposed by Akasofu and Meng (1969).

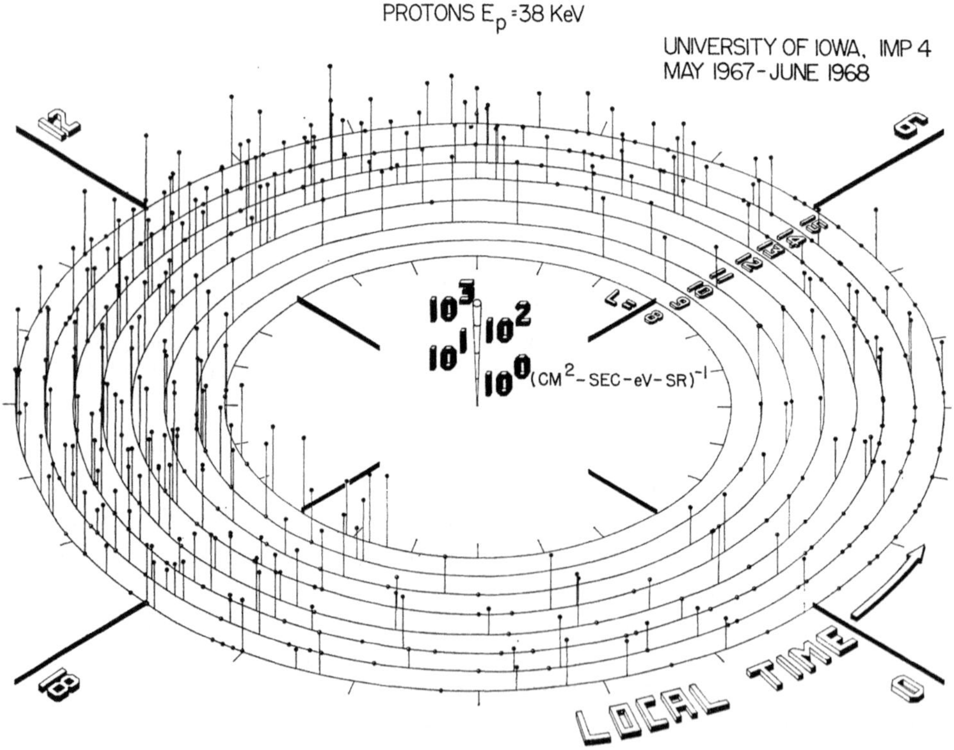

Fig. 3. Observed distribution of ring current protons constructed by Frank (1968).

There are at least two important features to be mentioned about the model. First of all, the ring current is greatly asymmetric with respect to the earth's dipole axis and is located in the day-evening sector. This geometry is in good agreement with the observed geometry of the low energy proton belt obtained by Frank (1968) as shown in Figure 3.

Secondly, the model implies that both the auroral electrojet and the asymmetric ring current are parts of a single current circuit. It is interesting to note in this connection that after estimating the intensity of the ring current, Chapman (1952) noted that it is "the same order of magnitude as the maximum auroral electrojet".

If the above implication is correct, the degree of the asymmetry of the ring current should be correlated with the intensity of the electrojet; this is indeed the case. Figures 4a and b illustrate two examples showing such a correlation. In particular, if the ring current field, however intense it may be (say, as large as 300γ), is symmetric, there is little indication of an auroral electrojet in the polar region. In Figures 4a and b

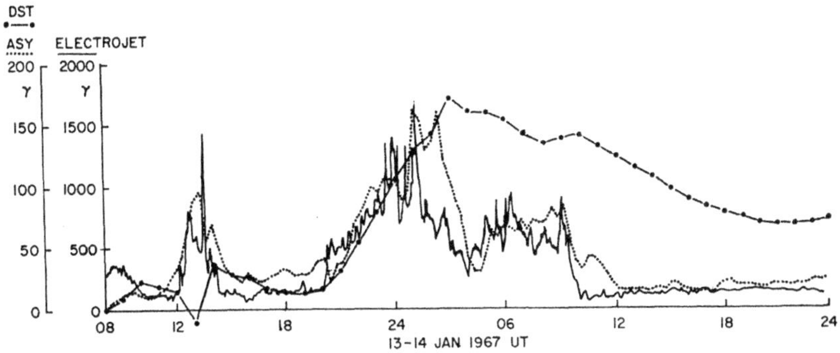

Fig. 4a. Relationships between the auroral electrojet index (AE), the ring current asymmetric index, and the Dst index for the January 13–14, 1967 storm.

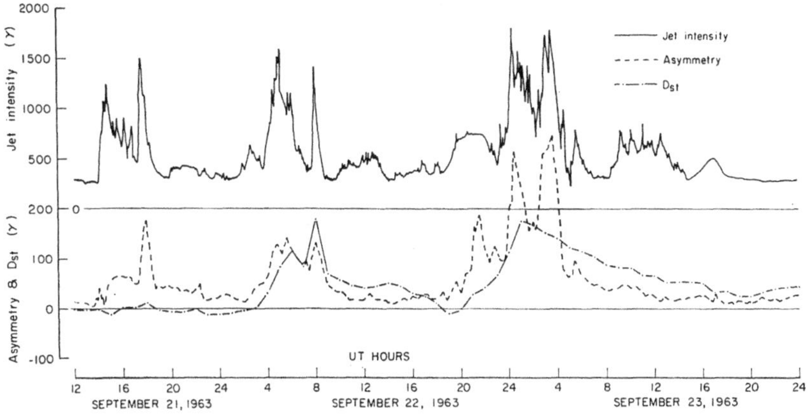

Fig. 4b. Relationships between the auroral electrojet index (AE), the ring current asymmetric index and the Dst index for the September 21–23 1963 storm.

the intensity of the auroral electrojet is given in terms of the AL index. The ring current asymmetry index was obtained by superposing magnetic records from six low latitude stations and then taking the *distance* between the upper and the lower envelopes; the principle behind such a procedure is that H component variations in low latitudes primarily result from those of the ring current field (Cummings and Coleman, 1968).

For a quantitative study of the proposed current system, a model by Kirkpatrick has been modified so as to better fit with the geometry of the proposed current system. The computer model current system thus constructed consists of:

A sheet current of intensity

$$I = 0, \qquad 0 \leqslant \phi \leqslant \pi$$
$$I = 2I_0 \sin 2\phi, \quad \pi \leqslant \phi \leqslant 2\pi \tag{1}$$

per radian of ϕ, flowing along the line of force which crosses the equatorial plane at a geocentric distance $r = 6.55\ R_E$, and ϕ denotes the azimuth angle reckoned from the 06 meridian in the dipole latitude and time coordinate.

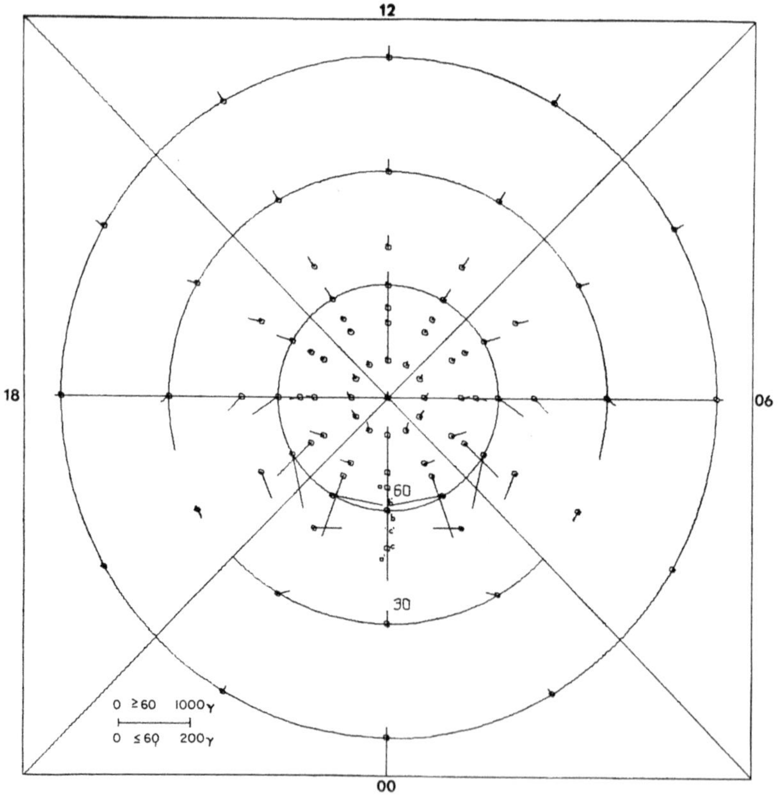

Fig. 5a. Computed distribution of geomagnetic disturbance vectors without including an eastward circular current in the equatorial plane (for details see the text).

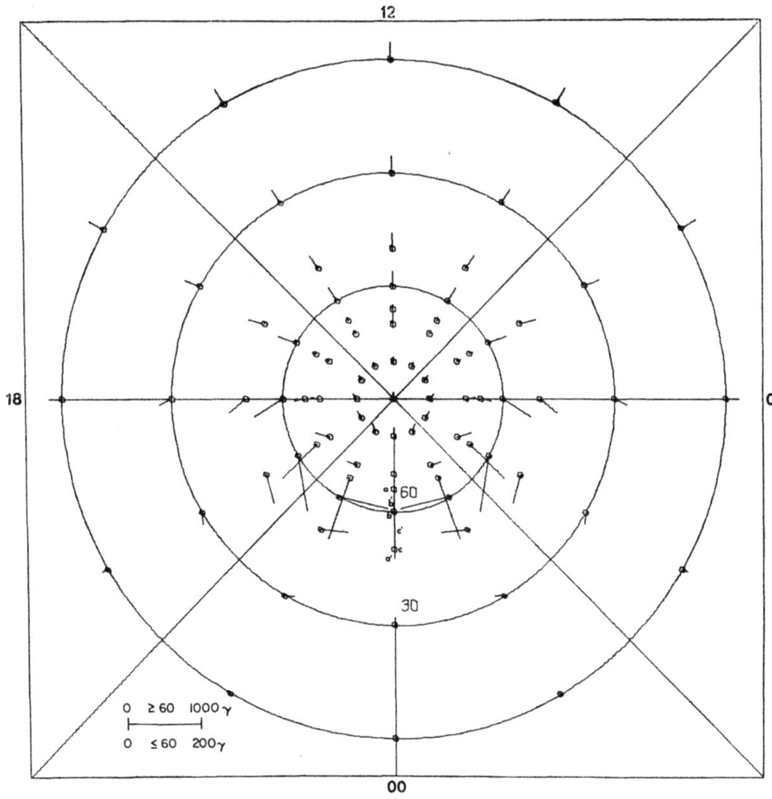

Fig. 5b. Computed distribution of geomagnetic disturbance vectors, including an eastward circular current in the equatorial plane (for details see the text).

A line current intensity

$$I = 0, \qquad 0 \leqslant \varphi \leqslant \pi$$
$$I = I_0 \sin \phi, \quad \pi \leqslant \varphi \leqslant 2\pi \tag{2}$$

which flows along the auroral zone at the 100 km level.

A line current of intensity

$$I = 2I_0 + I_0, \qquad 0 \leqslant \phi \leqslant \pi$$
$$I = 2I_0 \cos \phi + I_0, \quad \pi \leqslant \phi \leqslant 3\pi/2$$
$$I = -2I_0 \cos \phi + I_0, \quad 3\pi/2 \leqslant \phi \leqslant 2\pi \tag{3}$$

which flows in the equatorial circle of radius 6.55 R_E. The second term $(+I_0)$ in the above equations represents an additional eastward current all around the earth.

In Figure 5, the distribution of geomagnetic disturbance vectors is shown in a dipole latitude-time coordinate; it is computed without the second term $(+I_0)$ in Equation (3). It reproduces reasonably well the distribution shown in Figure 1, except for the poleward directed (positive of the H component) vectors in low latitudes in the dark sector. In Figure 1, these poleward directed fields were supposed to be produced by the eastward return current from the westward auroral electrojet.

A detailed comparison of simultaneous records from Honolulu and the synchronous satellite (ATS) demonstrated clearly that the positive change (or the so-called 'low latitude positive bay' is caused by a current system far above the ionosphere; Cummings and Coleman (1968) suggested that the positive change is caused by a reduction of the ring current particles in the dark sector. Therefore, in order to simulate the effect of the reduction of the ring current particle, and eastward '*equivalent*' ring current of intensity, $+I_0$ is added to the previous computer model. The implication of this reduction will be discussed later, when mechanisms which generate the magnetospheric substorm are discussed. The distribution of geomagnetic disturbance vectors thus obtained is shown in Figure 5b. It is in a good agreement with the case examined by Haerendel and Lüst (1970) in connection with their Ba cloud release in the outer magnetosphere.

As mentioned earlier, in Figure 1, the eastward directed disturbance vectors in the

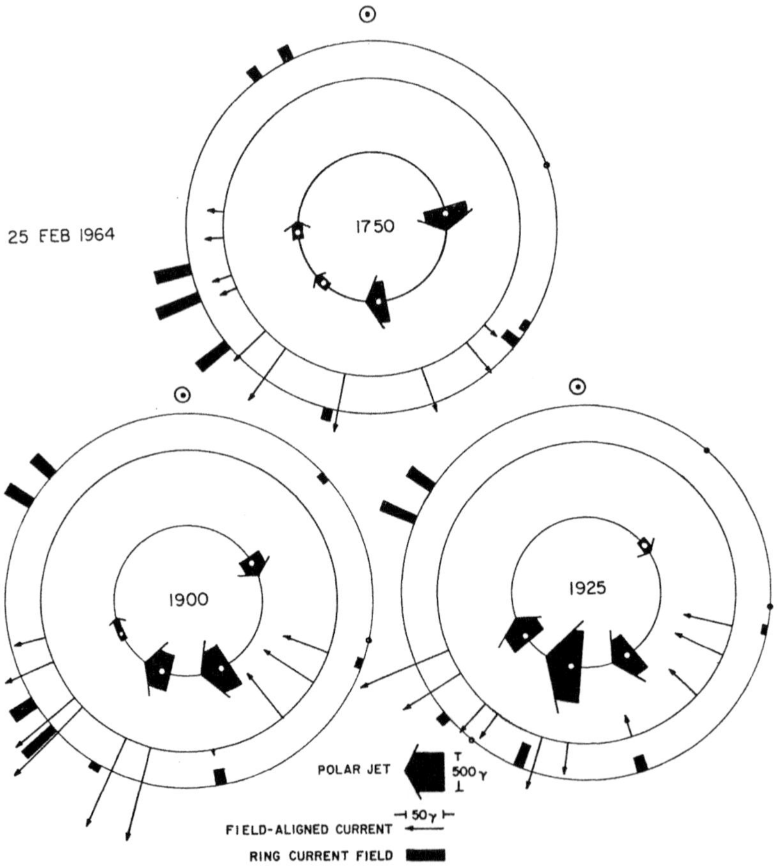

Fig. 6a. Three dimensional current systems at 1750, 1900, and 1925 UT on February 25, 1964.

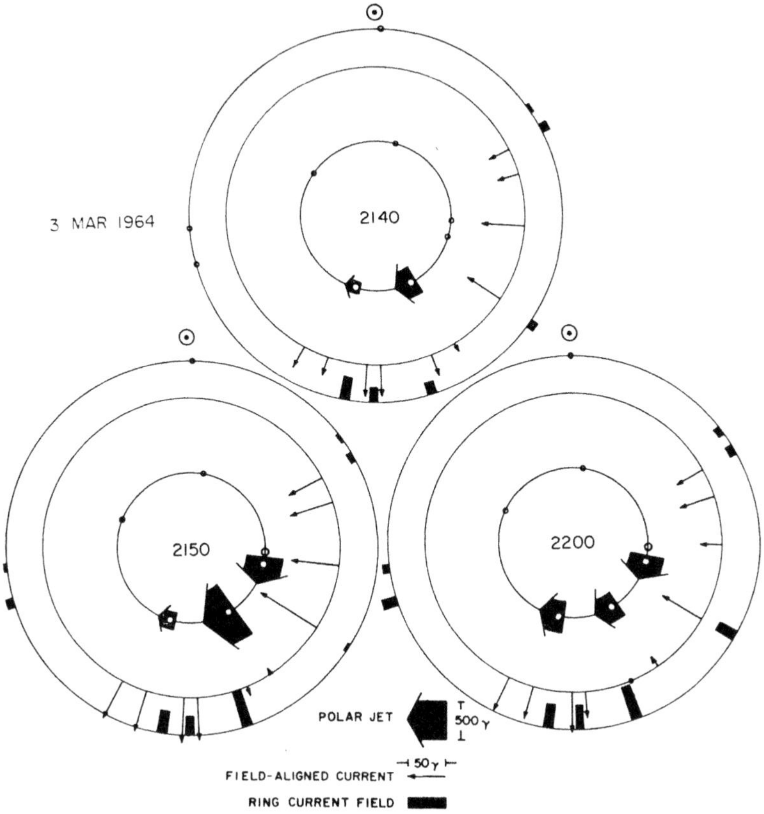

Fig. 6b. Three dimensional current systems at 2140, 2150, and 2200 UT on March 3, 1964.

evening sector in mid-latitudes were interpreted to be caused by the equatorward directed return current, and westward directed disturbance vectors in the morning sector in mid-latitudes by the poleward directed return current. In the proposed model, however, these disturbance vectors are produced by field aligned sheet currents, outward in the evening sector and inward in the morning sector. Thus, variations of the *D* component in mid-latitudes can be used to examine the direction and intensity of the field aligned currents. It was noted earlier that changes of the population of the ring current particles can be inferred from variations of the *H* component in low latitudes. It also is known that *H* component variations at stations along the auroral zone results from the auroral electrojet. Therefore, on the basis of analyses of magnetic records from polar, mid-latitudes, and low latitudes it is possible to infer qualitatively the geometry of three dimensional current systems. Figures 6a and b show examples of three dimensional current systems constructed by Meng and Akasofu (1969). Because of the semi-quantitative nature of the results, a particular form of presentation is devised by using three concentric circles.

The inner circle represents the auroral zone. The sign and the magnitude of the *H* component changes at auroral zone stations are assumed to represent the direction and magnitude, respectively, of the electrojet. A negative bay is assumed to be caused by a westward electrojet, represented by a westward pointing arrow whose width is proportional to the magnitude of the negative bay. The sign and magnitude of *D* component changes in mid-latitudes are assumed to represent the direction and magnitude, respectively, of the field aligned current. A positive (eastward) change should be caused by an outward current, and a negative (westward) change by an inward current. The currents are represented by arrows on the middle circle whose lengths are proportional to the magnitude of the *D* component changes, normalized to the value at dipole lat 50°. The sign and magnitude of *H* component changes at a low latitude station are assumed to represent changes, at least qualitatively, of the population of the ring current particles in the sector where the station is located. A negative change should be produced by an increase of particles, and a positive change by a decrease in the population of the particles. These are plotted on the outer circle; a negative change by an outward bar and a positive change by an inward bar; the lengths of the bars are proportional to the magnitudes of the changes.

An important implication of the proposed three dimensional current system is that the asymmetric ring current completes its circuit by driving the auroral electrojet; the auroral electrojet thus is generated when either (1) an asymmetric ring current grows, (2) a part of the circular ring circuit is broken or (3) both. Akasofu and Meng (1969) showed that in an early epoch of the magnetospheric substorm, the ring current particles appear in the sunlit sector while the dark sector is deserted at the same period. The eastward equivalent current discussed earlier may be associated with such a desertion of ring current belt particles and results in a positive change of the *H* component in low latitudes on the earth's surface, as well as within and inside the

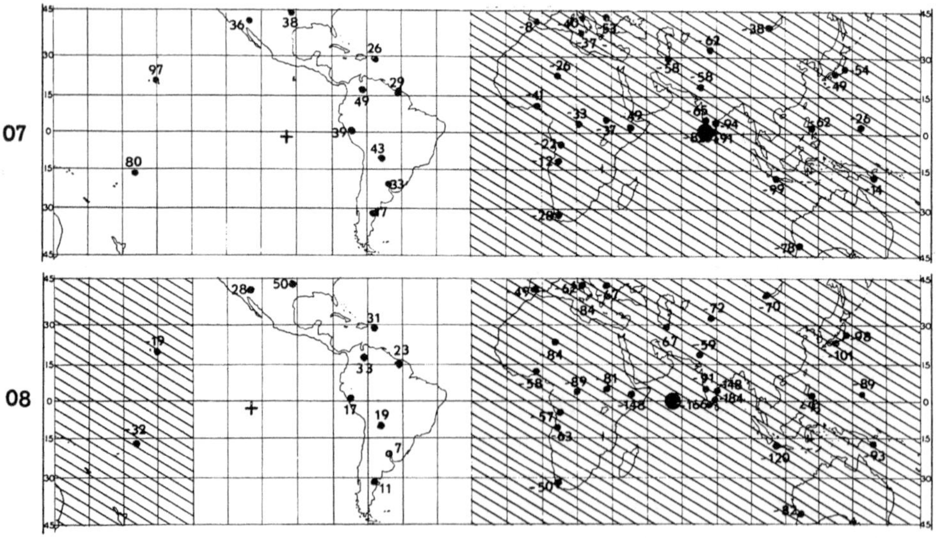

Fig. 7. Distribution of changes of the main phase field from the values at the onset time of an intense substorm (0600 UT) during the April 18, 1965, storm.

ring current (in the equatorial plane, as observed by the ATS). Figure 7 shows changes of the H component in mid and low latitudes after the intense substorm which began at 0600 UT, April 18, 1965. The region where a negative change occurred (that is, the sector in which an increase of ring current particles is observed) is shaded; the rest is the region where a positive change (that is, the sector in which a decrease of ring current particles) occurred. At 0800 UT, 2 hr after the onset of the substorm, the dark sector became populated by ring current particles. Though not shown, at 0900 UT, a negative change was recorded in all longitudes.

A very low density plasma in the magnetosphere must participate in conducting 10^6 amp to and from the ring current belt. Swift (1967) demonstrated by his detailed numerical study that the conduction occurs in a thin outer layer of the ring current belt. Let us suppose that the aurora is the projection of this thin layer onto the polar ionosphere. Then, a current of 10^6 amp must flow across a rather small cross sectional

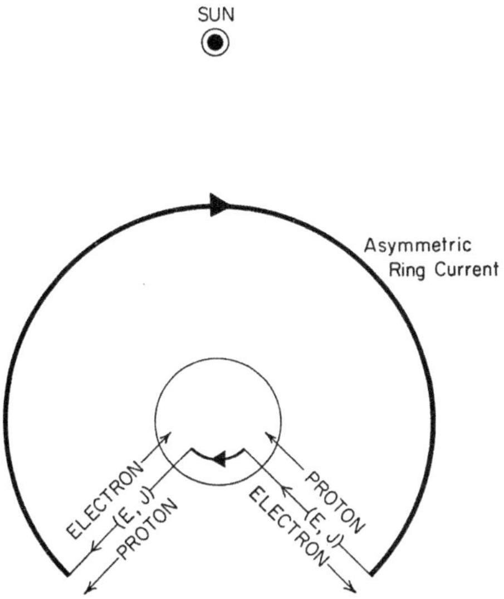

Fig. 8. Scheme of the acceleration of auroral particles associated with the proposed three dimensional current system.

area; for a typical auroral arc the N–S thickness is of the order of 1 km, and its E–W length is of the order of 4000 km. We can show then that in the equatorial plane the current density is of the order of 1.8×10^{-10} amp/cm^2. If the plasma density is of the order of 1/cm^3, electrons in the plasma must be accelerated to have a speed of the order of 1.1×10^9 cm/sec (corresponding to 340 eV) in order to carry the above current density. If the electron density is of the order of 0.1 cm^3, electrons must be accelerated to have an energy of order 34 keV. Swift (1965), and Alfvén and Carlqvist (1967) showed that well before such a stage is reached an instability of the plasma disrupts the current flow. The latter authors suggested that a solar flare results from an explosive conversion of the magnetic energy in the circuit into kinetic energy of particles by such disruption of electric current circuits in the solar atmosphere.

The scheme of the acceleration process associated with the proposed model current system is schematically shown in Figure 8 (Akasofu, 1969). In the morning sector, electrons are accelerated away from the earth, and in the evening sector towards the earth. Montgomery (1968) demonstrated that there is a far greater flux of energetic electrons in the morning half of the magnetospheric tail than in the evening half (Figure 9). On the other hand, bright auroral forms in the evening sector, such as

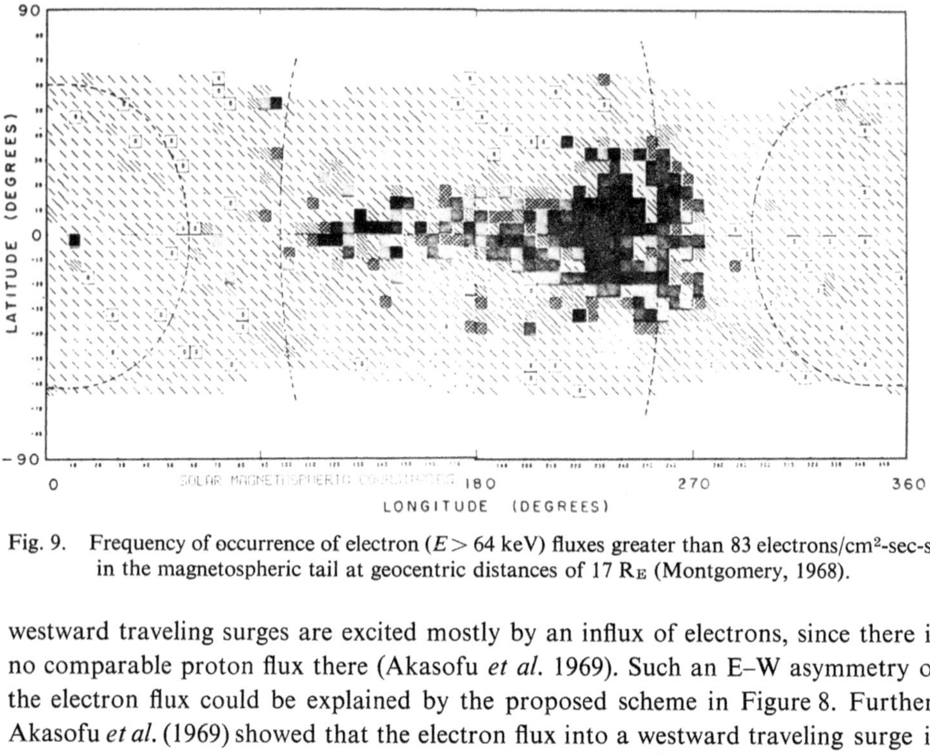

Fig. 9. Frequency of occurrence of electron ($E > 64$ keV) fluxes greater than 83 electrons/cm^2-sec-sr in the magnetospheric tail at geocentric distances of 17 R_E (Montgomery, 1968).

westward traveling surges are excited mostly by an influx of electrons, since there is no comparable proton flux there (Akasofu *et al.* 1969). Such an E–W asymmetry of the electron flux could be explained by the proposed scheme in Figure 8. Further, Akasofu *et al.* (1969) showed that the electron flux into a westward traveling surge is equivalent to an upward current of the order of 0.5×10^6 amp. An intense flux of electrons in the morning sector may be caused by a dumping (by VLF wave particle interactions) of electrons which are accelerated upwards in that sector.

The scheme presented in Figure 8 suggests that protons are accelerated upwards in the evening sector and downwards in the morning sector. Both ground observations of the Hα emissions and satellite observations of protons do not contradict with such an acceleration scheme.

It has been suggested by Piddington (1967) and others that the asymmetric proton belt is produced by an inward motion of protons in the tail plasma and their subsequent westward drift motion. However, it should be noted that it takes many hours for 10 keV protons to reach the afternoon sector. On the other hand, both detailed analyses of ground geomagnetic disturbance vectors and also a preliminary study of

ATS data indicate that the asymmetric distribution of ring current protons is established in an early epoch of the magnetospheric substorm.

Acknowledgments

The work reported here was supported in part by a grant from the National Aeronautics and Space Administration NGR-02-001-001, and in part by a grant from the National Science Foundation GA-1661.

References

Akasofu, S.-I.: 1968, in *Polar and Magnetospheric Substorms*, D. Reidel Publishing Company, Dordrecht, Holland.
Akasofu, S.-I.: 1969, *Nature* **221**, 1020.
Akasofu, S.-I. and Meng, C.-I. 1969, *J. Geophys. Res.* **74**, 293.
Akasofu, S.-I., Eather, R. H., and Bradbury, J. N.: 1969, *Planetary Space Sci.* 1409.
Alfvén, H. and Carlqvist, P.: 1967, *Solar Phys.* **1**, 220.
Chapman, S.: 1952, *Ann. Geophys.* **8**, 205.
Cummings, W. D.: 1966, *J. Geophys. Res.* **71**, 4495.
Cummings, W. D. and Coleman, P. J., Jr.: 1968, *Radio Sci.* **3**, 758.
Fejer, J. A.: 1961, *Can. J. Phys.* **39**, 1409.
Frank, L. A.: 1968, *Trans. Amer. Geophys. Union* **49**, 738.
Haerendel, G. and Lüst, R.: 1970, this volume, p. 213.
Meng, C.-I. and Akasofu, S.-I.: 1969, *J. Geophys. Res.* **74**, 1969.
Montgomery, M. D.: 1968, *J. Geophys. Res.* **73**, 871.
Piddington, J. H.: 1967, *J. Atmos. Terr. Phys.* **29**, 87.
Silsbee, H. C. and Vestine, E. H.: 1942, *Terr. Mag. Atmos. Elec.* **47**, 195.
Swift, D. W.: 1965, *J. Geophys. Res.* **70**, 3061.
Swift, D. W.: 1967, *Planetary Space Sci.* **15**, 835.

ON THE ORIGIN OF RADIATION BELT AND AURORAL PRIMARY IONS

W. I. AXFORD

Dept. of Physics, Dept. of Applied Physics and Information Science, and Institute for Pure and Applied Physics, University of California, San Diego, La Jolla, Calif., U.S.A.

Abstract. The possibilities that the energetic ions in the radiation belts and in the auroral primaries might originate in the ionosphere or in the solar wind are examined. It is argued that there is no clear evidence favoring one source over the other on the basis of the observations that have been carried out to date. An entirely different type of observation is proposed as a means of providing a cheap and effective way of deciding this question and of providing information concerning the rate of accretion of inert gases by the atmosphere.

1. Introduction

It is usually assumed that most of the energetic particles in the magnetosphere originate in the solar wind; however, the evidence is at best fragmentary and far from compelling. As far as the electrons are concerned there are good reasons for believing that energy sharing takes place in the outer magnetosphere (e.g., Kennel, 1969), and hence it is not possible to clearly identify any particular group of electrons as having originated in the solar wind or in the ionosphere. In the case of the ions it is possible to decide one way or the other by examining their composition and state of ionization, although it is not a simple matter to make the appropriate observations.

In this paper various aspects of this problem will be considered to emphasize its importance and to bring it to the attention of experimenters. The possibility of an ionospheric origin is discussed in Section 2, where particular attention is paid to the possibility that the particles forming the plasma sheet might be polar wind particles which have been trapped and accelerated on closed field lines in the tail of the magnetosphere. In Section 3 the possibility of a solar wind origin for the particles is discussed, noting that this has the attractive feature of providing particles with reasonably large initial energies, but also that their mode of access to the magnetosphere is not obvious. If the particles in fact do originate in the solar wind, there are some interesting implications concerning the atmospheric budgets of the inert gases which will be examined briefly. Finally, in Section 4 the experimental evidence as it exists at present is reviewed, and an experiment which has not previously been considered in this context, but which promises to yield definite results at a relatively low cost, is described also. The possibilities that cosmic ray neutron albedo or neutral atoms captured from the solar wind should be significant sources of magnetospheric energetic particles will not be discussed.

As a preliminary, we will estimate the number of particles that are likely to be involved in the radiation belts and in auroral precipitation. On quiet days the total

B. M. McCormac (ed.), Particles and Fields in the Magnetosphere.

energy of particles trapped in the radiation belts is $\sim 5 \times 10^{21}$ erg (corresponding to a surface magnetic field depression $\sim 10\gamma$) and the number of energetic particles involved is $\sim 3 \times 10^{29}$ assuming that they are mostly protons with an average energy of 10 keV. During the main phase of a large magnetic storm the number of trapped energetic protons could be as large as 3×10^{30} to 10^{31}, and these might have to be injected in a period as short as 10^4 sec, which requires peak injection rates of 3×10^{26} to 10^{27} protons $\sec^{-1}$. The trapped particles eventually leave the magnetosphere (during the recovery phase if a storm is involved) by convection, diffusion, charge-exchange with geo-coronal H atoms and precipitation into the atmosphere. In the case of electrons, auroral precipitation appears to be the dominant process, and as a rough guide it is reasonable to assume that most of the electrons enter the atmosphere during or shortly after being injected on the nightside of the magnetosphere, while most of the protons and other ions are temporarily trapped and thus form the ring current. The total precipitated electron flux varies from as little as $\sim 10^{16}$ erg $\sec^{-1}$ on extremely quiet days, to $\sim 10^{18}$ erg $\sec^{-1}$ and more during magnetic storms (Sharp and Johnson, 1968); it seems safe to conclude that the average flux is $\sim 10^{17}$ to 3×10^{17} erg $\sec^{-1}$. According to Evans *et al.* (1966), the average proton energy flux is about 14% of the electron energy flux, although Whalen and McDiarmid (1969) favor a higher value (30 to 50%). If a rough mean of these estimates is taken, we find that the average proton energy flux is $\sim 3 \times 10^{16}$ to 10^{17} erg $\sec^{-1}$, and hence if the average particle energy is taken to be 6 keV $\approx 10^{-8}$ erg, we find that the average proton flux into the atmosphere is 3×10^{24} to 10^{25} $\sec^{-1}$. In the following sections this result is used as a basis for estimating the fluxes of various other ions that might be involved in auroral precipitation.

2. The Ionosphere as a Possible Source of Energetic Particles

The possibility that many, if not most of the electrons that are precipitated into the atmosphere to produce aurora, originate in the ionosphere has been discussed by a number of authors (e.g. Block and Fälthammar, 1968; Dessler and Hanson, 1961; Kennel, 1969; Michel, 1968; Perkins, 1968; Speiser, 1969). On the whole the arguments advanced seem sound and it is reasonable to presume therefore that 'cold' ionospheric electrons are accelerated to high energies by one means or another. In the case of protons and other ions, however, the situation is less clear, although as described in this section it is possible to make a case for having the entire plasma sheet formed of ionospheric plasma.

According to the theoretical work on the polar wind that has been carried out to date it can be reasonably expected that the tail of the magnetosphere contains ionospheric plasma which is in the process of escaping from the earth (Dessler and Michel, 1966; Axford, 1968; Banks and Holzer, 1968, 1969a, 1969b; Banks, 1969a, 1969b; Holzer, 1969). The vertical speeds of ions observed at altitudes of the order of 1000 km over the polar regions are already 10 to 20 km $\sec^{-1}$ (Hoffman, 1968), and it can be argued that the ions continue to accelerate and eventually attain energies characteristic of photoelectrons (~ 15 eV). If this is so, the flux of polar wind ions in the tail of the magnetosphere should typically be 5×10^4 cm^{-2} sec^{-1}, corresponding to a number density $n = 0.02$ cm^{-3} and a bulk speed of 50 km sec^{-1}. The ions should be mostly protons, with the relative concentrations of other ions being as follows: $[^4He^+]:[H^+] \approx 10^{-2}$, $[^4He^{++}]:[H^+] \lesssim 10^{-4}$, $[^3He^+]:[H^+] \approx 10^{-8}$ and $[O^+, N^+, Ne^+, \text{etc.}]:[H^+] \lesssim 10^{-4}$ (e.g., Axford, 1969a).

Polar wind ions can be trapped in the plasma sheet if the magnetic field lines in the tail of the magnetosphere reconnect and thus close at large distances from the earth as indicated in Figure 1 (cf., Speiser, 1969). Observed plasma sheet number

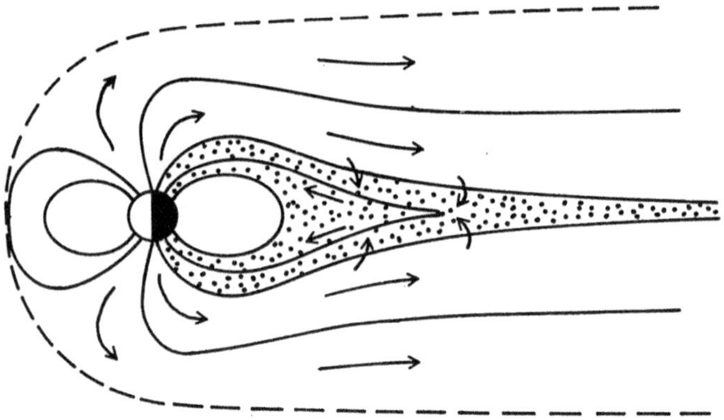

Fig. 1. A schematic diagram of the magnetosphere showing how polar wind plasma could become trapped on field lines in the tail which reconnect and contract towards the earth. The plasma sheet is indicated by the dotted region and the direction of plasma flow by the arrows.

densities are of the order of 1 cm^{-3} (Bame *et al.*, 1967; Vasyliunas, 1968; Gringauz, 1969), hence the degree of compression (associated with contraction of closed field lines) which must have taken place if the plasma is indeed associated with the polar wind, is a factor ~ 50. The corresponding increase of particle energy, assuming a simple adiabatic compression, is a factor $\sim 50^{2/3}$, so that the plasma sheet, if formed in this manner, would contain particles with energies of the order of a few hundred eV. The number density and average energy of the particles should be highest at the center (i.e., near the 'neutral' sheet), with both quantities decreasing towards the edge of the sheet, which should correspond to the last closed field lines. Further contraction of the magnetic field lines followed by convection deep within the magnetosphere could easily lead to particle energies of the order of tens of keV.

This description of the formation of the plasma sheet from polar wind (i.e., ionospheric) ions and electrons has a certain appeal despite its very qualitative nature. It should be noted that it does not necessarily defeat the purpose for which the polar wind was invented (viz., to explain the plasmapause and the loss of ^{4}He from the atmosphere), provided (a) only a fraction of the polar wind is trapped in this way before escaping, or (b) the particles are first convected into the magnetosphere and escape as they are convected out again, or (c) during polar substorms reconnection takes place deep within the plasma sheet so that large chunks of the latter tear off and are expelled from the tail of the magnetosphere. It is estimated that the average total polar wind flux is $\sim 3\times 10^{25}$ to 10^{26} ions $\sec^{-1}$ (i.e., a factor 10 larger than the average auroral proton flux) which suggests that in fact most of the particles concerned do ultimately escape.

It is possible of course that there are other ways in which ionospheric ions can be accelerated to relatively high energies (e.g., as a result of the viscous drag of the solar wind acting on the plasma just within the magnetopause). The point of the foregoing discussion is simply to show that an ionospheric origin for the energetic ions in the magnetosphere is not out of the question, and therefore that there is a good reason for carrying out studies of the composition and state of ionization of these ions.

3. Accretion of Particles from the Solar Wind

If the plasma sheet and other energetic particles in the magnetosphere originate mainly in the solar wind, there is no difficulty in explaining the particle energies involved since the solar wind can be expected to provide ~ 1 keV per particle initially. On the other hand it is by no means obvious on the basis of observational evidence how the particles can gain access to the magnetosphere, since the magnetopause seems on the whole to be rather impervious to the solar wind. However, it is important to note that we are only required to account for perhaps 10^{25} to 3×10^{25} protons $\sec^{-1}$ on the average (allowing for a 25% loss into the atmosphere and a 75% loss into space), which amounts to no more than $\sim 2\times 10^{-4}$ to 5×10^{-4} of the total solar wind flux incident on the magnetosphere. Thus a relatively slight diffusion of solar wind particles across the magnetopause would be sufficient to account for the necessary accretion rate. The energy carried by these particles can amount to only a few per cent at most of the total amount extracted from the solar wind, but of course the energy transfer is associated with processes which lead to the formation of the tail of the magnetosphere (e.g., Axford, 1967), and these may have nothing to do with particle transport.

It might be expected that solar wind particles can easily enter the magnetosphere along magnetic field lines that have undergone reconnection and thus directly link the geomagnetic and interplanetary magnetic fields. Indeed the high latitude aurora described, for example, by Lassen (1967) and by Eather and Akasofu (1969) could perhaps be regarded as evidence that direct entry does occur. Since we do not understand the details of the interaction between the solar wind and the magnetosphere associated with the reconnection process it is difficult to make any deductions concerning the rate at which solar wind particles can flow along 'open' magnetic field lines into the magnetosphere. However, it seems reasonable to surmise that if it occurs at all, the inflow takes place most easily on the upstream face of the magnetosphere (Figure 2), since further downstream where the solar wind is supersonic it should be relatively difficult for the particles to turn around and flow back towards the earth.

If the region of the magnetosphere near the magnetopause is sufficiently turbulent, solar wind particles might diffuse across the interface at a rate sufficient to give the required particle influx. Diffusion is most likely to occur in regions where the mean geomagnetic field strength is low and hence the relative amplitude of fluctuations $|\delta B/B|$ can easily be large. Thus the narrow bands along each side of the tail defined by the neutral sheet (Figure 2), are favorable regions for diffusion, especially since

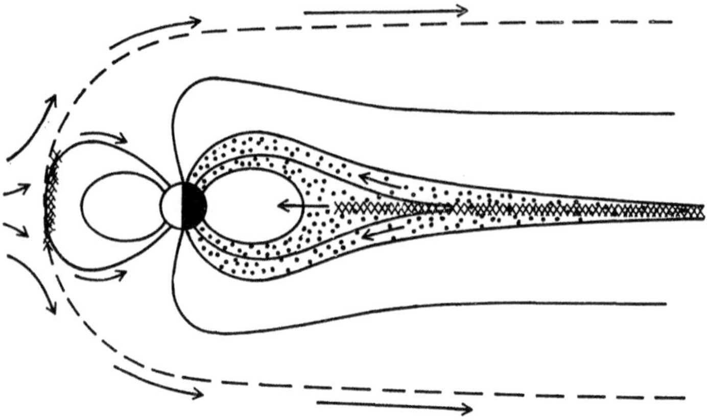

Fig. 2. A schematic diagram showing how solar wind plasma might enter the magnetosphere. The regions where entry takes place most easily are indicated by the hatching, the plasma sheet is indicated by the dotted region, and the direction of plasma flow is indicated by the arrows.

there are reasons for believing that the neutral sheet might be quite turbulent in any case. It should be noted that particles with large mass-to-charge ratios might have favored access to the magnetosphere in these circumstances.

If solar wind particles form a substantial fraction of the ions comprising the plasma sheet, the trapped radiation, and the auroral primaries, their presence should be detectable as a result of their characteristic composition and state of ionization. As a first approximation the composition should roughly correspond to that of the sun, and the state of ionization should correspond to thermal equilibrium at coronal temperatures. However, deviations from such a description of the plasma can be expected since (a) variations of composition are known to occur, (b) the state of ionization does not correspond exactly to thermal equilibrium, and (c) the processes which enable the particles to penetrate the magnetopause and the subsequent convection and acceleration mechanisms which act on the particles within the magnetosphere are likely to affect the various species of ion differently. Furthermore, the composition of the sun is not well established as far as some of the more interesting species are concerned (notably ^{2}H and ^{3}He) and even the relative abundance of ^{4}He is somewhat uncertain (e.g., Lambert, 1967).

In column (a) of Table I, the abundances in the solar system of selected species are listed (relative to that of Si taken as 10^6) according to Cameron (1968). For the heavier species we give the total abundance of all isotopes of each element and indicate only the isotopic number (A) of the most abundant. The state of ionization of each element (Z) corresponding to thermal equilibrium at 10^6 to 2×10^6 °K (e.g., Tucker and Gould, 1966) is given in column (b) of the table; the numbers indicate ions that are likely to account for 10% or more of the total. The approximate mass-to-charge ratio (A/Z) is given in column (c). The particle kinetic energies corresponding to an average bulk flow speed of 400 km sec^{-1} are listed in column (d). The heavier particles can be seen to have energies which approach those of low energy cosmic rays, if indeed it is a valid assumption that their mean velocity is that of the solar wind. The combination of high particle energies and high mass-to-charge ratios for these species has led Cameron (1963, 1964) to suggest that they might be very efficiently captured from the solar wind by the magnetosphere, and in particular that the atmospheric abundance of certain Xe isotopes might be explained in this way.

Estimates of the average fluxes of the various species in the solar wind are given in column (e). It should be noted that the observed flux of $^4He^{++}$ is less relative to $^1H^+$ than would be expected on the basis of photospheric abundances (see Hundhausen, 1968). This is presumably a result of diffusion or gravitational settling in the corona (e.g., Jokipii, 1966; Nakada, 1969). These effects are probably controlled by the mass-to-charge ratio of $^4He^{++}$ relative to $^1H^+$. Hence we have adjusted the estimated fluxes of other species to the observed average $^4He^{++}$ flux rather than the $^1H^+$ flux, and included a $<$ sign in cases where the mass-to-charge ratio should have a markedly unfavorable influence.

The average rates of accretion of solar wind ions by the atmosphere are given in column (f), assuming that there is no discrimination between species as far as access to and behavior within the magnetosphere are concerned. These rates have been calculated using the average auroral proton flux ($\sim 3 \times 10^{24}$ sec^{-1}) and the relative fluxes in the solar wind of each species with respect to protons from column (e). Estimates of the rates at which the various species are lost from the atmosphere as a result of polar wind and evaporative escape are listed in column (g). In the case of heavier particles the loss rates are very uncertain because the appropriate calculations of polar wind models have not been carried out yet. For the inert gases, the atmospheric content is listed in column (h) and in column (j) the time (τ) it would take to accumulate these amounts at the estimated present solar wind accretion rate assuming no losses.

At this point the table can be seen to contain a number of interesting implications, all based on the premise of course that the energetic protons and other ions in the magnetosphere and the aurora originate in the solar wind. For example, it is evident that on the average the earth loses ^1H and ^2H faster than it can accrete these isotopes from the solar wind. (Note that the ^2H losses are mainly associated with the polar wind.) It is not clear whether or not the earth gains more C, N and O than it loses because this requires a better knowledge of the polar wind losses than we have at present, but of course the amounts involved are trivial.

TABLE I

Element	(a) Abundance in sun	(b) Charge at coronal temperatures Z	(c) Mass to charge ratio A/Z	(d) Particle kinetic energy in solar wind (keV)	(e) Solar wind flux (cm^{-2} sec^{-1})
^{1}H	2.6×10^{10}	1	1	0.85	2×10^{8}
^{2}H	3.9×10^{6}	1	2	1.7	1.7×10^{4}
^{3}He	6.3×10^{5}	2	1.5	2.6	2.7×10^{3}
^{4}He	2.1×10^{9}	2	2	3.4	9×10^{6}
^{12}C	1.35×10^{7}	4–6	2.5	11	6×10^{4}
^{14}N	2.44×10^{6}	5–7	2.5	13	1.1×10^{4}
^{16}O	2.36×10^{7}	6–7	2.5	14	1×10^{5}
^{20}Ne	2.36×10^{6}	8	2.5	17	1×10^{4}
^{36}Ar	2.28×10^{5}	~ 14	2.5	34	1×10^{3}
^{84}Kr	64.4	~ 14	6	71	$< 3 \times 10^{-2}$
^{129}Xe	7.1	~ 14	9	110	$< 3 \times 10^{-3}$

The ^{3}He losses, which are due mainly to the evaporative escape of neutrals, are approximately in balance with solar wind accretion. It has been suggested previously (Axford, 1969a; Johnson and Axford, 1969) that this is in itself good evidence that the auroral primary ions originate in the solar wind. However, the argument has been weakened with the recent discovery of a possible primordial source of ^{3}He (Clarke and Craig, 1969). The rate of accretion of ^{4}He is only a few per cent of the rate at which it is produced by radiogenic processes; in this case it appears necessary to invoke the polar wind as the main loss mechanism if a condition of equilibrium exists (Axford, 1968).

Neon is an interesting case, since we seem to be within a factor 10^3 of being able to supply the entire atmospheric content (i.e., τ(Ne) is about 10^3 times the age of the earth). It is possible that effects associated with having $A/Z = 2.5$ could boost our estimate by this factor, but it is at least as reasonable to argue that the average solar wind flux has been higher in the past. In any case, the Ne abundance sets some sort of an upper limit on the amount of material that could have been accreted from the sun (MacDonald, 1963). Note however that MacDonald's estimates of the accretion flux are probably too high, since he simply assumed the capture cross section to be the geometrical cross section of the earth.

The atmospheric A is almost certainly of internal origin, being produced by the radioactive decay of ^{40}K; it is in any case not easy to argue that it might have been accreted. Krypton and Xe provide interesting but difficult problems; the present accretion rates fail by a factor $\sim 10^6$ to account for the atmospheric content, and if one follows Cameron's (1963, 1964) suggestion that the rate of accretion can be enhanced if A/Z is large the degree of enhancement must be an extremely steep function of this ratio. The best that can be done is to assume that the Ne content is entirely due to solar accretion, so that in the case of Xe the factor to be accounted for becomes of the order of $\tau(\mathrm{Xe})/\tau(\mathrm{Ne}) \approx 10^3$. Since the capture efficiency for

ospheric ɜtion rate -1)	(g) Atmospheric loss rate (sec^{-1})	(h) Atmospheric column content (cm^{-2})	(j) Accretion time τ (years)	(k) Auroral flux ($cm^{-2}\ sec^{-1}$)	(l) Amount collected by target foil (cm^3 STP)
10^{24}	2×10^{26}	–	–	3×10^7	4×10^{-5}
10^{20}	$\sim 10^{22}$	–	–	3×10^3	4×10^{-9}
$< 10^{19}$	4×10^{19}	1.41×10^{14}	7×10^5	350	4×10^{-10}
$< 10^{23}$	5×10^{24}	1.13×10^{20}	1.3×10^8	1.4×10^6	1.5×10^{-7}
10^{20}	–	–	–	9×10^3	1.5×10^{-8}
$< 10^{20}$	–	–	–	1.7×10^3	2×10^{-9}
$< 10^{20}$	–	–	–	1.5×10^3	1.5×10^{-9}
$< 10^{19}$	–	3.84×10^{20}	4×10^{12}	150	2×10^{-10}
$< 10^{18}$	–	1.93×10^{23}	2×10^{16}	15	2×10^{-11}
5×10^{14}	–	2×10^{19}	$> 7 \times 10^{15}$	$< 4.5 \times 10^{-3}$	$< 7 \times 10^{-15}$
5×10^{13}	–	1.6×10^{18}	$> 6 \times 10^{15}$	$< 4.5 \times 10^{-4}$	$< 7 \times 10^{-16}$

protons is 10^{-4} to 5×10^{-4}, it is evident that the magnetosphere must be almost 100% efficient in capturing Xe if Cameron's suggestion is correct. This is not impossible, but it is not entirely convincing, especially when it is remembered that the solar wind fluxes of the heavy ions quoted in column (e) are at best upper limits, and there is no allowance for discrimination against ions with large A/Z associated with processes occurring within the magnetosphere.

Magnetospheric particles with energies less than about 30 to 50 keV are almost certainly injected directly by convection (Axford, 1969b; Kennel, 1969). As far as the protons and other ions are concerned, the process is adiabatic to a first approximation, and thus it is reasonable to think in terms of 'forbidden regions' (Alfvén and Fälthammar, 1963). It is easily seen from the simplest theory of this effect that the size of the forbidden region for a particular species varies as $(A/Z)^{1/4}$ if the particles have the same energy per nucleon initially. Thus one should expect some discrimination against species with $A/Z > 1$ whenever convection is the dominant injection process. The effect is not very pronounced, however, unless A/Z is quite large. Protons typically penetrate to the plasmapause at $L \approx 4$, hence α-particles ($A/Z = 2$) should penetrate to $L \approx 5$ without difficulty and thus should normally be present in the aurora. Xenon ions with $A/Z \approx 9$ will have difficulty in penetrating closer than $L \approx 7$ in most conditions, which suggests that there might be some discrimination against accretion of Xe by the earth.

Nonadiabatic processes are more difficult to discuss since they require that we know the power spectrum of fluctuating fields reasonably well if we wish to estimate diffusion coefficients. To precipitate ions into the atmosphere it is probably necessary that there should be hydromagnetic turbulence with a small scale length in order to change the pitch angle by a resonant interaction of the particle gyration with Doppler shifted field fluctuations (e.g., Kennel, 1969; Tverskoy, 1969). Similarly, diffusion into and out of the magnetosphere is controlled by a resonant interaction with electric field fluctuations at the drift frequency (Fälthammar, 1965; Tverskoy, 1969). Clearly particles with differing A/Z will resonate with different parts of the fluctuation spectrum whatever process is involved, but unfortunately one cannot make any predictions concerning the relative ease of diffusion of different species without a detailed knowledge of the power spectrum of the fluctuations. It is important to note however, that there is no *a priori* reason for believing that the relative abundances of the different species in the solar wind should be preserved within the magnetosphere. One can only hope that the distortion is not so great as to make it difficult to distinguish between particles which originate in the ionosphere and those which originate in the solar wind. (Tverskoy (1965) and Hess (1966) have endeavored to predict the ratio of α-particle and proton abundances in the radiation belts, but results are open to the objections raised above and others.)

4. Observations

The first attempts to identify species other than protons and electrons in the energetic particle population of the magnetosphere involved the use of rocket borne nuclear emulsions (Freden and White, 1959, 1960, 1962; Armstrong *et al.*, 1961; Heckman and Armstrong, 1962; Naugle and Kniffen, 1963). The flights took place in the inner radiation belt and the shielding was such that only particles with energies exceeding about 10 MeV/nucleon were detected. Of more than 1700 tracks examined, all were due to protons except for six due to ^{2}H and five due to ^{3}H; no α-particles were detected. It is presumed that the ^{2}H and ^{3}H were produced by interactions between energetic trapped protons and atmospheric nuclei (Heckman and Armstrong, 1962). An upper limit for the flux of all nuclei heavier than protons was reported by Naugle and Kniffen (1963) to be 6×10^{-3} of the proton flux in the proton energy range 12 to 31 MeV. These results were confirmed by Fenton (1967) using data from the Chicago charged particle telescope on OGO 1. In this case the data sample included 15000 protons, and with no α-particles being observed, Fenton was able to put an upper limit of 6×10^{-4} on the α-particle to proton ratio in the range 26 to 85 MeV/nucleon.

More recently observations from satellite borne detectors at lower energies ($\gtrsim$0.5 MeV/nucleon) have shown that α-particles are present (Krimigis and Van Allen, 1967, 1968; Krimigis *et al.*, 1968; Paulikas and Blake, 1968; Fritz and Krimigis, 1969; Van Allen, 1969). According to Fritz and Krimigis (1969) the ratio of the intensities of α-particles and protons is of the order of 2×10^{-4} for particles with energies greater than 0.46 MeV/nucleon; the ratio increases if the comparison is made on the basis of equal energy per charge or equal total energy. [See also the article by Krimigis (1970).]

If one were to assume that the particles observed in the radiation belts originated in the solar wind, then these data provide useful if not conclusive information on the injection and loss processes. However, any measurement which involves having the particles pass through matter is unable to distinguish between $^4He^+$ and $^4He^{++}$ in the incident flux, and thus cannot give an answer to the problem under discussion. It will be recalled that theoretical studies of the solar wind (Banks and Holzer, 1969a, 1969b) suggest that the flux of $^4He^+$ should be about 1% of the proton flux, and that observations of low energy particles in the plasmasphere also indicate $[^4He^+]:[^1H^+] \approx 10^{-2}$ (Taylor *et al.*, 1965). Since the ratio of the concentration of ^4He ions to protons (i.e., $\{[^4He^+]+[^4He^{++}]\}:[^1H^+]$) in magnetospheric plasma originating in ionosphere has roughly the same value as in the solar wind there is no obvious reason for expecting vastly different concentrations of ^4He ions in the radiation belts if the particles originate from either of these sources. Experiments which can distinguish between singly and doubly ionized ^4He should of course be able to provide an immediate answer, but these are not easily constructed if they have to cope with more or less isotropic particle distributions. The superthermal ion detector on ATS 1 (Freeman, 1968) does have such a capability but since the particles normally observed have quite low energies ($\sim$1 eV) it is difficult to believe that they can be of other than ionospheric origin, and thus the ^4He ions should be $^4He^+$. Freeman (1968) has noted that $^4He^{++}$ seems to be present in the observed spectra with a relative concentration of a few per cent, however, this requires further confirmation since the uncertainty of the measurements is rather large in the region of interest.

A few attempts to measure the flux of energetic ^4He ions in the aurora have been reported. The first of these (Albert, 1967) has been heavily criticized (Albert, 1968; Reasoner *et al.*, 1968; Wax and Bernstein, 1968) and the results do appear to be rather dubious. Reasoner *et al.* (1968) have measured energy spectra of precipitating ions in the aurora using two types of detectors simultaneously. They have suggested that the observed spectra could be interpreted simply if α-particles were present having a flux of $1.6 \pm 0.7\%$ of the proton flux at the same energy per charge. Clear evidence for the presence of ^4He ions among the auroral primaries has been obtained by Whalen and McDiarmid (1968a, 1969) using a rocket borne magnetic spectrometer capable of separating $^4He^+$ ions from protons with energies greater than 50 keV. With a few reasonable assumptions it can be deduced from the results of this experiment that the ^4He ion to proton ratio is roughly 0.18 at a particle energy of about 50 keV, or roughly 0.02 at an energy per unit charge of about 50 keV (Reasoner, 1968; Whalen and McDiarmid, 1968b). However the ratio is determined it is apparent from this experiment that the auroral primaries do contain ^4He ions, although whether the incident beam contains mainly $^4He^+$ or $^4He^{++}$ ions is not known since at the altitudes attained the beam would have been approximately in charge equilibrium.

Attempts to detect ^4He line emissions (especially λ5876 HeI) in the aurora have usually been unsuccessful (Eather, 1968) although with the energetic ^4He ion fluxes determined in the experiments described above, the emission intensities must have been on the border line of detectability (Reasoner *et al.*, 1968). Stoffregen (1969), however, has reported two cases in which λ 5876 HeI was observed during very strong auroral events.

In summary, there is evidence for the existence of ^4He ions among the auroral primaries, but it is less certain that these are $^4He^{++}$ rather than $^4He^+$. The only way this can be shown definitively is to use detectors which discriminate on the basis of particle momentum as well as energy. A preliminary experiment using such a detector has been flown into the aurora (McDiarmid, 1969) but to date no results are available. (An experiment which can separate solar wind ions in a similar manner has been flown on Explorer 34 (Ogilvie *et al.*, 1968a, 1968b; Ogilvie and Wilkerson, 1969), but unfortunately it was designed only to cope with $^1H^+$ and $^4He^{++}$.)

It is evident from the foregoing discussion that alternative techniques for measuring the composition of magnetospheric and auroral ions should be examined. If only ^4He ions are to be considered in addition to protons, it is necessary to be able to discriminate between $^4He^+$ and $^4He^{++}$ if the origin of the particles is to be determined, and this is not a simple matter. Since it is difficult enough to detect ^4He it does not seem likely that other species can be detected with commonly used instruments. However, there is one relatively simple and cheap experiment which would allow us to detect most species of interest, as well as providing information on the rate at which they are accreted by the atmosphere. The experiment is essentially the same as that proposed for the solar wind by Signer *et al.* (1965) and carried out for the first time during the Apollo 11 landing on the moon. This consists of exposing a thin, high purity target foil for a suitable period in the medium to be investigated (i.e., the solar wind in the case of the Apollo 11 experiment), bringing the foil back to earth and analyzing the accumulated nuclides using various high sensitivity techniques that have been developed for analyzing meteorites.

In the case of magnetospheric and auroral particles, it would be necessary to carry the target foil on a satellite at a suitable altitude and expose it for several days before returning it to earth. A polar orbit would be most suitable for detecting auroral primaries and thereby measuring atmospheric accretion rates since in this case the satellite traverses the auroral zone four times during each orbit. If the satellite altitude is relatively low (say 300 to 400 km), the particles captured should be predominantly associated with the aurora. The optimum orbit for this purpose is one in which the orbital plane contains the sun, since the satellite will then pass through the breakup region twice during each orbit.

Let us suppose that a thin foil of area 10^3 cm^2 is carried on a satellite orbiting as described, and exposed for 5 days before being returned to earth. During this period the satellite spends about 10% of the time traversing the auroral zones. Thus the total exposure to the auroral primary beam is about 4×10^4 sec. Estimates of the fluxes of auroral primaries are given in column (k) and the total amount of each species accumulated in the target foil during the mission is given in column (1) of Table I.* These quantities are small, but, except for Kr and Xe can be detected by available high precision techniques. It should be remembered however that the amounts quoted could be over estimates because our calculations have not taken into account any

* The experiment could also be performed if it is part of a recoverable rocket payload. In this case the much shorter exposure time available ($\sim$ 200 sec) could be compensated for to some extent by firing into a very bright aurora. Under favorable conditions the quantities collected should be perhaps a factor 10 less then those listed in column 1.

possible discrimination against species with large A/Z taking place within the magnetosphere.

Since the experiment proposed here does not permit any evaluation of the charge of the incident ions, it is necessary to be careful in interpreting any results since the relative amounts of the species captured, in most cases would not be vastly different if the ions originated in the ionosphere rather than the solar wind. In fact it appears that only ^3He can provide a definitive result since the amount of ^3He one expects to be accumulated in the target is at least one hundred times greater for the case of solar wind ions than for ionospheric ions. It is an advantage that $^3He^{++}$ ions are the least likely of all species heavier than ^1H to be discriminated against in the solar wind and in the magnetosphere.

Finally, it is worth noting that if the auroral primaries are shown to be of solar wind origin, it should be possible to measure the rates of accretion of Ne and A, and to test the suggestion due to Cameron (1963, 1964) that the large A/Z ratio of solar wind Xe ions might result in a great enhancement of the accretion rate. If the atmospheric Xe is indeed accreted from the solar wind, our estimates of the amounts of Xe (and presumably Kr) that can be accumulated in the target are too small by a factor $\approx 10^3$.

Acknowledgments

This work was supported in part under NASA Grant No. NGR-05-009-081, and in part by the Advanced Research Projects Agency (Project DEFENDER) and was monitored by the U.S. Army Research Office, Durham, under contract DA-31-124-ARO-D-257.

References

Albert, R. D.: 1967, *Trans. Amer. Geophys. Union* **48**, 866.
Albert, R. D.: 1968, *J. Geophys. Res.* **73**, 4453.
Alfvén, H. and Fälthammar, C.-G.: 1963, *Cosmical Electrodynamics: Fundamental Principles*, Oxford University Press, New York.
Armstrong, A. H., Harrison, F. B., Heckman, H. H., and Rosen, L.: 1961, *J. Geophys. Res.* **66**, 351.
Axford, W. I.: 1967, *Space Sci. Rev.* **7**, 149.
Axford, W. I.: 1968, *J. Geophys. Res.* **73**, 6855.
Axford, W. I.: 1969a, in *Atmospheric Emissions* (ed. by B. M. McCormac and A. Omholt), Van Nostrand and Reinhold Company, New York, p. 317.
Axford, W. I.: 1969b, *Rev. Geophys.* **7**, 421.
Bame, S. J., Asbridge, J. R., Felthauser, H. E., Hones, E. W., and Strong, I. B.: 1967, *J. Geophys. Res.* **72**, 113.
Banks, P. M.: 1969a, in *Atmospheric Emissions* (ed. by B. M. McCormac and A. Omholt), Van Nostrand and Reinhold Company, New York, p. 533.
Banks, P. M.: 1969b, in *Production and Maintenance of the Polar Ionosphere* (ed. by G. Scovli), to be published.
Banks, P. M. and Holzer, T. E.: 1968, *J. Geophys. Res.* **73**, 6846.
Banks, P. M. and Holzer, T. E.: 1969a, *J. Geophys. Res.* **74**, 6304.
Banks, P. M. and Holzer, T. E.: 1969b, *J. Geophys. Res.* **74**, 6317.
Block, L. P. and Fälthammar, C.-G.: 1968, *J. Geophys. Res.* **73**, 4807.

Cameron, A. G. W.: 1963, *Icarus* **1**, 314.
Cameron, A. G. W.: 1964, in *The Origin and Evolution of the Atmosphere and Oceans* (ed. by P. J. Brancazio and A. G. W. Cameron), John Wiley and Sons, Inc., New York, p. 235.
Cameron, A. G. W.: 1968, in *Origin and Distribution of the Elements* (ed. by L. H. Ahrens), Pergamon Press, Oxford, p. 125.
Clarke, W. B. and Craig, H.: 1969, *Trans. Amer. Geophys. Union* **50**, 209.
Dessler, A. J. and Hanson, W. B.: 1961, *Astrophys. J.* **134**, 1024.
Dessler, A. J. and Michel, F. C.: 1966, *J. Geophys. Res.* **71**, 1421.
Eather, R. H.: 1968, *Space Res.* **8**, 201.
Eather, R. H. and Akasofu, S.-I.: 1969, *J. Geophys. Res.* **74**, 4794.
Evans, J. E., Joki, E. G., Johnson, R. G., and Sharp, R. D.: 1966, *Space Res.* **6**, 773.
Fälthammar, C.-G.: 1965, *J. Geophys. Res.* **70**, 2503.
Fenton, K. B.: 1967, *J. Geophys. Res.* **72**, 3389.
Freden, S. C. and White, R. S.: 1959, *Phys. Rev. Letters* **3**, 9.
Freden, S. C. and White, R. S.: 1960, *J. Geophys. Res.* **65**, 1377.
Freden, S. C. and White, R. S.: 1962, *J. Geophys. Res.* **67**, 25.
Freeman, J. W., Jr.: 1968, *J. Geophys. Res.* **73**, 4151.
Fritz, T. A. and Krimigis, S. M.: 1969, *J. Geophys. Res.* **74**, 5132.
Gringauz, K. I.: 1969, *Rev. Geophys.* **7**, 339.
Heckman, H. H. and Armstrong, A. H.: 1962, *J. Geophys. Res.* **67**, 1255.
Hess, W. N.: 1966, in *Radiation Trapped in the Earth's Magnetic Field* (ed. by B. M. McCormac), D. Reidel Publishing Company, Dordrecht-Holland, p. 352.
Hoffman, J. H.: 1968, *Trans. Amer. Geophys. Union* **49**, 253.
Holzer, T. E.: 1969, in *Production and Maintenance of the Polar Ionosphere* (ed. by G. Scovli), to be published.
Hundhausen, A. J.: 1968, *Space Sci. Rev.* **8**, 690.
Johnson, H. E. and Axford, W. I.: 1969, *J. Geophys. Res.* **74**, 2433.
Jokipii, J. R.: 1966, in *The Solar Wind* (ed. by R. J. Mackin, Jr. and M. Neugebauer), Pergamon Press, New York, p. 215.
Kennel, C. F.: 1969, *Rev. Geophys.* **7**, 379.
Krimigis, S. M.: 1970, this volume, p. 364.
Krimigis, S. M. and Van Allen, J. A.: 1967, *J. Geophys. Res.* **72**, 5779.
Krimigis, S. M. and Van Allen, J. A.: 1968, in *Earth's Particles and Fields* (ed. by B. M. McCormac), Reinhold Publishing Corporation, New York, p. 127.
Krimigis, S. M., Van Allen, J. A., and Burns, A. L.: 1968, *Trans. Amer. Geophys. Union* **49**, 267.
Lambert, D. L.: 1967, *Nature* **215**, 43.
Lassen, K.: 1967, in *Aurora and Airglow* (ed. by B. M. McCormac), Reinhold Publishing Corporation, New York, p. 453.
MacDonald, G. J. F.: 1963, *Rev. Geophys.* **1**, 305.
McDiarmid, I. B.: 1969, private communication.
Michel, F. C.: 1968, *Trans. Amer. Geophys. Union* **49**, 740.
Nakada, M. P.: 1969, *Solar Phys.* **7**, 302.
Naugle, J. E. and Kniffen, D. A.: 1963, *J. Geophys. Res.* **68**, 4065.
Ogilvie, K. W., McIlwraith, N., and Wilkerson, T. D.: 1968a, *Rev. Sci. Inst.* **39**, 441.
Ogilvie, K. W., Burlaga, I. F., and Wilkerson, T. D.: 1968b, *J. Geophys. Res.* **73**, 6809.
Ogilvie, K. W. and Wilkerson, T. D.: 1969, *Trans. Amer. Geophys. Union* **50**, 302.
Paulikas, G. A. and Blake, J. B.: 1968, paper presented at International Symposium on the Physics of the Magnetosphere, Washington, D.C.
Perkins, F. W.: 1968, *J. Geophys. Res.* **73**, 6631.
Reasoner, D. L., Eather, R. H., and O'Brien, B. J.: 1968, *J. Geophys. Res.* **73**, 4185.
Reasoner, D. L.: 1968, *J. Geophys. Res.* **73**, 6387.
Sharp, R. D. and Johnson, R. G.: 1968, *J. Geophys. Res.* **73**, 969.
Signer, P., Eberhardt, P., and Geiss, J.: 1965, *J. Geophys. Res.* **70**, 2243.
Speiser, T. W.: 1969, in *Atmospheric Emissions* (ed. by B. M. McCormac and A. Omholt), Van Nostrand and Reinhold Company, New York, p. 337.
Stoffregen, W.: 1969, in *Production and Maintenance of the Polar Ionosphere* (ed. by G. Scovli), to be published.

Taylor, H. A., Jr., Brinton, M. C., and Smith, C. R.: 1965, *J. Geophys. Res.* **70**, 5769.
Tucker, W. H. and Gould, R. J.: 1966, *Astrophys. J.* **144**, 244.
Tverskoy, B. A.: 1965, *Geomagn. Aeron.* **5**, 617.
Tverskoy, B. A.: 1969, *Rev. Geophys.* **7**, 219.
Van Allen, J. A.: 1969, *Rev. Geophys.* **7**, 233.
Vasyliunas, V. M.: 1968, *J. Geophys. Res.* **73**, 2839.
Wax, R. L. and Bernstein, W.: 1968, *J. Geophys. Res.* **73**, 4452.
Whalen, B. A. and McDiarmid, I. B.: 1968a, *J. Geophys. Res.* **73**, 2307.
Whalen, B. A. and McDiarmid, I. B.: 1968b, *J. Geophys. Res.* **73**, 6388.
Whalen, B. A. and McDiarmid, I. B.: 1969, in *Atmospheric Emissions* (ed. by B. M. McCormac and A. Omholt), Van Nostrand and Reinhold Publishing Company, New York, p. 93.

MATHEMATICAL MODELS OF MAGNETOSPHERIC CONVECTION AND ITS COUPLING TO THE IONOSPHERE

VYTENIS M. VASYLIUNAS

Dept. of Physics and Center for Space Research, Massachusetts Institute of Technology, Cambridge, Mass., U.S.A.

1. Introduction

The phrase 'magnetospheric model' can have several different meanings. Here I am using the term 'model' in the sense described by Parker (1968): "We construct idealized and simplified theoretical models for the purpose of demonstrating how the basic laws of physics lead to a certain observed effect." A model in this sense is a solution of the equations that describe the system under consideration. Obtaining an exact solution of the equations governing a system as complex as the magnetosphere is clearly impossible, and to construct a theoretical model the equations must be simplified (often drastically) to the point of tractability. The aim is to isolate those aspects of the physical situation that are essential to the particular phenomenon one is attempting to understand. One thus proceeds by solving the basic equations under a variety of simplifying assumptions and noting what assumptions are required to reproduce the essential features of the phenomenon under study. Of course, no model will predict in precise quantitative detail all the features of the observations, but then our primary goal is understanding, not forecasting.

In this paper I discuss theoretical models (in the sense described above) of some aspects of magnetospheric convection and its effects on charged particle populations. Starting with the classic work of Axford and Hines (1961), the concepts and consequences of magnetospheric convection have been discussed widely in qualitative form (e.g., Axford, 1969; Kennel, 1969); the task now is to express some of these ideas in quantitative form as appropriate solutions of the basic equations. First I will formulate the mathematical problem of calculating a model of steady state convection and the associated particle distribution. Then I will discuss solutions of several parts of the problem (a solution of the complete problem has not yet been obtained) and their relation to observed phenomena.

2. Magnetospheric Convection as a Mathematical Problem

In principle, magnetospheric convection could be treated with MHD equations, relating the flow velocity to the stresses present in the plasma; this approach has been highly successful in treating the solar wind. The flow speeds associated with convection, however, are expected to be very small compared to propagation speeds of waves in the magnetospheric medium, and the characteristic flow times thus should be very long compared to the travel time of a signal between the ionosphere and the equa-

B. M. McCormac (ed.), Particles and Fields in the Magnetosphere.

torial plane (except possibly in the region of 'open' field lines deep within the magnetotail). Thus, we expect the plasma at any point of a magnetic flux tube within the magnetosphere to remain continually 'adjusted' to the boundary conditions at the feet of that flux tube on the ionosphere. With the expected close coupling between the magnetosphere and the ionosphere, two media with very different dynamical properties, it proves convenient to formulate the convection problem not in terms of the dynamical concepts of flow and stress but in terms of electric field and current (fundamentally, of course, these two modes of treating the problem are equivalent, since in a plasma there is a close connection between the flow and the electric field and between the stress and the electric current).

The electric field and the charged particle distribution in the magnetosphere can be simultaneously calculated from a closed self-consistent chain of equations as first

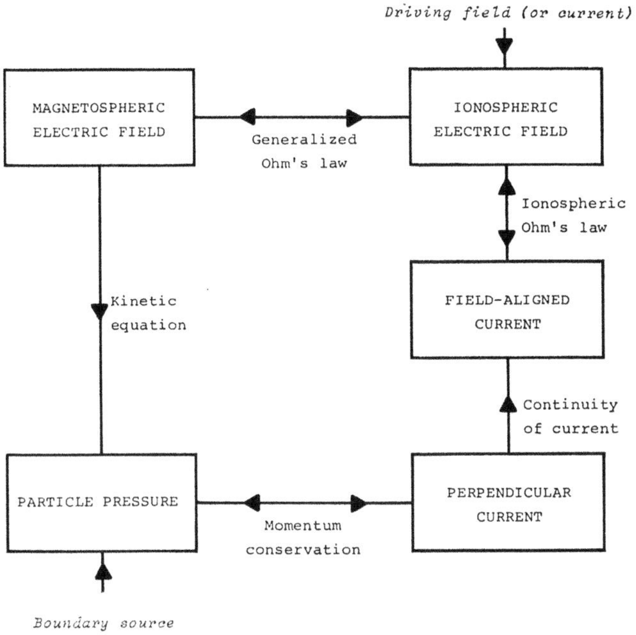

Fig. 1. Outline of the self-consistent calculation of magnetospheric convection. The principal quantities being calculated are shown enclosed in boxes. Each line joining two boxes is labeled with the physical principle or equation that provides the link between the two quantities; the arrow points to that one of the two which can be calculated from the other. Quantities written in italics are boundary conditions that must be specified. ('Kinetic equation' is used as a general term to include whatever method for computing the particle distribution is deemed to be appropriate, e.g., Liouville's theorem, Boltzmann–Vlasov equation and its transformations, etc.)

introduced by Fejer (1964) and later used by Swift (1967, 1968). The complete calculation is outlined in Figure 1. Since it is a closed loop, we can break into it at any point; it is convenient to start with the electric field. Let us provisionally suppose, then, that we know the electric field configuration in the magnetosphere:

(a) First link: with knowledge of the electric field, we calculate the motion and distribution of protons and electrons in the magnetosphere, and hence in particular the total plasma pressure at any point;

(b) Second link: from the plasma pressure gradients we calculate the components of the electric current perpendicular to the magnetic field.

(c) Third link: by calculating the divergence of the perpendicular current and averaging over each flux tube, we obtain the field aligned currents flowing between the magnetosphere and the ionosphere.

(d) Fourth link: from the requirement that these field aligned currents be closed by perpendicular ohmic currents in the ionosphere, we obtain the configuration of the electric field in the ionosphere.

(e) Fifth and final link: the ionospheric electric field can be mapped into the magnetosphere, and the requirement that it agree with the magnetospheric electric field assumed at the outset determines the field and thus closes the system of equations.

Let us now examine in more detail the individual links of the chain.

(a) Calculation of the charged particle distribution in a given configuration of electric (and magnetic) fields is an extensively studied and familiar process. The simplest method, if only adiabatic processes are considered, is to calculate single particle trajectories (e.g., Alfvén, 1939; Taylor and Hones, 1965; Kavanagh *et al.*, 1968) and apply Liouville's theorem; or one may develop, by suitable transformations of the Boltzmann equation, a transport equation (e.g., Siscoe, 1964; Vasyliunas, 1969), with the capability of including nonadiabatic scattering, acceleration, and loss processes. As a boundary condition, it is necessary to specify the particle population at the boundary of the region under study, at those points of the boundary where particles can move into the region. Usually the region studied is the magnetosphere out to some 10 to 15 R_E on the night-side and the electric field is such that particles move from the magnetotail toward the earth; then it is necessary to specify the particle population at the near earth end of the magnetotail, and the observed plasma sheet population is a reasonable choice.

(b) The calculation of the perpendicular current from the total pressure can be looked at in two equivalent ways. We may write down the momentum conservation equation for the plasma, which in the approximation of flow speed very small compared to particle thermal speed (valid throughout the magnetosphere beyond the plasmapause, except possibly within the neutral sheet) reduces to

$$\nabla \cdot \boldsymbol{P} = \frac{1}{c} \mathbf{J} \times \mathbf{B} \tag{1}$$

where $\boldsymbol{P}$ is the total (proton plus electron) pressure tensor and the rest of the symbols have their usual meaning (Gaussian units). Solving for the perpendicular component of the current density yields, if the pressure is isotropic

$$\mathbf{J}_\perp = \frac{c\mathbf{B} \times \nabla P}{B^2} \tag{2}$$

and if the pressure is anisotropic but field aligned

$$\mathbf{J}_\perp = \frac{c\mathbf{B} \times [\nabla P_\perp + (P_\parallel - P_\perp) \nabla B/B]}{B^2 + 4\pi (P_\perp - P_\parallel)}. \tag{3}$$

Equivalently, we may sum the currents due to all the single particle gradient, curvature, and magnetization drifts obtaining the same result (e.g., Parker, 1957). Note that $\mathbf{J}_\perp$ depends only on the total pressure and the magnetic field and has no *explicit* dependence on the electric field or the detailed energy spectra of the particles. Given only the configuration of the magnetic field and the total pressure, however obtained, the perpendicular current is uniquely determined. (It is assumed here that the general configuration of the magnetic field is known from observation or field models, and that the detailed magnetic effects of **J** can be neglected.)

(c) The current density given by Equation (2) or (3) in general has a nonzero divergence. The total current density, however, must have zero divergence even if the convection is assumed to be highly nonsteady. It is readily shown that under magnetospheric conditions the charge accumulation implied by the divergence of $\mathbf{J}_\perp$ alone would lead to electric field changes on a time scale of milliseconds (not a surprising result if one recalls that allowing $\nabla \cdot \mathbf{J} \neq 0$ is the same as keeping the displacement current term in Maxwell's equations). Hence, in all magnetospheric processes other than high frequency plasma waves $\nabla \cdot \mathbf{J} = 0$ and the divergence of $\mathbf{J}_\perp$ must be cancelled by the divergence of a magnetic field aligned current density $J_\parallel$. From the divergence of Equation (2) or (3) thus we obtain the derivative of $J_\parallel$ along the magnetic field direction:

$$\frac{\partial}{\partial l}\left(\frac{J_\parallel}{B}\right) = \frac{2c(\mathbf{B} \times \nabla B) \cdot \nabla P}{B^4} \tag{4}$$

for the isotropic pressure, or

$$\frac{\partial}{\partial l}\left(\frac{J_\parallel}{B}\xi\right) = \frac{c(\mathbf{B} \times \nabla B) \cdot \nabla (P_\perp + P_\parallel)}{\xi B^4} \tag{5}$$

where $\xi \equiv 1 + 4\pi(P_\perp - P_\parallel)/B^2$ for the general case. As expected from a simple picture of particle drifts, field aligned currents are present whenever the pressure gradient has a component along the direction of the ∇B drift. If, for example, we have a region of enhanced pressure within a narrow longitude sector, a field aligned current flows into the ionosphere at its western edge and out of the ionosphere at its eastern edge. The total current density flowing in or out of the ionosphere at any point must be found by integrating Equation (4) or (5) along the field line; with the isotropic pressure, which must be constant along the field line, Equation (4) can be integrated explicitly (Vasyliunas, 1969), yielding

$$J_\parallel = \frac{B_m}{2B_e} c\left(\nabla P \times \frac{\mathbf{B}_e}{B_e}\right) \cdot \nabla \int \frac{\mathrm{d}l}{B} \tag{6}$$

where $\mathbf{B}_e$ is the magnetic field at the equator and B_m the field magnitude at the ionosphere ('mirror field'). The gradients are with respect to the equatorial coordinates of the field line, the integration is along the entire length of the field line above the ionosphere, and $J_{\|}$ is per unit area of the *ionosphere* and positive if *into* the ionosphere.

(d) Continuity of current in the ionosphere requires that

$$\nabla \cdot \mathbf{I} = J_{\|} \sin \chi \tag{7}$$

where **I** is the height integrated horizontal current density, χ is the inclination of the magnetic field, and $J_{\|}$ is the field aligned current density flowing into the ionosphere. If the electric field **E** at the top of the ionosphere has no horizontal variation except on a sufficiently long scale ($\gtrsim$10 km), **I** can be related to **E** by the equation

$$\mathbf{I} = \boldsymbol{\Sigma} \cdot \left(\mathbf{E} + \frac{\mathbf{V}_{\mathrm{n}}}{c} \times \mathbf{B} \right) \tag{8}$$

where $\mathbf{V}_{\mathrm{n}}$ is the velocity of the neutral gas in the ionosphere (commonly assumed to corotate with the earth) and $\boldsymbol{\Sigma}$ is the familiar height integrated conductivity tensor (e.g., Fejer, 1953, 1964). Inserting Equation (8) into (7) provides a differential equation from which **E** may be obtained if $J_{\|}$ is known. ($\partial \mathbf{B}/\partial t = -c\nabla \times \mathbf{E}$ must be known or set to zero.) Note that given the conductivity and the neutral gas motion **E** is uniquely determined by $J_{\|}$ (or vice versa) and does not depend *explicitly* on any other magnetospheric variables such as pressure gradients.

(e) The electric field at the top of the ionosphere and the electric field within the magnetosphere are related by the generalized Ohm's law, and thus either can be calculated from the other. For quasi-steady convection, in which the electric field can be derived from a potential, one needs to calculate the potential drop along the field line from the ionosphere to the equatorial plane. A common assumption that appears to be at least a good first approximation is that the conductivity along the field lines is high enough so that this potential drop can be neglected and the potential simply mapped out from the ionosphere into the magnetosphere. Anomalous resistivity caused by instabilities (e.g., Kennel, 1970) may produce significant potential drops which, in principle, can be calculated if $J_{\|}$ is known.

This entire calculation should provide a reasonably complete model of the processes associated with convection within the 'closed' part of the magnetosphere, in which the magnetic field lines connect from one hemisphere to the other. The 'open' field lines that extend far out into the magnetotail must ultimately enter either the interplanetary medium, as in the models of Dungey (1961) and his followers, or/and a magnetospheric boundary layer within which a viscous drag (Axford and Hines, 1961; Axford, 1964) or some other tangential stress responsible for maintaining the magnetotail occurs (Axford *et al.*, 1965; Siscoe, 1966). On these magnetic field lines, the electric field is determined by processes associated with the field line merging or the tangential stress, whose physics at present is largely unknown; it is this electric field, however, that sets the entire convection system going and thus it may appropriately be called the *driving field.* Equations (7) and (8) are still applicable and given the driving field serve to determine the field aligned current density $J_{\parallel}$ in the polar cap region. Alternatively, $J_{\parallel}$ may be fixed by the field line merging or tangential stress processes in which case it is the *driving current* setting up the convection; Equations (7) and (8) then determine the associated electric field. Finally, if the calculation is to be restricted only to the closed field line region, the driving field must still be partially specified as the boundary condition on **E** now required to solve Equations (7) and (8).

In summary, understanding the origin of the driving electric field (or current) and the origin of the plasma sheet is a prerequisite for a complete model of magnetospheric convection, and such understanding we do not now possess. However, the properties of the plasma sheet particles are well known from observation and, as will appear shortly, crude but useful models of the driving field can be guessed. Given both of these factors the set of equations described here allows one to construct a model of the convection within the closed part of the magnetosphere, including the effects of interactions between the plasma, the electric fields, and the ionosphere; this now requires nothing more than solution of a well defined computational problem! Even with drastic simplifications the problem is extremely complicated and I have not yet obtained a solution of the complete self-consistent chain. Individual links of the chain, however, have been studied and useful results obtained. Models of the first link, particle distribution in given electric fields, have been constructed by Taylor and Hones (1965) and Kavanagh *et al.* (1968) from a strictly adiabatic viewpoint. I have studied the same problem with the inclusion of nonadiabatic effects; the calculation is published elsewhere (Vasyliunas, 1969), and the basic physical ideas and some results have been described by Kennel (1969). In the following section I consider simple models of another link: the connection between field aligned currents and electric fields in the ionosphere.

3. A Model Electric Field and Current System in the Ionosphere

The basic assumption of the convection models first introduced by Axford and Hines (1961) and Dungey (1961) is that magnetic field lines from the polar caps (assumed to extend into interplanetary space or a viscous drag boundary layer) move, at least partially, with the solar wind. This implies an electric field over the polar cap (the driving field) directed in a general way from dawn to dusk; thus on the polar cap boundary the electric potential (assuming now that any variations in the convection pattern are slow enough so that induction fields can be neglected) is positive on the dawn side and negative on the dusk side. To obtain a very simple model, we approximate the polar cap boundary by a circle of constant latitude $\lambda = 72°$ and assume the potential on the boundary

$$\Phi(\lambda, \phi) = \Phi_0 \sin\phi \quad \text{at} \quad \lambda = 72° \tag{9}$$

where ϕ is the longitude measured eastward from the midnight meridian and Φ_0 is a constant. Since we are now interested in ionospheric effects on convection, we neglect particle pressure gradient effects and solve Equations (7) and (8) with $J_{\|}=0$ to find the potential at lower latitudes with Equation (9) as the boundary condition. The solutions are obtained in a frame of reference rotating with the earth and $\mathbf{V}_n$ in Equation (8) is set to zero. For simplicity, symmetry between the two hemispheres is assumed and day-night variations in the conductivity are neglected (some of the effects neglected here are included in a similar calculation recently done independently by Wolf (1969). The conductivity model of Fejer (1953, 1964) is used, with a value of 3.5 for the ratio of height integrated Hall–Pedersen conductivities.

The resulting potential configuration in the ionosphere is shown in Figure 2. It

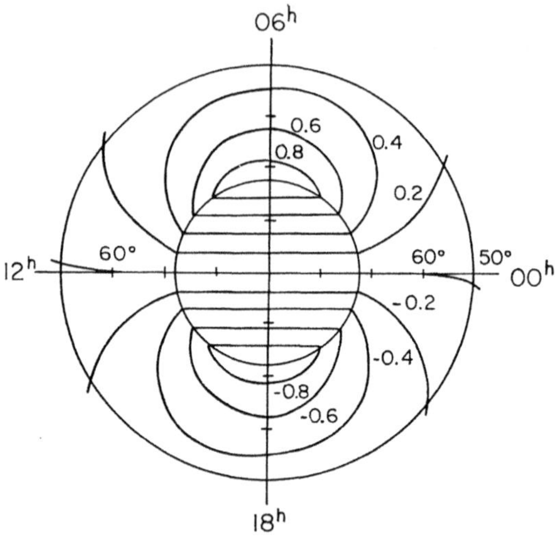

Fig. 2. Calculated equipotential contours in the ionosphere labeled by values in units of Φ_0 (see text). Only the region above 50° latitude is shown (the calculation, however was carried down to the equator and all magnetic field dip angle effects included). The outer circle is $\lambda = 50°$; the inner $\lambda = 72°$.

looks rather similar to the configuration inferred from DP2 magnetic variations (Obayashi and Nishida, 1968). To obtain a solution within the polar cap it was assumed that $J_{\|}=0$ also over the polar cap for lack of any better choice. The field aligned currents (which necessarily must be present (cf., Vasyliunas, 1968) are confined then to the polar cap boundary. For the case shown in Figure 2, the current density is

$$J_{\|}(\lambda, \phi) = \alpha(\Phi_0 \Sigma_p / R^2)\, \delta(\lambda - \lambda_0) \sin(\phi - \phi_0) \tag{10}$$

where $\lambda_0 = 72°/57.3°$, $\alpha = 6.75$, $\phi_0 = 0.45°$, $R = 1$ R_E, Σ_p is the height integrated Pedersen conductivity, and Φ_0 is the same as in Equation (9). Instead of postulating a driving field (Equation (9)) one could have obtained Figure 2 by postulating a driving current instead (Equation (10)); this was effectively the approach of Iwasaki and Nishida (1967), who obtained models similar to the present one.

Some magnetic observations recently reported by Behannon (1969) indicate that such field aligned currents do exist near the outer boundary of the plasma sheet (which may reasonably be taken to be the boundary of 'open' polar cap field lines). Behannon found a broad depressed field region within the magnetotail that appears to coincide with the plasma sheet; he also found that field lines in the magnetotail diverge away from the midnight meridian (interpreted as an expansion of the tail) and noted that this divergence is *larger inside the depressed field region* than outside. The latter observation implies that the current flowing on the boundary of the depressed field region has a field aligned component, and working through the geometry one finds that this field aligned current flows toward the earth in the dawn side of the tail and away from the earth in the dusk side, exactly in the directions required by Figure 2 and Equation (10).

The potential of Figure 2 was calculated assuming that the height integrated conductivities are independent of latitude. Figure 3 shows the effect on the potential

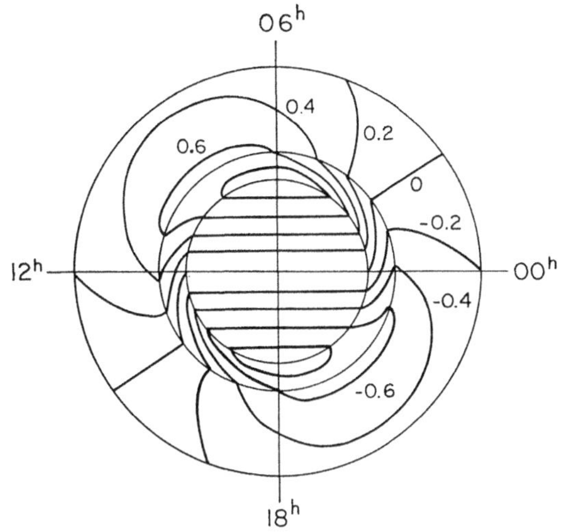

Fig. 3. Same as Figure 2 but including effects of enhanced conductivity between $\lambda = 65$ and $72°$. The three circles are $\lambda = 50$, 67, and $72°$, respectively.

if the conductivity is enhanced at latitudes just below the polar cap, corresponding to the auroral zone. (Models with enhanced conductivity rings have been considered by Iwasaki and Nishida (1967), Wolf (1969), and in a somewhat different context by Gottlieb and Fejer (1967), and Swift (1968).) Specifically, both the height integrated Hall–Pedersen conductivities were assumed to be increased by a factor of 10 between $\lambda = 67°$ and $\lambda = 72°$ maintaining the same ratio. The potential over the polar cap is unaffected (since Equation (9) is still assumed to hold), while the potential pattern at latitudes below the enhanced conductivity ring is rotated eastward by $\sim 37°$. The amount of this rotation (for which a simple physical explanation can be given) is primarily governed by the Hall–Pedersen conductivity ratio within the enhanced ring.

The field aligned current density is given by Equation (10) with $\alpha = 72.85$, $\phi_0 = 37.2°$, and Σ_p the low latitude (not enhanced) value.

If the potential distribution of Figure 3 is mapped out to the equatorial plane, the transformation from a corotating to an inertial frame (i.e., the addition of the corotation electric field) will produce a region of closed flow lines near the earth which, following Nishida (1966) and Brice (1967), is to be identified with the plasmasphere. The maximum radial extent of this region here will occur at a local time $\sim 37°$ east of the dusk meridian, in qualitative agreement with the observed 'bulge' of the plasmasphere (Carpenter, 1966). (In the absence of an enhanced conductivity ring the maximum would occur on the dusk meridian itself (cf., Figure 2).)

It is of interest to consider the ionospheric currents driven by the potential of Figure 3 and their magnetic effects. With the assumptions made all current components are sinusoidal functions of longitude with latitude dependent amplitudes and phases. The amplitudes of the height integrated E–W and N–S current density components

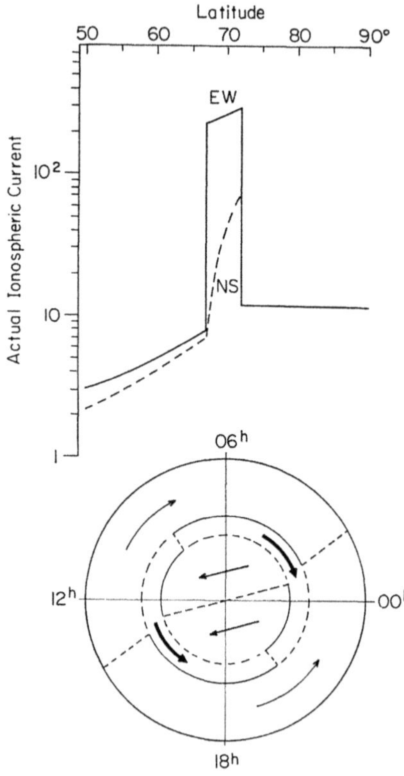

Fig. 4. Actual height integrated ionospheric current driven by the potential of Figure 3. Top: amplitude of the E–W and N–S components of the current density in units of $\Phi_0\ \Sigma_p/R$ (see text; Σ_p is the low latitude value). Above $\lambda = 72°$ the E–W and N–S components have the same amplitude and differ in phase by 90°. Bottom: phase of the E–W component; the dashed line indicates where the current changes from eastward to westward and the arrows indicate the current direction. Solid circles (or portions thereof) are at $\lambda = 50$, 67, and 72°.

and the phase of the E–W component are shown in Figure 4. The presence of the enhanced conductivity ring has produced two electrojets that bear a rather strong resemblance to the observed auroral electrojets. The uniform current over the polar cap is directed slightly east of the solar direction, as expected from the Hall current (along the equipotentials, toward the sun) and the Pedersen current (dawn to dusk). This is in disagreement with the 'equivalent' current inferred from polar cap magnetic disturbances (e.g., Silsbee and Vestine, 1942; Fairfield, 1963) which generally points well to the west of the sun.

However, *it is meaningless to ask about the magnetic effects of a current system that is not closed*, and hence the ionospheric current **I** shown in Figure 4 cannot be compared in any way with 'equivalent' currents constructed from observed magnetic variations. The current density **I**, nevertheless, can be uniquely decomposed into a 'source free' current $\mathbf{I}_{SF}$ that is confined within the ionosphere and a current that serves merely to close the field aligned currents: we write

$$\mathbf{I} = \mathbf{I}_{SF} + \nabla\tau \qquad (11)$$

where by assumption $\nabla \cdot \mathbf{I}_{SF} = 0$; the function τ can be calculated by inserting Equation

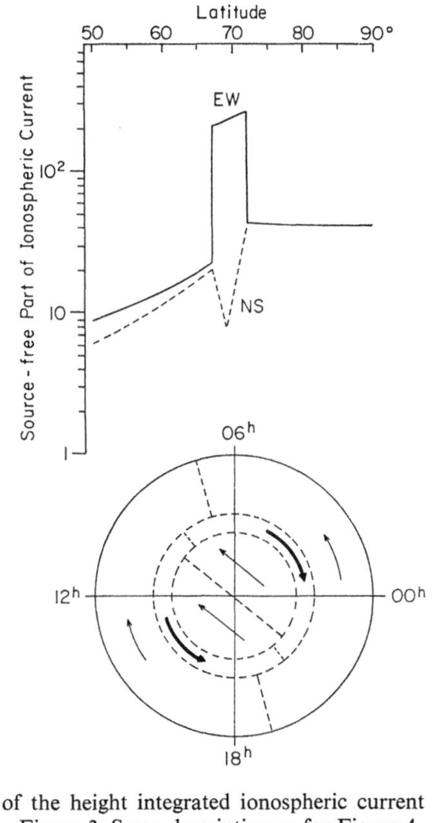

Fig. 5. Source free part of the height integrated ionospheric current driven by the potential of Figure 3. Same description as for Figure 4.

(11) into (7)

$$\nabla^2 \tau = J_{\parallel} \sin \chi \tag{12}$$

and $\mathbf{I}_{SF}$ can be obtained then from Equation (11). (In the absence of horizontal conductivity gradients, $\mathbf{I}_{SF}$ is just the Hall current.) The ground level magnetic disturbance then can be considered as the sum of two contributions: (a) the magnetic field of the closed purely ionospheric current system $\mathbf{I}_{SF}$; and (b) the magnetic field of the closed current system consisting of the ionospheric currents $\nabla\tau$, the field aligned currents and their associated magnetospheric currents, taken all together. (A third contribution from closed currents flowing entirely within the magnetosphere, e.g., the symmetric part of the ring current, adds only a field essentially uniform over the earth.) If it is assumed that the *ground level* magnetic field from the current system (b) can be neglected, as suggested for instance by Boström (1964) and Kern (1966) (and proved in the approximation of a plane ionosphere and straight field lines by Fukushima (1968)), then the current $\mathbf{I}_{SF}$ is identical with the conventional 'equivalent' current.

Figure 5 shows the source free part $\mathbf{I}_{SF}$ of the ionospheric current driven by the potential of Figure 3. As can be seen, it bears a striking resemblance to the equivalent current of Silsbee and Vestine (1942), and in particular predicts the observed westward tilt over the polar cap (the tilt is a consequence of the enhanced conductivity ring, as can also be demonstrated by a simple qualitative argument). It should be understood clearly that in this model the equipotentials over the polar cap are still directed along the solar direction and the *actual* ionospheric current flows to the east of the sun, but a magnetometer on the polar cap observes a disturbance that, if attributed solely to an overhead sheet current, would imply a current direction tilted to the west. Only the source free part of the actual overhead current contributes to the magnetic field, the rest assumed to be cancelled by effects of field aligned and magnetospheric currents. These predictions of the model have been confirmed by the results of a polar cap Ba cloud release reported at this conference by Wescott (1970), who observed that the plasma motion was approximately in the antisolar direction while the simultaneously observed ground level magnetic disturbance corresponded to an overhead current directed $\sim 55°$ west of the sun.

In summary, a very simple model of magnetospheric convection has been constructed, postulating a primary flow away from the sun over the polar cap and considering only the effects of the ionosphere. With the inclusion of an enhanced conductivity region at the auroral zones, this model predicts the observed location of the plasmasphere bulge and, with the assumption that 'equivalent' currents constructed from magnetic disturbances primarily reflect the source free component of the ionospheric current, it reproduces the main features of the DS current system, including in particular the westward tilt over the polar cap that had long been a problem for convection theories.

Acknowledgments

I am grateful to A. Nishida and S. Olbert for valuable discussions. This research was supported by the National Aeronautics and Space Administration under grant NGR 22-009-015.

References

Alfvén, H.: 1939, *Kungl. Sv. Vet-Akademiens Handl.* III, **18**, 3.
Axford, W. I.: 1964, *Planetary Space Sci.* **12**, 45.
Axford, W. I.: 1969, *Rev. Geophys.* **7**, 421.
Axford, W. I. and Hines, C. O.: 1961, *Can. J. Phys.* **39**, 1433.
Axford, W. I., Petschek, H. E., and Siscoe, G. L.: 1965, *J. Geophys. Res.* **70**, 1231.
Behannon, K. W.: 1969, GSFC preprint X-616-69-146.
Boström, R.: 1964, *J. Geophys. Res.* **69**, 4983.
Brice, N. M.: 1967, *J. Geophys. Res.* **72**, 5193.
Carpenter, D. L.: 1966, *J. Geophys. Res.* **71**, 1675.
Dungey, J. W.: 1961, *Phys. Rev. Letters* **6**, 47.
Fairfield, D. H.: 1963, *J. Geophys. Res.* **68**, 3589.
Fejer, J. A.: 1953, *J. Atmos. Terr. Phys.* **4**, 184.
Fejer, J. A.: 1964, *J. Geophys. Res.* **69**, 123.
Fukushima, N.: 1968, "Equivalence in Ground Geomagnetic Effect of Chapman–Vestine's and Birkeland–Alfvén's Electric Current Systems for Polar Magnetic Storms", preprint.
Gottlieb, B. and Fejer, J. A.: 1967, *J. Geophys. Res.* **72**, 239.
Iwasaki, N. and Nishida, A.: 1967, *Rept. Ionosph. Space Res. Japan* **21**, 17.
Kavanagh, L. D., Freeman, J. W. and Chen, A. J.: 1968, *J. Geophys. Res.* **73**, 5511.
Kennel, C. F.: 1969, *Rev. Geophys.* **7**, 379.
Kennel, C. F.: 1970, this volume, p. 257.
Kern, J. W.: 1966, *J. Geomag. Geoelectr.* **18**, 125.
Nishida, A.: 1966, *J. Geophys. Res.* **71**, 5669.
Obayashi, T. and Nishida, A.: 1968, *Space Sci. Rev.* **8**, 3.
Parker, E. N.: 1957, *Phys. Rev.* **107**, 924.
Parker, E. N.: 1968, in *Earth's Particles and Fields* (ed. by B. M. McCormac), Reinhold Publishing Corporation, New York, p. 357.
Silsbee, H. C. and Vestine, E. H.: 1942, *Terr. Mag. Atmos. Elec.* **47**, 195.
Siscoe, G. L.: 1964, Ph.D. Thesis, Massachusetts Institute of Technology, Cambridge, Mass., unpublished.
Siscoe, G. L.: 1966, *Planetary Space Sci.* **14**, 947.
Swift, D. W.: 1967, *Planetary Space Sci.* **15**, 835.
Swift, D. W.: 1968, *Planetary Space Sci.* **16**, 329.
Taylor, H. E. and Hones, E. W.: 1965, *J. Geophys. Res.* **70**, 3605.
Vasyliunas, V. M.: 1968, *J. Geophys. Res.* **73**, 5805.
Vasyliunas, V. M.: 1969, to be published.
Wescott, E.: 1970, this volume, p. 229.
Wolf, R. A.: 1969, "Theoretical Model for Convective Flow of Plasma in the Magnetosphere", preprint.

PART II

THE BOW SHOCK

SOLAR WIND STIMULATION OF THE MAGNETOSPHERE

S. J. BAME

University of California, Los Alamos Scientific Laboratory, Los Alamos, N.M., U.S.A.

The manner in which the solar wind stimulates geomagnetic activity has been a subject of interest since spacecraft measurements of the solar wind and interplanetary magnetic field became available. It was demonstrated with Mariner 2 measurements that a rough relationship exists between solar wind bulk speed and geomagnetic activity (Snyder *et al.*, 1963). The average size of the magnetosphere *is less at times of high* K_p, suggesting that the solar wind pressure is greater then (Gosling *et al.*, 1967). Sudden commencements and impulses in the geomagnetic field are produced by interplanetary shocks and discontinuities as reported by a number of groups (e.g., Gosling *et al.*, 1968). Increases in the solar wind proton density and α-particle abundance which occur following interplanetary shocks, and at about the time of the main phase development of geomagnetic storms, have also been reported (e.g., Bame *et al.*, 1968; Hirshberg *et al.*, 1970). The importance of a southward directed magnetosheath magnetic field for the production of disturbances has been reported by Fairfield and Cahill (1966). Numerous other studies have been published but space precludes a discussion of them here.

It has been argued that time variations in solar wind parameters are the principal

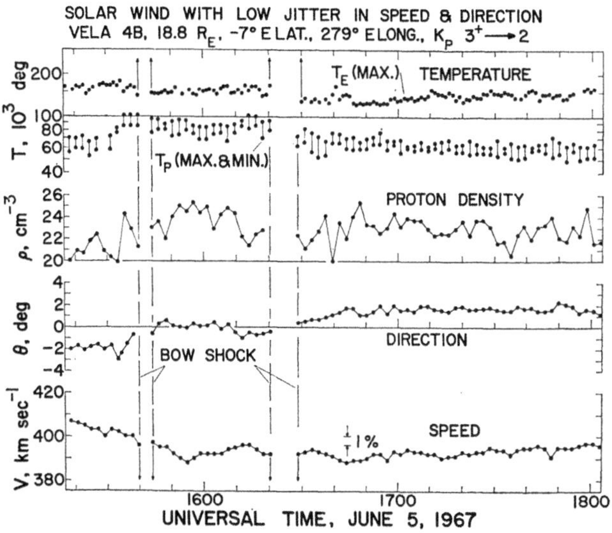

Fig. 1. Vela 4B solar wind parameters on June 5, 1967, prior to the 1915 UT passage of an interplanetary shock. Data gaps occur during times when the magnetosheath advanced to engulf the spacecraft. In this example fluctuations in bulk speed, direction, and temperatures are lower than usual, but the density fluctuations are fairly high and are probably partly responsible for the K_p value of 2_0 recorded during this period.

B. M. McCormac (ed.), Particles and Fields in the Magnetosphere.

agent responsible for geomagnetic activity, as characterized by the K_p index (e.g., Patel and Dessler, 1966). The present study investigates that possibility by comparing 3 hr average values of solar wind parameters and their deviations for periods of low ($K_p < 1_0$) and high ($K_p > 3_0$) geomagnetic activity.

Measurements used here were obtained with electrostatic analyzers on the Vela 3A, 3B, and 4B spacecraft in circular orbits at $\sim 18\ R_E$. Discussions of data analysis techniques and some results on solar wind protons and electrons can be found in Hundhausen *et al.* (1967) and Montgomery *et al.* (1968).

Examples of the behavior of the solar wind during a period of relative calm, preceding an interplanetary shock on June 5, 1967, and during a time of greater

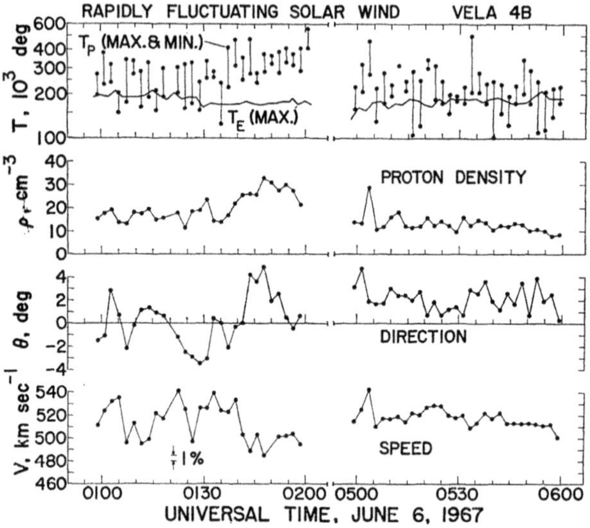

Fig. 2. Vela 4B solar wind parameters after the shock passage on June 5. During the early time the parameters are significantly disturbed compared to average fluctuations and a K_p value of 8_0 was recorded. At later times the fluctuations decrease to more normal values while the K_p value falls from 5_0 to 2_0.

TABLE I

		ϱ cm^{-3}	$\Delta\varrho$ cm^{-3}	V km sec^{-1}	ΔV km sec^{-1}	θ deg	$\Delta\theta$ deg	T 10^3 deg	ΔT 10^3 deg
	max	12.1	4.1	532	26	+ 4.1	3.9	202	52
$K_p < 1_0$	min	0.9	0.2	259	1	− 6.7	0.2	11	1
	ave	6.5	1.1	354	7	− 2.9	1.1	38	11
	max	29.4	8.4	682	45	+ 6.3	4.7	404	135
$K_p > 3_0$	min	2.1	0.3	325	2	−10.5	0.1	32	3
	ave	8.8	1.7	476	12	− 1.0	1.7	155	35

Maximum, minimum, and average values of Vela 3 3-hr averages of solar wind proton density ϱ, bulk speed V, unaberrated direction approximately in the ecliptic θ, average proton temperature T, and their rms deviations for two ranges of K_p.

disturbance following the shock when K_p reached 8_0, are shown in Figures 1 and 2. Table I presents 3 hr averages of Vela 3 measurements averaged over times when K_p was $< 1_0$ and $> 3_0$ between July 1965 and December 1967. As expected the average speed was higher during times of high K_p, but in addition the density was 35% higher, and the proton temperature was a factor of 4 higher. A frequently observed direction swing from east toward west of the sun during times of disturbance is also reflected in the table. Fluctuations in these parameters are significantly higher during times of high K_p.

An interesting feature shown in the table is the prevalence of high proton temperatures during disturbed times. Figure 3 shows distributions of the 3 hr averages of

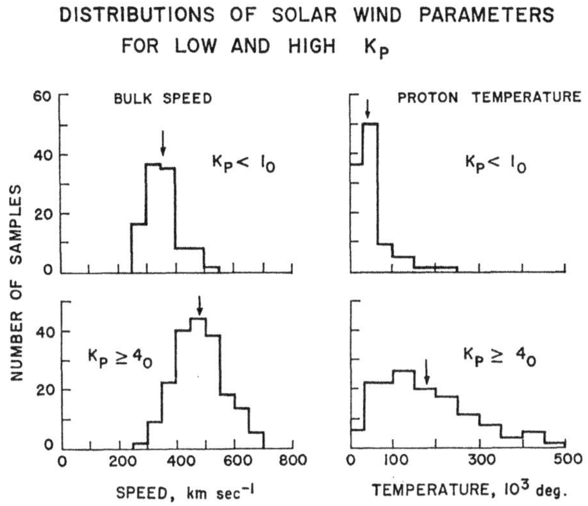

Fig. 3. Vela 3 distributions of 3-hr averages of solar wind speed and average proton temperature for 2 K_p ranges. Average values are given by the arrows.

bulk speed and temperature for quiet and disturbed times. The correlation of speed and temperature evident in the figure has been noted by Neugebauer and Snyder (1966) and others.

In conclusion, it is apparent that geomagnetic activity can be stimulated by a complex of changes in the solar wind. Study of individual events shows, as might be expected, that increasing density and speed together produce the most dramatic examples of activity. However, in some events either the density or speed changes alone play the dominant role. Changes in direction also generate activity, but usually in concert with other changes. Fluctuations in density, speed, and direction which increase at higher levels of density and speed also must contribute to the activity.

Some examples of activity occur when the solar wind plasma changes and fluctuations do not appear sufficient to drive the activity. It seems likely that at those times the interplanetary magnetic field changes (not measured on Vela) may have played the dominant role (see Fairfield, 1968).

The correlation of high proton temperature with high geomagnetic activity levels seems best explained on the basis of the temperature being an indicator of the disturbance level in the solar wind. High plasma temperatures are somehow produced by solar wind density, speed, and directional changes and fluctuations on their journey

TABLE II

Maximum, minimum, and average values of Vela 3 3-hr averages of solar wind parameters for two ranges of proton temperature

		ϱ cm^{-3}	$\Delta\varrho$ cm^{-3}	V km sec^{-1}	ΔV km sec^{-1}	θ deg	$\Delta\theta$ deg
	max	14.2	7.4	454	12.0	+0.6	2.8
$T \leqslant 2 \times 10^4$	min	0.9	0.1	249	0.4	−9.1	0.1
	ave	6.3	0.9	312	3.5	−3.3	0.6
	max	30.9	10.6	618	29.4	+8.4	4.7
$T \geqslant 2 \times 10^5$	min	3.9	0.2	370	3.3	−8.5	0.4
	ave	9.9	1.8	497	14.0	+0.9	1.9

from the sun, and in turn these continuing fluctuations and changes disturb the magnetosphere. In support of this hypothesis, Table II shows maximum, minimum, and average values of the Vela 3 3-hr averages of the various parameters and their deviations for two temperature ranges. It is clear that high temperature is to be associated generally with a disturbed solar wind.

Acknowledgments

This research was done as part of the Vela Nuclear Test Detection Satellite program, jointly administered by the Advanced Research Projects Agency of the Department of Defense and the U.S. Atomic Energy Commission, and managed by the U.S. Air Force.

References

Bame, S. J., Asbridge, J. R., Hundhausen, A. J., and Strong, I. B.: 1968, *J. Geophys. Res.* **73**, 5761.
Fairfield, D. H.: 1968, *Space Research* VIII, p. 107.
Fairfield, D. H. and Cahill, Jr., L. J.: 1966, *J. Geophys. Res.* **71**, 155.
Gosling, J. T., Asbridge, J. R., Bame, S. J., and Strong, I. B.: 1967, *J. Geophys. Res.* **72**, 101.
Gosling, J. T., Asbridge, J. R., Bame, S. J., Hundhausen, A. J., and Strong, I. B.: 1968, *J. Geophys. Res.* **73**, 43.
Hirshberg, J., Alksne, A., Colburn, D. S., Bame, S. J., and Hundhausen, A. J.: 1970, *J. Geophys. Res.* **75**, 1.
Hundhausen, A. J., Asbridge, J. R., Bame, S. J., Gilbert, H. E., and Strong, I. B.: 1967, *J. Geophys. Res.* **72**, 87.
Montgomery, M. D., Bame, S. J., and Hundhausen, A. J.: 1968, *J. Geophys. Res.* **73**, 4999.
Neugebauer, M. and Snyder, C. W.: 1966, *J. Geophys. Res.* **71**, 4469.
Patel, V. L. and Dessler, A. J.: 1966, *J. Geophys. Res.* **71**, 1940.
Snyder, C. W., Neugebauer, M., and Rao, U. R.: 1963, *J. Geophys. Res.* **68**, 6361.

SHOCK WAVES IN THE SOLAR WIND

A. J. HUNDHAUSEN

University of California, Los Alamos Scientific Laboratory, Los Alamos, N.M., U.S.A.

Since the advent of *in situ* observations of the solar wind, a sufficient number of interplanetary shock waves has been studied to suggest some general conclusions regarding their properties. While some of these conclusions agree with inferences drawn earlier from indirect evidence, a disparity appears to exist which is of interest in relating shock associated geomagnetic activity to solar activity. The directly measured propagation speeds of interplanetary shocks imply transit times from their assumed solar origin to 1 AU which are longer than those often inferred from solar flare-SC correlation studies.

For references to papers describing the direct observations see Wilcox (1969) and Hundhausen and Gentry (1969). These observations have shown that the typical interplanetary shock is of intermediate strength, moving at 100 to 200 km/sec^{-1} through the plasma or at $\sim$500 km sec^{-1} relative to a stationary observer. As further evidence for this conclusion, Figure 1 shows a shock observed by the Vela 4B satellite on June 5,

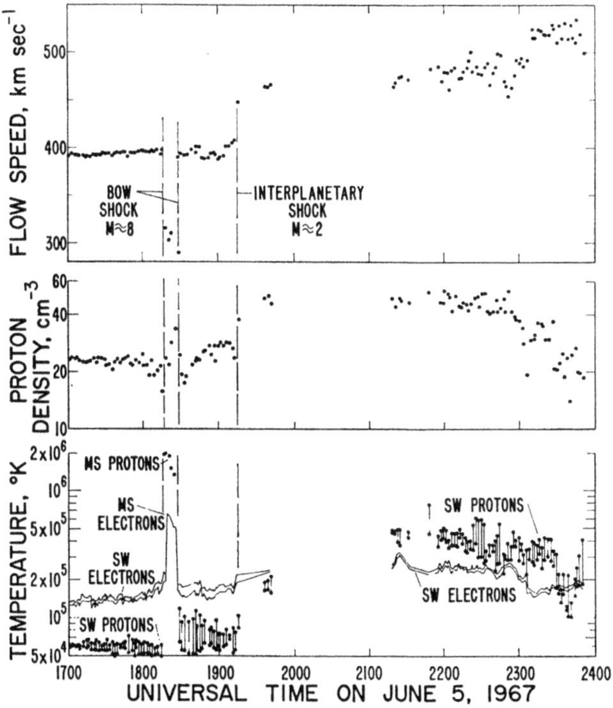

Fig. 1. Flow speed, proton density, and proton and electron temperatures measured by the Vela 4B satellite on June 5, 1967.

B. M. McCormac (ed.), Particles and Fields in the Magnetosphere.

1967. The flow speed, proton density, and the proton and electron temperatures (the latter has not been measured in previous shock observations) are shown as functions of time. At 1815 UT the satellite, previously in the solar wind, crossed the earth's bow shock into the magnetosheath. The bow shock is fairly strong with the flow speed of the incoming material (viewed in the shock frame) about 8 times the sound speed. The large changes in proton and electron temperatures by factors of ~30 and ~3, respectively, are characteristic of a strong shock. The satellite moved back into the solar wind at 1825 UT. The abrupt increases in flow speed, density, and particle temperatures at 1925 UT signified the passage of an interplanetary shock wave. In this case, the proton temperature increased by a factor of ~3, the electron temperature by only ~15%. The contrast between these moderate changes and those observed an hour earlier at the bow shock graphically illustrates the weaker nature of the interplanetary shock. The shock propagation speed is calculated to be 510 km sec^{-1} by inserting the measured densities and flow speed into the mass continuity equation (assuming propagation radial from the sun). The ambient material was flowing into the shock at only 1.8 times the speed of sound.

Figure 2 shows the relationship between transit time from the sun and shock speed

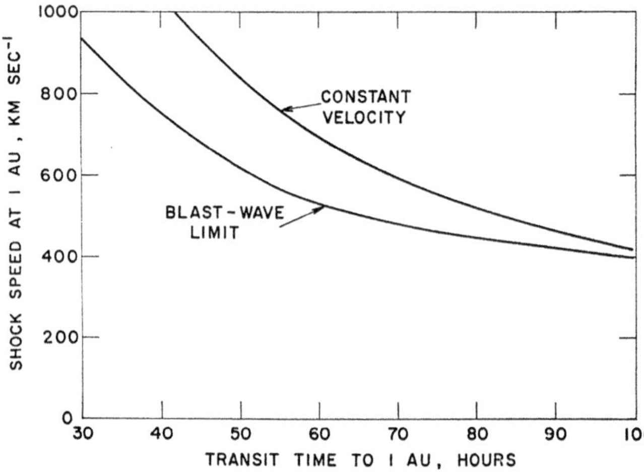

Fig. 2. The relationship between transit time to 1 AU and shock speed at 1 AU assuming (1) constant shock propagation speed, and (2) an impulsive generation of the disturbance (i.e., a blast wave) in an ambient solar wind moving at 400 km sec^{-1}.

at 1 AU for two limiting cases, constant propagation speed and the 'blast wave' case treated by Hundhausen and Gentry (1969). The latter applies to shock waves originating in impulsive (short duration) disturbances. Such disturbances have high propagation speeds near the sun and slow to the 1 AU speed due to momentum transfer to the ambient material swept up in transit. The flow speed continues to rise following the shock shown in Figure 1, suggesting that this disturbance is closer to the constant velocity case than the blast wave case (see Hundhausen and Gentry, 1969). The expected transit time for this shock is then 70 to 80 hr. Solar flares as large as class 2 were reported at 0847 UT on June 2, at 0800 UT on June 3, and at 0954 UT on June 4, i.e., at 82.6, 59.5, and 33.5 hr, respectively, before the shock observation. The transit time estimate based on the measured shock speed favors the earliest of these flares as the source of the disturbance, but it is not clear that this association would be chosen if the flow speed were not available.

If 500 km sec^{-1} is accepted as a typical interplanetary shock speed, Figure 2 gives an expected typical transit time from 65 to 80 hr. The transit times obtained by unambiguous flare associations with directly observed shocks generally fall in the 50 to 70 hr range (see Bame *et al.*, 1968; Taylor, 1969; Hirshberg *et al.*, 1970), in reasonable agreement with this prediction. These transit times are longer than the canonical value implied by flare-SC correlation studies (e.g., Akasofu and Yoshida, 1967). This may suggest that some of the SC's included in the correlation studies were not produced by flare associated shocks, but by recurrent streams or other types of discontinuities in the solar wind (e.g., Gosling *et al.*, 1967; Burlaga and Ogilvie, 1969).

Acknowledgments

This research was carried out as part of the Vela nuclear test detection satellite program, jointly administered by the Advanced Research Projects Agency of the Department of Defense and the Atomic Energy Commission, and managed by the U.S. Air Force. The author wishes to thank Drs. S. J. Bame and M. D. Montgomery for providing the previously unpublished shock observation given above.

References

Akasofu, S.-I. and Yoshida, S.: 1967, *Planet. Space Sci.* **15**, 39.
Bame, S. J., Asbridge, J. R., Hundhausen, A. J., and Strong, I. B.: 1968, *J. Geophys. Res.* **73**, 5761.
Burlaga, L. F. and Ogilvie, K. W.: 1969, *J. Geophys. Res.* **74**, 2815.
Gosling, J. T., Asbridge, J. R., Bame, S. J., Hundhausen, A. J., and Strong, I. B.: 1967, *J. Geophys. Res.* **72**, 3357.
Hirshberg, J., Alksne, A., Colburn, D. S., Bame, S. J., and Hundhausen, A. J.: 1970, *J. Geophys. Res.*, **75**, 1.
Hundhausen, A. J. and Gentry, R. A.: 1969, *J. Geophys. Res.* **74**, 2908.
Taylor, H. E.: 1969, *Solar Phys.* **6**, 320.
Wilcox, J. M.: 1969, to be published in *Proceedings of COSPAR Symposium on Solar Flares and Space Research*.

HYDROMAGNETIC OBSERVATIONS IN THE SOLAR WIND

K. W. OGILVIE and L. F. BURLAGA
Laboratory for Space Sciences, NASA-Goddard Space Flight Center, Greenbelt, Md., U.S.A.

1. Introduction

Experiments are conducted in the solar wind to determine the velocity distribution function of the plasma which is a function of time, space, and the species of particles studied. As has been pointed out (Vasyliunas, 1969) the length scale of the instruments used is so much less than any of the characteristic lengths of the plasma, that these instruments are just devices for the analysis and measurement of charged particle beams, and do not exploit directly any of the cooperative properties of the plasma. What is done is to determine the numbers of particles $(dn)/(dv)\ dv$ in various differential regions of velocity space by counting them, and then either to fit to an assumed function, or use these values of differential density to approximate the unknown function. There are thus two problems which have been fully discussed by Vasyliunas:

(a) The conversion of observed particle counts to values of $(dn)/(dv)\ dv$ (the unfolding problem), which requires a good knowledge of the characteristics of the instrument; and

(b) The approximation to the distribution function.

The fluid approximation characterizes the state of the plasma by a few parameters density, velocity, and temperature which describe the distribution function and are analogous to statistics. These quantities, supplemented by magnetic field measurements, give a complete hydromagnetic description of the plasma. These can be determined by fitting the differential densities to a *known* form of distribution function or by using the method of moments (Ogilvie *et al.*, 1967). The hydromagnetic approach is appropriate to study a situation in which there are scalar pressures for ions and electrons, large lengths, disturbances with rather low frequencies, and very large electrical conductivity.

It has been shown by Hundhausen *et al.* (1968) that the mean coulomb collision time in the solar wind becomes of the order of the expansion scale time at 0.1 AU so that collisions between charged particles cannot maintain an isotropic maxwellian distribution function beyond 0.1 AU. Similarly, collisions of ions with neutral H which may be present are also unimportant. The result is that the temperature parallel to the magnetic field tends to become greater than the perpendicular temperature $T_{\perp}$ as the plasma moves away from the sun. However, observations show that $T_{\parallel}/T_{\perp} \sim 2$ at 1 AU (Hundhausen *et al.*, 1967) where the magnetic field usually makes a 45° angle with the earth-sun line, so an experiment which measures the temperature T along the earth-sun line gives a temperature which is between $T_{\parallel}$ and $T_{\perp}$ and usually

B. M. McCormac (ed.), Particles and Fields in the Magnetosphere.

differs from these by $< 50\%$. Since the temperature varies from $< 10^4$ K to $\sim 10^6$ K a factor of 2 is not important in many cases, and the fluid can be considered to be isotropic with temperature T to zeroth approximation. Presumably, this near isotropy is maintained in the absence of collisions by instabilities (Kennel and Scarf, 1968).

The Goddard-University of Maryland plasma experiment which formed part of the experiment complement of the satellite Explorer 34 had an acceptance angle of 2.5° in aximuth and $\pm 9°$ in elevation, which was parallel to the satellite spin axis and perpendicular to the ecliptic plane. The time resolution is 3 min. Each spin, corresponding to observation at one energy per unit charge, was divided up into 16 equal angular intervals of $22\frac{1}{2}°$ width. This angular resolution is useful in the study of the flow of plasma around the magnetosphere, but does not allow us to detect the thermal anisotropy of the plasma. In this way, Burlaga and Ogilvie (1968) examined the change in bulk speed across the earth's bow shock and the flow direction behind the shock along the flanks of the magnetosheath; the results support the fluid model.

2. Discontinuities in the Solar Wind

Burlaga (1968) showed that there are simultaneous discontinuities in the plasma and magnetic field parameters in the solar wind, and found several signatures which are consistent with those for tangential discontinuities but not for shocks. To prove that such a discontinuity is tangential, one must show that three necessary conditions are satisfied: (a) it does not propagate relative to the solar wind, (b) the total pressure does not change across the surface, and (c) the magnetic fields $\mathbf{B}$ and $\mathbf{B}'$ on either side of the discontinuity surface are perpendicular to the surface normal $\hat{\mathbf{n}}$. Condition (a) was discussed by Fairfield (1968), and Burlaga (1968) showed a cascade of discontinuities which satisfied condition (b). Explorer 34 plasma data were used together with magnetic field data from Explorers 33, 34, and 35 to show that some discontinuities satisfy condition (c) (Burlaga and Ness, 1969).

Several very large discontinuities in the bulk speed (> 60 km/sec in < 3 min) were found in the Explorer 34 data and were attributed to relative motions along the surface of a hydromagnetic tangential discontinuity (Burlaga, 1969a). Each of these discontinuities was associated with a directional discontinuity in the magnetic field, and the flow direction changed across at least two of the glide planes. The change in the magnetic field direction w tended to be near 90° and was never less than 46°; this is somewhat surprising, since Burlaga (1969b) showed that directional discontinuities tend to have small w. It was found that the large velocity discontinuities with w near 90° were consistent with the conditions for stability with respect to the Kelvin-Helmholtz mode (Sen, 1963, 1964), but that similar discontinuities with w near 180° would be unstable. Thus, the unusual w distribution could indicate that large velocity discontinuities with small w disintegrated by the K-H instability.

Hydromagnetic shocks have been studied by several groups, but only the work of Sonett *et al.* (1966), and Ogilvie and Burlaga (1969) is based on simultaneous magnetic field and plasma data. The latters' work on six propagating shocks shows that the density jump, and in most cases the temperature jump, is given within experimental errors by the Rankine-Hugoniot conditions for an isotropic single fluid plasma. The values of shock velocities, density ratios, and Mach numbers so obtained indicate that at 1 AU the typical interplanetary shock is not strong (Hundhausen and Gentry, 1969).

Knowledge of the direction of the shock normal is necessary in order to apply the Rankine-Hugoniot conditions. We found the normal directions of three of the shocks using magnetic field observations from more than one satellite.

3. Interaction of Discontinuities with the Earth

A geomagnetic sudden commencement or sudden impulse can, in principle, be caused by a shock in the interplanetary medium whose front intersects the magnetosphere, or by a tangential discontinuity convected past the magnetosphere with the bulk speed of the plasma.

Gold (1955) suggested that SC's were caused by shock waves propagating through the interplanetary medium from the sun. Subsequently, Sonett *et al.* (1966) reported direct interplanetary observations of a shock-like discontinuity which was moving from the sun and was associated with a SC that was observed by 51 ground stations. Other observations of propagating events thought to be shocks, for which both plasma and magnetic field information was available, have been described by Ogilvie and Burlaga (1969). Nishida (1964) suggested that SC's could also be produced by a nonshock mode, presumably a hydromagnetic wave or a tangential discontinuity.

Nishida also suggested that the nonshock mode discontinuity must be the cause of SI, and Sonett and Colburn (1963) proposed that the SI^- is generally due to a reverse shock. Gosling *et al.* (1967) presented interplanetary observations which showed a discontinuous decrease in density and a gradual increase in temperature at the time of a world-wide discontinuous decrease in the earth's field, thus establishing that an SI^- could be caused by a nonshock mode discontinuity. Ogilvie *et al.* (1968) examined simultaneous interplanetary plasma and magnetic field data associated with a similar SI^- and observed a discontinuous decrease in density. Also observed was a discontinuous increase in magnetic field intensity and no appreciable change in temperature thus showing that the SI^- was caused by a hydromagnetic discontinuity whose signature was a tangential discontinuity.

We have also shown that positive impulses SI^+ sometimes are caused by hydromagnetic discontinuities. Gosling *et al.* (1967) reported an observation of a discontinuous decrease in temperature at the time of an SC, showing that the event was not caused by a forward shock but not ruling out a reverse shock. It is clear from these observations that both SI and SC can be produced in a variety of ways.

Taylor (1968) examined the causes of thirty-six SC events using interplanetary magnetic field data from Explorer 28. He concluded that eight of these events were due to tangential discontinuities, and that twenty-six were possible shocks, which caused the 'larger' SC events. Nishida (1964) suggested that the rise time of the impulse is small ($\lesssim$2 min) for the events caused by shocks and large ($\gtrsim$2 min) for events caused by thicker nonshock mode disturbances which propagate slowly or not at all.

A study using both interplanetary plasma and magnetic field data has been carried out to see if it is possible to predict, using only ground observations, which type of interplanetary structure is responsible for a given SI or SC (Burlaga and Ogilvie, 1969).

A theory of the interaction of the solar wind with the earth shows that the change ΔH in the horizontal component of the earth's magnetic field should be proportional

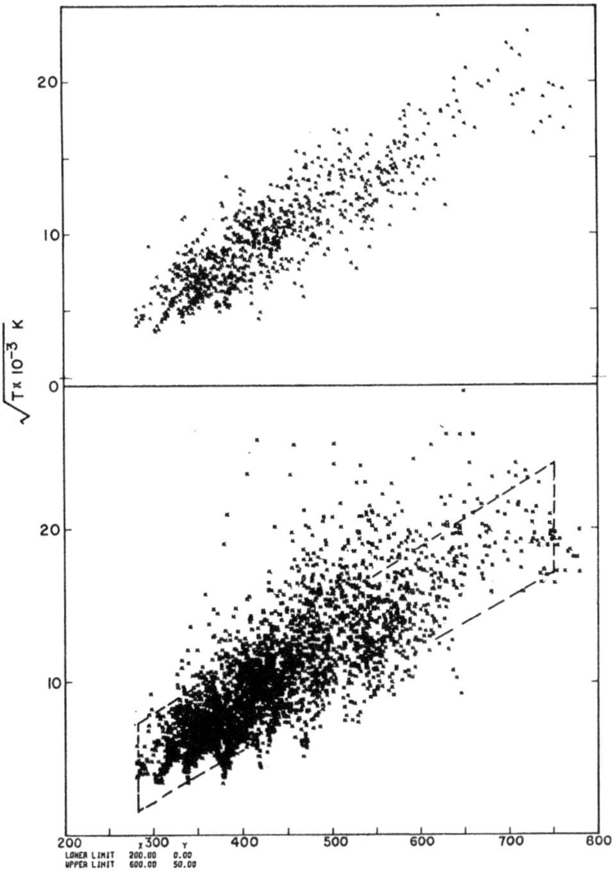

Fig. 1. Lower: $T^{\frac{1}{2}}$ vs. bulk speed, hourly values. Upper: $T^{\frac{1}{2}}$ vs. bulk speed, 3-hourly values.

to the change in momentum flux. Siscoe *et al.* (1968), from a study of 13 SI^+ events, showed that from this theory one can calculate the change in momentum flux using ΔH and an empirical constant of proportionality which possibly varies with time. Ogilvie *et al.* (1968) showed a similar empirical result for both SI and SC events.

Clearly, ΔH alone is not sufficient to distinguish between a shock and a tangential discontinuity as the cause of a given event.

Magnetic field observations by Ness and Fairfield and plasma observations by the present authors on Explorer 34 from June to December 1967 were used and the principal results from a study of 19 SC's and SI's will be discussed.

All the events were associated with abrupt changes in the state of the solar wind near the earth. These changes were either shocks (≤ 12), tangential discontinuities (5), or small dense regions of dimension $\simeq 0.005$ AU producing pulses in the terrestrial magnetogram. There is a tendency for shocks to be identified as SC's and tangential discontinuities as SI's, but this relation is not good enough to be used for prediction. Events which were identified by nearly all observatories as SC were due to shocks and the two SI^- events were caused by tangential discontinuities across which the density decreased.

In general, the thickness of a discontinuity in the interplanetary medium is not related to the rise time of the corresponding event, and one cannot associate the rise time of a SC with the velocity of the causative shock.

4. Heating in the Solar Wind

We now consider the temperature of the solar wind and its heating mechanism. Figure 1 shows a distribution of the square root of hourly (lower) and 3 hourly (upper) average values of $T^{\frac{1}{2}}$ plotted against bulk speed V. Although a good deal of scatter is evident a functional relation is suggested. When these data are replotted by com-

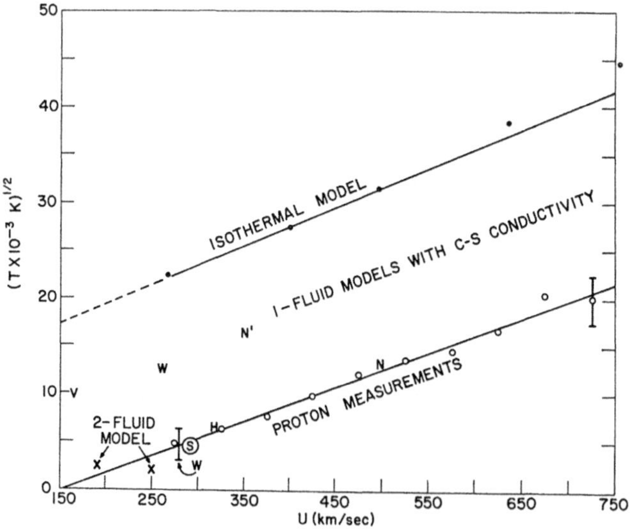

Fig. 2. Values of $T^{\frac{1}{2}}$, computed from 3-hourly averages, plotted as a function of bulk speed V for intervals of 50 km sec^{-1}. The open dots are Explorer 34 results, and the solid dots are Parker's solution to Bernoulli's equation for an isothermal corona $T \propto V^2$.

puting $T^{\frac{1}{2}}$ from 3 hr averages for bulk speed intervals 250 km/sec^{-1} $< V <$ 300 km/sec^{-1} etc. in order to take out short period fluctuations in T and V we see the result in Figure 2. Here the present observations are shown as open dots and the variability is shown by the bar on the uppermost point. Various theoretical predictions are shown on this diagram (Parker's original isothermal model, and the results of various single fluid models) as well as certain other experimental results. The point marked W is from Wolfe's observations (Wolfe and McKibbin, 1968) on December 15, 1965. The average values of $T^{\frac{1}{2}}$ and V for the flight of Mariner 2 in 1962 are given by the point N (Neugebauer and Snyder, 1966). The point H represents observations by Hundhausen referring to quiet times during 1962–67. It appears that the solar wind properties are represented by a point on this line whose position varies, but the slope of the line remains constant.

The functional form of the relationship,

$$\sqrt{T} = (0.036 \pm 0.003)\, V - (5.54 \pm 1.5) \tag{1}$$

is empirical (Burlaga and Ogilvie, 1970). The relation $V^2 = a + bT$, which is suggested by the Bernoulli equation, gives a somewhat worse fit.

It has been thought that the predictions of the two fluid model of Sturrock and Hartle (1966) are inconsistent with the observations, giving temperatures which are too low. Figure 2 shows that the predictions of Hartle and Sturrock (1968) are consistent with the extrapolated observations. Their two fluid model is for the very quiet solar wind and does not apply to nonquiet conditions. Although the two fluid model does give lower than average temperatures, it also gives lower than average speeds. Since $k\,(T_e + T_i) \ll mV^2$ the low speed is a more serious problem than the low temperature.

Parker (1963) showed that if heat were added to the plasma such that its temperature remained constant out to tens of solar radii with no heating beyond, high wing velocities could be obtained. Table I shows calculations of T and V at 1 AU, assuming

TABLE I

R_1 ($\times 10^6$ km)	V (km sec^{-1})	T (K)	T (observed, K)
5.4	260	6×10^3	1.2×10^4
8	320	1.2×10^4	3.6×10^4
20	410	5×10^4	8.5×10^4
40	460	1.4×10^5	1.3×10^5

$T = 10^6$ K to radius R_1 using Parker's model. Such a crude calculation gives velocity and temperature values which are not too different from those observed at nonquiet times, suggesting that an accurate calculation could account for the observed values of V and T by the use of a two fluid model with heating out to some characteristic distance R_1 and no heating in the interplanetary medium. Some preliminary calculations by Hartle and Barnes (1969) support this hypothesis.

Given the *T-V* relation it is possible to identify regions having anomalously high temperatures. These regions usually occur during large positive gradients in bulk speed, consistent with heating due to the collision of a fast stream with slower plasma. Heating does not occur at most negative bulk speed gradients, which suggests that the Kelvin-Helmholtz instability is not an important heating process. Heating by the action of hydromagnetic shocks is not an important heating process due to the relative rarity of these events at 1 AU (Burlaga and Ogilvie, 1970).

5. Helium Observations

The first long term observations of the solar wind made by Neugebauer and Snyder (1966) on Mariner 2, showed that a second peak in the energy per unit charge spectrum was often present. This occurred at twice the energy per unit charge that characterized the H^+ peak and was ascribed to H_e^{++}, an identification which has been confirmed

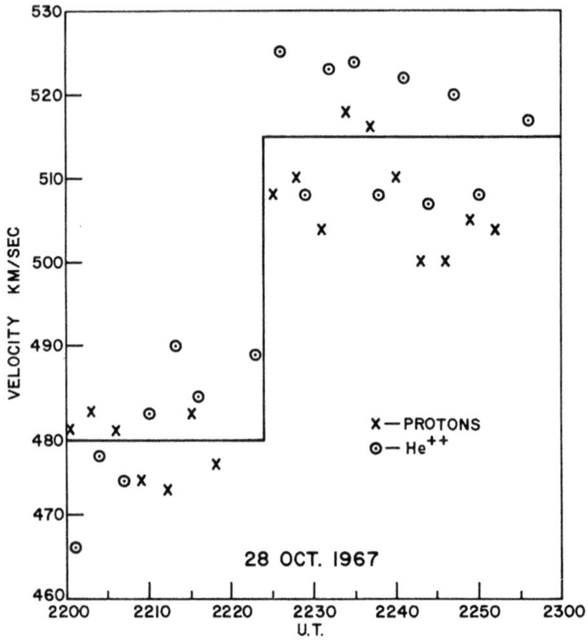

Fig. 3. This diagram shows how the bulk speeds, calculated separately, for the H and He ions undergo a coincident discontinuous increase.

by the present experiment. Thus, the solar wind should be considered a three component fluid – protons, electrons, and H_e^{++}.

Explorer 34 observations (Ogilvie and Wilkerson, 1969) show that the bulk speeds of H^+ and H_e^{++} are equal on the average. This is illustrated in Figure 3, which shows the independently determined H and He velocities for 50 min when they increased by

8%. The accuracy of determination of the He velocity, of course, is less than that of the H velocity but the two are equal to within the combined errors.

The Explorer 34 experiment separates H_e^{++} and H^+ by both a mass per charge and energy per charge analysis and in principle can obtain H_e^{++}/H^+ for any temperature. Due to an instrumental background it was often not possible to determine a value for the abundance n_α/n_p when this quantity was low simultaneously with the proton density n_p and temperature. There were times of high density and temperature when H_e^{++} densities could be measured, so observations which confirm and extend earlier results have been obtained.

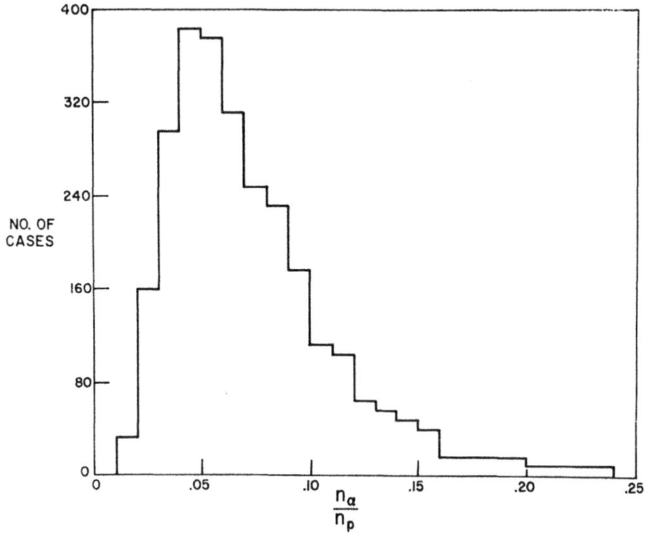

Fig. 4. This histogram shows the results of the measurements of N_α/n_p for the period of operation of Explorer 34.

In Figure 4 we show the frequency distribution of all observations of n_α/n_p obtained during the life of the experiment. The number of measurements giving a relative abundance of zero (that is, no He detected) is not shown. Due partly to the instrumental background this distribution is skewed and is not a normal distribution as it would be if there was a unique constant relative abundance and the variability was the result of measurement errors. We can conclude from this distribution that the average value of n_α/n_p is approximately 0.05 ± 0.015 and that there is considerable variability. The average value is somewhat too high due to the exclusion of periods of time when the relative abundance of He is especially low. The variability is a real physical effect associated with solar conditions, and has time scale of the order of hours or tens of hours. There exist many continuous subsets of the data, lasting between a half day and two days and containing of the order of 500 consecutive readings whose distributions are quite distinct from one another. The average value of $n_\alpha/n_p=$

(0.051) compares with the result of Neugebauer and Snyder (1966), $0.046 \pm .038$, and of Strong *et al.* (1967) 0.046.

If the plasma temperature is low, the bulk speed must coincide with a differential velocity channel or no counts will be recorded by the experiment. If the temperature is high, enough counts must fall into each of the differential velocity channels covered by the distribution to exceed the background. For times when the proton density was high ($> 10\ \text{cm}^{-3}$) the apparent absence of He always coincided with low temperatures and bulk speeds falling appropriately between instrumental channels. Thus, this biased sample does not provide evidence for an occasional complete absence of He from the plasma.

Periods when the ratio n_α/n_p is greater than normal sometimes closely follow geomagnetic storms. Although it is probably correct to assume the He rich plasma occurs as a result of flare activity on the sun, often as part of the driver of a shock, our data are consistent with a random association. In view of this and of known difficulties of associating storms and flares, we must regard this case as not entirely proven.

The ratio of the He temperature T_α to that of the protons T_p is of particular interest from the point of view of understanding heating processes in the solar wind. If $T_\alpha = T_p$, the two species are in thermodynamic equilibrium whereas if $T_\alpha = 4T_p$ they may be characterized by the same velocity distribution function. As pointed out by Jokipii and Davis (1968), a process which causes a change in the velocity distribution function depending upon only the particle velocity will give ions the same velocity distribution and temperatures proportional to their masses. Thus, observations showing that $T_\alpha/T_p \simeq 4$ are consistent with heating by hydromagnetic waves, while equal temperatures would be consistent with collisional heating leading to thermal equilibrium. Such observations cannot tell us, however, where the heating took place since adiabatic expansion conserves the ratio.

When more than two energy channels for the He ions register counting rates above the background, it is possible to determine independent values of the temperatures and other fluid parameters for each species by the curve fitting method normally used for the protons. Examples of the results of this procedure are shown in Table II. The periods of time indicated in the second column are those for which conditions remained constant; the fluid parameters set out are averages of those determined every 3 min during these periods. The two independent determinations of the bulk velocity of the plasma in the last two columns show remarkably good agreement. The ratio T_α/T_p is found to be always greater than 1 and usually less than 4, suggesting that some form of hydromagnetic wave heating mechanism was operative. The He observations thus agree well with the hypothesis of heating close to the sun but cannot provide a decisive test of it.

6. Flow behind Shocks

Shocks occurred on August 11, June 25, and June 26, 1967, when both Explorer 34 and Mariner 5 were measuring solar wind parameters at different locations. Exami-

TABLE II

Day	Time (UT)	n_p (cm^{-3})	n_α (cm^{-3})	n_α/n_p	T_p (K)	T_α (K)	U_p (km sec^{-1})	U_α (km sec^{-1})	T_α/T_p
May 30	2258–2334	17.8	1.42	0.08	1.3×10^5	3.7×10^5	593	580	2.8
June 5	1922–1953	38.5	1.6	0.042	1.9×10^5	5.6×10^5	456	461	2.9
June 8	0638–0651	9.6	1.1	0.11	9×10^4	2.8×10^5	426	431	3.1
June 25	0306–0358	27.4	1.3	0.047	8.4×10^4	2×10^5	348	353	2.4
	0515–0626	18.6	1.3	0.07	1×10^5	2×10^5	351	347	2.0
	0653–0801	15.8	1.2	0.076	1×10^5	2.4×10^5	355	346	2.4
	0930–1007	20.5	1.1	0.054	9×10^4	2×10^5	357	351	2.2
Aug. 11	1136–1151	8.2	0.8	0.098	3.6×10^5	5.6×10^5	540	538	1.6
Sept. 7	1400–1538	17.7	1.1	0.064	8.4×10^4	2.4×10^5	386	388	2.9
Sept. 20	0342–0548	15.5	0.75	0.048	2×10^5	5.2×10^5	519	530	2.6
Oct. 28	2149–2216	11.0	1.05	0.095	8×10^4	4×10^5	479	478	5.0
Oct. 29	1730–1748	19	1.1	0.058	1.6×10^5	4×10^5	578	577	2.5

nation of the fluid parameters behind the shocks showed a high degree of correlation between the measurements made at both separations. This work will be fully discussed by Lazarus *et al.* (1970).

Hourly averages of the solar wind parameters observed on Mariner 5 were used to predict the parameters seen at earth by making the following adjustments: (a) the number density was reduced by assuming an inverse square dependence on distance from the sun; and (b) the convection speed V was assumed to remain constant and the arrival time was obtained by dividing the difference in distance from the sun by V. These adjustments were applied to each of the hourly averages. Note that we assumed an advancing 'front' of plasma perpendicular to the earth-sun line. During the event of August 11, Mariner 5 was essentially on the earth-sun line and 1.6×10^7 km from the earth. During the events of June 25 and 26 the distance separating the two spacecraft was less than 1.7×10^6 km.

The flow behind the two shocks on June 25 and 26 was very complicated, possibly due to the short interval between them. During the quiet periods immediately before the shock on June 25 and around 1200 hr on June 26 agreement between the two plasma instruments was very good. The velocities and most probable thermal speeds agreed well at other times, and the disagreement between the density observations for the period 1800 on June 25 to 0600 on June 26 is taken to indicate that the scale of density fluctuation was as small at 0.01 AU.

The bulk speed increases monotonically during the period of observation, and the mean thermal speed which is equivalent to the temperature shows three peaks, one immediately after each shock and one between them. The increasing velocity and temperature behind the shocks is inconsistent with their being blast waves, behind which the models predict monotonically decreasing density, velocity, and temperature. This indicated that the shocks probably are driven but the driving mechanism cannot be determined by the present experiment.

For the August 11 event both spacecraft were situated on the earth-sun line, and separated by a distance of 1.6×10^7 km. The agreement between hourly averages of bulk speed, density, and most probably thermal speed is striking, except between 0600 and 1800 on August 11. This indicates that both spacecraft were studying the same material which was moving with little distortion over the length scale of 1.6×10^7 km or ~ 0.1 AU.

These data are consistent with the material seen before the velocity maximum being ambient gas heated by the passage of the shock and piled up in front of the driving gas. This period has been identified as a region of local heating (Burlaga and Ogilvie, 1970). The shock is then to be interpreted as 'standing off' the advancing high speed stream. A similar event occurred on September 20, 1967 (Burlaga and Ogilvie, 1969). The observations qualitatively resemble the results by Hundhausen and Gentry (1969) for driven shocks. If we identify the stand-off distance with the product of the bulk speed and the time between the shock and the velocity maximum, this distance is of the order of 0.15 AU at earth. This value is similar to, or is somewhat smaller than, stand-off distances given by Hundhausen and Gentry for spherically symmetric shocks. A small stand-off distance and a weak shock can be achieved simultaneously by postulating that the advancing driver gas is not spherically symmetrical. This is consistent with observations of Akasofu and Yoshida (1967), Hirschberg (1968), and also with shock observations by Taylor (1968), who noted departures from spherical symmetry in shock fronts.

7. Conclusion

We have interpreted approximately 3000 hr of satellite observations with reference to a hydromagnetic fluid model of the solar wind. We find that this model describes the main features of the observations and that many hydromagnetic phenomena are exhibited, including shocks, heating due to waves, and discontinuities, and also that the solar wind has a variable chemical constitution.

Acknowledgments

The results discussed in this paper could not have been obtained without the generous cooperation of the experimenters of the GSFC magnetometer Drs. D. H. Fairfield and N. F. Ness. Many useful discussions with Dr. T. D. Wilkerson also are acknowledged. The comparison of the Mariner 5 and Explorer 34 results was made with the cooperation of Dr. A. Lazarus.

References

Akasofu, S. E. and Yoshida, S.: 1967, *Planetary Space Sci.* **15**, 39.
Burlaga, L. F.: 1968, *Solar Phys.* **4**, 67.
Burlaga, L. F.: 1969a, *Solar Phys.* **7**, 72.
Burlaga, L. F.: 1969b, *Solar Phys.* **7**, 54.
Burlaga, L. F. and Ness, N. F.: 1969, *Solar Phys.* **9**, 467.
Burlaga, L. F. and Ogilvie, K. W.: 1968, *J. Geophys. Res.* **73**, 6167.
Burlaga, L. F. and Ogilvie, K. W.: 1969, NASA-GSFC, X-616-69-142.
Burlaga, L. F. and Ogilvie, K. W.: 1970, *Astrophys. J.* **159** (in press).
Fairfield, D. H.: 1968, *J. Geophys. Res.* **73**, 6179.
Gold, T.: 1955, in *Gas Dynamics of Cosmic Clouds*, North-Holland Publishing Company, Amsterdam, Holland.
Gosling, J. T., Asbridge, J. R., Bame, S. J., Hundhausen, A. H., and Strong, I. B.: 1967, *J. Geophys. Res.* **72**, 3357.
Hartle, R. E. and Sturrock, P. A.: 1968, *Astrophys. J.* **L51**, 1155.
Hartle, R. E. and Barnes, A.: 1969, to be published.
Hirschberg, J.: 1968, *Planetary Space Sci.* **16**, 309.
Hundhausen, A. J. and Gentry, R. A.: 1969, *J. Geophys. Res.* **74**, 2908.
Hundhausen, A. J., Bame, S. J., and Ness, N. F.: 1967, *J. Geophys. Res.* **72**, 5265.
Hundhausen, A. J., Gilbert, H. E., and Bame, S. J.: 1968, *Astrophys. J.* **152**, L3.
Jokipii, J. R. and Davis, L. R.: 1968, Enrico Fermi Institute Preprint No. 68-68.
Kennel, C. F. and Scarf, F. L.: 1968, *J. Geophys. Res.* **73**, 6149.
Lazarus, A. J., Ogilvie, K. W., and Burlaga, L. F.: 1970, submitted to *Solar Physics*.
Neugebauer, M. and Snyder, C. W.: 1966, *J. Geophys. Res.* **71**, 4469.
Nishida, A.: 1964, *Rep. Ion. Space Res. Japan*, **18**, 295.
Ogilvie, K. W. and Burlaga, L. F.: 1969, *Solar Phys.* **8**, 422.
Ogilvie, K. W., Burlaga, L. F., and Richardson, H.: 1967, NASA-GSFC, 612-67-503.
Ogilvie, K. W., Burlaga, L. F., and Wilkerson, T. D.: 1968, *J. Geophys. Res.* **73**, 6809.

Ogilvie, K. W. and Wilkerson, T. D.: 1969, *Solar Phys.* **8**, 435.
Parker, E. N.: 1963, in *Interplanetary Dynamical Processes*, Interscience Publishers, New York.
Sen, A. K.: 1963, *Phys. Fluids* **6**, 1154.
Sen, A. K.: 1964, *Phys. Fluids* **7**, 1293.
Siscoe, G. L., Formisano, V., and Lazarus, A. J.: 1968, *J. Geophys. Res.* **73**, 4869.
Sonett, C. P. and Colburn, D. S.: 1963, *Planet. Space Sci.* **13**, 675.
Sonett, C. P., Colburn, D. S., and Briggs, B. R.: 1966, in *The Solar Wind*, Pergamon Press, New York.
Strong, I. B., Asbridge, J. R., Bame, S. J., and Hundhausen, A. J.: 1967, NASA-SP150.
Sturrock, P. A. and Hartle, R. E.: 1966, *Phys. Rev. Letters* **16**, 628.
Taylor, H. E.: 1968, NASA-GSFC X-616-68-239.
Vasyliunas, V. M.: 1969, 'Methods of Experimental Physics'. IX Plasma Physics.
Wolfe, J. H. and McKibbin, D. C.: 1968, *Planet. Space Sci.* **16**, 953.

PLASMA MEASUREMENTS NEAR THE EARTH'S BOW SHOCK: VELA 4

MICHAEL D. MONTGOMERY
University of California, Los Alamos Scientific Laboratory, Los Alamos, N.M., U.S.A.

1. Introduction

The supersonic flow of solar wind plasma past the earth's magnetic field is now known to produce a well defined, essentially permanent standing shock wave (Ness *et al.*, 1964; Gosling *et al.*, 1967; Burlaga and Ogilvie, 1968). Because collisional mean free paths in the solar wind are much larger than the shock's dimensions and therefore cannot play a direct role in the shock's formation, collective (wave particle) interactions must provide the necessary dissipation. Measurements of the thickness of the magnetic transition (Heppner *et al.*, 1967) indicate that the shock is too thin (much less than an ion cyclotron radius) to result from magnetic wave turbulence such as Alfvén wave turbulence suggested by Kennel and Sagdeev (1967). Tidman (1967) indicated that a likely alternative would be electrostatic ion wave turbulence driven by interpenetrating ion beams, and measurements of high frequency electric and magnetic field fluctuations published by Fredricks *et al.* (1968) and Scarf *et al.* (1970) show that electrostatic turbulence is likely to be the dominant dissipation mechanism. However, a necessary requirement for the instability of electrostatic waves with respect to small currents or interpenetrating ion beams is $T_e \gg T_i$, and recent observations (Montgomery *et al.*, 1969; Montgomery *et al.*, 1970) show that the bow shock remains well defined for T_e/T_i as small as 0.6. This paper briefly presents measurements of electron and positive ion velocity distributions that provide evidence for electron preheating near the leading edge of the shock. The resulting increase in T_e/T_i may be large enough to cause significant destabilization of ion waves in the region where ion heating is observed to occur.

2. Observations

The measurements were made with a hemispherical plate electrostatic analyzer flown on the Vela 4B satellite. The satellite was launched in April 1967 into a near circular orbit with a radius of ≈ 19 R_E, inclination (with respect to ecliptic plane) $\sim 60°$, and an orbital period of 4.7 days. The spin period of the satellite is ~ 64 sec and its spin axis is continuously oriented along the satellite earth line. Descriptions of a similar analyzer flown on previous Vela flights have been published in Singer (1965), Bame *et al.* (1967) and Hundhausen *et al.* (1967).

The polar acceptance angle of the analyzer is $\pm 55°$ (measured from the spacecraft's equatorial plane) and the resolution in azimuth ϕ is 2.8° FWHM forming a 110° by 2.8° fan shaped solid angle. During each satellite rotation period a set of sixteen

B. M. McCormac (ed.), Particles and Fields in the Magnetosphere.

differential energy spectrums (each requiring ≈ 0.375 sec) spaced 4 sec apart is obtained for both electrons and positive ions. Since the instrument integrates over $\pm 55°$ in polar angle θ, only a two dimensional approximation $F(v, \phi)$ to the velocity distribution function $f(\mathbf{v})$ is obtained, where $F(v, \phi) \equiv \int_{-\pi/2}^{\pi/2} f(\mathbf{v}) \cos\theta \, d\theta$. The plots which will be used to describe $F(v, \phi)$ are differential energy spectrums converted to F (see Montgomery *et al.*, 1970 for details), and thus represent slices through F at constant ϕ.

Figure 1 presents Vela observations of plasma fluid parameters derived from F as

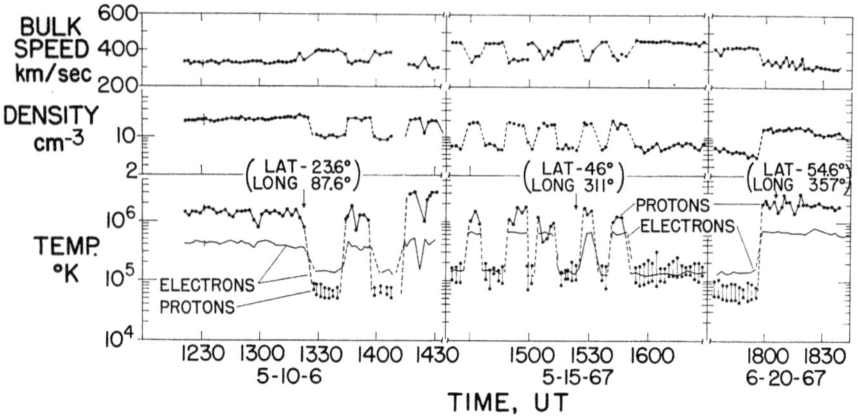

Fig. 1. Plasma measurements during bow shock crossings. The maximum and minimum solar wind proton temperatures are indicated by the points at the top and bottom of the short vertical lines. Note the strong proton heating indicated by the large jumps in proton temperature. In the magnetosheath, where both the electron and positive ion distributions are very non-Maxwellian, the temperature is a measure of the average energy of the distributions (see text).

defined above. Changes in density, bulk speed, electron temperature, and ion temperature during a number of representative crossings are shown. The temperature used here, unless otherwise stated, is defined by $T = m\langle v^2 \rangle / 3k$ and therefore is a measure of the internal energy of the *entire* distribution. This figure which is discussed elsewhere (Montgomery *et al.*, 1970) is included here to show typical jump conditions and the general nature of crossings as observed by the Vela 4B satellite.

A crossing which provides a clear example of electron preheating will now be discussed in detail. The crossing occurred at 1817 UT on June 5, 1967, when the satellite was located at 285° solar ecliptic longitude and $-14.6°$ solar ecliptic latitude. It was chosen because its features were separated in time more than usual, resulting in effectively better time resolution for the measurements. The above average time separation is probably the result of a smaller than average shock-satellite relative velocity at the time of the crossing. This seems reasonable because another crossing back into the solar wind occurred at 1827 UT only 10 min later. Thus, the relative velocity must have gone to zero at some point between the crossings and, in addition, typical magnetosheath positive ion distributions usually observed far downstream

(Montgomery *et al.*, 1970) were not seen, implying that the satellite did not entirely traverse the shock structure.

The solar wind plasma parameters at the time of the crossing can be summarized as follows: The bulk speed V_B was 398 km sec^{-1}, the proton temperature T_p was 7×10^4 K, and the particle density n was 16 cm^{-3}. Figure 2 shows values of the

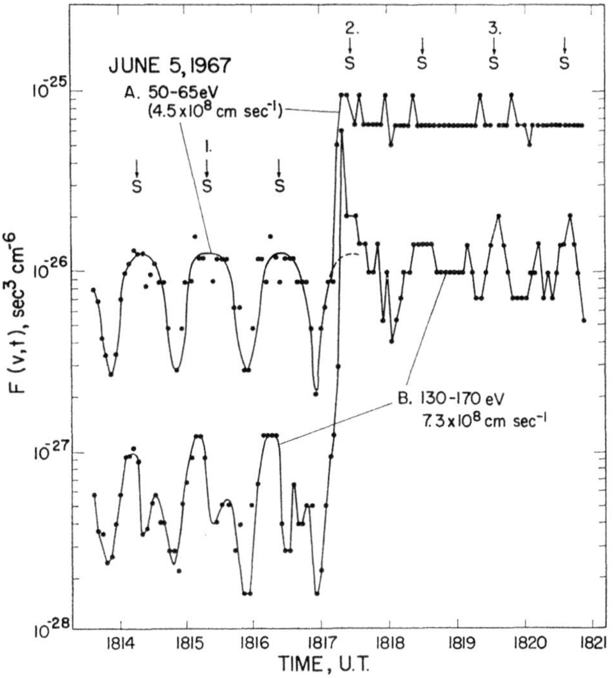

Fig. 2. Fixed velocity samples of the electron distribution plotted versus time during a bow shock crossing. The S symbols indicate when the analyzer acceptance fan looks toward the sun, and the numbers 1, 2, and 3 indicate the approximate times the distributions appearing in Figures 3 and 4 were obtained. The data points are 4 sec apart.

electron velocity distribution at two fixed velocities plotted vs. time. As is usually observed, most of the electron heating occurred during the ~3.6 sec gap between measurements (at about 1817:14 UT). In fact, electron heating has been observed to occur in ~0.03 sec (Montgomery *et al.*, 1970). The S arrows indicate when the instrument's acceptance fan was looking toward the sun, and the points labeled 1, 2, and 3 mark the times when the distributions presented by Figures 3 and 4 were obtained. Modulation of the curves at the spin frequency is caused by the plasma bulk speed V_B and the complicated shape of the 7.3×10^8 cm sec^{-1} (curve B) in the solar wind is due to the combined effects of a small electron temperature anisotropy and a skewing of the distribution due to thermal conduction (Montgomery *et al.*, 1968).

Figure 3 shows three branches of the electron distributions at different positions with respect to the shock (refer also to Figure 2). These distributions were all measured at a spacecraft azimuthal angle ϕ of 340°, i.e., 20° before the acceptance fan passed the sun. Curve 1 was obtained approximately 2 min before the shock was encountered and is a typical solar wind electron distribution yielding a temperature T_E of 2.1×10^5 K. Curve 2 represents a distribution obtained between 5 and 8 sec after electron heating

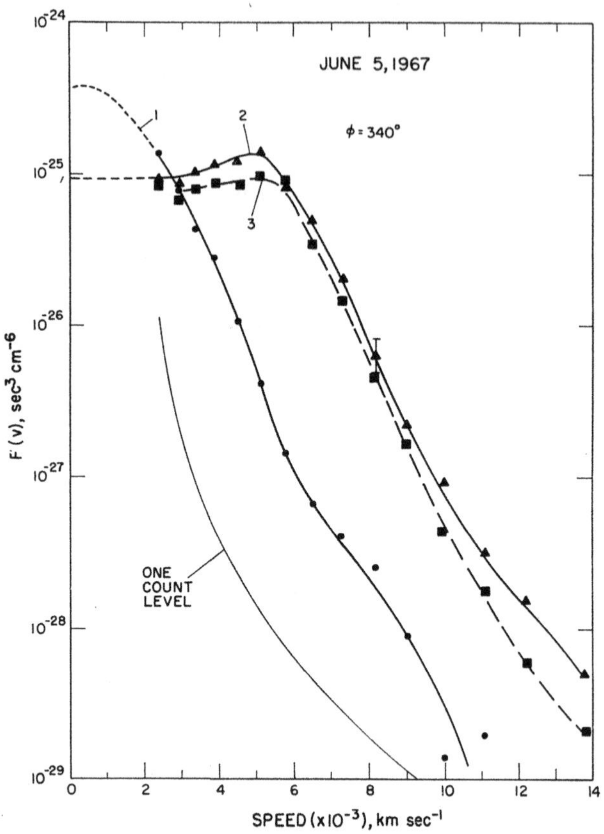

Fig. 3. Branches of the electron velocity distribution obtained at the three positions near the shock indicated in Figure 2. The 'one count level' is the distribution that would be obtained if all data samples contained one count. Thus, the vertical separation between the one count level and the curves indicates the relative counting rate. Adjacent data points were obtained 0.016 sec apart.

and is characterized by $T_E = 6.5 \times 10^5$ K. The flat topped and sometimes concave nature of this distribution is discussed in Montgomery *et al.* (1970). Curve 3 was obtained 2 min after Curve 2 and somewhat further downstream. It is little changed from Curve 2 except for some reduction of the small peak near 5000 km sec^{-1}, and is characterized by a temperature of 6.4×10^5 K. The dashed part of Curve 1 is the result of a bi-Maxwellian least squares fit (Montgomery *et al.*, 1968); the dashed part of Curves 2 and 3 is simply a horizontal line drawn from the lowest speed points to the ordinate axis (no measurements were made below about 2400 km sec^{-1}).

The corresponding positive ion distributions are shown in Figure 4. Note that since the angular width of the positive ion beam is small, positive ion measurements are available only when the detector is looking in roughly the sunward direction (noted by the S's on Figure 2). Curve 1 is a slice through the peak of the solar wind proton velocity distribution at $\phi = 4.6°$ obtained 4 sec after Curve 1 of Figure 3 was obtained. A bi-Maxwellian fit to this distribution gives the upstream parameters $V_B = 389$ km

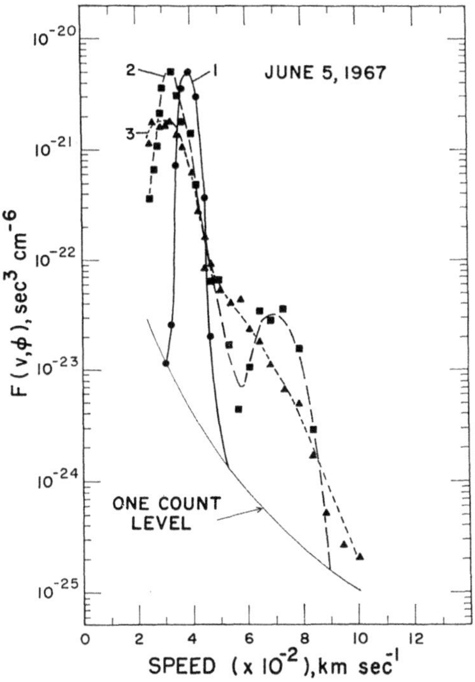

Fig. 4. Cuts through the positive ion distribution obtained at the three positions near the shock indicated in Figure 2. Adjacent data points were obtained 0.008 sec apart. Notice that the beam first slows down forming an enhanced high energy tail and then broadens. The α-particle contribution (1.5% in particle density) has been eliminated from Curve 1.

sec^{-1}, $T_p = 7 \times 10^4$ K, and $n = 16$ cm^{-3}. The temperature anisotropy ratio $T_{p\,max}/T_{p\,min}$, is 1.24.

Curve 2 shows a cut through the positive ion distribution at $\phi = 10°$ obtained 6 sec after Curve 2 of Figure 3 (11 to 14 sec after the sudden electron heating shown in Figure 2). A substantial deceleration of the proton beam (from 398 km sec^{-1}) has occurred with little change in the shape of the distribution near the peak. This deceleration of the ion beam has been discussed by Neugebauer and Snyder (1969) and seems to be a consistently observed feature. However, the amount of deceleration varies rapidly in time and the position of the peak is observed to rapidly fluctuate in speed by ~50 to 60 km sec^{-1}, changing considerably during the 1 sec interval between solar oriented positive ion measurements. Higher velocity particles are observed to appear in the tail of the distribution that commonly result in a fluctuating secondary peak between 1.5 and 2 V_B and formation of an enhanced high energy tail. If the temperature of the protons is taken from the shape of the distribution near the peak, the temperature ratio T_e/T_p becomes $6.5 \times 10^5/8 \times 10^4 = 8$ which may be high enough to account for the destabilization of electrostatic waves in this region. The shape of the electron distribution may also contribute to destabilization, but is not measured in the important region near the ion thermal speed. The increase in internal energy of the ions at this point is substantial but cannot be accurately calculated because rapid fluctuations in the distribution prevent the combined use of spectra from different angles.

Curve 3 represents the ion distribution further downstream and was obtained 4 sec after Curve 3 of Figure 3 at $\phi = 4.7°$. Considerable heating has now taken place near the peak and the final downstream bulk speed has essentially been reached. The width and position of the peak give a temperature and bulk speed of 6×10^5 and 310 km sec^{-1} while integration over the entire distribution yields the following values: $V_B = 325$ km sec^{-1}, $T_p = 1.1 \times 10^6$ K, and $n = 34$ cm^{-3}. The strong heating which has occurred near the peak of the distribution in this part of the shock may be the result of non-linear Landau damping of the ion waves observed by Scarf *et al.* (1970).

3. Summary

Evidence has been presented which indicates that electron preheating occurs in the bow shock. The resulting enhancement of the electron-ion temperature ratio T_e/T_i may be sufficient to destabilize the electrostatic waves observed by Fredricks *et al.* (1968) and Scarf *et al.* (1970). Evidence is also seen for the possible strong interaction of the ion velocity distribution with these waves resulting in strong ion heating. A possible mechanism for fast preheating of the electrons may be the Buneman (1959) instability (Kennel, 1970) driven by large electron drift speeds. The possible existence of electron drift speeds greater than the electron thermal speed is indicated by the steepness of the magnetic field gradient within the shock (Fredricks *et al.*, 1968; Scarf *et al.*, 1970). Because of the discontinuous nature of the ion measurements, clear indications of electron preheating are rare, and it is not possible at this time to establish that it is a permanent feature.

Acknowledgments

This research was done as part of the Vela Nuclear Test Detection Satellite Program which is jointly sponsored by the Advanced Research Projects Agency of the Department of Defense and U. S. Atomic Energy Commission. I acknowledge helpful discussions with Drs. D. W. Forslund, R. W. Fredricks, and C. F. Kennel.

References

Bame, S. J., Asbridge, J. R., Felthauser, H. E., Hones, Jr., E. W., and Strong I. B.: 1967, *J. Geophys. Res.* **72**, 113.
Buneman, O.: 1959, *Phys. Rev.* **115**, 503.
Burlaga, L. F. and Ogilvie, K. W.: 1968, *J. Geophys. Res.* **73**, 6167.

Fredricks, R. W., Kennel, C. F., Scarf, F. L., Crook, G. M., and Green, I. M.: 1968, *Phys. Rev. Letters* **21**, 1761.
Gosling, J. R., Asbridge, J. R., Bame, S. J., and Strong, I. B.: 1967, *J. Geophys. Res.* **72**, 101.
Heppner, J. P., Sugiura, M., Skillman, T. L., Ledley, B. G., and Campbell, M.: 1967, *J. Geophys. Res.* **72**, 5417.
Hundhausen, A. J., Asbridge, J. R., Bame, S. J., Gilbert, H. E., and Strong, I. B.: 1967, *J. Geophys. Res.* **72**, 78.
Kennel, C. F.: 1970, this volume, p. 257.
Kennel, C. F. and Sagdeev, R. Z.: 1967, *J. Geophys. Res.* **72**, 3303.
Montgomery, M. D., Bame, S. J., and Hundhausen, A. J.: 1968, *J. Geophys. Res.* **73**, 4999.
Montgomery, M. D., Asbridge, J. R., Bame, S. J., and Hundhausen A. J.: 1969, *Trans. Geophys. Union* **50**, 460.
Montgomery, M. D., Asbridge, J. R., and Bame, S. J.: 1970, *J. Geophys. Res.* (in press).
Ness, N. F., Scearce, C. S., and Seek, J. B.: 1964, *J. Geophys. Res.* **69**, 3531.
Neugebauer, M. and Snyder, C. W.: 1969, *Trans. Geophys. Union* **50**, 460.
Scarf, F. L., Fredricks, R. W., and Kennel, C. F.: 1970, this volume, p. 102.
Singer, S.: 1965, *Proc. IEEE* **53**, 1935.
Tidman, D. A.: 1967, *J. Geophys. Res.* **72**, 1799.

AC ELECTRIC AND MAGNETIC FIELDS AND COLLISIONLESS SHOCK STRUCTURES

F. L. SCARF, R. W. FREDRICKS, and C. F. KENNEL*

Space Sciences Laboratory, TRW Systems Group, Redondo Beach, Calif., U.S.A.

Abstract. Instruments on board Pioneer 8 and 9, and OGO 5 have provided high time resolution plasma diagnostics from which the gross and fine structure of collision-free shocks can be deduced. We have found several types of shock structures in the high beta ($\beta \approx 1$) high Mach number ($M_A \approx 10$) flows of solar wind plasma impinging upon the magnetospheric obstacle and in the low Mach number interplanetary shocks. By far the most common structure is a large amplitude MHD pulse structure having a characteristic length of the initial gradient and trailing wavetrain corresponding to a few times c/ω_{pe}, the electron inertial length. The dissipation mechanism in such shock structures is provided by electrostatic wave turbulence arising apparently from current driven electron-proton or proton-proton two stream instabilities which saturate nonlinearly. Nonlinear whistler mode waves also occur in some shock structures, but provide much less efficient dissipation or proton thermalization compared with that due to the electrostatic turbulence.

1. Introduction

The concept of a collisionless shock transition between the magnetosphere and the free streaming solar wind was originally introduced on the grounds that the interplanetary magnetic field, somehow, would bind the collisionless plasma into a conventional fluid. Details of the dissipation mechanisms were avoided, and aerodynamic analogies were used to compute the location of the shock boundary, assuming that the relevant Mach number is (u/V_A), where u is the wind speed and V_A is the Alfvén wave speed (Axford, 1962; Kellogg, 1962; Spreiter and Jones, 1963). To an amazing extent, these rough kinematical techniques do provide an accurate statistical prediction for the location of the mean proton boundary, or magnetopause, and the mean shock boundary, or position where the solar wind is first disturbed (Ness *et al.*, 1964). However, it has been clear for several years that MHD concepts such as generation of Alfvén waves could not be used alone as basic building blocks for theories of collisionless shocks. The particle dissipation occurred so rapidly that microscopic phenomena associated with plasma instabilities and particle scattering from high frequency waves had to be invoked to provide explanations for the shock transitions.

The first direct observations of high frequency ($f > f_c^+$) wave phenomena in the bow shock region were made on OGO 1 and OGO 3. The JPL-UCLA search coil experiments usually detected strong magnetic fluctuations associated with the bow shock in the frequency range 0.5 to 500 Hz (Olson *et al.*, 1969), corresponding to electron whistler mode waves. However, a number of plausible theories suggested that the dominant dissipation mechanisms would involve electrostatic waves, which are readily stimulated by two stream interactions. In recent years it has been possible to examine

* Permanent address: Department of Physics, University of California, Los Angeles, California 90024.

directly ac electric fields beyond the magnetopause, and it has been verified that electrostatic turbulence provides the dominant positive ion dissipation mechanism for a large class of collisionless shocks. Here we display some of the new results and summarize the initial interpretations.

2. Observations

The first solar wind and magnetosheath measurements of low frequency ($f \geqslant 100$ Hz) electric fields were carried out with the very simple TRW electric field experiment on Pioneer 8, launched in December 1967. Figure 5 of the initial report on the Pioneer 8 experiment (Scarf *et al.*, 1968) shows that significantly enhanced electric field noises are associated with detection of a collisionless shock in the magnetosheath; the abrupt increase in broadband noise by about 10 mV between 0215 and 0230 UT on December 15, 1967, corresponded to an associated increase in peak solar wind speed from $u_1 \simeq 260$ km/sec to $u_2 \simeq 360$ km/sec (Wolfe and Intriligator, 1969). Many other low Mach number interplanetary shocks have been detected on Pioneer 8 (the probe is presently about 35° ahead of the earth, in the sense that corotating solar wind features encounter Pioneer 8 before they arrive at earth), and in almost all cases enhanced electric field noise with $f \geqslant 100$ Hz is observed at the shock boundary. Similar interplanetary results are now being obtained from the Pioneer 9 experiment (launched on November 8, 1968, and presently about 40° behind the earth), but no measurements of the standing bow shock were made on either of these Pioneer probes.

The OGO 5 spacecraft was launched March 4, 1968, into a highly eccentric earth orbit with a geocentric apogee of 24 R_E. The TRW plasma experiment includes five electric dipole and three magnetic loop (search coil) sensors mounted on a 22 ft boom. Three short wire grid electric dipole sensors are mounted orthogonally, so that all three electric field vector components are measured in time sequence. The output is fed into narrow band (15%) filters with center frequencies at 0.56, 1.3, 3.0, 7.35, 14.5, 30.0, and 70.0 kHz. In a given frequency band, each directional component is sampled for 9.2 sec sequentially; after 27.6 sec the center frequency is advanced to the next channel. A complete axis and frequency scan sequence requires 3.2 min. In addition, the wave form on one axis is continuously monitored by a separate broadband (1 to 22 kHz) analog telemetry system. The remaining two boom mounted dipoles are colinear. Their output is monitored through a 200 Hz center frequency filter for about 2 sec every 9.2 sec. The three wire grid electric dipoles do not operate as resistively coupled Langmuir probes, but rather they are capacitively coupled to the plasma (Crook *et al.*, 1969) since the dipole dimensions (50 cm) are much smaller than the Debye length, and the current collecting area is small. These dipoles measure potential gradients induced across them by ambient electric fields. Three electrostatically shielded magnetic loops are also boom mounted orthogonal to the main electric dipoles, and they are sequentially sampled by axis and by frequency through 0.56 and 70 kHz filters. The 0.56 kHz magnetic output is phase shifted, mixed with the 0.56 kHz electric output, and monitored through a zero crossing correlator, so that electrostatic and electromagnetic waves may be separated. In addition to all of these diagnostics, the OGO experiment has a broadband (200 Hz to 70 kHz) peak electric field detector with a charging time constant of 10 μ sec. The output is stored in a sample and hold circuit and read out every 9.2 sec.

The OGO 5 experiment is extremely comprehensive compared to those flown on Pioneer 8, 9. The latter are equipped with unbalanced electric field dipoles, and they have only two bandpass channels [(400 Hz and 22 kHz (on Pioneer 8) or 30 kHz (Pioneer 9)] in addition to the qualitative broadband measurement. Moreover, on OGO the digital telemetry leads to as many as 56 samples per second, while on Pioneer at the highest rate we obtain only one sample per seven seconds. Therefore, all analysis of the shock fine structure is based on OGO observations.

Figure 1 shows how large amplitude electric fields are detected at the shock fronts,

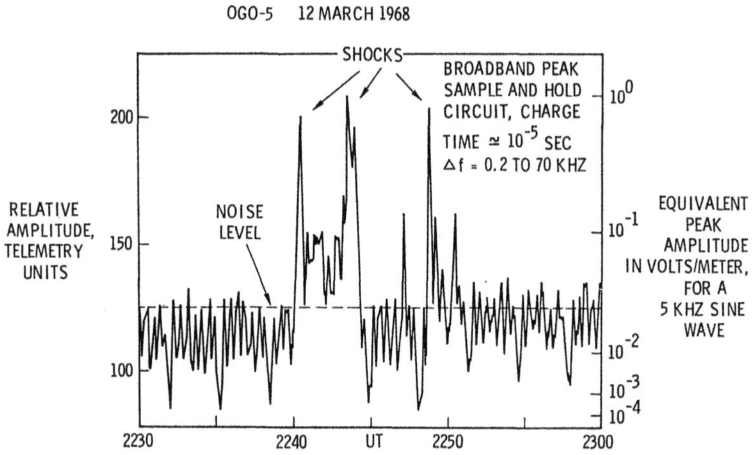

Fig. 1. Peak electric fields detected in the region of multiple bow shock crossings using the OGO 5 broadband sample and hold channels. The peak to average ratio in the wave is generally much greater than unity, but this ratio depends in detail on the noise spectrum. The broadband peak values shown here are typical for the strongest electrostatic shocks, when the maximum rms values detected in the 15% bandwidth channels are near 20–40 mV/m.

with a rapid decline in wave excitation as the spacecraft moves downstream in the magnetosheath. For this series of multiple crossings on March 12, 1968, OGO 5 was outbound with $r \simeq 18.6\ R_E$ at 0740 LT. Between 2230 and 2240 to 2241 UT, the spacecraft was in the solar wind with $|B| < 10\gamma$ and with a directional flux of 0 to 600 eV protons $< 5 \times 10^4\ cm^{-2}\ sec^{-1}$. (The proton measurements from the Lockheed light ion mass spectrometer were supplied by G. Sharp and K. Harris. The magnetic field data from the UCLA fluxgate were supplied by P. Coleman, Jr. and C. Russell.) Similar interplanetary conditions prevailed between about 2244 and 2249 UT, with the final outbound shock crossing occurring just before midnight. Between 2240:30 and 2244 and after 2249 the magnetometer and spectrometer reported magnetosheath conditions ($|B| \simeq 19-25\gamma$, directed proton flux $\geqslant 10^7\ \text{cm}^{-2}\ \text{sec}^{-1}$). It is clear that electrostatic shocks of this type involve the production of intense VLF electric fields, although the peak detector gives little information about the type of wave excited.

A more detailed example of such a structure is shown on Figure 2, which contains

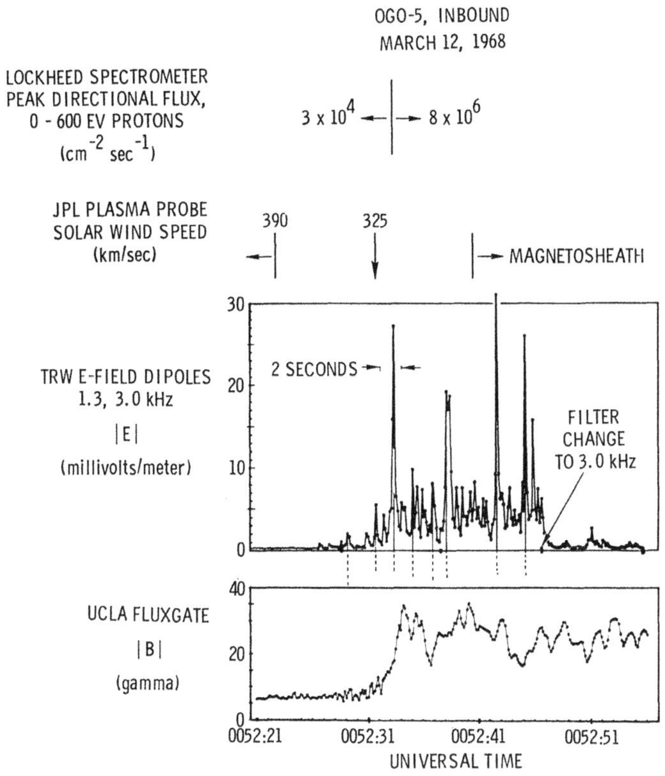

Fig. 2. Production of intense electrostatic turbulence (with $\Delta B <$ a few mγ) in a current layer having scale length near c/ω_{pe}. The slowing down of the solar wind protons just in front of the shock is due to the charge separation electric field that momentarily transfers energy to the electrons, producing a current instability. The sudden increase in low energy proton flux indicates that the VLF electric fields thermalize the positive ions.

a profile of the magnitude of the magnetic field and the response of the TRW plasma wave detector in the 1.3 kHz bandpass channel. In this shock, which has been discussed elsewhere (Fredricks *et al.*, 1968; Fredricks and Coleman, 1969), the compression of $|\mathbf{B}|$ is carried almost exclusively by the two components of $\mathbf{B}$ lying in the plane of the shock front. The component of $\mathbf{B}$ along the shock normal direction manifests much smaller variation (most of which probably is due to uncertainty in determination of the shock normal $\mathbf{n}_s$).

The upstream Alfvén Mach number for this shock was $M_A = U_0/C_A \sim 10$ with an energy ratio $\beta_+ \sim 1$, $\beta_- \sim 2$, which implies a high beta, high Mach number shock of the nearly 'perpendicular' type ($\mathbf{B}_0 \cdot \mathbf{n}_s \approx 0$). The bulk energy of upstream protons was near 850 eV with a temperature near 15 eV, so that the Lockheed spectrometer measured only background levels. However, at 00:52:30 UT, at the first large gradient in $|\mathbf{B}|$, the electric field in the 1.3 kHz bandpass rose in association with the magnetic profile variation, to two orders of magnitude above the threshold level measured in the upstream solar wind. During this time, a jump in proton flux of about two orders of magnitude is also observed in the energy range 0 to 600 eV. This reveals that significant energy transfer within the proton distribution function has occurred. We interpret this as the manifestation of the randomization or dissipation mechanism in this shock. Indeed, the energy distribution measured near to, but downstream from the front at 00:52:30, can be fit to a quasi-thermal distribution corresponding to some 100 eV proton temperature (Sharp, 1969). By the time this near-equilibrium has occurred, the electric field amplitudes in the 0.56 and 1.3 kHz channels have dropped to levels about one order of magnitude less than the peak values shown on Figure 2.

This type of shock we interpret as the high beta, high Mach number analog of the simple $\beta=0, M_A \approx 1$ fluid shock having $\mathbf{B}_0$ perpendicular to $\mathbf{n}_s$ (Davis *et al.*, 1958). In such a shock, the current density and magnetic gradient are related via

$$\frac{4\pi}{c}\mathbf{J} = \nabla \times \mathbf{B}$$

which for one dimensional shocks is just

$$\frac{4\pi}{c} Ne(\langle U_+\rangle - \langle U_-\rangle) = -\frac{\partial B}{\partial X}$$

where $\langle U_\pm\rangle$ are the bulk speeds of ions and electrons transverse to the shock normal, $\mathbf{n}_s = (n_s, 0, 0)$ and also transverse to $\mathbf{B}$. It is easily shown that $\langle U_-\rangle$ predominates, and one achieves a magnetic profile with characteristic length c/ω_{p-}. Meanwhile, the bulk speed associated with the current density, $\langle U_-\rangle$ greatly exceeds $c_s = (\kappa T_-/M_+)^{1/2}$, the ion acoustic wave phase speed. If $T_- \gtrsim T_+$, an instability of the ion acoustic wave can grow to large amplitudes in the shock front, thus yielding ion acoustic wave turbulence (Sagdeev, 1958). The effect of this nonlinear turbulence should be the randomization, or heating, of ions by nonlinear ion Landau damping (Kadomtsev, 1965). The whole wave particle interaction can be described macroscopically by an effective collision frequency which in turn implies an effective conductivity. There will then be an effective Joule heating J^2/σ in the shock layer over a characteristic length $\sim c/\omega_{p-}$.

The broadband telemetry allows us to examine the waveform in great detail, but this analysis does not *directly* confirm the above interpretations. One serious problem faced in trying to make specific mode identifications in regions such as the bow shock is illustrated in Figure 3. Here the magnetic field profile for the April 5 outbound shock is shown on a moderate time scale while the electric field ampligrams, made once every second, are used to investigate the fine structure in great detail. It can be seen here that the VLF E field spectrum changes drastically over time periods as short as 50 msec. Since these waves are quite strong ($E_{max} \simeq 20$ to 50 mV/m in a 15% bandwidth channel), they imply that large amplitude perturbations in the plasma properties occur over short time scales (e.g., 50 msec) or short distances. At 1325 UT on April 5, OGO 5 was moving outward with a speed of 1 m/msec with respect to the earth; if the shock of Figure 3 was stationary, and if wave convection was unimportant,

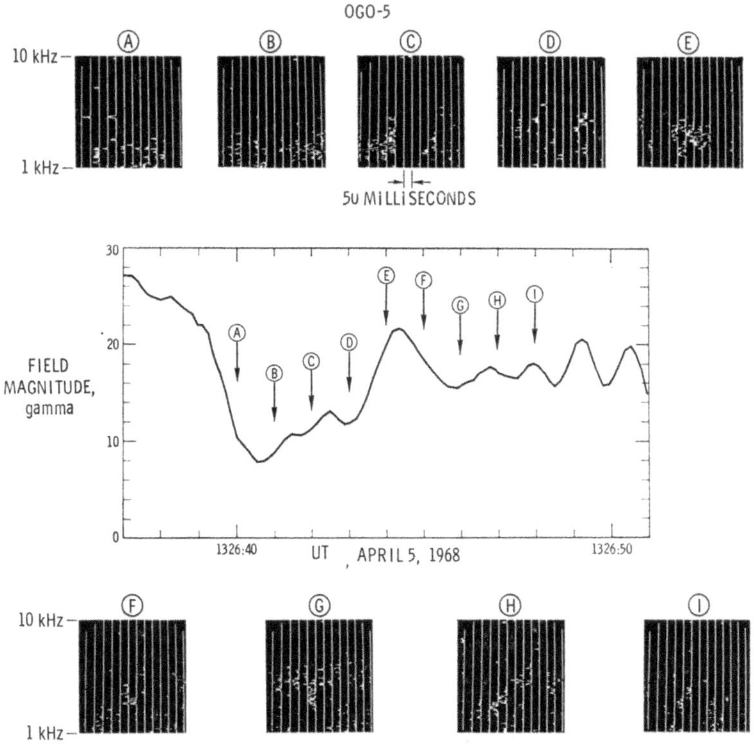

Fig. 3. High resolution spectrograms of the ac electric fields developed within the bow shock. At this point the spacecraft speed with respect to the magnetosphere was 1 m/msec.

then large changes in the *E* spectrum occurred over distances scale as small as 50 m or (2 to 3) Debye lengths. This is compatible with detection of short wavelength ion acoustic waves which have low plasma speeds and undergo Doppler shifts, but since the fastest plasma probes on OGO 5 have sequence times of the order of 2 to 9 sec, rather than 50 msec, it is not possible to follow directly the fastest fluctuations in the plasma parameters.

3. Conclusions

The majority of OGO 5 shocks examined to date are of the form discussed here, although electromagnetic or whistler mode shocks have been detected (Fredricks and Coleman, 1969). We are continuing our analyses and hope to be able to confirm the tentative identifications of the wave modes and the plasma instability mechanisms in the near future.

Acknowledgments

We are extremely grateful to Dr. P. J. Coleman, Jr. and to Dr. C. Russell for their continuing cooperation and their generous permission to display data from the UCLA fluxgate magnetometers. We also thank Dr. G. Sharp, Dr. M. Neugebauer, Dr. J. H. Wolfe and Dr. D. Intriligator for furnishing plasma probe data from the various spacecraft experiments. This work was supported in part by the National Aeronautics and Space Administration under Contracts NAS 5-9278, NAS 2-4673 and NASW-1598.

References

Axford, W. I.: 1962, *J. Geophys. Res.* **67**, 3791.
Crook, G. M., Scarf, F. L., Fredricks, R. W., Green, I. M., and Lukas, P.: 1969, *IEEE Trans. Geosci. Electronics*, **GE-7**, **2**, 120.
Davis, L., Lüst, R., and Schlüter, A.: 1958, *Z. Naturforsch.* **13a**, 916.
Fredricks, R. W., Kennel, C. F., Scarf, F. L., Crook, G. M., and Green, I. M.: 1968, *Phys. Rev. Letters* **21**, 1761.
Fredricks, R. W. and Coleman, P. J., Jr.: 1969, *Proceedings of the International Conference on Plasma Instabilities in Astrophysics*, (in press).
Kadomtsev, B. B.: 1965, in *Plasma Turbulence* (ed. by B. B. Kadomtsev), Academic Press, New York.
Kellogg, P. J.: 1962, *J. Geophys. Res.* **67**, 3805.
Ness, N. F., Scearce, C. S., and Seek, J. B.: 1964, *J. Geophys. Res.* **69**, 3531.
Olson, J. V., Holzer, R. E., and Smith, E. J.: 1969, *J. Geophys. Res.* 4601.
Sagdeev, R. Z.: 1958, in *Plasma Physics and the Problem of Controlled Thermonuclear Research* (ed. by L. A. Artsimovich), Pergamon Press, London, p. 454.
Scarf, F. L., Crook, G. M., Green, I. M., and Virobik, P. F.: 1968, *J. Geophys. Res.* **73**, 6665.
Sharp, G. W.: 1969, private communication.
Spreiter, J. R. and Jones, W. P.: 1963, *J. Geophys. Res.* **68**, 3555.
Wolfe, J. and Intriligator, D.: 1969, private communication.

PART III

MAGNETOSPHERIC PARTICLES

ENERGETIC PARTICLE PHENOMENA IN THE EARTH'S MAGNETOSPHERIC TAIL

JAMES A. VAN ALLEN
Dept. of Physics and Astronomy, University of Iowa, Iowa City, Ia., U.S.A.

Abstract. Solar electrons of kinetic energy $E \gtrsim 50$ keV are used as test particles to study the gross magnetic topology of the earth's distant magnetospheric tail and electric fields therein. The observations have been made with similar systems of detectors on the earth orbiting Explorer 33 and the moon orbiting Explorer 35. Based on study of electron shadowing by the moon, of the simultaneous intensity of solar electrons in interplanetary space and within the magnetotail, and of the very short time delay in access of impulsively emitted solar electrons into the magnetotail, the following conclusions are proposed:

(1) The gross magnetic topology of the distant magnetotail is an "open" one (i.e., dynamic interconnection of geomagnetic field lines to those in the interplanetary medium).

(2) There are no closed electrical equipotential surfaces in the magnetotail at downstream distances greater than 64 R_E.

(3) Solar electrons enter the magnetotail at downstream distances between 64 and 900 R_E.

(4) Support is given to the idea that the *motional electromotive force* caused by the motion of the interplanetary magnetic field past the earth is responsible for driving magnetospheric plasma convection and electrical currents in the polar ionosphere.

1. Introduction

This paper is concerned with the gross magnetic topology of the earth's distant magnetospheric tail and with electric fields therein.

The relevant observational work employs as *test particles* electrons $E_e \gtrsim 45$ keV and protons $E_p \gtrsim 320$ keV that are of solar origin. Three pieces of equipment are used:

(a) A system of three thin window Geiger-Mueller tubes and one multi-channel solid state detector on Explorer 33 in a loose earth orbit since July 1, 1966;

(b) A similar system of detectors on Explorer 35 in orbit about the moon since July 22, 1967; and

(c) A spherical rock, radius 1738 km (the moon), used as an occulting disc.

The University of Iowa equipment on the two GSFC/NASA spacecraft Explorers 33 and 35 has been described in previous papers (Van Allen and Ness, 1967; Armstrong and Krimigis, 1968; Van Allen and Ness, 1969).

2. Electron Shadowing by the Moon

Study of the shadowing effect of the moon on solar electrons first became possible with the injection of Explorer 35 into lunar orbit on July 22, 1967, and is continuing at the present date. Two sets of detectors on this spacecraft, one of the University of Iowa and one of the University of California, are suitable for the relevant observations. Both comprise systems of thin window Geiger-Mueller tubes though the detailed arrangements are quite different. The respective three axis flux gate magnetometers of the Goddard Space Flight Center (Ness *et al.*, 1967) and of the Ames Research Center (Colburn *et al.*, 1967) measure the magnetic vector **B** at the spacecraft.

The earliest reports on the shadowing of energetic electrons by the moon during solar events in July and August 1967 were given by Van Allen and by Lin and Anderson at the Midwest Cosmic Ray Conference in Iowa City, March 1 to 2, 1968 and soon thereafter at the meeting of the American Geophysical Union in Washington, D.C. on April 9, 1968 (Van Allen, 1968; Lin, 1968a).

The two basic papers on the subject have been published (Lin, 1968b; Van Allen and Ness, 1969, hereafter referred to as VN).

The observational results of the two groups are generally concordant, though only the latter authors have studied the detailed magnetic topology associated with lunar shadowing and the angular distribution of particles.

Subsequent discussion refers only to shadow events that have been observed within the magnetospheric tail of the earth, and, more specifically, to only those that occur on the sunward (and earthward) side of the moon and exhibit $\approx 100\%$ shadowing of solar electrons having energies of the order of 20 to 60 keV.

Two altogether different interpretations of such events were suggested originally by the two respective groups:

INTERPRETATION A (LIN AND ANDERSON)

"These observations indicate that the solar flare electrons enter the tail at a point beyond the lunar orbital radius of $\sim 64\ R_E$ (see also Lin and Anderson, 1966). The electrons then stream up the tail toward the earth and presumably are lost close to the earth" (Lin, 1968b). This view was further emphasized by Lin (1968a). In the accompanying discussion at the April 1968 A.G.U. meeting, Anderson agreed with Lin and asserted that he wished to "go on record" with the interpretation that the results required that electrons be streaming up the tail (i.e., toward the earth) with a "highly anisotropic" angular distribution at points outside of the shadow.

INTERPRETATION B (VAN ALLEN AND NESS)

It was shown by VN, however, "that essentially 100% shadowing of 50 keV solar electrons occurs on the sunward side of the moon when it is within the magnetotail *and when the unshadowed intensity is isotropic*. The inference of Lin (1968b) that 'inward streaming' of electrons at the point of observation in the magnetotail is a necessary condition for such shadowing is thus shown to be invalid" (VN, p. 93). Further, defining a quantity D the "impact parameter" with respect to the center of the moon of the rectilinear projection of the magnetic vector **B** at the spacecraft, we found that "the edges of electron shadows occur accurately at points defined by $D = 1.00$" (VN, p. 88). (A number of examples are cited and shown graphically in Figures 9, 10, 11, 12, 13, and 14 of VN.) "The accuracy of this statement (i.e., $D = 1.00$ at the edges of electron shadows) is measured by the maximum departure from rectilinear projection that is permitted by the electron shadowing observations. This is approximately $0.1\ R_M$ in $2\ R_M$ or 3°" (VN, p. 88). Later (VN, p. 91) we give our estimate of the error in rectilinear projection of **B** as of the order of 2°. Since the observational results establish $D = 1.00$ as a necessary and sufficient condition for the occurrence of edges of electron shadows (i.e., a null result on trans **B** drift), the only remaining question is the sensitivity of the technique for detecting departures from this condition. This question was answered by considering the observed sharpness of the edges and the comparative locations of the edges in the several detectors that accept particles of different pitch angles and senses of motion (cf. Figure 17 of VN). We adopted the rather generous upper limit of 200 km ($= 0.1\ R_M$). Further (pp. 88 and 89 of VN):

"We suggest the following interpretation of observations such as those just described: When the moon intercepts a tube of electrons, it isolates the segment of the tube between itself and the earth. Electrons within this tube that have just passed the moon before it intercepted the tube and are inbound toward the earth will be reflected by the much stronger field near the earth and will return within ~5 sec to strike and be absorbed by the moon as they move outward. Thus, if there are no significant trans **B** drifts or diffusion due to any cause (curvature drift, $\mathbf{E} \times \mathbf{B}$ drift, $\mathbf{B} \times \nabla \mathbf{B}$ drift, or non adiabatic scattering by fine scale 'noise' in the magnetic field) and further if there are no local sources, the entire segment of the tube of force between the moon and the earth will be drained of electrons within ~5 sec. The observed fact is that the edges of electron shadows are about as sharp as is physically possible (~1 gyroradius). Hence, leakage of electrons through the side walls of the shadowed tube of force cannot be occurring at a velocity greater than about 200 km/5 sec $= 40$ km sec^{-1} in a direction parallel to the ecliptic plane or, since the shadows are often noncentral (e.g., Figure 14), at a velocity greater than about 100 km sec^{-1} in any direction orthogonal to **B**.

"*Second conclusion.* The preceding discussion leads to the conclusion that electron shadowing data provide no *direct* information on the region of access of interplanetary electrons into the magnetotail but do provide an upper limit on the trans **B** diffusion velocity $v_\perp$ of electrons due to all causes. In the central portion of the magnetotail during magnetically quiet conditions and for radial distances $\lesssim 64\ R_E$, $v_\perp \lesssim 100$ km sec^{-1} for 50 keV electrons. This upper limit requires, among other possibilities, that the transverse electric field

$$E_\perp (= Bv_\perp/c) \leqslant 5 \times 10^{-4}\ \text{Vm}^{-1}$$

unless the (exceedingly improbable) vector cancellation of the electric drift by other drifts occurs. The actual value of $E_\perp$, of course, may be much less than the upper limit quoted here."

In a subsequent paper, Anderson and Lin (1969) have abandoned Interpretation A and have adopted Interpretation B as the basis for a discussion of the magnetic topology in the magnetospheric tail, as revealed by a variety of lunar shadowing events. They conclude that the topology of the distant tail is usually open (i.e., interconnection to the magnetic field in the interplanetary medium).

3. Further on the Electric Field in the Magnetospheric Tail

During the prolonged solar electron event of November 10 to 22, 1967, the earth-moon system was bathed in an isotropic homogeneous beam of solar electrons whose ntensities were slowly varying functions of time. A large body of simultaneous observations by University of Iowa equipment on the earth orbiting Explorer 33 and the lunar orbiting Explorer 35 is available for this period (VN).

In Figure 1 are shown relevant segments of the orbits of Explorer 33 and of the

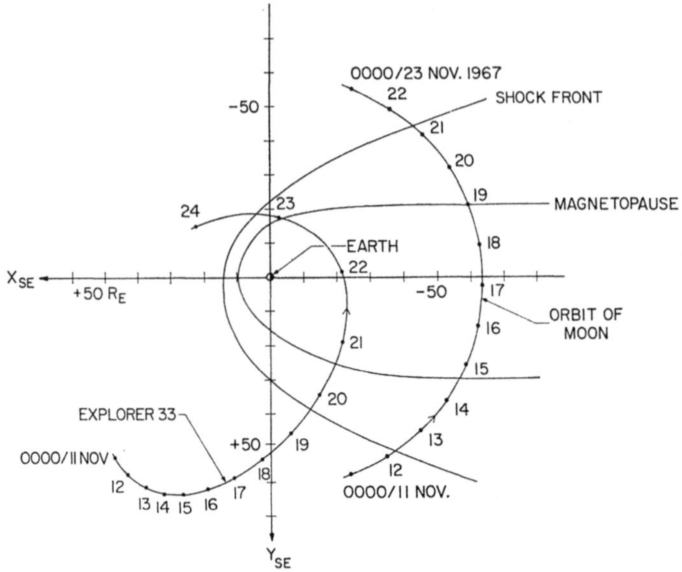

Fig. 1. Projection on the ecliptic plane of segments of the orbits of Explorer 33 and the moon during the November 10 to 22, 1967 solar electron event. Explorer 35 is in orbit about the moon. The curves labeled 'shock front' and 'magnetopause' are after Behannon (1968).

moon projected on to the ecliptic plane. The orbit of Explorer 35, for the present purpose, is represented adequately by that of the moon. The curves labeled 'Shock Front' and 'Magnetopause' are after Behannon (1968).

Simultaneous hourly averages of the counting rates of two nominally identical, thin window Geiger-Mueller tubes on Explorer 33 and Explorer 35 are plotted in Figure 2. The energy thresholds of both of these integral detectors are $E_e \approx 48$ keV and $E_p \approx 750$ keV. The counting rates shown are raw absolute data and have not been mutually normalized in any way. Hourly averages of GM3 data from Explorer 35 that included all or a significant fraction of a lunar shadow period have been deleted. The upper solid curve is drawn through the Explorer 33 data (solid circles). The Explorer 35 data (the $\times$'s) are plotted independently.

The lower solid curve through the triangles represents on the same scale the estimated 'background' contribution to the counting rates of detectors GM3 by galactic

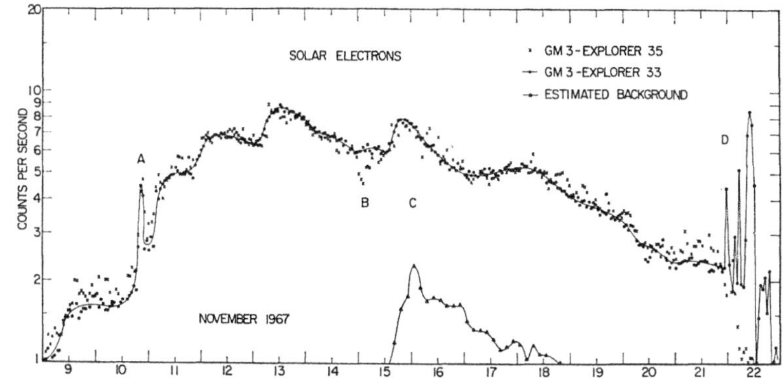

Fig. 2. Hourly averaged counting rates of nominally identical Geiger-Mueller tubes GM3 on Explorer 33 (solid circles) and GM3 on Explorer 35 (×'s) plotted independently on an absolute basis with no arbitrary normalization. The upper solid curve is drawn through the Explorer 33 points, ignoring the Explorer 35 points. The lower solid curve through the triangles shows the estimated 'background' contribution to the counting rates by solar protons and galactic cosmic rays.

cosmic rays and solar protons, as derived from simultaneous measurements by other detectors on Explorer 35.

For the most part the counting rates of GM3 on Explorer 33 and of GM3 on Explorer 35 are indistinguishably different. The accurate temporal tracking throughout the event (note especially that in the vicinity of the brief peak labeled A) bespeaks the homogeneity of electron intensity over distances of the order of 100 R_E irrespective of the relative locations of the two observing points. Perhaps the only significant departure of the counting rate of GM3 on Explorer 35 from that in interplanetary space (GM3 on Explorer 33) occurs during passage through the magnetopause in the region labeled B in Figure 2 (cf. Figure 1 and VN). The large fluctuations of the counting rate of GM3 on Explorer 33 in the region labeled D are caused by magnetospheric electrons.

I take the data of Figure 2 to establish that the intensities of solar electrons at Explorer 33 and at Explorer 35 are identical ($\pm 10\%$) *throughout the segments of their orbits shown in Figure 1.*

It has been shown in the previous paper (VN) that the intensities are accurately isotropic at both observational positions and throughout the event.

As an indication of the level of geomagnetic activity, the following values of the K_p Daily Sum for the 13 days during November 10 to 22 inclusive are noted: 8, 17+, 28+, 25+, 19−, 17−, 17, 3−, 5−, 6+, 3, 7+, and 20.

Observations made by Explorer 33 (designated position A) up to November 20 will be considered to have been in interplanetary space, as though the earth's system did not exist.

The trajectories of solar electrons from interplanetary space into the magnetospheric tail (e.g., to Explorer 35 designated position B) and to the polar caps of the earth are not known. (The letters *A*, *A'*, and *B* as used hereafter are not related to the *A*, *B*, *C*, and *D* used as labels in Figure 2.) Nonetheless, it will be shown that general dynamical principles provide a significant inference from the observations of the previous section.

It is first necessary to estimate the dimensions of the region over which the electron intensity is sensibly constant. The more-or-less spiral interplanetary magnetic field structure which corotates with the sun is presumed to control the particle populations to a first approximation. The circumferential velocity of this structure with respect to the earth is about 400 km sec^{-1}, or 226 R_E hr^{-1} or 5425 R_E day^{-1}. The observed electron intensity at A is isotropic and is a slowly varying function of time over periods of the order of several hours (Figure 2). Hence it is reasonable to suppose that the intensity is homogeneous over a region having dimensions of the order of hundred R_E in the vicinity of the earth. There is also substantial direct support for this supposition on the basis of essentially identical solar electron intensities at Explorers 33 and 35 for events (not shown here) during which both spacecraft are in interplanetary space and are separated by distances of the order of 100 R_E.

Electrons that pass through A, in general, will not pass through B, and indeed may not enter the magnetotail at all. However, those that arrive at B in a specified direction and with a specified energy must have passed through *some point* A' in interplanetary space, just outside of the influence of the earth's magnetic system. For all possible directions and energies there will be a family of such 'source points' A' distributed perhaps quite widely but likely within a region of dimensions of the order of several hundred R_E. There will be a different family of source points for each different position of B.

Let it be supposed then:

(a) that the unidirectional differential spectral intensity is isotropic and homogeneous throughout the source region, and

(b) that point A is within the source region.

Irrespective of the complexity of the physical trajectories of particles in the arbitrary electric and magnetic fields that they traverse in going from A' to B, one is assured by Liouville's theorem that the density D of their representative points in the six-dimensional (ordinary) phase space is constant along a dynamical trajectory (Swann, 1933), that is,

$$D = \text{constant along a dynamical trajectory.} \tag{1}$$

From this basic principle, it follows (Jánossy, 1948) that

$$\frac{m}{p^3}\left(\frac{\mathrm{d}j}{\mathrm{d}p}\right) = D = \text{const.} \tag{2}$$

wherein p denotes the scalar momentum, m the relativistic mass, and $\mathrm{d}j/\mathrm{d}p$, the unidirectional differential spectral intensity of particles (i.e., number of particles per unit momentum at p crossing a unit area perpendicular to $\mathbf{p}$ per unit solid angle per unit time).

In terms of the more convenient experimental parameter E, the *kinetic energy* of the particles:

$$\frac{1}{p^2}\left(\frac{\mathrm{d}j}{\mathrm{d}E}\right) = D = \text{const.} \tag{3}$$

or

$$\frac{1}{(E^2 + 2m_0c^2E)}\left(\frac{\mathrm{d}j}{\mathrm{d}E}\right) = \text{const.} \tag{4}$$

In Equation (4), m_0 is the rest mass and c is the speed of light.

Let the kinetic energy of each electron be changed by an amount ΔE (as by an electric field) between points A' and B. Then from Equation (4),

$$\frac{\left(\dfrac{\mathrm{d}j_B}{\mathrm{d}E}\right)_{E+\Delta E}}{\left(\dfrac{\mathrm{d}j_{A'}}{\mathrm{d}E}\right)_E} = \left[1 + \frac{2(E + m_0c^2) + \Delta E}{(E + 2m_0c^2)}\frac{\Delta E}{E}\right] \tag{5}$$

(cf. Lundquist, 1954; Webber, 1967).

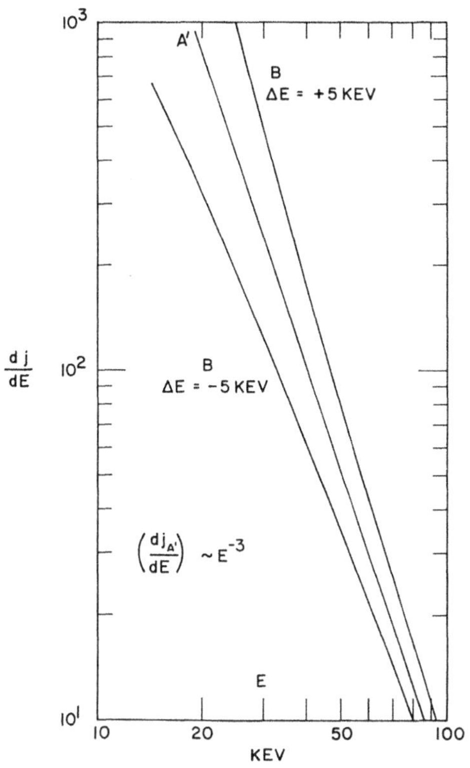

Fig. 3. The effect on the differential energy spectrum $(\mathrm{d}j_{A'}/\mathrm{d}E) \sim E^{-3}$ at point A' by a change of kinetic energy $\Delta E = +5$ keV and -5 keV between A' and B (cf. Equation (5) of the text).

The application of this formula to the present situation is illustrated in Figure 3 for a differential energy spectrum $\mathrm{d}j/\mathrm{d}E = kE^{-\gamma}$ with $\gamma = 3$. Previous work has established that γ is typically >3 and usually $\geqslant 4$ for solar electrons.

The observations yield

$$J(E_0) = \int_{E_0}^{\infty} \left(\frac{\mathrm{d}j}{\mathrm{d}E}\right) \mathrm{d}E$$

where E_0 is the effective energy threshold of the detector. It follows from Equation (5) (Van Allen, 1970) that, approximately,

$$\frac{\Delta J}{J} = \gamma \frac{\Delta E}{E_0}. \tag{6}$$

In the present experiment it has been established that

$$\left|\frac{\Delta J}{J}\right| \leqslant 0.10. \tag{7}$$

Hence for $\gamma \geqslant 3$,

$$\left|\frac{\Delta E}{E_0}\right| \leqslant 0.03$$

or

$$|\Delta E| \leqslant 1.5 \text{ keV} \tag{8}$$

along a trajectory from A' to B. It is clear that the exact value of γ is not important to the significance of the result.

Since a change of kinetic energy can be caused only by an electric field, this inference can be written in the more significant form

$$\left|\int_{A'}^{B} \mathbf{E} \cdot \mathbf{ds}\right| \leqslant 1.5 \text{ kV} \tag{9}$$

wherein **E** is the vector electric field, from any cause, at vector element **ds** of the physical trajectory and the line integral is taken along this trajectory from the source point A' to the observing point B.

It is possible, of course, that stochastic electric fields may have large intensities point-by-point. Also, no statement can be made concerning the magnitude of the component of **E** perpendicular to particle trajectories.

Nonetheless, Equation (9) prevails for points B throughout the magnetospheric tail at the orbit of the moon and over a wide range of geomagnetic activity. It appears to have fundamental significance.

4. Access of Solar Electrons to the Magnetospheric Tail

As shown by Van Allen and Ness (1969) and recapitulated above, 50 keV electrons diffuse *across* the magnetospheric tail at downstream distances $\lesssim 64\ \mathrm{R_E}$ at the remarkably slow rate $v_{\perp} \lesssim 100\ \mathrm{km\ sec^{-1}}$, this value being a generous upper limit. In contrast, such electrons pass from the interplanetary medium into the tail with an unobservably small delay. The situation is examined in the most definitive way by simultaneous observations of impulsive solar electron events (Van Allen and Krimigis, 1965) with two spacecraft, one outside and one inside the magnetospheric tail.

A valuable case occurred on August 14, 1968 (Van Allen, 1970). The geocentric solar ecliptic coordinates X_{SE}, Y_{SE}, and Z_{SE} for Explorer 33 and the moon at the onset of the event were -56.25, -12.80, -10.20 and -17.96, -60.07, $+2.33$ R_E, respectively. Hence, Explorer 35 was clearly outside of the magnetospheric tail and Explorer 33 was inside. The fact that Explorer 33 was deep within the tail was further certified in a direct observational sense by the prevalence of sporadic clouds of magnetospheric electrons.

The simultaneous electron observations with approximately identical detectors on the two spacecraft are shown in Figure 4. In this figure, the counting rates of G1AV/35

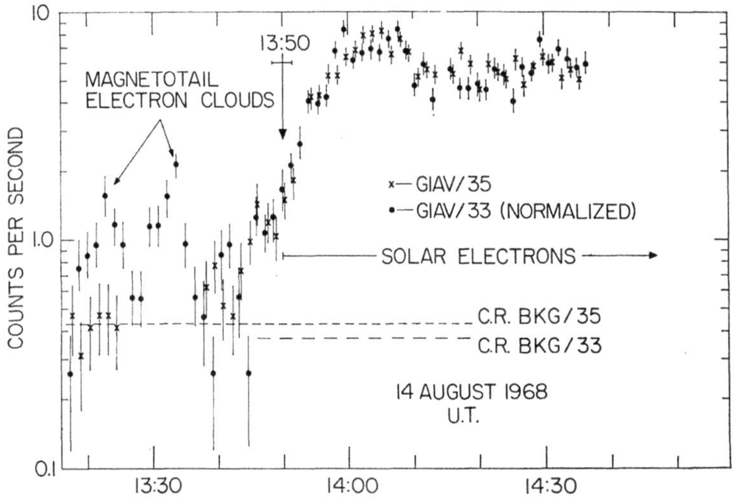

Fig. 4. Time dependences of electron intensities as observed by G1AV on Explorer 33 in the magnetotail and G1AV on Explorer 35 in interplanetary space during the early phase of the impulsive solar electron event of August 14, 1968. Note the prevalence of magnetotail electron clouds at Explorer 33. The counting rates of G1AV/35 are plotted on an absolute basis. Those of G1AV/33 have been normalized to those of G1AV/35 by equating the average values of their differences above cosmic ray background during 1410 to 1436 UT. The horizontal bar through the vertical time marker at 1350 UT represents the ± 100 sec that I have adopted as the outer bounds on the lack of simultaneity of the two intensity time curves. Each point represents a counting rate averaged over 25.565 sec. There is an interval of 81.808 sec between observations.

are plotted on an absolute basis. Those of G1AV/33 have been normalized to those of G1AV/35 by equating the average values of their differences above cosmic ray background during the plateau period 1410 to 1436 UT. There is a small precursor peak at both spacecraft in the time interval 1344 to 1350 UT (probably high energy solar electrons).

The principal event begins at 1350 UT. There is no discernible delay in arrival of electrons at the two spacecraft. Even the temporary standstill at ≈ 1355 UT is observed at both. I have adopted ± 100 sec as generous outer bounds on the lack of simultaneity of the two intensity time curves.

Measured from the onset of the associated X-ray flare at 1326 UT, the total flight time of the earliest ≈ 50 keV electrons to arrive near the earth could not have exceeded $14+8$ min $=1320$ sec. This yields an apparent rectilinear velocity of 1.14×10^5 km sec^{-1}, which is the rectilinear velocity of a 40 keV electron. As observed in other cases, it is probable that the first electrons to arrive had higher energy, having *not* traveled in a straight line. The angular distribution was significantly anisotropic for only a few minutes at the onset of the event. Adopting the above apparent velocity, the observed maximum delay of 100 sec corresponds to a distance of 0.076 AU $=1780$ R_E. This is an approximate upper limit on the difference ΔL between the two paths from the sun to the respective spacecraft. The observations are consistent with $\Delta L=0$. As shown above, the average drift velocity of 50 keV electrons transverse to the magnetospheric tail $v_\perp \leqslant 100$ km sec^{-1} in the region closer than 64 R_E to the earth.

The *lower* limit on the drift time across 10 R_E (a value appropriate to the position of Explorer 33 during the August 14 event) is therefore 640 sec, a value much greater than the *upper limit* on the observed delay.

Hence, the evidence appears to require that the primary access of solar electrons into the magnetospheric tail occurs at downstream distances between 64 *and* 900 R_E, the latter figure being the upper bound on $\Delta L/2$.

It may be of importance to note that August 14, 1968 was a magnetically disturbed day with $K_p=5^-$ for 1200 to 1500 UT and with $\sum K_p=32^-$ for the day.

5. Interpretative Remarks

Another aspect of the foregoing evidence favoring an 'open' magnetic topology in the magnetotail is developed as follows:

If the topology of the magnetotail is *closed*, solar electrons can gain access throughout its interior only by diffusion *across* magnetic lines of force. Further, if there are differences in electrical potential between various points across the tail of the order of 50 kV, as are apparently required by a large body of evidence on magnetospheric convective effects, trans-**B** diffusion (by magnetic gradient drift, curvature drift, and nonadiabatic scattering but not by $\mathbf{E} \times \mathbf{B}$ drift) would imply diffusion across *closed* electrical equipotential surfaces and would result in changes of kinetic energy by an amount very much greater than the upper limit $|\Delta E| \leqslant 1.5$ keV inferred in an earlier section.

The results of the work reported herein therefore give strong support to the widely discussed possibility of an *open* magnetic topology in the distant magnetospheric tail, at least for $-X_{SE} \gtrsim 64$ R_E (i.e., dynamic interconnection of geomagnetic field lines to those in the interplanetary medium).

Moreover, they support the idea that the 'electrostatic' fields that drive magnetospheric current systems are in fact attributable to the *motional electromotive force* that results from the movement of the interplanetary magnetic field (frozen into the conducting solar wind plasma) with respect to the earth at $\sim 400 \text{ km sec}^{-1}$ (cf., Sonett and Colburn, 1968). Note that the electric field in the earth's frame of reference

$$\mathbf{E} = -(\mathbf{v} \times \mathbf{B})/c \approx 1.4 \times 10^{-3} \text{ Vm}^{-1}$$
$$= 10 \text{ kV}/\text{R}_\text{E}$$ (in a direction more or less perpendicular to the ecliptic plane).

The embarrassment of the correspondingly large value of the difference in potential across the diameter of the magnetotail ($\sim 40 \text{ R}_\text{E}$) may be relieved by supposing that the tail becomes flattened along the $\pm Z_{SE}$ axis at values of $-X_{SE} \gg 64 \text{ R}_\text{E}$ (Burrows, 1969; Fritz, 1969; Vasyliunas, 1969).

The role of the motional electromotive force in magnetospheric phenomena is apparently a common concept among theorists but it is rarely mentioned or discussed explicitly. The paper of Rostoker and Fälthammar (1967) is a noteworthy exception.

References

Anderson, K. A. and Lin, R. P.: 1969, *J. Geophys. Res.* **74**, 3953.
Armstrong, T. P. and Krimigis, S. M.: 1968, *J. Geophys. Res.* **73**, 143.
Behannon, K. W.: 1968, *J. Geophys. Res.* **73**, 907.
Burrows, J. R.: 1969, private communication.
Colburn, D. S., Currie, R. G., Mihalov, J. D. and Sonett, C. P.: 1967, *Science* **158**, 1040.
Fritz, T. A.: 1969, private communication.
Jánossy, L.: 1948, *Cosmic Rays*, Clarendon Press, Oxford, p. 266.
Lin, R. P.: 1968a, *Trans. Amer. Geophys. Union* **49**, 235.
Lin, R. P.: 1968b, *J. Geophys. Res.* **73**, 3066.
Lin, R. P. and Anderson, K. A.: 1966, *J. Geophys. Res.* **71**, 4213.
Lundquist, S.: 1954, *Tellus* **6**, 260.
Ness, N. F., Behannon, K. W., Scearce, C. S., and Cantarano, S. C.: 1967, *J. Geophys. Res.* **72**, 5769.
Rostoker, G. and Fälthammar, C.-G.: 1967, *J. Geophys. Res.* **72**, 5853.
Sonett, C. P. and Colburn, D. S.: 1968, *Phys. Earth Planet. Interiors* **1**, 326.
Swann, W. F. G.: 1933, *Phys. Rev.* **44**, 224.
Van Allen, J. A.: 1968, *Trans. Amer. Geophys. Union* **49**, 235.
Van Allen, J. A.: 1970, *J. Geophys. Res.* **75** (in press).
Van Allen, J. A. and Krimigis, S. M.: 1965, *J. Geophys. Res.* **70**, 5737.
Van Allen, J. A. and Ness, N. F.: 1967, *J. Geophys. Res.* **72**, 935.
Van Allen, J. A. and Ness, N. F.: 1969, *J. Geophys. Res.* **74**, 71.
Vasyliunas, V. M.: 1969, private communication.
Webber, W. R.: 1967, *Handbuch der Physik* XLVI/2, *Kosmische Strahlung II*, Springer-Verlag, Berlin, 181.

ANISOTROPIC DISTRIBUTIONS OF ENERGETIC ELECTRONS IN THE EARTH'S MAGNETOTAIL AND MAGNETOSHEATH

S. SINGER and S. J. BAME

University of California, Los Alamos Scientific Laboratory, Los Alamos, N.M., U.S.A.

1. Introduction

It is now well established (Anderson, 1965; Bame *et al.*, 1966, 1967; Montgomery *et al.*, 1965; Murayama, 1966; Montgomery, 1968) that the earth's magnetotail and magnetosheath support a population of energetic electrons at distances well beyond the boundary of stable trapping, and that the spatial and temporal characteristics of the population depend on whether the particles are observed in the plasma sheet, the high latitude magnetotail, or in the magnetosheath. The directional characteristics of the electrons have been investigated less extensively. Early Vela results (Bame *et al.*, 1964, 1965; Asbridge *et al.*, 1965; Singer *et al.*, 1967) reported streaming and non-streaming anisotropies, although at least some of the streaming events were due to instrumental effects. At radial distances up to 16.5 R_E, both Serlemitsos (1966) and Haskell (1969) have reported trapped electrons whose pitch angles are nearly parallel to the magnetic field, a condition which Haskell attributes to L shell splitting (Mead, 1966; Roederer, 1967). Haskell also detected streaming electrons in the magnetosheath near the magnetopause, and interpreted them in terms of particles leaking out of the magnetosphere.

The results are presented of an analysis of 6 mo of data from the Electrostatic Analyzer and Energetic Particle experiments on the Vela 4 satellites in which the observations of electron anisotropies are extended to $\sim$18.5 R_E and to high (60°) ecliptic latitudes. The major conclusions may be summarized as follows:

(a) Streaming energetic electrons are unambiguously detected in the magnetosheath not far from the magnetopause. Their spatial distribution shows that they are not necessarily found close to the solar magnetospheric equatorial plane, and have never been observed *inside* the magnetopause. The streaming direction is usually antisolar and often approximately parallel to the magnetopause.

(b) Large directional anisotropies in plasma-sheet energetic electrons have been detected that do not arise from streaming but instead from symmetric anisotropic pitch angle distributions. The spatial distribution of these occurrences shows that they are never seen outside the magnetopause, and that they are far more likely to be seen at magnetospheric longitudes near $\sim$240° than at other locations. Anisotropic distributions are almost never seen near the midnight meridian despite the fact that isotropic energetic electrons are frequently detected there. Near the neutral sheet, the directional anisotropies seem to be consistent only with a nearly vertical magnetic field and with particle pitch angles that peak near 90°.

B. M. McCormac (ed.), Particles and Fields in the Magnetosphere.

2. Instrumentation

The general features of the instruments used have been described elsewhere (Montgomery *et al.*, 1968; Singer *et al.*, 1968); for clarity, we summarize here only those aspects which are pertinent to the observations being reported.

The energetic electron data are taken from the four GM counters of the Vela Electron-Proton Spectrometer (EPS) and from the Energetic Particle (ST) experiments. The GM counters (~40 keV threshold) are located at spacecraft colatitudes of 22.5°, 67.5°, 112.5°, and 157.5° in a single spacecraft meridian plane. The ST semiconductor electron sensor, which detects electrons in the energy range 33 keV to 2.25 MeV, is placed in a different meridian plane at a colatitude of 90°. The electron fluxes are sampled by each sensor once each 4 sec, so that during the course of a single spacecraft rotation (64 sec), 80 samples of data are obtained that define the energetic electron angular distribution over the 4π solid angle.

Although the Vela satellites have no magnetometer, the spacecraft location (magnetosheath, plasma sheet, or high latitude magnetotail) can be determined unambiguously by comparing the measured characteristics of the low energy ions and electrons (as inferred from the EPS) with those established (Coon, 1966; Bame, 1968) for the two particle species in the various spatial regions. For example, a transition from the plasma sheet into the magnetosheath would be identified by the sudden appearance of positive ions with a well defined velocity vector, and a change in electron spectral shape wherein the intensity of ~0.1 keV electrons is enhanced relative to that of the >1 keV electrons.

3. Streaming Electrons in the Magnetosheath

In Figure 1, the electron and ion signals from the EPS and the ST electron sensors are plotted for a 40 min period near 1501 UT on July 8, 1967. Prior to ~4 min from reference time (R+4), the spacecraft was in the plasma sheet as indicated by the absence of low energy ions and the low value of the 63 to 78 eV electron flux. At that time, the energetic electrons had a nonstreaming type of anisotropic pitch angle distribution to be described below. After R+4, the sudden appearance of low energy ions and the increase in the low energy electron flux indicate that the spacecraft crossed the magnetopause and entered the magnetosheath. The character of the energetic electron population underwent a sudden change: strong spin modulation was still present, indicating anisotropic directional characteristics; however, there was only one counting rate peak per spacecraft rotation. Since the peaks occurred when the detector was pointing most nearly at the sun, these signals may be interpreted as a beam of electrons streaming in the antisolar direction. The streaming is apparent in both the >33 and >57 keV channels, and its appearance is sporadic, i.e., it appears briefly, disappears, and reappears again repeatedly. The sporadicity may persist for several hours; it is rare that a single episode of streaming lasts for more than 10 to 15 min.

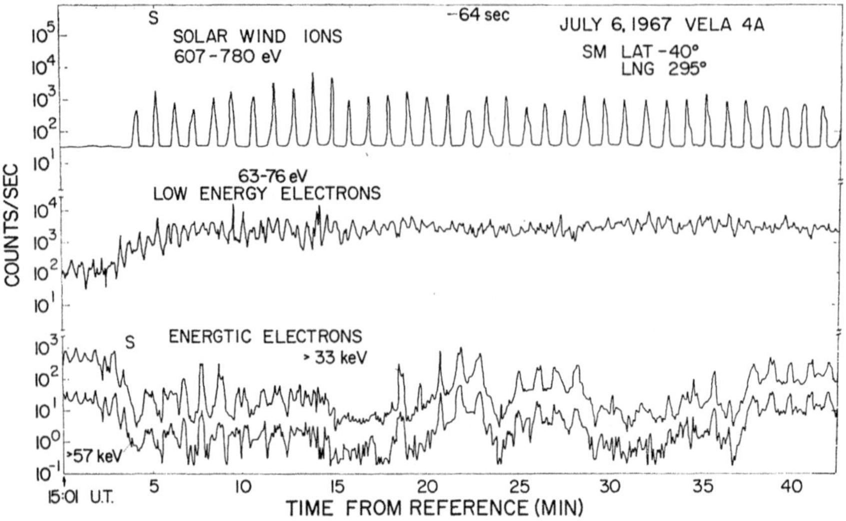

Fig. 1. Streaming Energetic Electrons in the Magnetosheath. The spacecraft entered the magnetosheath at ~ 1505 UT, as determined from the low energy and ion spectrums. Streaming electrons (showing one counting rate peak per spacecraft rotation) are seen at 1506–1510 UT, 1520–1524 UT, 1526–1530 UT, and 1538–1543 UT. The small tics labeled '*S*' denote the instant during each spacecraft rotation when the detecting instrument is most nearly looking at the sun. The spacecraft location in all figures is given in solar magnetospheric coordinates.

A more striking example of streaming electrons is shown in Figure 2. Here the EPS electron and ST energetic electron counting rates are plotted for a 20 min period after 0937 UT on December 3, 1967. During this period the spacecraft was in the magnetosheath, and just after 0937 electron streaming in the antisolar direction commenced. After R+12 the streaming ceased although it reappeared intermittently and frequently for the next 2 hr. Despite an anisotropy ratio (ratio of maximum to minimum counting rate during a single spacecraft rotation) of ~30, the minima in the counting rates during the streaming interval were in general agreement with the isotropic counting rate after R+12. This situation further supports an interpretation in terms of a directed beam of particles superimposed on a weak isotropic flux. A detailed study of the period before R+12 shows (a) the streaming direction is not precisely parallel to the satellite sun line; and (b) the widths of the peaks (in terms of angle) are substantially greater than the angular resolution of the detector, indicating that the streaming electrons can have sizable pitch angles relative to the magnetic field direction. Furthermore, the low energy electrons show no obvious signs of streaming, and the peak energetic electron counting rate (~150/sec, corresponding to a flux of ~3000/cm^2-sec-sr) is rather low compared to the other types of electron events seen by the Vela satellites. All of these observations appear to be typical of the several dozen streaming events under consideration.

In the examples described above, the streaming events were observed to occur in the magnetosheath. A better conclusion can be drawn from a study of the Vela 4A

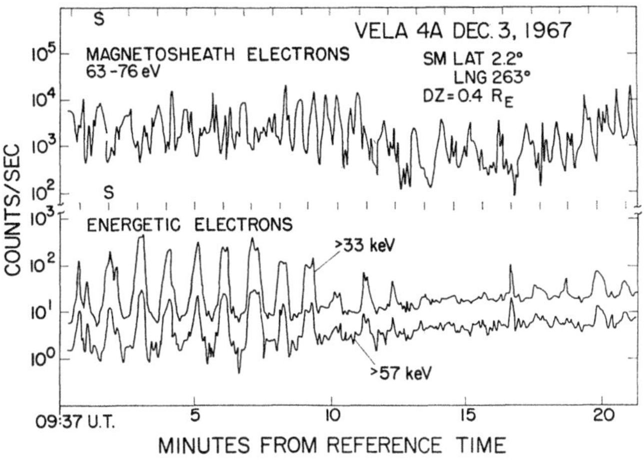

Fig. 2. Streaming Energetic Electrons in the Magnetosheath. The streaming electrons are seen at 0937–0947 UT and are identified by the single counting rate peak per satellite rotation. *S* denotes the time in each rotation at which the detector most nearly points at the sun. The peaks in the counting rate are not precisely solar oriented, and the width of the peaks exceed the instrumental resolution. No evidence for streaming of low energy electrons can be seen.

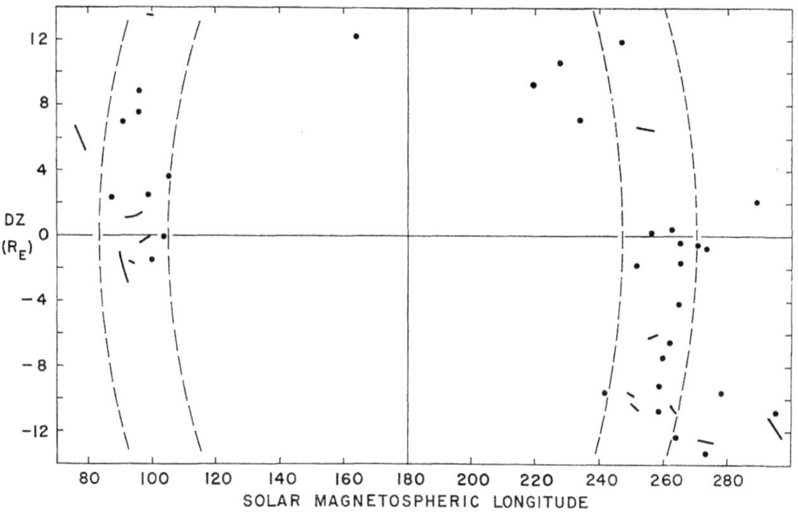

Fig. 3. Spatial Distribution of Streaming Events. Spacecraft trajectories from May–November 1968 are plotted here in terms of distance from the extended neutral sheet (DZ) vs. solar magnetospheric longitude, and are made visible only when a streaming electron event was detected. The pairs of dashed lines correspond to the average innermost and outermost locations of the magnetopause as detected by the Vela satellites.

data from May to November 1967, when the satellite passed through the magnetotail once every $4\frac{1}{2}$ day orbital period. The results are shown in Figure 3, where the spatial distribution is displayed in terms of the spacecraft distance from the neutral sheet (estimated using the results of Russell and Brody (1967)) vs its solar magnetospheric longitude. The spacecraft trajectory has been made visible only when a streaming event occurred; ordinarily, the duration of the event was so short that the trajectory appears as a single dot. For reference, the average outermost and innermost magnetopause locations (Bame *et al.*, 1967) are shown as dashed lines; crossings of the magnetopause are almost always encountered between these sets of limits. It is apparent that the location of nearly all of the streaming events lie within the limits. The four observations of streaming inside the inner boundary have been examined carefully, and the EPS data clearly indicate that the spacecraft was indeed in the magnetosheath. Presumably, the magnetosphere was grossly distorted at these times.

4. Discussion

In most streaming events, the flow is approximately in the antisolar direction, so it is important to consider whether the sun is a source of background that masquerades as a beam of particles. Although GM counters are known to respond to solar radiations, the ST semiconductor detector counts X rays only when extremely intense solar X ray events occur. We have concluded that the streaming events cannot be due to a solar background for the following reasons: (a) the solar aspect angle at which the maximum streaming electron counting rate occurs is at times detectably different from 0° (the angle at which maximum response to X rays occurs). On several occasions the electron rates peaked at angles ~45°. (b) The semiconductor detector shows streaming at electron energies in excess of 100 keV; even during intense X ray flares, the solar background cannot be detected at those energies. (c) Solar X ray monitors on the spacecraft normally detect no changes in solar X ray emission during streaming events. (d) The shape of the streaming event pitch angle distribution is usually inconsistent and unreconcilable with the 'pitch angle' distribution obtained when solar X rays are known to be the cause of detector counting rates.

5. Nonstreaming Angular Anisotropies in the Plasma Sheet

A second kind of directional asymmetry is often seen when studying the energetic electrons associated with the plasma sheet in the earth's magnetotail and a typical example is shown in Figure 4. There EPS electron and ion data are plotted together with ST energetic electron detector data from a 4A pass near the dawn meridian on May 19, 1967. From R+4 to R+15, the spacecraft was in the magnetosheath; before and after this interval, it was in the plasma sheet. Near R+9 min, the ion intensity decreased briefly and the low energy electron spectrum hardened, an occurrence which usually implies the very near presence of the magnetopause. The energetic electrons show a strong spin modulation when the spacecraft is in the plasma sheet, whereas the fluxes in the magnetosheath are essentially isotropic. The abrupt transition from anisotropy to isotropy is observed to occur at the magnetopause. Near R+9, one lobe of spin modulation appears, indicating that the spacecraft was within a cyclotron radius of the magnetopause for ~30 sec. The magnetopause thus acts as a barrier that effectively isolates the anisotropic from the isotropic fluxes.

The nature of the anisotropy seen in the plasma sheet is different from that discussed above in connection with streaming electrons. In the plasma sheet, the distributions possess *two* peaks of equal amplitude separated by 180° in azimuth, and no net flux of electrons can be detected streaming in a direction parallel to the line joining the

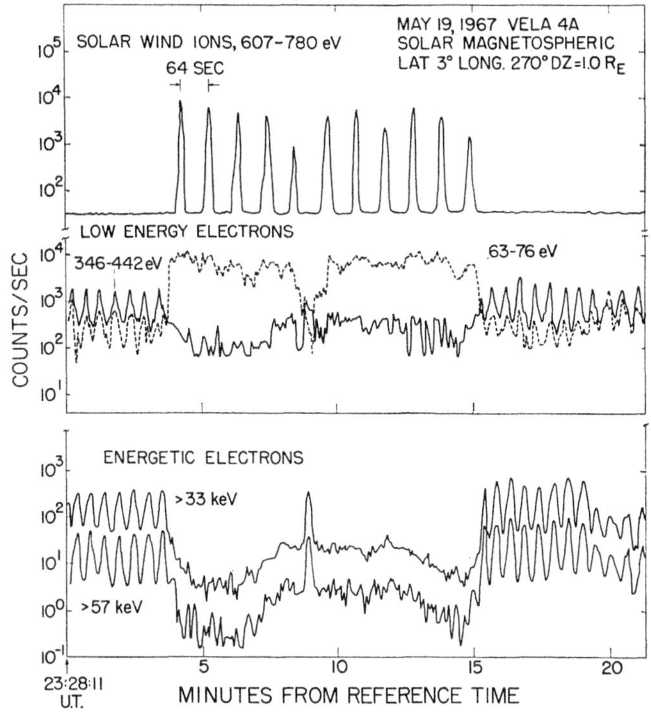

Fig. 4. Nonstreaming Directional Anisotropies in Plasma Sheet Electrons. The low energy electron-ion spectra show that the spacecraft was in the magnetosheath from ~2332–2343 UT, and otherwise was in the plasma sheet. A close approach to the magnetopause occurred at 2337 UT. The non-streaming anisotropies (2 counting rate peaks per satellite rotation) are seen only in the plasma sheet, and possibly during the close approach to the magnetopause. The anisotropy can be seen also in the low energy electrons, and the anisotropy ratio generally increases with electron energy.

directions of maximum counting rate. The anisotropy therefore must be caused by an anisotropic but symmetric pitch angle distribution which is similar to the distribution possessed by particles trapped in the earth's field. Although the anisotropy ratio itself may undergo slow amplitude modulations, it is common to see an anisotropy persist as long as 1 hr, in contrast to the sporadic nature of the streaming events. It is interesting to note that the anisotropy ratio often increases with the electron energy. For example, it is possible to observe low energy electrons that are isotropic, >33 keV electrons that are just detectably anisotropic, and >100 keV electrons that have an anisotropy ratio of 20:1. In Figure 4, for example, it is apparent that the anisotropy ratio is greater for the >57 keV electrons than for the >33 keV electrons. Occasionally, the *low energy* electrons also possess directional anisotropies; the event shown in Figure 4 illustrates such an occurrence. When the spin modulation extends to the low energies, changes in the anisotropy ratio usually occur simultaneously for both sets of particles.

The spatial distribution of the nonstreaming directional anisotropies has been

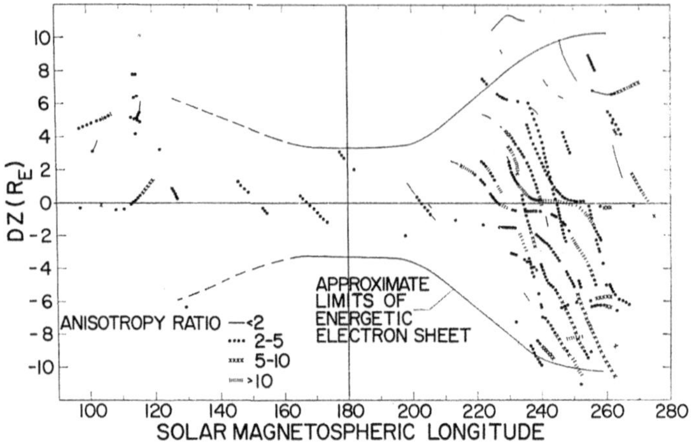

Fig. 5. Spatial Distribution of Nonstreaming Anisotropy Events. The coordinate system employed in Figure 3 is used here to show the frequency of observation of nonstreaming anisotropic electrons. The orbit trajectories are made visible with a set of plotting symbols that denote the electron anisotropy ratio. The probability of observing a nonstreaming anisotropy is greatest at longitudes of 235–260° and least near the midnight meridian.

derived from the series of passes through the magnetotail in May to November 1967. From it, we conclude that these events are observed in the plasma sheet and *never* in the magnetosheath. The longitude distribution of the nonstreaming events is shown in Figure 5, where the estimated spacecraft distance from the neutral sheet has been plotted vs spacecraft solar magnetospheric longitude for all of the orbits in the relevant time period. However, the trajectories are rendered visible only when *anisotropic angular distributions* of energetic electrons are detected. Different plotting symbols are employed to denote the anisotropy ratio in the range <2 to >10. Electron fluxes which are isotropic (i.e., anisotropy ratio $\lesssim 1.25$) are not plotted. The limits of the energetic electron sheet associated with the plasma sheet (Montgomery, 1968) are also indicated. It is apparent that the bulk of the observations is contained within the energetic electron sheet and between longitudes 230 to 260° (0330 to 0530 LT). For longitudes less than 230°, the likelihood of detecting anisotropic distributions decreases drastically, and few observations are made near the midnight meridian. The paucity of observation near the dusk meridian is probably related to the dawn-dusk asymmetry in the plasma sheet energetic electron population (Montgomery *et al.*, 1965; Bame *et al.*, 1967; Montgomery, 1968). The near absence of anisotropic distributions near the midnight meridian is surprising, since there is no lack of observation of isotropic electron fluxes there.

6. Discussion

If no important streaming or drift velocities are involved, the plasma sheet electrons undergo simple guided center motion and the magnetic field vector is a line of symmetry of the particle motion. With the 80 point set of directional distribution samples from the Vela sensors, the symmetry axis direction can be determined as follows: for each point i ($i=1,\ldots 80$) of the directional distribution, a unit mass point S_i is defined whose distance from the origin is proportional to the ith counting rate and whose radius vector has the same direction as that associated with the ith directional distribution sample. The S_i form a surface of unit mass density whose products of inertia can be readily computed, and whose principal axes can be obtained through the application of the principal axis theorem. The axis of symmetry is that principal axis for which the surface S depends on the polar angle measured relative to the principal axis only. The inertia eigenvalue associated with the symmetry axis offers some insight into the nature of the pitch angle distribution: If the eigenvalue has the maximum value, the pitch angles tend to be peaked near 90° (pancake distribution), if it has the minimum value the particles have near zero pitch angles (cigar shaped distribution).

The analysis described above has been applied to a few observations of nonstreaming anisotropy near the neutral sheet at longitudes between 230 to 270°. For one such event, Figure 6 shows the pitch angle distribution for the two choices of the axis of symmetry. In Figure 6a the minimum eigenvalue axis was assumed to be coincident with the magnetic field direction. However, the data are not consistent with that choice: The pitch angle distribution for the GM counters peaks at 90 and 0° (not merely 0°), whereas the distribution for the ST energetic electrons is strongly peaked toward angles $<70°$. The internal inconsistency merely reflects the dependence of the distribution on both the polar and azimuthal angles (in the principal axis coordinate system). In Figure 6b, the maximum eigenvalue principal axis, which was nearly perpendicular to the solar magnetospheric equatorial plane, was assumed to be the symmetry axis. It is apparent that particles have pitch angles preferentially near 90° and that the distribution depends on polar angle only.

Analyses of several similar events (i.e., where the spacecraft is near the neutral sheet) give the same result, and we conclude that at such times the local magnetic field is nearly perpendicular to the solar magnetospheric equatorial plane, and that electrons have pitch angles which are predominately 90° (as opposed to 0° as seen by Haskell (1969) and Serlemitsos (1966)). It is not clear that these particles are 'trapped'. It is conceivable instead that a local acceleration mechanism may populate field lines preferentially with 90° pitch angles; or, particles may be injected isotropically and their loss cone may somehow be substantially larger than the few degrees one would estimate.

It may be possible to explain the curious lack of observation of anisotropic fluxes near the midnight meridian in terms of the field line curvature near the neutral sheet. Near the midnight meridian, field lines which cross the neutral sheet form a cusp

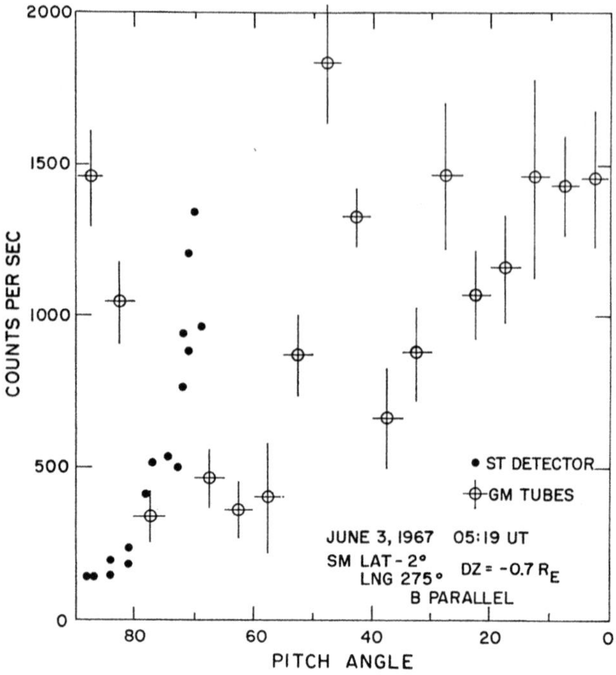

6a.

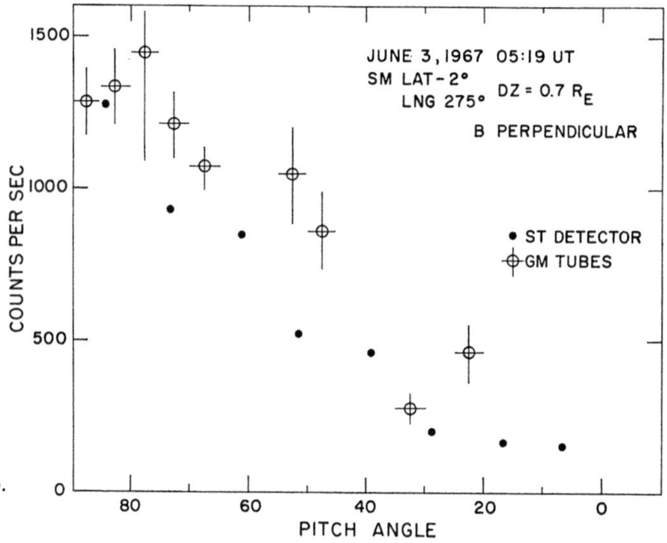

6b.

Fig. 6. Pitch Angle Distributions of Nonstreaming Electrons. Plotted here are the electron counting rates vs. pitch angle for two choices of the magnetic field direction. The data (from one rotation of the spacecraft) were averaged over intervals of 2.5°. The statistical uncertainties in the GM counters are indicated by the vertical strokes. a. Magnetic field direction parallel to minimum eigenvalue axis, corresponding to particles having mostly small pitch angles. The data are inconsistent with this choice, because the counting rates depend on both the polar and azimuthal angles in the principal axis system. For this choice, the minimum eigenvalue happened to be nearly parallel to the neutral sheet and resembled a stretched field line. b. Magnetic field direction parallel to maximum eigenvalue axis, corresponding to particles with mostly 90° pitch angles. This axis is nearly perpendicular to the neutral sheet. The data confirm that it is indeed the symmetry axis of the particle motion.

shaped curve near the crossover point, whereas Fairfield (1968) has shown that near 0300 to 0600 LT the field is nearly perpendicular and the curvature at the crossover point is much smaller than at the midnight meridian. If the plasma sheet anisotropic pitch angle distributions reported here imply pseudo-trapping of electrons, it is possible that the midnight meridian field line curvature at the crossover is too great to permit particles to make the turn (Taylor, 1969) and hence a quasi-stable particle population cannot be built up. Closer to the dawn meridian, however, the curvature is smaller and may permit a sizable population of quasi-trapped particles to exist. Presumably those particles which are equatorially mirroring in the plasma sheet would be lost at the magnetopause (Mead, 1966; Roederer, 1967).

Acknowledgment

This research was done as part of the Vela Nuclear Test Detection Satellite Program which is jointly sponsored by the Advanced Research Projects Agency of the Department of Defense and U.S. Atomic Energy Commission.

References

Anderson, K. A.: 1965, *J. Geophys. Res.* **70**, 4741.
Asbridge, J. R., Bame, S. J., Felthauser, H. E., Olson, R. A., and Strong, I. B.: 1965, *Trans. Amer. Geophys. Union* **46**, 142.
Bame, S. J.: 1968, in *Earth's Particles and Fields* (ed. by B. M. McCormac), Reinhold Publishing Corporation, New York, p. 373.
Bame, S. J., Asbridge, J. R., Felthauser, H. E., Olson, R. A., and Strong, I. B.: 1964, *Trans. Amer. Geophys. Union* **45**, 624.
Bame, S. J., Asbridge, J. R., Felthauser, H. E., Olson, R. A., and Strong, I. B.: 1965, *Trans. Amer. Geophys. Union* **46**, 142.
Bame, S. J., Asbridge, J. R., Felthauser, H. E., Olson, R. A., and Strong, I. B.: 1966, *Phys. Rev. Letters* **16**, 138.
Bame, S. J., Asbridge, J. R., Felthauser, H. E., Hones, E. W., and Strong, I. B.: 1967, *J. Geophys. Res.* **72**, 113.
Coon, J. H.: 1966, in *Radiation Trapped in the Earth's Magnetic Field* (ed. by B. M. McCormac), D. Reidel Publishing Company, Dordrecht-Holland, p. 231.
Fairfield, D. H.: 1968, *J. Geophys. Res.* **73**, 7329.
Haskell, G. P.: 1969, *J. Geophys. Res.* **74**, 1740.
Mead, G. D.: 1966, in *Radiation Trapped in the Earth's Magnetic Field* (ed. by B. M. McCormac), D. Reidel Publishing Company, Dordrecht-Holland, p. 481.
Montgomery, M. D.: 1968, *J. Geophys. Res.* **73**, 871.
Montgomery, M. D., Singer, S., Conner, J. P., and Stogsdill, E. E.: 1965, *Phys. Rev. Letters* **14**, 209.
Montgomery, M. D., Bame, S. J., and Hundhausen, A. J.: 1968, *J. Geophys. Res.* **73**, 4999.
Murayama, T.: 1966, *J. Geophys. Res.* **71**, 5547.
Roederer, J. G.: 1967, *J. Geophys. Res.* **72**, 981.
Russell, C. T. and Brody, K. I.: 1967, *J. Geophys. Res.* **72**, 6104.
Serlemitsos, P.: 1966, *J. Geophys. Res.* **71**, 61.
Singer, S., Conner, J. P., and Stogsdill, E. E.: 1967, *Trans. Amer. Geophys. Union* **48**, 161.
Singer, S., Aiello, W. P., Conner, J. P., and Klebesadel, R. W.: 1968, *IEEE Trans. Nucl. Sci.* **NS–16**, 336.
Taylor, H. E.: 1969, to be published.

TRAPPED AND POLAR PARTICLES DURING THE JUNE 9, 1968 MAGNETIC STORM

P. L. ROTHWELL, V. H. WEBB, and L. KATZ
Air Force Cambridge Research Laboratories, Bedford, Mass., U.S.A.

Abstract. During June 7 to 26, 1968, simultaneous observations of protons and electrons (0.1 to 1.0 MeV) were made. The proton fluxes were consistent with radial diffusion conserving the first two adiabatic invariants. However, the electron fluxes showed a preferential hardening around $L = 4.0$ consistent with a weak acceleration mechanism. The presence of both electrons and protons outside the trapping boundary was observed for LT = 2100 h. Pitch angle distributions are presented.

1. Introduction

Since the end of the last decade extensive studies of the radiation belts have mapped out the magnetic structure and energetic particle population of the earth's immediate environment. It is now quite clear that solar produced particles find their way into the radiation belts, however, studies to date have not been able to show how this is accomplished. In this paper we present data that we hope will aid in the understanding of this problem.

During the June 9, 1968 magnetic storm, we observed the growth at local noon of energetic electron fluxes well inside the trapping boundary. The appearance of energetic electron fluxes inside the trapping boundary during magnetic bays has been observed previously by Williams *et al.* (1968), Rao (1969), and Parsignault *et al.* (1969). We observe this electron growth as being concentrated around $L = 4.0$ which is reflected in the e-fold energy as well as the flux distribution. It will be shown that this growth is consistent with a weak local acceleration.

In contrast to the large variations observed in the electron population during the storm, very little changes are seen in the proton population. The proton fluxes show only mild changes and the e-fold energy varies with L in a manner consistent with radial diffusion that conserves the first two adiabatic invariants.

Also, during the storm low energy electrons and protons were observed beyond the trapping boundary at high latitudes on the nightside of the earth. These polar particle fluxes were not observed on the dayside and, as one would expect, are short lived.

Finally, pitch angle distributions for both protons and electrons are considered.

2. Instruments and Magnetic Aspect

Electron analysis and detection is accomplished by the use of a 180° focusing magnetic spectrometer with an array of solid state detectors. A uniform magnetic field of 1000 G is provided by permanent magnets to measure 0.1 to 1.0 MeV electrons. The detector array consists of eight 300 μ solid state detectors. Each detector is collimated so as to

B. M. McCormac (ed.), Particles and Fields in the Magnetosphere.

define an energy bin with an energy resolution of 15% FWHM and is backed up by a discriminator with a 90 to 300 keV pulse height window to reject protons. The geometrical factor for the spectrometer varies from 1.3×10^{-2} cm^2 sr at channel 1, which is the lowest energy bin, to 0.3×10^{-2} cm^2 sr at channel 8. During calibration the spectrometer was irradiated with 100 MeV protons and was determined to have a sensitivity of 6% or less per channel.

Proton detection and energy identification is accomplished by pulse height analysis of pulses from a 100 μ thick solid state detector with, as for the magnetic spectrometer, the pulses being separated into eight different energy bins. The bin widths vary from approximately 31% at channel 3 to 15% at channel 8. Immediately following and operating in anticoincidence with the 100 μ detector, is a 300 μ thick solid state detector which eliminates protons of energy between 3.5 and 60 MeV which could otherwise

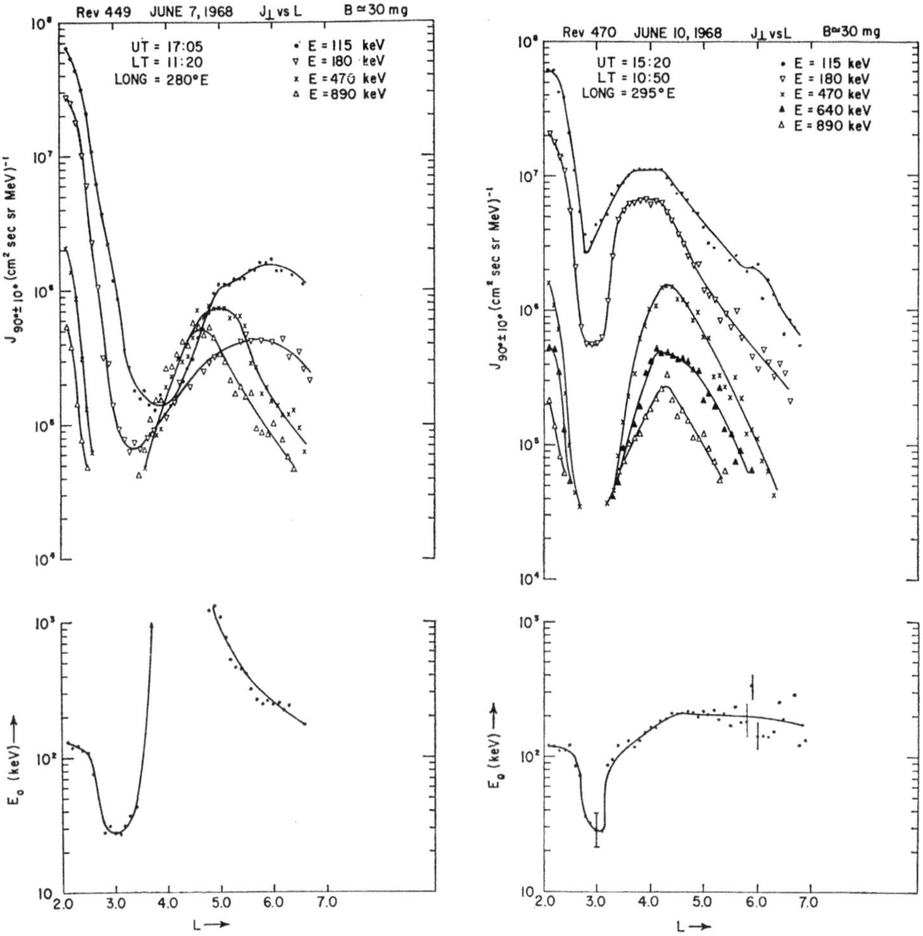

Fig. 1. Electron fluxes during magnetic quiet and at the beginning of a magnetic storm. (Note 115 keV cutoff at $L = 9$–10.)

be counted in the first detector as superfluous events. Low energy electrons which could be erroneously counted as protons are eliminated by a 1000 G sweeping magnet which is mounted into the detector shielding. This electron shield removes all electrons with energies less than about 2 MeV. This spectrometer has a geometrical factor of 0.53×10^{-2} cm^2 sr.

Magnetic aspect information is supplied by two sets of triaxial magnetometers; one set of 100 mG range and the other of 600 mG range. Each axis of both magnetometer sets is examined twice each second. A least squares fit of the functional form $A + B \cos \omega t$ is made every 12 sec to the data for the X and Z axis. The satellite is spin stabilized about the Y axis; hence, the Y axis data are not modulated. From the fits the orientation of the spectrometers with respect to the B field is determined each time a count-rate meter reading (every 40 msec) is taken.

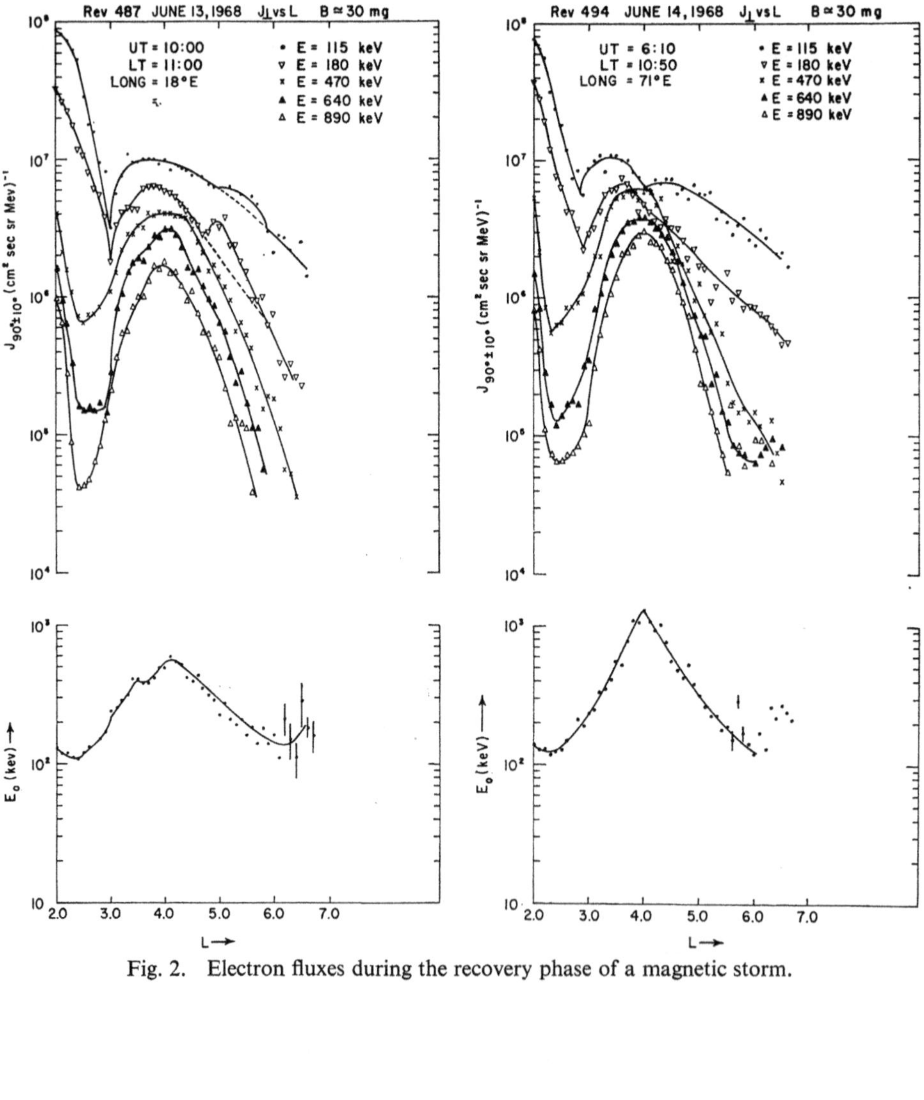

Fig. 2. Electron fluxes during the recovery phase of a magnetic storm.

3. Observations

The orbit of Air Force satellite OV1 13 was polar with an apogee of 9300 km, a perigee of 560 km, a 100° inclination, and a period of 200 min. This orbit allowed the sampling of the outer belt particle population at the same local times. In this paper we concentrate on: (a) electron flux, energy spectra, and pitch angle distributions at LT=1100 h, (b) proton energy spectra at LT=1100 h, and (c) particle fluxes observed at high latitudes (polar particles) at LT=2100 h.

June 1968 was magnetically quiet up to the 9th, 2200 UT (ESSA, 1968). A series of storms, with a major one occurring on the June 10 (2150 UT), kept the Dst curve depressed to June 11 0600 UT when it reached a minimum value of $-70\ \gamma$. The Dst

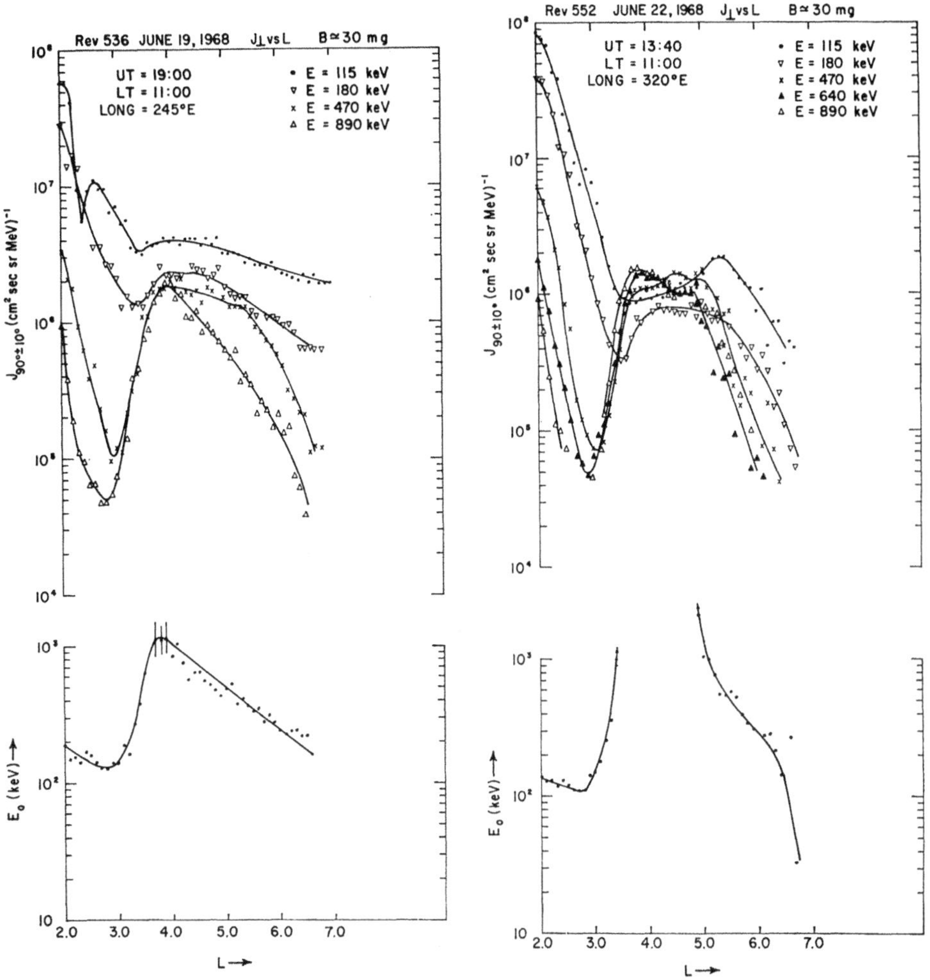

Fig. 3. Electron fluxes after the magnetic storm.

recovered to a positive value by June 15, 1968, and except for mild fluctuations it remained positive until the end of June.

Figure 1 shows the flux curves and energy spectra for mirroring electrons at LT= 1100 h. Rev 449 was preceded by two of the five quietest days in June (Lincoln, 1968) and represents the pre-storm quiet time condition. The main phase of the storm is characterized by a decrease of the 1.0 MeV electrons accompanied by a large increase of the 100 to 200 keV fluxes (Rev 470). This increase is well inside the belt and probably represents ring current particles.

The 1.0 MeV particles deplete further until during the recovery phase an enhancement of 1.0 MeV particles is observed. This enhancement occurs around $L=4$, is also seen from the e-fold energy, and is consistent with the location predicted by Williams *et al.* (1968). The hardening of the spectrum continued at $L=4.0$ until on June 14, 0610 UT it had a value of 1.2 MeV (Figure 2).

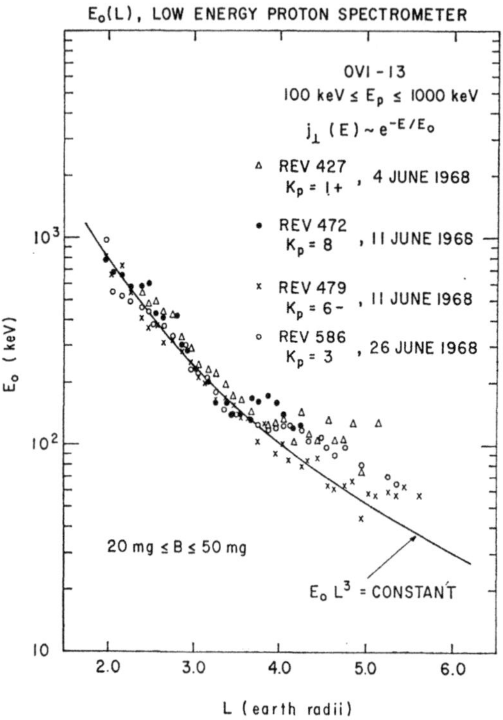

Fig. 4. The exponential energy parameter E_0 as a function of L before, during, and after the magnetic storm.

Following the storm (Figure 3) the 100 to 200 keV channel fluxes decay rapidly in the neighborhood of $L=4.0$ creating a post-storm peak at $L=6.0$. Also at $L=4.0$ the 1.0 MeV post-storm flux is higher than the 470 keV flux. One feature of the radiation belts is that the 470 to 890 keV fluxes have the same values on the inner edge of

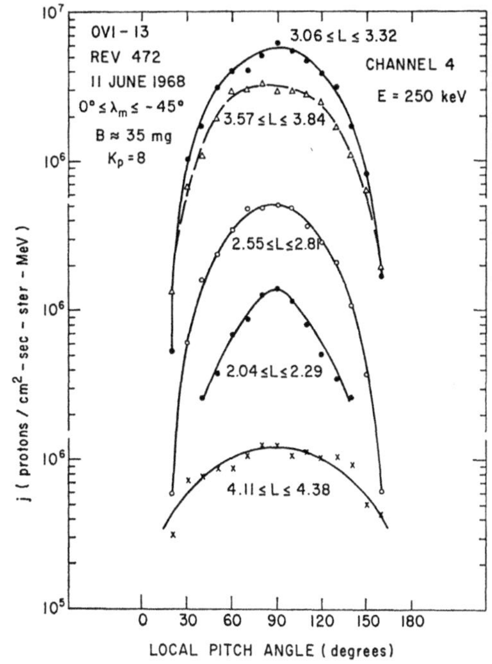

Fig. 5. Sample proton local pitch angle distributions.

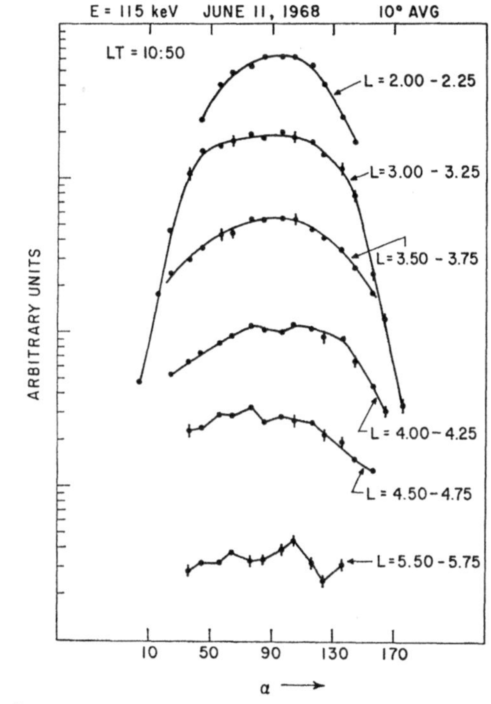

Fig. 6. Sample electron local pitch angle distributions.

the outer belt before and after the storm. The quiet time trapping boundaries on the dayside were located at $L=9$ (115 keV), $L=7$ (470 keV), and $L=6.5$ (890 keV).

Figure 4 shows the proton energy parameter E_0 as a function of L for the same time period. Very little variation is observed from a L^{-3} dependence which is consistent with adiabatic conserving diffusion.

The local pitch angle distributions (10° average, Figures 5 and 6) were fitted to a form $j=A\sin^N\alpha$, where j is the flux and α is the local pitch angle. For electrons N varied linearly from 3.0 at $L=2.0$ to $N=0$ (isotropic) for $L\geqslant 5$. For protons N varied linearly from 4.0 at $L=2.0$ to $N=0$ at $L\geqslant 5.0$. No variation in energy was observed for either protons or electrons. These observations were taken during a period of high K_p. However, the flattening out of the local pitch angle distributions at higher values of L is a general feature independent of magnetic activity.

On two passes during June 10 to 14, isotropic high particle fluxes were encountered

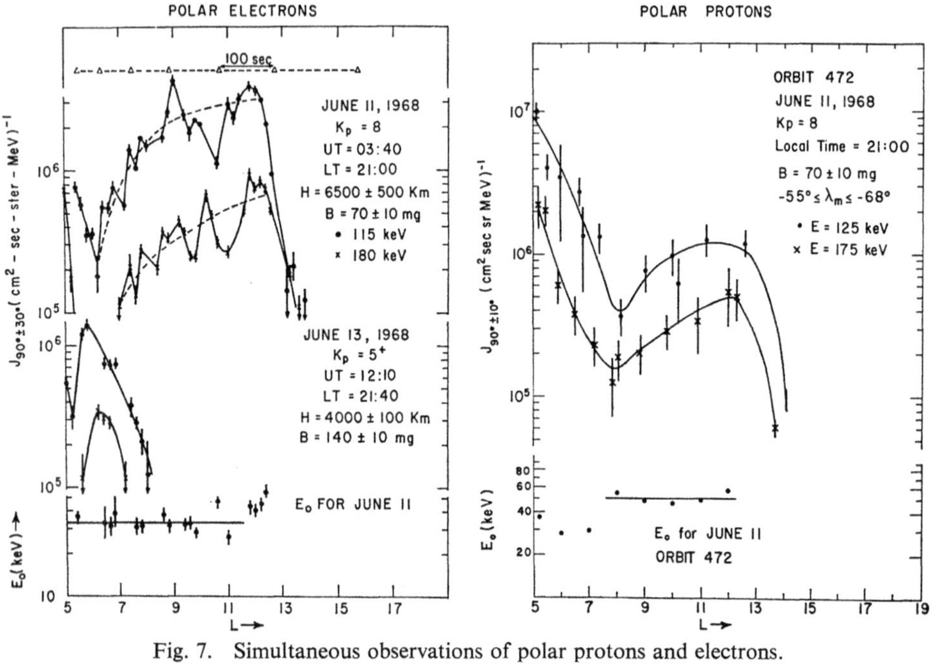

Fig. 7. Simultaneous observations of polar protons and electrons.

beyond the trapping boundary at LT = 2100 h (Figure 7). (90° ± 50° in pitch angle was sampled in this region.) The trapping boundary for 115 keV electrons on the nightside varied between $L=5$ and 7. The high latitude particle fluxes were first observed on June 11, 0300 UT ($K_p=8$). Both proton and electron fluxes had an e-fold energy of 30 to 50 keV. High latitude electron fluxes have also been observed by McDiarmid and Burrows (1965) and Vampola (1969).

A later pass on June 14, 1200 UT (LT = 2100 h) showed a much lower electron flux suggesting a remnant of the June 11 flux. Noise in the proton detector prevented positive identification of a real signal. A pass taken on June 10, 1500 UT during which the flare protons were arriving at close to peak intensity (ESSA, 1968) showed no measurable fluxes at high latitudes.

4. Discussion

The response of electrons to a magnetic storm in the outer belt is very energy dependent. From the dayside observations it is very suggestive that the higher energy electrons have been accelerated from lower energies at $L=4.0$. A separate exponential flux distribution was assumed below and above $E_e=300$ keV. The average energy of

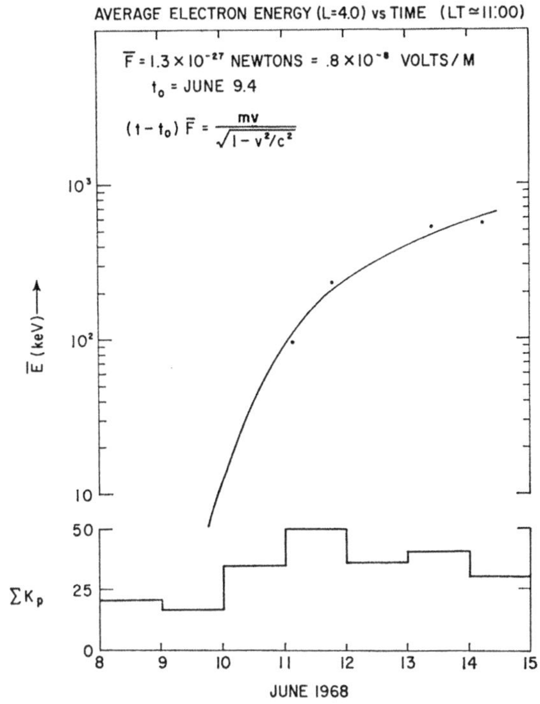

Fig. 8. Average electron energy at $L=4.0$ as a function of time. A constant force acceleration is assumed.

the total distribution was plotted as a function of time (Figure 8) and fitted for a constant average force. The fit is consistent with 40 keV particles having been picked up during the main phase of the storm and accelerated to the observed energies.

We have assumed the force (electric field $=10^{-8}$ V/m) is operative uniformly for the entire period (10^5 sec). However, this is equivalent to a field of 1 mV/m accelerating the electron on the order of 1 sec. Therefore, the observed acceleration is not inconsistent with resonant interactions with whistlers.

Because of their presence at local night the polar particles suggest a tail source.

This source could be plasma sheet particles that, upon moving in during a magnetic storm (Vasyliunas, 1968), are accelerated in the auroral tubes of force (Hones *et al.*, 1968). It is unlikely that these are solar particles brought in by an electric field in the tail (Taylor and Hones, 1965) because of the presence of protons as well as electrons.

References

ESSA: 1968, *Solar-Geophysical Data*, No. 287.
Hones, E. W., Jr., Singer, S., and Rao, C. S. R.: 1968, *J. Geophys. Res.* **73**, 7339.
Lincoln, V.: 1968, *J. Geophys. Res.* **73**, 6868.
McDiarmid, I. B. and Burrows, J. R.: 1965, *J. Geophys. Res.* **70**, 3031.
Parsignault, D. R., Paolini, F. R., Rothwell, P. L., and Katz, L.: 1969, *Proceedings of the Twelfth Plenary Meeting of COSPAR*, to be published.
Rao, C. S. R.: 1969, *J. Geophys. Res.* **74**, 794.
Taylor, H. E. and Hones, E. W.: 1965, *J. Geophys. Res.* **70**, 3605.
Vampola, A. L.: 1969, *J. Geophys. Res.* **74**, 1254.
Vasyliunas, V. M.: 1968, *J. Geophys. Res.* **73**, 2839.
Williams, D. J., Arens, J. F., and Lanzerotti, L. J.: 1968, *J. Geophys. Res.* **73**, 5673.

SOLAR PARTICLE OBSERVATIONS OVER THE POLAR CAPS

G. A. PAULIKAS, J. B. BLAKE, and A. L. VAMPOLA
Space Physics Laboratory, Aerospace Corporation, El Segundo, Calif., U.S.A.

1. Introduction

The increasing incidence of energetic particle emission from the sun since 1966 has provided a number of opportunities to measure the spatial, angular and energy distribution of these particles at high latitudes with instruments flown aboard low altitude polar orbiting satellites. Such measurements may provide insight into the structure of the distant magnetosphere and complement the measurements carried out by high altitude spacecraft.

The phenomenology exhibited by solar particles observed over the polar caps becomes increasingly complex as one studies lower energy particles. Of special interest, because of their small gyroradii, are solar protons and α-particles with energies less than about 10 MeV/nucleon. Solar electrons with energies down to 40 keV also have been observed over the polar caps and provide an additional means (a very low rigidity particle) for studying solar particle access to the polar caps. This paper will summarize some of the recent measurements of spatial (latitude) distributions and angular distributions. We will also compare solar proton and electron data and discuss the connection between the limits of trapping and the access and cutoff latitudes for solar particles. The bulk of the data to be discussed was obtained aboard the USAF-OAR satellites 1966-70A (OV3-3), 1967-72D (OV1-12), and 1968-59A (OV1-15).

2. Penetration of Solar Electrons to the Polar Caps

It has been found that solar electrons (Vampola, 1969) in addition to the well documented low energy protons and alphas have ready access to the polar caps. The measurements now have been extended down to electron energies of 40 keV; a representative electron spectrum obtained on April 13, 1969, is shown in Figure 1. The salient features of the solar electron observations over the polar caps are summarized:

(a) Electrons have access to the polar caps with high efficiency. Measurements outside the magnetosphere have been compared to measurements over the polar caps and show a transmission efficiency near unity. At the present time it is not clear whether the turnover in the electron spectrum (Figure 1) is a magnetospheric effect or associated with the properties of the source on the sun.

(b) In contrast to the low energy solar proton fluxes whose latitude profiles are often characterized by spatial variations in the flux (Blake *et al.*, 1968; Williams and Bostrom, 1969) most solar electron observations show featureless flux vs. latitude profiles; solar electrons can be distinguished from the trapped electron population

B. M. McCormac (ed.), Particles and Fields in the Magnetosphere.

on the fringes of the trapping region by examining the latitude intensity profiles. Such nonuniformities as have been detected, in most cases appear to be associated with motions past the satellite of the trapping boundary containing terrestrial electrons.

(c) Diurnal asymmetry in the polar electron flux is sometimes observed – usually

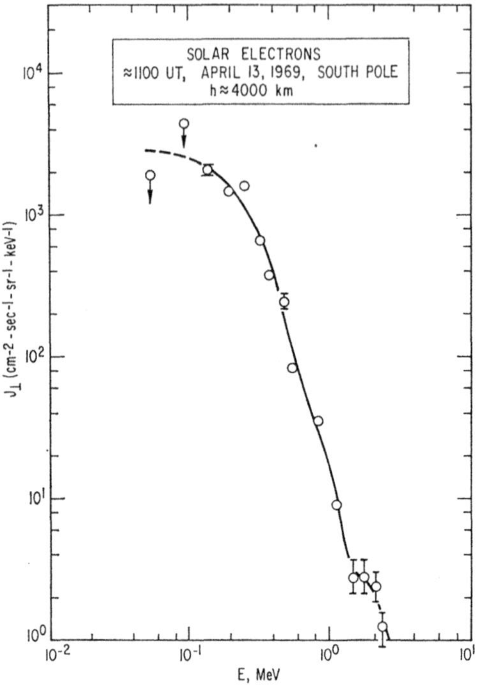

Fig. 1. Electron spectrum obtained on April 13, 1969, onboard the polar orbiting OV1-19 (1969-27D). A flux of electrons illuminated the polar cap uniformly.

in the early stages of the polar particle event. The polar electron fluxes merge abruptly with the trapped outer zone electron fluxes (as determined by a change in the pitch angle distribution) near the local midnight meridian; in contrast, near the noon meridian a more gradual connection between the solar flux and the outer zone electrons is found. A drop in the solar electron flux spatially coincident with the beginning of an anisotropic pitch angle distribution for solar protons on the day-side hemisphere is often observed (Figure 2). Measurements of electron pitch angle distributions here show a peaking of the flux perpendicular to the magnetic field. There is no latitude gap between the domain of solar electrons and the domain of terrestrial electrons.

3. Latitude and Pitch Angle Distributions

Latitude profiles of low energy (> 10 MeV) solar proton fluxes observed over the polar cap exhibit an amazing range of behavior. An indication of the possibilities has

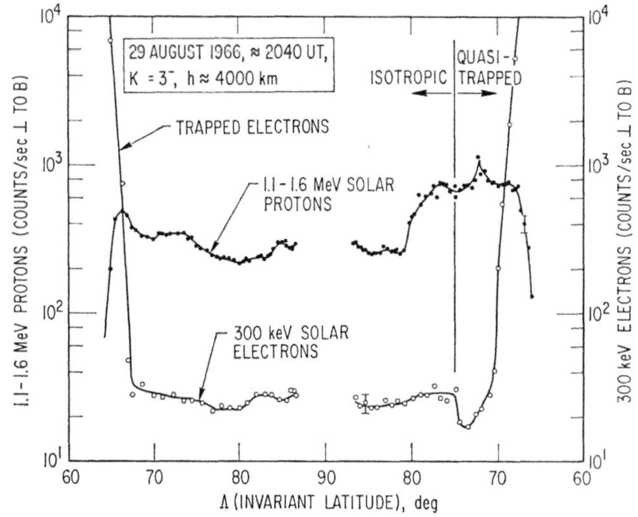

Fig. 2. Comparison of solar electrons (225 keV $\leqslant E_e \leqslant$ 375 keV) and solar proton (1.1 MeV $\leqslant E_p$ 1.6 MeV) on August 29, 1966. Data were obtained onboard OV3-3 (1966-70A). Note the alignment of the electron flux depression and the beginning of the quasi-trapped region for electrons and protons. These data were obtained over the south polar cap. The dayside flux enhancement contains, in part, an anisotropic pitch angle distribution.

been presented in Figure 2. Compilation of all our data obtained since 1966 indicates that spatial nonuniformity is the rule for solar protons but is the exception for solar electrons (Table I).

Frequently one observes an enhancement in the solar proton flux in a latitude

TABLE I

Solar particle flux profile summary*

	August 1966 June 1967		July 1967 Dec. 1967	July 1968 Nov. 1968
	Protons 1.1–1.6 MeV	Electrons 225–375 keV	Protons 3–10 MeV	Protons 0.75–1.7 MeV
Non-uniform profiles	46	4	21	15
Flat profiles	14	55	2	7

* These observations are weighted (because of satellite operational considerations and instrument counting statistics) toward the later phases of large solar particle events.

interval at higher latitudes than the cutoff. Examples of such enhancement are found at all local times. Measurements for protons and electrons made at altitudes > 1000 km and away from local midnight show anisotropic pitch angle distributions with the particle flux peaking in a direction perpendicular to the local field. The width of this region of pitch angle anisotropy approaches 10° in latitude near local noon at high (≈4000 km) altitudes. These distributions indicate temporary trapping of particles and have been discussed for protons by various authors (Taylor, 1967; Gall *et al.*, 1968). A simple way to describe the generation of an anisotropic solar particle flux on the dayside is by inverting the arguments used by Roederer (1967) to describe the motion of trapped particles; drift shells of particles with low altitude mirror points near local noon connect to open field lines in the magnetic tail. The motion of electrons (in the range of these measurements) is adiabatic in the absence of wave particle interactions throughout most of the magnetosphere (Gall *et al.*, 1968). If the tail is filled with a flux of solar particles, then field lines closed near local noon are populated by particles drifting in longitude. Note that no particle scattering need be invoked. Similar arguments also apply to ≈1 MeV protons, although their motion does not satisfy the adiabatic criterion and protons may suffer from scattering by cyclotron radius sized field curvatures in the limit of almost open field lines. The boundary between isotropic and quasi-trapped fluxes, occurring at latitudes between 73° and about 78°, delineates the separation of locally closed field lines from field lines which extend far into the magnetotail.

4. Connection Between Trapping Boundary and the Cutoff Latitude

The idea of the last closed drift shell (the last family of lines upon which a low rigidity particle may be permanently trapped) may be used as a definition of the limits of access of extraterrestrial particles. The concept can be experimentally tested by comparing a trapping boundary of low energy electrons with the cutoff latitude of low energy protons. We have performed this experiment and illustrate the results in Figure 3a for the midnight meridian and in Figure 3b for the noon meridian. In the case of near midnight, the disappearance of the trapped electrons is directly connected with the appearance of solar protons. Studies of trapped particle trajectories in a Mead-Williams model of the magnetosphere using Roederer's (1967) program SPLIT indicate that near local midnight the last closed field line also defines the last closed drift shell for particles mirroring at high latitude. Near local noon, however, solar protons appear deeper in the electron trapping region and in general exhibit a very ill-defined cutoff. During disturbed times solar protons merge with the trapped proton boundary. Figure 4 summarizes the comparison between the trapping termination and cutoff latitude near local midnight for quiet as well as disturbed times. The tracking of the Λ_p with Λ_e is evident; the small difference $\Delta\Lambda$ (about 1.5°) can be quantitatively connected to the nonzero gyroradius of the protons on these field lines near the magnetic equator. Near local noon definition of the cutoff becomes more difficult. Solar protons penetrate 3 to 4° past the apparent limit of electron trapping. If the electron boundary at local noon can truly be identified with the last *closed* drift shell, then proton scattering and hence radial diffusion because of shell splitting need be invoked in order to account for the experimental results.

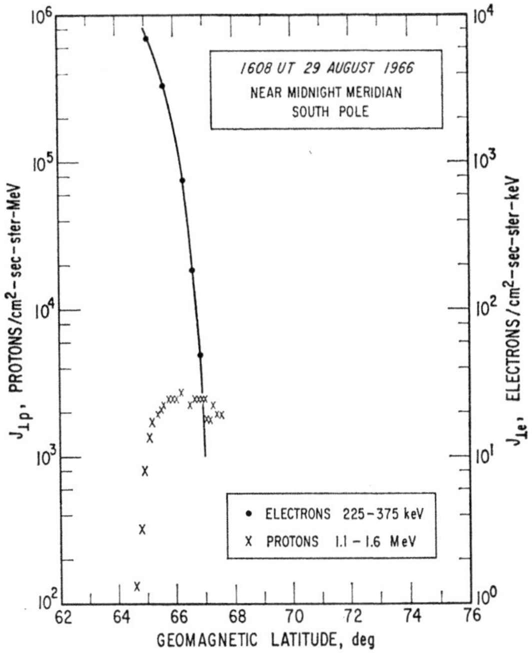

Fig. 3a. Boundary of trapped electrons and cutoff of 1.1–1.6 MeV solar protons. Data taken by OV3-3 satellite near local midnight.

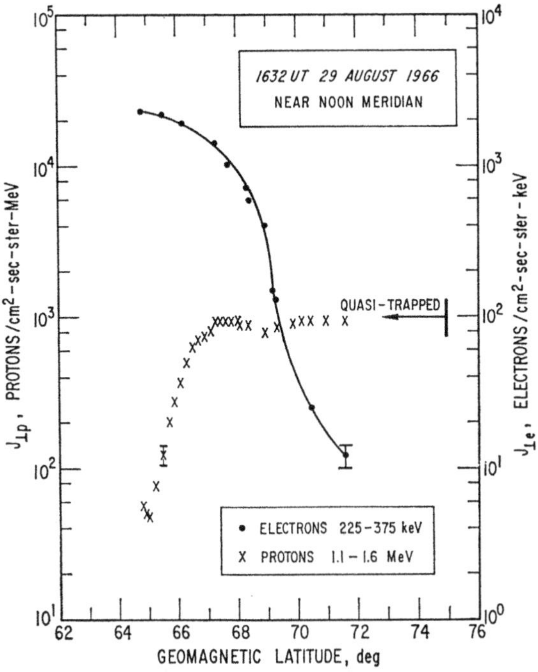

Fig. 3b. Boundary of trapped electrons and cutoff of solar protons near local noon. Quasi-trapped region (pitch angle distributions peak perpendicular to magnetic field) is accessible to solar protons by longitudinal drift.

5. Conclusions

The results of this work combined with others serve to define some of the properties of the magnetosphere for the transmission of solar particles.

Montgomery and Singer (1969), as well as Kane *et al.* (1968), find that solar proton temporal variations and angular distribution anisotropies, although not eliminated, are

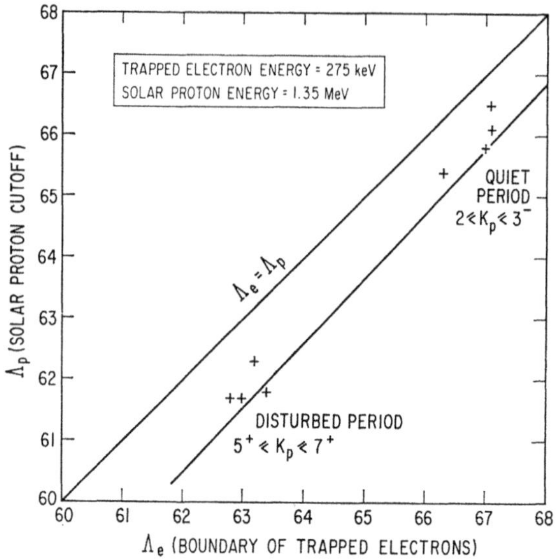

Fig. 4. Comparison of proton cutoff and outer zone boundary for magnetically quiet as well as disturbed periods. Data were obtained in the August 28–September 3, 1966, interval near local midnight. The lower solid line indicates cutoff latitude shift assuming protons penetrate one cyclotron radius (in a 20γ equatorial field) past the last closed field line.

smoothed in transit to the interior of the magnetosphere near 18 to 20 R_E. Assuming adiabatic motion for simplicity, remaining spatial gradients in the particle angular distribution function across the magnetotail will be translated into nonuniform latitude intensity profiles across the polar cap which is the usual observation. The uniformity of electron latitude intensity profiles in the domain of open field lines over the pole suggests that, in contrast, the magnetotail is filled very uniformly with solar electrons. Anderson and Lin (1969) have demonstrated that the magnetotail is filled by solar electrons traveling along field lines directly connected to the interplanetary field rather than strong diffusion.

The very sharp and temporally stable separation as seen by solar electron fluxes of the region of open field lines from locally closed field lines and the abrupt changes in pitch angle distributions suggest that diffusion by solar electrons in the region near the earth where open field lines are in contact with closed field lines is slow (less than ≈ 70 km per longitudinal drift period). Solar protons penetrate deeper into the region of closed field lines than pure trajectory tracing permits. We postulate that scattering of protons by (<1 cps) hydromagnetic waves is the mechanism which transports protons radially.

One of the persistent observations, ours as well as that of others (Blake *et al.*, 1968; Williams and Bostrom, 1969), has been the enhancement of the solar proton flux at latitudes just above the cutoff, i.e., on open field lines. This can be interpreted as indicating at times diffusive entry of solar protons into the magnetosphere (Williams and Bostrom, 1969). The time development of the latitude intensity profile required for this explanation is one in which a flat latitude profile evolves with time. Observations exist where such evolution did not proceed although flux enhancement near the cutoffs persisted throughout a proton event. A possible alternative explanation of the higher fluxes near cutoff is that this region is fed by the magnetotail as well as by entry plus subsequent scattering of protons into the quasi-trapped region from the flanks of the magnetosphere.

We conclude that some scattering of low energy solar protons, but not electrons, is required in order to account for the experimental observations of solar particles over the polar caps.

Acknowledgment

This work was performed under contract AF-4706-69-C-0066.

References

Anderson, K. A. and Lin, R. P.: 1969, *J. Geophys. Res.* **74**, 3953.
Blake, J. B., Paulikas, G. A., and Freden, S. C.: 1968, *J. Geophys. Res.* **73**, 4927.
Gall, R., Jimenez, J., and Camacho, L.: 1968, *J. Geophys. Res.* **73**, 1593.
Kane, S. R., Winckler, J. R., and Hofmann, D. J.: 1968, *Planet Space Sci.* **16**, 1381.
Montgomery, M. D. and Singer, S.: 1969, *J. Geophys. Res.* **74**, 2869.
Roederer, J.: 1967, *J. Geophys. Res.* **72**, 981.
Taylor, H. E.: 1967, *J. Geophys. Res.* **72**, 4467.
Vampola, A. L.: 1969, *J. Geophys. Res.* **74**, 1254.
Williams, D. J. and Bostrom, C. O.: 1969, *J. Geophys. Res.* **74**, 3019.

THE REACTION OF THE PLASMAPAUSE TO VARYING MAGNETIC ACTIVITY

C. R. CHAPPELL, K. K. HARRIS, and G. W. SHARP
Lockheed Palo Alto Research Laboratory, Palo Alto, Calif., U.S.A.

1. Introduction

Concentrations of H^+, He^+, and O^+ ions have been measured from an initial perigee height of ~300 km to an apogee of ~23 R_E by the light ion mass spectrometer aboard OGO 5. The instrument has a sensitivity of less than 1 ion/cm^3 and a range of 8 orders of magnitude. With these capabilities the instrument was able to measure ion concentrations from within the plasmasphere through the plasmapause and the 'trough' region of the outer magnetosphere. In this paper we are concerned with the position and density profile of the plasmapause which is generally located in the region $\sim 3 \leqslant L \leqslant 6$.

The details of the instrument design and calibration and the method of data analysis have been described elsewhere (Harris and Sharp, 1969, 1970). The data to be presented have been studied as a function of L and local time. This choice of coordinates is to facilitate comparisons with other plasmapause measurements (cf., Carpenter, 1966; Binsack, 1967; Brinton *et al.*, 1968; Vasyliunas, 1968, and others). The inclination of the satellite was such that it covers geomagnetic latitudes of about -25 to $+5°$ on the inbound plasmapause crossings and about 18 to 35° on outbound crossings. By using the L values and local time we are essentially plotting the magnetic equatorial crossing of the field line on which the plasmapause is detected. Our approach has been to examine the concentration of each orbit from approximately $L = 10$ inbound through perigee to $L = 10$ outbound and to locate the plasmapause in L and local time.

2. Plasmapause Boundary

A typical plasmapause crossing displaying concentration vs. L value is shown in Figure 1. This plot shows the rise in H^+, He^+, and O^+ as the vehicle passes inbound through the plasmapause. The H ion concentration rises from less than 1 ion/cm^3 in the 'trough' region to $>10^3$ ions/cm^3 inside the plasmasphere.

The plasmapause itself is a loosely defined region but has generally been described as the region at which the ion (or electron) concentration rapidly falls off by greater than an order of magnitude, generally from >100 ion/cm^3 to about 1 ion/cm^3 (Carpenter, 1966).

The locations of the plasmapause for 100 ions/cm^3 are shown in Figure 2. These data display the inbound and outbound plasmapause crossings for each of the first twenty orbits of the satellite. The data cover local times of ~1000 ± 2 hr inbound

B. M. McCormac (ed.), Particles and Fields in the Magnetosphere.

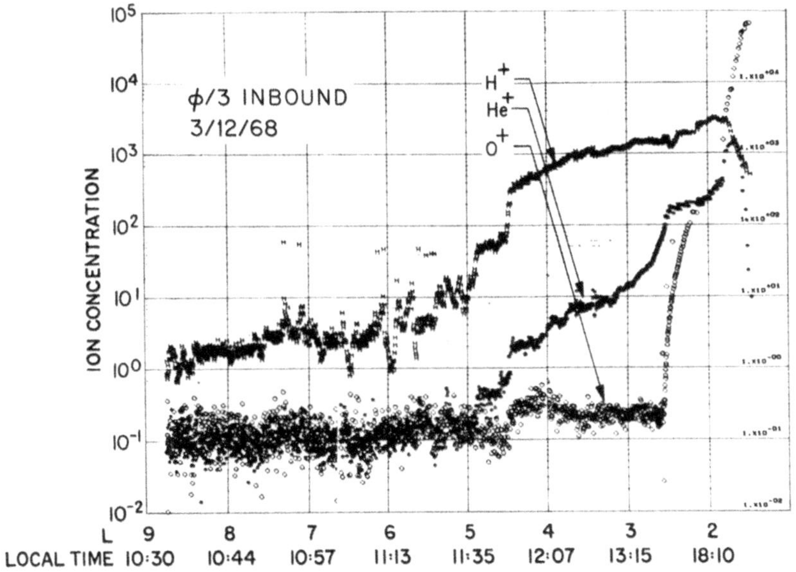

Fig. 1. A typical inbound plasmapause crossing showing the H^+, He^+, and O^+ ion concentrations as a function of L.

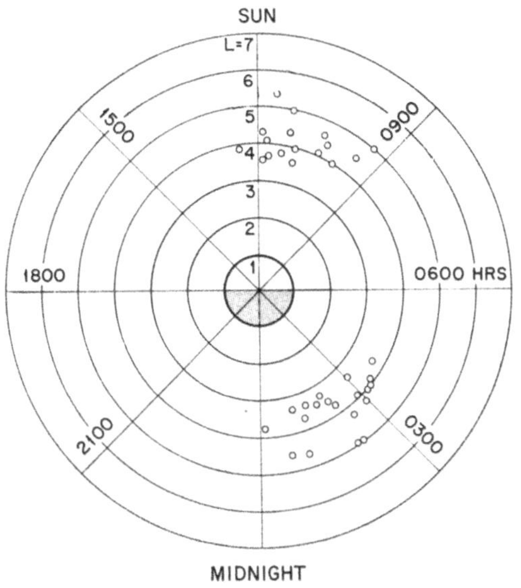

Fig. 2. Location of the plasmapause crossings given in L and local time for the inbound and outbound passes of the first twenty orbits. The inbound passes are in the 1000 ± 2 hr LT region and the outbound are in the 0200 ± 2 hr LT region. The L parameter represents the equatorial crossing distance in R_E of the field line on which the plasmapause was measured.

and ~0200 ±2 hr outbound. From Carpenter's (1966) measurements it is known that there is a local time variation in the position of the plasmapause which ranges from a maximum of about $L=6$ at the bulge on the evening side to a minimum of about $L=3.5$ at dawn. This variation is small in the 4 hr region around 1000 and hence will be ignored while the same local time variation in the 0200 region is more substantial and is considered below. With the local time variation taken into account, the remaining plasmapause movement will be assumed to be due to processes connected with magnetic activity.

The first problem with which one is confronted in attempting to correlate plasmapause location with the level of magnetic activity is the unknown time lag between the onset of the activity and the resulting influence on the plasmapause. Based on whistler measurements of the electron concentration, Carpenter (1967) has estimated a lag of approximately 6 hr, but it is not at all clear that this estimate is always applicable. We have studied this time lag by plotting average K_p and maximum K_p vs. the L value of the plasmapause position over several different time periods of up to 24 hr immediately preceding the time that our measurement of the plasmapause was made. These studies have shown that for the inbound passes in the 1000 LT region there is a general decrease in plasmapause radius with increasing magnetic activity but there is no particular time lag which appears more favorable than the others examined in this local time region.

The 0200 LT region was examined in the same manner as the 1000 LT region with the addition of two further considerations. First, as we have stated, there is a local time variation in this region from midnight to 0400. This local time variation was taken into account by using a normalizing technique based on Carpenter's (1966) measurement of the average position of the plasmapause as a function of local time. Second, we should realize that the location of the plasmapause is not always explicitly given by the point at which the concentration falls below 100 ions/cm^3. This is particularly evident in curve a of Figure 3 where the plasmapause covers the region from $L=4$ to $L=6$ but actually reaches the low concentrations characteristic of the trough at $L=6$. Although the plasmapause concept is usually discussed in terms of the position of the steep gradient in the density profile rather than the position of a certain level of concentration, it would be convenient to arrive at some decision on a particular concentration level since even the position of the gradient is sometimes ambiguous (Figure 3, curve a). In many cases, as in Figure 3, curve b, there is a negligible difference in locating the plasmapause if one were to use either the <100 ions/cm^3 or <10 ions/cm^3, however, in cases such as Figure 3, curve a, we feel that the concentration level of 10 ions/cm^3 more accurately defines the edge of the plamasphere since it more nearly represents the boundary between values characteristic of the plasmasphere and those of the trough. It is still necessary to be cautious in applying this new definition to all local time cases. In Figure 3, curve c, which is in the 1000 LT region, for example, the concentrations in the trough region are sometimes quite high and exceed 10 ions/cm^3 in several places; these places should not be confused with the plasmapause position which is at about $L=4.4$. However, for the

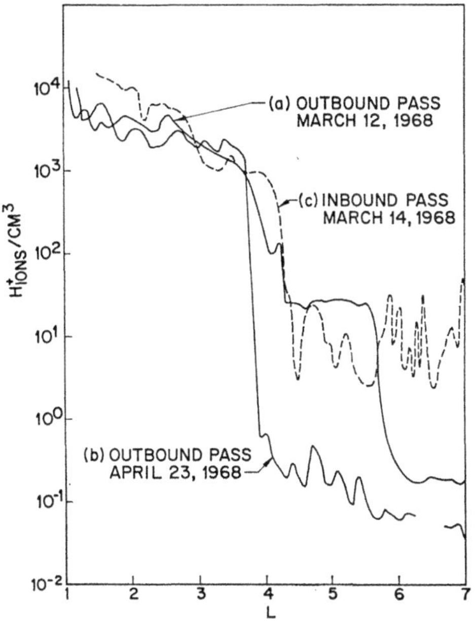

Fig. 3. Examples of a plasmapause crossing as shown by plotting the concentration of H ions vs *L*. a. This March 12, 1968, outbound pass is an example of a case in which the plasmapause position currently is not defined clearly. b. This outbound pass, April 23, 1968, shows a well defined plasmapause position. c. This inbound pass, March 14, 1968, case shows high concentrations in the trough region outside the plasmapause.

outbound passes in the midnight to 0400 LT region the 10 ions/cm^3 point seems to define the outer edge of the plasmasphere quite well.

We can now use the position of the <10 ions/cm^3 concentration level as the outer boundary of the plasmasphere in this midnight to 0400 LT region and correct it for local time variations by the method described above. If this position is now plotted versus K_p for the same time intervals mentioned above we find that the correlation is by far the best for the 6 hr previous to the measurement. In conclusion, for this particular local time region between midnight and 0400 LT the lag time between the onset of magnetic activity and the plasmapause response is about 2 to 6 hr.

This response time is substantiated by Figure 4. Here we have grouped the twenty outbound passes according to the average magnetic activity in this 6 hr time period before the measurement was made. The four cases shown in Figure 4 are representative of the four different groups of magnetic activity in the previous time period. The maximum spread in each group is shown by the error bars at the 10 ion/cm^3 value. We attribute this spread to a combination of two things – the local time variations and the difference in each case of the magnetic history over the preceding 2 days. The effect of this difference in magnetic activity over longer preceding time periods than the 6 hr which we have adopted will be examined in more detail in a later study. Several characteristics of the plasmapause shape should be discussed further. First we note that the sharpness of the knee increases as the magnetic activity increases.

Second, the maximum spread in each group decreases as magnetic activity increases Third, the levels of concentration inside the plasmasphere and outside in the trough remain about the same as the plasmapause changes position. Finally, the drop in H ion density at the plasmapause is about 3 orders of magnitude from about 10^3 ions/cm^3 to about 1 ion/cm^3. It should be emphasized that Figure 4 represents only the region

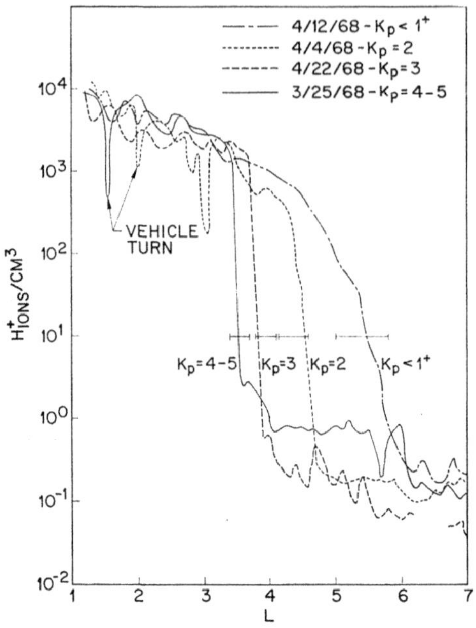

Fig. 4. A composite of several typical plasmapause crossings representing different levels of magnetic activity. These plots of H ion concentration vs. L represent all of the outbound passes in the local time region from midnight to 0400.

of local times from midnight to 0400 and consequently the characteristics of this region should not be readily generalized to other local times. It should be pointed out also that of the twenty orbits under consideration there was one which did not fit into the group in which its magnetic activity would have placed it. Thus, one must be aware of the possibility of the occasional influence of other events which have not been considered.

We have also tried to apply the 2 to 6 hr response time to the 1000 LT region given by the inbound passes. If we group the passes according to their average magnetic activity in this period we find that although the passes seem to group themselves better in this 6 hr period, there still remains a sizable spread within each group which virtually obscures the influence of the magnetic activity on the plasmapause position. This large spread is not unexpected because of the other processes in this region which can possibly influence the shape and position of the plasmapause. These processes such as the influence of the magnetopause movement on the plasmapause position will be examined in more detail elsewhere.

3. Summary

In examining the connection between magnetic activity, as given by the K_p index and the location of the plasmapause, we have found that there exists a fair correlation between an increase in the activity and a decrease in the plasmapause position within ± 2 hr of 1000 and of 0200 LT. This correlation agrees very well with the same comparison made by Carpenter (1967) using whistler data. In examining different response times between change in magnetic activity and corresponding change in plasmapause radius for the two local time regions we find that a delay of 2 to 6 hr applies quite well to the 0200 region while the 1000 region shows no striking difference in degree of correlation for periods of 6, 12, or 24 hr. The 2 to 6 hr response time of the 0200 LT region is reinforced by Figure 4 in which we have assumed this lag time and have been able to group the different plasmapause locations quite well according to the magnetic activity in this previous time interval. The plasmapause reaction to increasing magnetic activity as shown in this figure can be considered the typical reaction in this local time region. It is characterized by an increasing sharpness of the knee with increasing activity while the concentration levels inside and outside the knee remain about the same. The precise shape of the plasmapause, however, can differ from these typical shapes depending on the particular sequence of magnetic activity changes and the possible presence of other influencing processes. One possible key to the further study of the plasmapause location is the use of an index other than the magnetic activity index which may not directly reflect the effect of all processes connected with the plasmapause formation.

Acknowledgments

The authors would like to thank Mr. A. J. Fernandez for his work in developing the computer programs for data processing and Mr. C. W. Gilbreth for his development of the electronic circuitry of the instrument. This work was funded by National Aeronautics and Space Administration Contract No. NAS 5-9092.

References

Binsack, J. H.: 1967, *J. Geophys. Res.* **72**, 5231.
Brinton, H. C., Pickett, R. A., and Taylor, H. A., Jr.: 1968, *Planet. Space Sci.* **16**, 899.
Carpenter, D. L.: 1966, *J. Geophys. Res.* **71**, 693.
Carpenter, D. L.: 1967, *J. Geophys. Res.* **72**, 2969.
Harris, K. K. and Sharp, G. W.: 1969, *Trans. IEEE Geoscience* **GE-7**, 93.
Harris, K. K. and Sharp, G. W.: 1970, *J. Geophys. Res.*, to be published.
Vasyliunas, V. M.: 1968, *J. Geophys. Res.* **73**, 2839.

PART IV

MAGNETIC AND ELECTRIC FIELDS IN THE MAGNETOSPHERE

MAGNETIC FIELDS IN THE EARTH'S TAIL

KENNETH W. BEHANNON

Laboratory for Space Sciences, NASA Goddard Space Flight Center, Greenbelt, Md., U.S.A.

1. Introduction

The initial mapping of the magnetic field in the geomagnetic tail out to a distance of 31 R_E by IMP 1 established the basic structure of the tail. In addition to the generally steady well collimated nature of the field in the tail, a radius of approximately 20 R_E out to a geocentric distance of 30 R_E in the antisolar direction was deduced and the existence of a relatively thin ($\leqslant 1\ R_E$) neutral sheet separating oppositely directed fields was established (Ness, 1965; Speiser and Ness, 1967). Explorer 14 measurements showed a broad region of low field magnitude surrounding the field reversal region at a distance of 10 to 12 R_E (Cahill, 1966). Observations of electrons and protons with $E > 100$ eV by the Vela satellites at geocentric distances between 15.5 and 20.5 R_E showed that the plasma sheet surrounding the neutral sheet is often 4 to 6 R_E thick near the midnight meridian, increasing to approximately twice that thickness near the dusk and dawn boundaries (Bame *et al.*, 1967).

Observations in the distant geomagnetic tail by Explorer 33 during 1966 showed that the tail is well defined to a distance of 80 R_E (Ness *et al.*, 1967a), and Pioneer 7 showed that it may extend as far as 1000 R_E (Ness *et al.*, 1967b). It was found that the field magnitude decreases with distance from the earth with a gradient that can be described by an inverse power law (Behannon, 1968; Mihalov *et al.*, 1968). The initially observed gradient has shown no statistically significant secular change (Mihalov and Sonett, 1968). It has been suggested that the tail gradient must be the result of either reconnection of field lines across the neutral sheet or an increase in the radius of the tail with distance, or a combination of the two mechanisms (Behannon, 1968; Mihalov *et al.*, 1968). The early Explorer 33 data further suggested that the tail is not circular in cross section but rather is elliptical with the major axis in the solar magnetospheric Z direction (Behannon, 1968). All of these basic features of the tail field topology have been reviewed in detail by Ness (1969) and by Behannon and Ness (1969).

During July 1967 to August 1968 additional features of the tail field topology were mapped by Explorer 33 and by Explorer 35 in orbit about the moon (Behannon, 1969). Those data provided evidence that

(a) Tail field lines are slightly diverging with distance from the earth;

(b) The neutral line is generally located beyond the orbit of the moon; and

(c) On average the magnetotail is aberrated by $3.1° \pm 1.5°$.

In addition, the Pioneer 8 spacecraft has shown that the geomagnetic tail at approximately 500 R_E downstream is similar in character to the tail observed by Pioneer 7 at 1000 R_E (Mariani and Ness, 1969).

B. M. McCormac (ed.), Particles and Fields in the Magnetosphere.

2. Divergence of the Magnetotail Field

Mihalov *et al.* (1968) reported that the geomagnetic tail field observed by Explorer 33 in 1966 tended to have a component toward the dawn side of the tail, both above and below the field reversal region. This skewing of the tail field was explained in terms of a twisting of field lines. From an analysis of later Explorer 33 data in addition to measurements by Explorer 35, Behannon (1969) found evidence that angular deviations of tail field lines from the sun–earth direction are produced both by a spreading of the tail field with distance from the earth and an aberration of the tail due to the earth's motion about the sun.

Shown in Figure 1 are 1262 hourly average values of the field azimuthal angle in

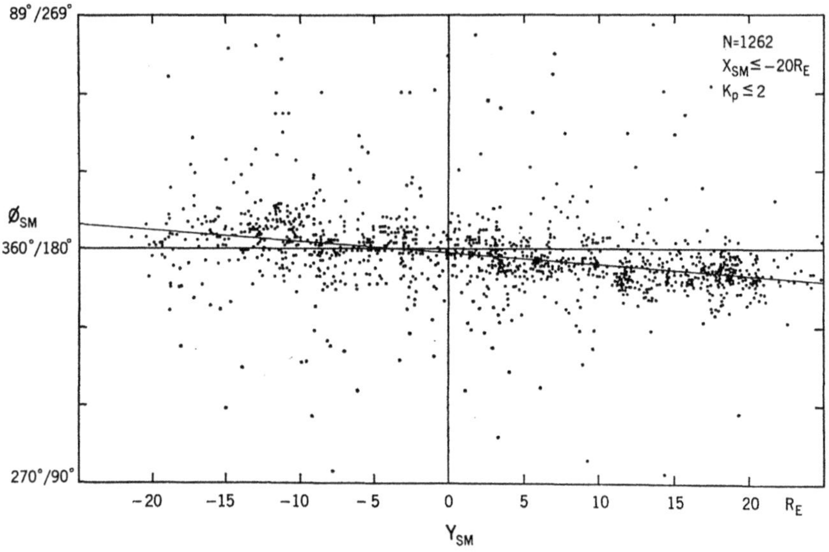

Fig. 1. Variation of the solar magnetospheric azimuthal angle of the tail field with distance from the XZ plane, with the polarity of field lines suppressed. The tilted orientation of the linear fit to the distribution and the displacement of the point at which that line has a value of 360/180° to the negative side of $Y_{sm} = 0$ demonstrate both the divergence of the tail field with distance from the earth and the aberration of the field.

solar magnetospheric coordinates φ_{sm} as a function of Y_{sm} across the tail. Data from both below the field reversal region and above it are folded together so that the angle between the field and the earth–sun line in the $X_{sm}Y_{sm}$ plane is plotted whereas the polarity of the field either toward or away from the earth is not considered. An appreciable number of cases are seen in which the field made a large angle with the 360/180° direction. These points are from the plasma sheet region where the direction is found to fluctuate widely. Even with the inclusion of those points, the general trend of the data is unmistakable.

A least squares fit to the plotted data is shown. The negative slope tilt of the distribution can be interpreted as resulting from the divergence of the field away from the tail axis with distance from the earth. The same result was obtained when the data were separated into measurements below the field reversal and those above it. Thus the distance of field lines from the $X_{sm}Z_{sm}$ plane is increasing with distance from the earth on both sides of that plane and both above and below the neutral sheet.

Using a solution for the magnetospheric boundary that includes pressure balance along the boundary of the magnetotail, Spreiter and Alksne (1969) have predicted that the tail radius will increase with distance from the earth up to a distance d_{max} and will remain constant thereafter. They have calculated that $R_{max} = 2.5\ D$ and is reached at a distance $d_{max} = 13\ D$, where D is the standoff distance of the magnetopause on the sunward side of the magnetosphere. For a value of $D = 10$ it is seen that $R_{max} = 25\ R_E$ and $d_{max} = 130\ R_E$. The observations by Explorers 33 and 35 are consistent with these predictions.

3. Location of the Neutral Line

Figure 2 demonstrates the existence of a broad region of depressed magnetic field magnitude in the geomagnetic tail near the solar magnetospheric equatorial plane. The data consist of 1346 hourly averages for which $-70 \leqslant X_{sm} \leqslant -20\ R_E$ and $K_p \leqslant 2$.

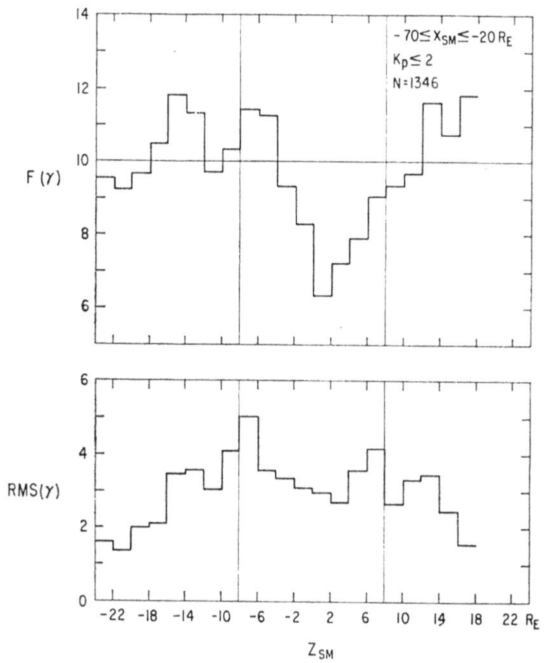

Fig. 2. Average variation of tail field magnitude with distance from the solar magnetospheric XY plane (upper curve). The lower curve shows the scatter in the hourly average magnitude values which were averaged to obtain the upper curve. The magnitude data show a broad region of depressed fields centered above $Z_{sm} = 0$.

Plotted in the top portion of the figure are averages of the hourly magnitude values for each 2 R_E interval of Z_{sm}. The rms magnitude deviations corresponding to each of the 2 R_E magnitude averages are shown in the lower part of the figure.

Vertical lines at $Z_{sm} = \pm 8$ R_E suggest that the depressed region is asymmetric with respect to $Z_{sm} = 0$. The average depressed region is approximately 12 R_E thick, which is consistent with the plasma sheet thickness near the dawn and dusk boundaries of the tail (Bame *et al.*, 1967). The relatively flat distribution of rms values suggests that the depressed magnitude region in the data is not produced merely by the constant motion of the relatively thin magnetic neutral sheet throughout the broader region.

An important feature of the depressed field region is the tendency for the B_Z component of the field to be larger within the region than outside. This is illustrated in Figure 3 where the elevation angle of the field θ_{sm} is shown as a function of the Z_{sm}

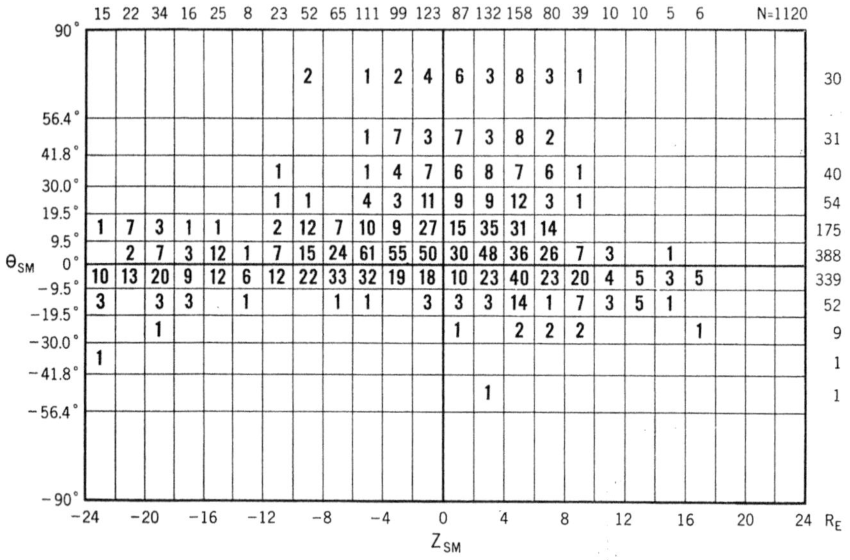

Fig. 3. Summary of tail field elevation angle in solar magnetospheric coordinates as a function of distance Z_{sm} from the XY plane. The numbers of hourly values falling within each of the indicated intervals of φ_{sm} and Z_{sm} are tabulated.

position of the measurement. The figure summarizes 1120 hr of data. The $\pm 90°$ range of θ_{sm} has been subdivided into 12 intervals corresponding to equal area latitude regions of a spherical surface. The number of cases of hourly values of θ_{sm} falling within each angular interval as well as within 2 R_E intervals of Z_{sm} have been tabulated, and the θ_{sm} and Z_{sm} marginal distributions are also given.

Figure 3 illustrates not only the thickness of the region in which the B_Z component of the field is a significant fraction of the total field but also the essentially positive character of the B_Z component. The average solar magnetospheric B_Z component in the depressed field region was found to decrease with distance from the earth. The average B_Z component was negative only 25% of the time while outside the region it was negative 67% of the time. Since 72% of the data used correspond to distances down the tail of greater than 50 R_E, these data support the view that in the Z_{sm} direction the field is converging and reconnecting within the plasma sheet and diverging outside that region, with the neutral line generally beyond the orbit of the moon. These results do not support the suggestion by Dessler (1968) that the neutral line is as near to the earth as 20 R_E since on the average the B_Z component is found to be generally positive in the plasma sheet region out to 60 R_E.

Individual cases can be found during which the neutral line must be closer to the earth than the lunar orbital distance. Such a case has been identified by Anderson and Lin (1969) using Explorer 35 observations of shadowing of magnetotail particles by the moon on November 17, 1967. This case is supported by the Explorer 35 magnetic field observations. The average solar magnetospheric B_Z component was southward and had a magnitude of 2.5γ, one third of the total field strength.

4. Aberration of the Magnetotail

A feature of the data shown in Figure 1 is the tendency for field lines with direction 360/180° to be located at a negative value of Y_{sm} (the point where the linear fit to the data crosses the 360/180° line). This reflects the aberration of the tail axis west of the antisolar direction as viewed from the earth. An average aberration of 2.7° is obtained from this analysis.

Another approach to obtaining a value for the average aberration is through analysis of the tail boundary data. This also can yield information on the shape and size of the tail cross section. Figure 4 summarizes such traversal data for Explorers 33 and 35 during 1967–68. The line segments represent periods during which a series of multiple crossings were observed. A nonlinear least squares fit of a circular cross section to the mean traversal points results in a radius of 23.8 R_E at the distance of the moon and an average angle between the tail axis and the earth–sun line of $3.1 \pm 1.5°$.

Walters (1964) suggested that the oblique orientation of the average interplanetary magnetic field relative to the earth–sun line would lead to a deflection of the plasma flow in the magnetosheath and hence to a tilt of the axis of symmetry of the magnetosphere away from the earth–sun line. For a radial solar wind of the average velocity of 440 km sec^{-1} observed by Explorer 34 during 8 mo of 1967 (Burlaga and Ogilvie, 1969) an aberration of the magnetotail of 3.9° results due to the earth's motion about the sun. The average deflection of the tail axis at the distance of the moon thus is even smaller than that due to aberration by a radially flowing plasma. If the observed deflection is entirely due to aberration, then these results would require either that the solar wind was radial with a velocity of 551 km sec^{-1} or that the solar wind had a lower velocity and an azimuthal component. Assuming the average velocity of 440 km sec^{-1} to be typical of the period of the tail geometry observations, then the observations suggest that the solar wind was on average from $0.8 \pm 1.5°$ east of the sun during that period.

On the basis of Vela satellite observations, Strong *et al.* (1967) reported that from

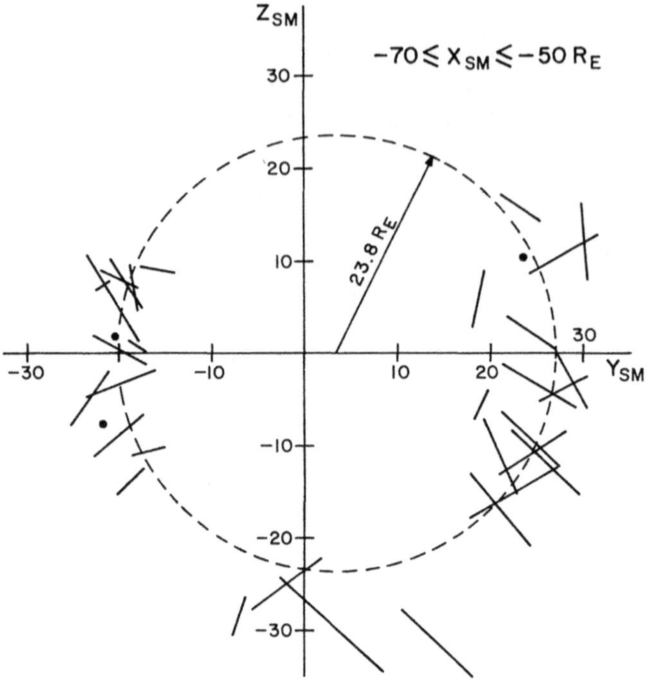

Fig. 4. Projections on the solar magnetosphere YZ plane of trajectory intervals when Explorers 33 and 35 observed the magnetotail boundary near the orbital distance of the moon. Reference circle shows aberration of the tail and the mean radius at which traversals along the dawn and dusk flanks occurred. Explorer 33 measurements near $Z_{sm} = 30$ R_E suggest an elliptical cross section for the tail.

July 1964 to July 1965 the solar wind velocity was on the average directed from 1.4° east of the sun. From an analysis of 604 photographic observations of type I comets, Brandt (1967) found that on average the solar wind was from 1.3° east of the sun near the orbit of the earth. However, from an analysis of IMP 1 solar wind measurements during November 1963 to February 1964, Egidi *et al.* (1969) have concluded that the solar wind came from west of the sun on 72% of the cases with the average being 1.5° west.

A solar wind from 1.4° east of the sun would permit a contribution of 0.6° by the oblique interplanetary field to the deflection angle for a 440 km sec^{-1} solar wind. Any larger contribution would require a smaller average total velocity. Thus it is unlikely that the effect suggested by Walters (1964) contributes more than approximately 1° to the tail axis deflection at the distance of the moon.

5. Length of the Magnetotail

The geomagnetic tail was observed intermittently during September 19 to 30, 1966, by the Pioneer 7 spacecraft as it passed behind the earth at a geocentric distance of approximately 1000 R_E (Ness *et al.*, 1967a; Wolfe *et al.*, 1967; Fairfield, 1968).

Magnetic field observations similar in character to those of Pioneer 7 were obtained by Pioneer 8 during January 1968 as it passed through the extended geomagnetic tail region 470 to 580 R_E downstream of the earth (Mariani and Ness, 1969). Tail observations at these large distances differ from measurements nearer the earth primarily in

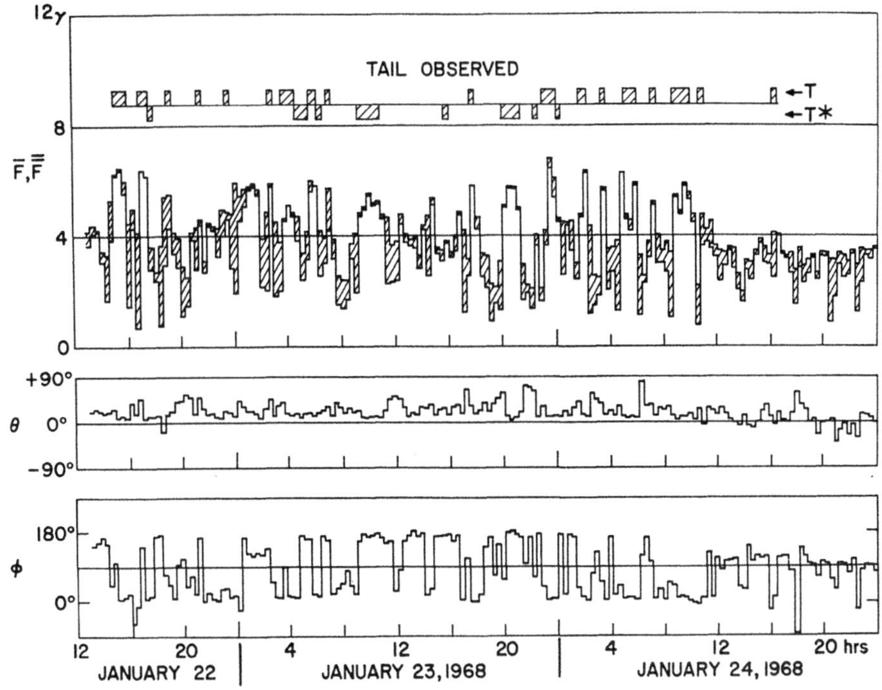

Fig. 5. Twenty minute averages of the magnetic field showing a section of data in the tail region, from 1200 UT on January 22, 1968, to 0000 UT on January 24. In the upper portion the time intervals of tail observations are indicated.

that the intervals of observation are much shorter in the far downstream region and the plasma measured in the far downstream tail does not disappear to the extent that it does in the near earth tail (Wolfe *et al.*, 1967).

Twenty minute averages of the magnetic field observed by Pioneer 8 are shown in Figure 5. The field intensity $\bar{F}$ has been computed from the averages of the field components while the average intensity $\bar{\bar{F}}$ is obtained by averaging the field intensities of individual data points. Also shown are the solar ecliptic azimuthal angle of the field φ and inclination angle θ. The time intervals when the observed field had an orientation close to that expected for the tail field, i.e., parallel or antiparallel to the earth–sun direction are indicated in Figure 5 by T or T^*, respectively. Identification of the tail is also made on the basis of a relatively high field intensity $\bar{F}$ (between 4 to 10γ), a small variance of the field (less than 1.0γ) and an inclination θ near to zero. Several crossings of a neutral sheet can be identified by a field intensity near zero as the azimuth changes rapidly from 0 to 180° or vice versa.

At 500 R_E the tail is not well ordered as it appeared in the Explorer 33 observations at 80 R_E. The tail appears to have been swept back and forth across the spacecraft trajectory, a consequence of temporal variations in the direction of the solar wind velocity. It is also possible that at 500 R_E the magnetotail is filamentary in structure due to penetration of the tail by the magnetosheath plasma or connection between the field in the tail and the interplanetary field. The tail was observed for a higher percentage of time by Pioneer 8 when in the region of an extended, aberrated tail than by Pioneer 7. At 1000 R_E Pioneer 7 observed the tail approximately 7% of the time over 11 days while Pioneer 8 near 500 R_E observed the tail approximately 25% of the time during a 9 day interval. It seems likely that the magnetotail generally ceases being a well ordered entity at some distance between 100 and 500 R_E downstream of the solar wind interaction with the geomagnetic field.

References

Anderson, K. A. and Lin, R. P.: 1969, Space Sciences Laboratory, University of California, Berkeley, Preprint, March 3.
Bame, S. J., Asbridge, J. R., Felthauser, H. E., Hones, E. W., and Strong, I. B.: 1967, *J. Geophys. Res.* **72**, 113.
Behannon, K. W.: 1968, *J. Geophys. Res.* **73**, 907.
Behannon, K. W.: 1969, NASA-GSFC Preprint X-616-69-146.
Behannon, K. W. and Ness, N. F.: 1969, in *Physics of the Magnetosphere* (ed. by R. L. Carovillano, J. F. McClay and H. R. Radoski), D. Reidel Publishing Company, Dordrecht-Holland, p. 409.
Brandt, J. C.: 1967, *Astrophys. J.* **147**, 201.
Burlaga, L. F. and Ogilvie, K. W.: 1969, NASA-GSFC Preprint X-616-69-142.
Cahill, L. J.: 1966, *Space Research IV*, 662.
Dessler, A. J.: 1968, *J. Geophys. Res.* **73**, 209.
Egidi, A., Pizzella, G. and Signorini, C.: 1969, *J. Geophys. Res.* **74**, 2807.
Fairfield, D. H.: 1968, *J. Geophys. Res.* **73**, 6179.
Mariani, F. and Ness, N. F.: 1969, NASA-GSFC Preprint X-616-69-54.
Mihalov, J. D. and Sonett, C. P.: 1968, *J. Geophys. Res.* **73**, 6837.
Mihalov, J. D., Colburn, D. S., Currie, R. G., and Sonett, C. P.: 1968, *J. Geophys. Res.* **73**, 943.
Ness, N. F.: 1965, *J. Geophys. Res.* **70**, 2989.
Ness, N. F.: 1969, *Rev. Geophys.* **7**, 97.
Ness, N. F., Behannon, K. W., Cantarano, S. C., and Scearce, C. S.: 1967a, *J. Geophys. Res.* **72**, 927
Ness, N. F., Scearce, C. S., and Cantarano, S. C.: 1967b, *J. Geophys. Res.* **72**, 3769.
Speiser, T. W. and Ness, N. F.: 1967, *J. Geophys. Res.* **72**, 131.
Spreiter, J. R. and Alksne, A. Y.: 1969, *Planet. Space Sci.* **17**, 233.
Strong, I. B., Asbridge, J. R., Bame, S. J., and Hundhausen, A.: 1967, in *The Zodiacal Light and the Interplanetary Medium* (ed. by J. L. Weinberg), NASA SP-150, NASA Technical Information Division, Washington, D.C., p. 365.
Walters, G. K.: 1964, *J. Geophys. Res.* **69**, 1769.
Wolfe, J. H., Silva, R. W., McKibbin, D. D., and Mason, R. H.: 1967, *J. Geophys. Res.* **72**, 4577.

MAGNETIC FIELD OBSERVATIONS IN HIGH β REGIONS OF THE MAGNETOSPHERE

M. SUGIURA, T. L. SKILLMAN, B. G. LEDLEY, and J. P. HEPPNER
NASA Goddard Space Flight Center, Greenbelt, Md., U.S.A.

1. Approximate Distribution of ΔB

For a description of the distortions of the geomagnetic field in the magnetosphere ΔB provides a useful measure. Figure 1 shows approximate equi-ΔB contours for the two 90° sectors centered at the noon and midnight meridian half planes, based on a preliminary sampling of a much greater body of OGO 3 and 5 rubidium magnetometer data.

For this preliminary presentation the data for the Southern Hemisphere were combined with those for the Northern Hemisphere assuming symmetry with respect to the geomagnetic equator in spite of the fact that the tail neutral sheet does not

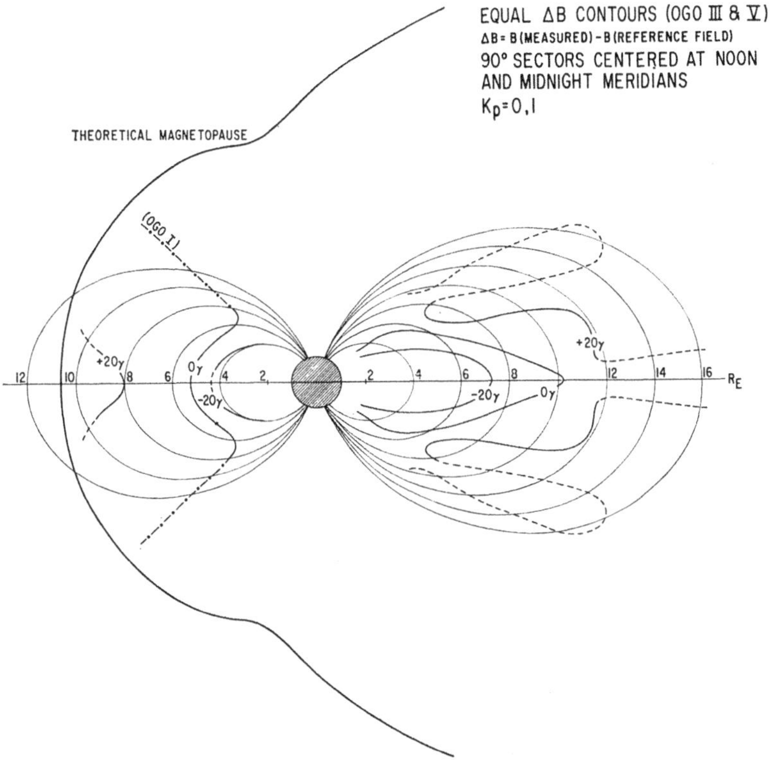

Fig. 1. Equi-ΔB contours for 90° sectors centered at the noon and midnight meridian half planes, in geomagnetic dipole coordinates; $K_p = 0$ and 1.

B. M. McCormac (ed.), Particles and Fields in the Magnetosphere.

necessarily coincide with the geomagnetic equator. The data used in Figure 1 were taken during periods when K_p was 0 or 1.

The negative ΔB values in the equatorial region between $L=2$ and 6 represent the field depression from the quiet time ring current. The maximum field decrease is approximately 35γ. The $\Delta B>0$ regions at higher latitudes on the nightside reflect the combined effects of the magnetic field inflation due to the equatorial ring current, the neutral sheet current in the tail, and the boundary surface current. There is an indication that a region of slightly lower ΔB exists at about 20 to 30° latitude and between 8 and 6 R_E, but it is not certain whether this is real or from a seasonal effect that is

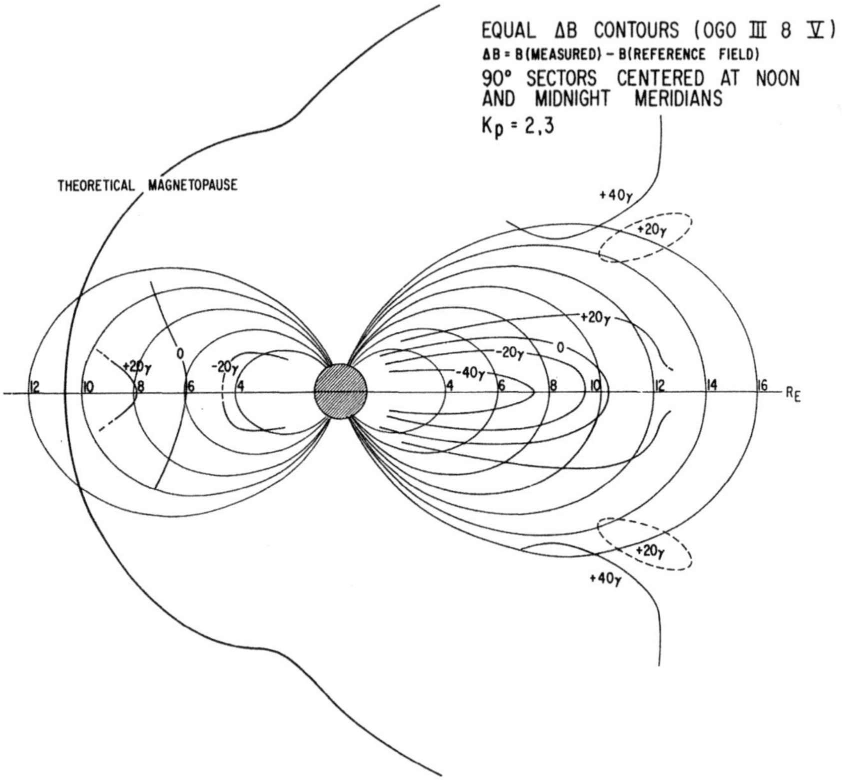

Fig. 2. Equi-ΔB contours for 90° sectors centered at the noon and midnight meridian half planes, in geomagnetic dipole coordinates; $K_p=2$ and 3.

not taken into account when averaging ΔB. It is noted, however, that there is a similar tendency in the ΔB contour map for $K_p=2$ and 3 shown in Figure 2.

On the front side, the $\Delta B=0$ line drawn at low latitudes with the present data matches very well the corresponding line at higher latitudes determined by Heppner *et al.* (1967) from the OGO 1 results. The change of the sign for ΔB across this line is mainly from the boundary surface current.

An equi-ΔB contour map based on the data taken during slightly disturbed periods with $K_p=2$ or 3 is presented in Figure 2. This ΔB map differs from that shown in Figure 1 for magnetically quiet periods in several respects: (a) the ring current is intensified; (b) the tail field is increased; (c) ΔB in the high ΔB region (above $\sim 30°$ latitude) becomes larger; and (d) the $\Delta B=0$ line on the front side changes its shape.

Parallelism between Dst (mainly representing the ring current field) and ap (a measure for polar substorm activity) has been noted earlier (Sugiura, 1964), but a comparison of Figures 1 and 2 clearly demonstrates that even during relatively weak substorms the quiet time ring current is intensified. Increases in the tail field during high K_p periods have been discussed by Behannon and Ness (1966) and Ness and Williams (1966). Calculations based on the boundary surface current model of Mead (1964) and the tail current model of Williams and Mead (1965) show that the high ΔB values exceeding 40γ or even 50γ observed by OGO 1, 3, and 5 above $\sim 30°$ latitude cannot be explained by these models, confirming the earlier conclusion reached by Heppner *et al.* (1967) from the OGO 1 observations. Still undetermined is whether the magnetic field inflation caused by the ring current is enough to account for such high ΔB values and if not, whether there is an additional pressure exerted from the polar region of the magnetosphere. In a later, more comprehensive analysis the possibility that the ΔB distribution depends on not only instantaneous but also previous magnetic activity will be examined.

2. Spatial Discontinuities in the Near Tail Region

The high β region in the nightside magnetosphere is referred to as the 'near tail' region. The dynamics of this region appears to be a key to the understanding of magnetospheric disturbance phenomena. In this section, magnetic field structures in the near tail region under relatively quiet conditions are discussed.

Figure 3 shows the variation in B on an inbound pass of OGO 3 near midnight. The steady tail field is abruptly interrupted by a sudden decrease in B at 11.3 R_E and at 12.7° geomagnetic latitude. Such a sudden decrease in B followed by irregular field variations is a feature repeatedly observed by OGO 3 on its nightside low latitude passes. When the satellite crosses the geomagnetic equator near 10 R_E the sudden drop in B is dramatic as seen in the example shown in Figure 4. In contrast to the sharpness of the variation in its magnitude the direction of the magnetic field does not generally indicate any abrupt change.

By estimating the kinetic energy density of the plasma based on the electron flux measurement by Frank (1967, 1968a, 1968b) it is found that the plasma kinetic energy density abruptly increases at the time a sharp decrease is observed in B. Plasma measurements in the tail by the Vela group indicate that the proton kinetic energy density is greater than (and probably several times) the electron kinetic energy density (Bame *et al.*, 1967). If this is so then it can be concluded that the abrupt decrease in B corresponds to a traversal of a discontinuity between a $\beta \ll 1$ region and a $\beta \gtrsim 1$ region. Sometimes two or more step type discontinuities are observed on one pass. Thus, the near tail region appears to have shell like structures with a discontinuity between successive shells filled with plasmas of different kinetic energy densities.

3. Tail Field Changes During Substorms

The OGO 1 and 3 magnetic field measurements in the tail indicated that in association with polar substorms on the earth, magnetic field disturbances are observed in the tail only near the meridional plane of the negative bay onset on the earth, that the tail field changes are delayed relative to the bay onset time on the ground by several to

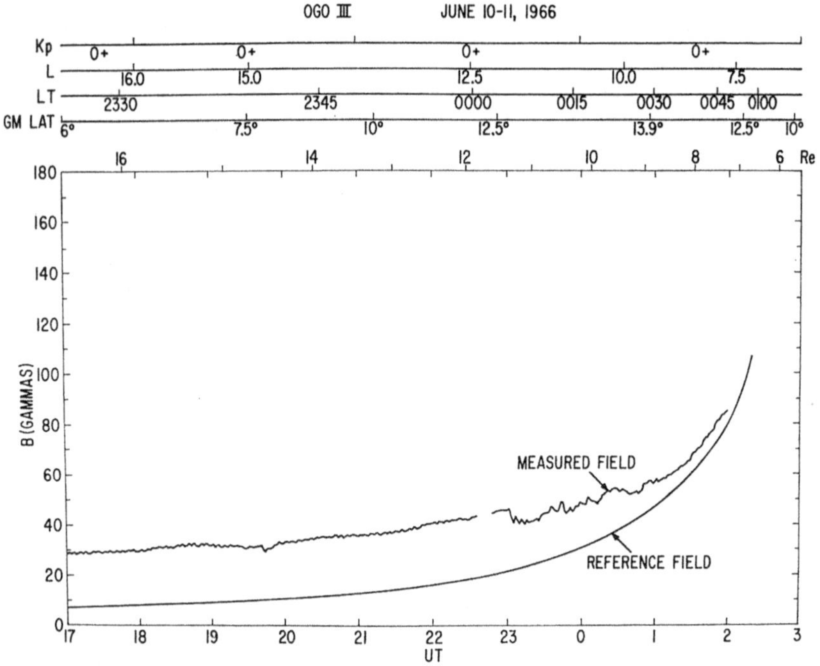

Fig. 3. *B* in the low latitude near tail region. K_p, *L*, LT, geomagnetic dipole latitude, and geocentric distance in R_E are indicated on the top.

15 min and that the tail field changes are such as to approach the dipolar configuration (Heppner *et al.*, 1967; Sugiura *et al.*, 1968). It can be pictured thus that at the onset of a negative bay the plasma pressure is released abruptly in the near tail region in the vicinity of the magnetic midnight meridian because of a highly channeled, rapid convection of the plasma (Heppner, 1969) or a drainage of the plasma triggered by the bay onset. The tail magnetic field beyond this region collapses and process proceeds tailward in a relatively narrow region along the magnetic midnight meridian. The electron behavior in the tail's plasma sheet observed by Vela 2 during magnetic bays (Hones *et al.*, 1967) seems to support this picture.

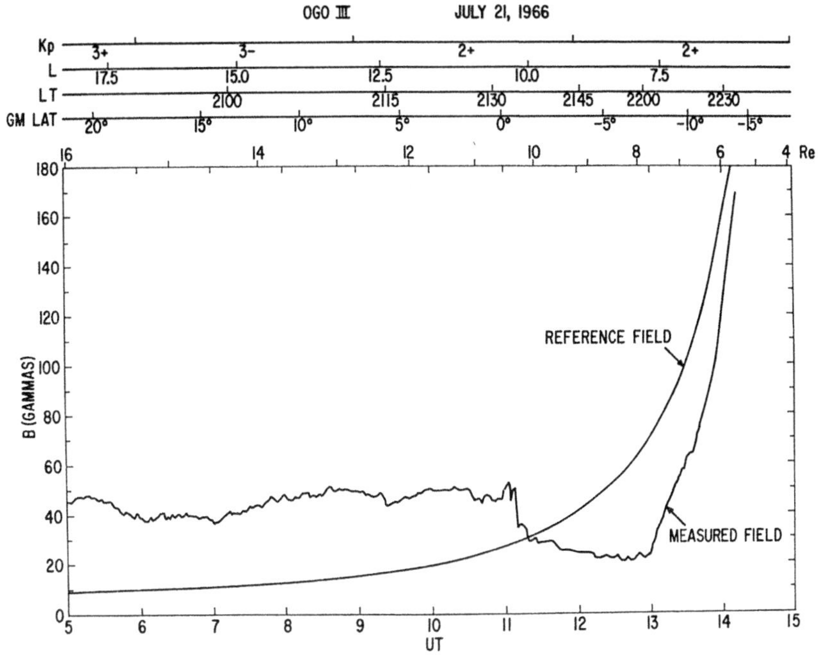

Fig. 4. B in the equatorial near tail region.

4. High β Region on the Dawn Flank of the Magnetosphere

The OGO 1 magnetic field measurements have shown (a) that at low geomagnetic latitudes near the dawn meridian the average gradient in B is essentially zero between $\sim 11\ R_E$ and the magnetopause; (b) that frequently there is very little contrast between magnetospheric and magnetosheath fields near the magnetopause, making boundary identification difficult using the magnetic field data alone; and (c) B in the magnetosheath is sometimes greater than B immediately inside the magnetopause (Heppner *et al.*, 1967). Based on these observations it was concluded that β must be approximately equal to, or greater than, unity in the low field-gradient region near the dawn meridian.

The more recent OGO 5 observations have confirmed the OGO 1 results and have provided examples in which this equatorial high β region extends to near local time 9 hr. In addition, whereas the previous argument for the presence of high β plasma in the region concerned was based solely on the magnetic field observations, the OGO 5 electric field measurements show that the total plasma flux in this equatorial region is often comparable to, if not greater than, the plasma flux in the adjacent magnetosheath.

5. Summary

The following aspects of the distortion of the geomagnetic field in the magnetosphere are discussed on the basis of the magnetic observations by the OGO 1, 3, and 5 satellites:

(a) Approximate distributions of ΔB (the magnitude of the measured field minus the magnitude of the reference field) are given in the noon and midnight sectors of the magnetosphere for magnetically quiet ($K_p=0$ or 1) and slightly disturbed ($K_p=2$ or 3) conditions. The field depression in the region of $L \lesssim 6$ caused by the quiet time ring current is shown to increase even during weak disturbances. The existence of regions of large ΔB above about 30° ML cannot be accounted for by the existing theoretical models for the magnetopause surface current and the neutral sheet current of the tail, and is unlikely to be explainable by the inflation due to the ring current. Large ΔB values in these regions may be a result of an additional pressure exerted from the polar regions of the magnetosphere.

(b) A sudden decrease in the magnetic field followed by irregular variations was often observed near midnight on low latitude inbound orbits of OGO 3. Comparing with the plasma observations this sudden decrease is interpreted as a traversal of the boundary between a low β region of the tail and an equatorial region of $\beta \gtrsim 1$ (i.e., the plasma sheet).

(c) Following the onset of a negative bay disturbance on the ground the magnetic field in the tail collapses near the meridional plane of the bay onset. Thus, the bay associated tail field change is such as to make the field configuration more dipolar.

(d) On the dawn side of the magnetosphere the magnetic field gradient is near zero in the equatorial region between about 11 R_E and the magnetopause. In this region the magnetic field in the magnetosheath is often comparable to, or even greater than, the field immediately inside the boundary. These observations suggest presence of a high β plasma in this region.

References

Bame, S. J., Asbridge, J. R., Felthauser, H. E., Hones, E. W., and Strong, I. B.: 1967, *J. Geophys. Res.* **72**, 113.

Behannon, K. W. and Ness, N. F.: 1966, *J. Geophys. Res.* **71**, 2327.

Frank, L. A.: 1967, *J. Geophys. Res.* **72**, 185.

Frank, L. A.: 1968a, in *Earth's Particles and Fields* (ed. by B. M. McCormac), Reinhold Publishing Corporation, New York, p. 67.

Frank, L. A.: 1968b, in *Physics of the Magnetosphere* (ed. by R. L. Carovillano, J. F. McClay and H. R. Radoski), D. Reidel Publishing Company, Dordrecht-Holland, p. 271.

Heppner, J. P.: 1969, in *Atmospheric Emissions* (ed. by B. M. McCormac and A. Omholt), Van Nostrand Reinhold Company, New York, p. 251.

Heppner, J. P., Sugiura, M., Skillman, T. L., Ledley, B. G., and Campbell, M.: 1967, *J. Geophys. Res.* **72**, 5417.

Hones, E. W., Jr., Asbridge, J. R., Bame, S. J., and Strong, I. B.: 1967, *J. Geophys. Res.* **72**, 5879.

Mead, G. D.: 1964, *J. Geophys. Res.* **69**, 1181.

Ness, N. F. and Williams, D. J.: 1966, *J. Geophys. Res.* **71**, 322.

Sugiura, M.: 1964, *Ann. IGY* **35**, 9.

Sugiura, M., Skillman, T. L., Ledley, B. G., and Heppner, J. P.: 1968, *J. Geophys. Res.* **73**, 6699

Williams, D. J. and Mead, G. D.: 1965, *J. Geophys. Res.* **70**, 3017.

FLUCTUATIONS IN THE DISTANT GEOMAGNETIC FIELD DURING SUBSTORMS: ATS 1

PAUL J. COLEMAN, Jr. and ROBERT L. McPHERRON

Dept. of Planetary and Space Science and Institute of Geophysics and Planetary Physics, University of California, Los Angeles, Calif., U.S.A.

1. Introduction

The initial data from the ATS 1 magnetometer led to the discovery of variations in the magnetic field in the night sector of the synchronous orbit that are characteristic of substorms (Cummings and Coleman, 1968; Cummings *et al.*, 1968). In this paper we will review these previously reported observations and present some results of more recent work.

2. Trajectory and Coordinate Systems

ATS 1 was launched into the synchronous equatorial orbit on December 6, 1966, from Cape Kennedy, Florida. The spacecraft is spin stabilized at a nominal rotation frequency of 1.67 Hz. The spin axis is parallel to that of the earth.

The geocentric range of the synchronous orbit is a constant 6.6 R_E. ATS 1 is stationed in the equatorial plane at longitude 150.3 °W. Thus, its coordinates are held fixed at geocentric range 6.6 R_E, latitude 0°, and longitude 150.3 °W.

The magnetic field, hereafter, will be described usually in terms of its vector components in one of two coordinate systems denoted by XYZ and DHV. The XYZ system has its origin at the spacecraft, the Z axis is parallel to the earth's axis and positive northward, the X axis is in the direction of the projection of the spacecraft-to-sun line on the plane perpendicular to the spacecraft spin axis, which is also the equatorial plane, and the Y axis completes the right handed system. The DHV system is also earth centered and it rotates with the earth. Thus, H is in the direction of the earth's spin vector, V is positive upward along the local vertical or radially outward from the earth, and D, which is positive eastward, completes the right handed system.

The ATS 1 orbit and the two coordinate systems are shown in Figure 1. The DHV system is particularly useful in comparing magnetograms from ATS 1 with those from surface sites. Comparisons with magnetograms from Honolulu (latitude 21° 19′ N, longitude 158° 00′ W) and College (64° 52′ N, 147° 50′ W) will be discussed in some detail here.

3. Instrumentation

The UCLA ATS 1 magnetometer is a biaxial, closed loop fluxgate magnetometer which has been described in some detail by Barry and Snare (1966). The instrument measures the component of the magnetic field parallel to the spin axis of the space-

B. M. McCormac (ed.), Particles and Fields in the Magnetosphere.

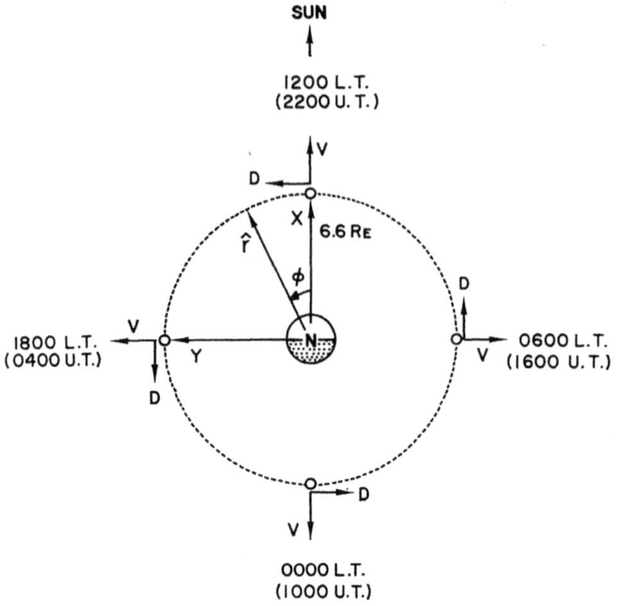

Fig. 1. View from the north of the ATS 1 orbit in the equatorial plane showing the *X* and *Y* directions of the nonrotating *XYZ* system and the *V* and *D* directions of the rotating *DHV* system. The *H* and *Z* directions are normal to the equatorial plane and positive northward.

craft and a single component transverse to the spin axis. The direction of this component is fixed relative to the spacecraft and therefore rotates in inertial space. The parallel component is sampled at a rate of 3.12 sec^{-1} and the transverse component is sampled at a rate of 6.24 sec^{-1}. The phase of the transverse component is measured relative to the pulse from a sun sensor with a field of view that sweeps across the sun during each revolution of the vehicle.

4. The Quiet Day Field

A description of the quiet day field at the synchronous, equatorial orbit will serve as a background for our discussion of the magnetic activity in the region. The ATS 1 data have been used to establish the properties of the quiet day field during the first 6 mo of 1967. K_p was used to determine the five quietest days in a given month. The days used in the quiet day study together with their corresponding K_p values, are listed in Table I.

The ATS 1 record includes the five quietest days of every month except April. There are three gaps in Table I. An SC magnetic storm began on 2348 UT on February 15. Thus, the last three hour value of K_p for February 15 was 5– and the ATS 1 data for this interval have been omitted. The blank spaces for April 13 and 28 indicate intervals during which the ATS 1 magnetometer was turned off.

The data in Table I show that the quiet days of some months were much more

TABLE I

Values of K_p and ΣK_p for the geomagnetically quiet days of the first 6 mo of 1967

Month	Day	K_p Three-Hour Range Indices 1	2	3	4	5	6	7	8	Sum
Jan.	4	0	0+	0+	0	1−	0+	1	1	4−
Jan.	5	1+	0+	0+	1−	1−	1−	0+	1−	5
Jan.	24	1	0+	0	0	0+	0+	0+	1+	4−
Jan.	30	0+	1−	1	1−	1−	0+	1−	1−	5
Jan.	31	0	0+	1−	0+	0+	1	1−	1−	4
Feb.	2	0	0	1−	1−	0+	0+	1+	1−	4
Feb.	3	0	0	0	2−	0+	0	1	1	4
Feb.	12	0+	0	0+	1−	1−	1−	0+	1+	4+
Feb.	13	1−	0	0+	1−	0+	2	2−	2−	7+
Feb.	15	1+	1	0+	0	0	0	1−		
March	2	0	1	1−	1+	2−	1+	1−	2−	8+
March	11	0	0	1	0+	1−	0+	0	0+	3−
March	12	0+	0	1−	1−	1−	1−	0+	0+	4−
March	24	1−	0+	0	0+	2+	1+	0+	0+	6−
March	25	0+	1	1+	1−	1+	1−	1	1−	7
April	13					0+	0+	1−	1+	
April	14	1−	0+	1+	1−	1−	1	1−	1−	6
April	15	1	0+	1−	0+	1	1	2−	1+	7+
April	26	0+	1	1−	1	1	0+	0+	0+	5
April	27	1	1−	1−	0+	0+	1−	1−	1	5+
April	28	0+	1	0+	0+	1−				
May	8	1	1	1	1	2	2	2	1+	11+
May	9	1	1+	2−	1+	1+	1+	1−	1	10−
May	15	3+	1+	0	1	1+	1	2+	2	12+
May	16	2	1+	1+	2−	1+	1	1+	2+	12+
May	21	1	1	0+	1	1−	1	3+	2+	11−
June	18	3−	1	1	0+	0+	1−	0+	1−	7
June	20	1−	1+	2	1−	1	1−	1−	1−	8−
June	21	0+	1	2	1	1	1	1	1+	9−
June	23	1−	1+	1	0+	0+	1	1	2	8−
June	24	1−	1−	1+	1−	1−	0+	1	2−	7

magnetically active than the quiet days of others. For example, the values of ΣK_p for May were more than twice the values for January.

In this quiet day study, 6 min averages of *D*, *H*, and *V* were used as the 'raw' data. For each month the magnetic fields for the five quiet days listed in Table I were averaged to yield an average quiet day for the month. The *D* components of these average quiet days are shown in Figure 2a as solid curves. The shaded regions in which the solid curves are centered represent the one standard deviation envelope, i.e., the width of the shaded region at any time point is equal to two standard

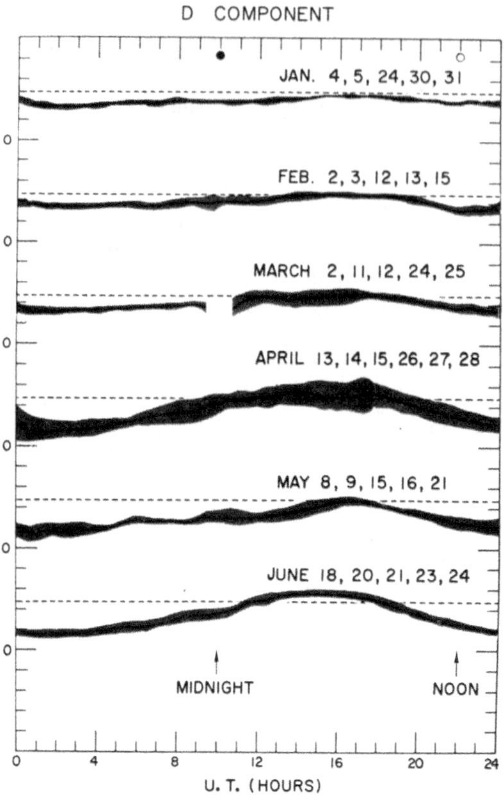

Fig. 2a.

Fig. 2. Mean values of D, H, and V for the quiet days of the indicated months. The shading indicates the envelope of one standard deviation. These results are derived from 6 min averages.

deviations $2\sigma_D$ where

$$\sigma_D = \sqrt{\sum_{i=1}^{n} \frac{(D_i - \langle D \rangle)^2}{n-1}}$$

and

$$\langle D \rangle = \frac{1}{n} \sum_{i=1}^{n} D_i$$

Usually, $n = 5$, but since some short gaps occur in the data n is occasionally less than 5. Similar curves and envelopes are shown for H and V in Figures 2b and 2c, respectively.

The magnitude of the diurnal variation in D increased from 7γ in January to 23γ in June. Throughout most of the 6 mo period the diurnal variation in D showed a peak shortly before dawn and a somewhat broader minimum shortly before dusk.

The magnitude of the diurnal variation in V increased from 5γ in January to 20γ

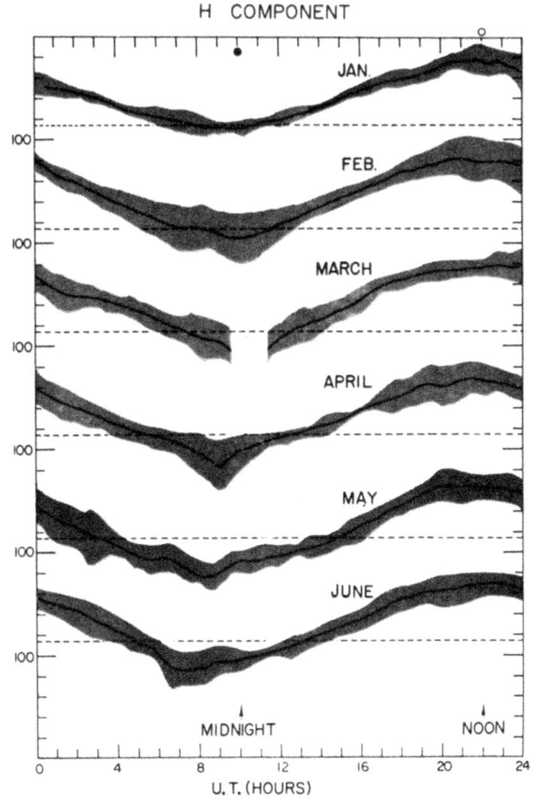

Fig. 2b.

in June. Note that, except for May, the quiet day monthly mean value of V at local noon was nearly constant over the entire 6 mo interval. Thus, the seasonal change in the field over the night sector produced the increase in the amplitude of the diurnal variation from January to June.

The H component had a maximum near 2200 UT, which is local noon at ATS 1, and a minimum near or shortly before 1000 UT, which is local midnight. The total excursion was 35 to 45γ. Here again the value of H at local noon on quiet days remained essentially constant during the 6 mo interval while the magnitude at local midnight decreased by about 10γ. Thus, the diurnal variation in the quiet day H increased by this amount.

The gap in the data that occurs in March is due to the eclipse of the sun by the earth near the spring equinox. The ATS 1 magnetometer data have been omitted for those approximately 1 hr eclipse intervals.

The earth's main field, as extrapolated from surface measurements (cf., Kahle, 1967), would appear as horizontal lines in Figures 2a, b, and c. This constant field has components $D_E = 23.5\gamma$, $V_E = -6.8\gamma$, and $H_E = 107.0\gamma$. The difference between the

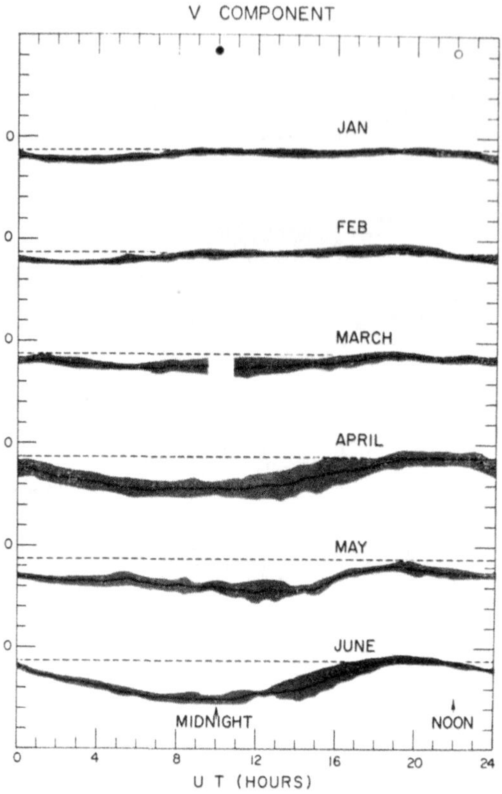

Fig. 2c.

constant field and the total field can be used to determine the field of the magnetospheric currents.

Cummings *et al.* (1969a) have used these quiet day data to estimate the equatorial component of the force of the interaction between the solar wind and the earth. They have concluded that the quiet day magnetic field indicates a tilt of the symmetry axis by an amount greater than can be accounted for by the effect of the earth's orbital velocity relative to the solar wind. This confirms a similar conclusion based on preliminary ATS data and Vela data (Roederer *et al.*, 1968; Roederer, 1969) and is consistent with Vela 3 measurements of the plasma flow direction in the magnetosheath for the same time interval (Hundhausen *et al.*, 1968). Walters (1964) has predicted an additional easterly deflection of the solar wind as it crosses the standing hydromagnetic shock upstream of the magnetosphere. For typical values of the interplanetary plasma and magnetic field parameters Walters computed deflection angles between 5 and 20°.

The cause of the January to June changes in the diurnal variation has not been established definitely. Olson and Cummings (1968) have found that a solar wind flow inclined southward by 8° and constant over the year could produce such an effect.

They assumed that the solar wind was specularly reflected at the magnetopause. Alternatively, since there was an increase in quiet day activity of the interval studied the changes may be associated with this increase.

5. Geomagnetic Substorms

Studies of the magnetic field at ATS 1 during geomagnetic substorms have added significantly to the body of empirical knowledge about this apparently fundamental magnetospheric phenomenon. The results of earlier phases of these studies have been reported by Cummings and Coleman (1968) and by Cummings *et al.* (1968).

An ATS 1 magnetogram, consisting of 6 min averages of *D*, *H*, and *V* for December 25, 1966, (UT) is shown in Figure 3. Negative bays associated with the

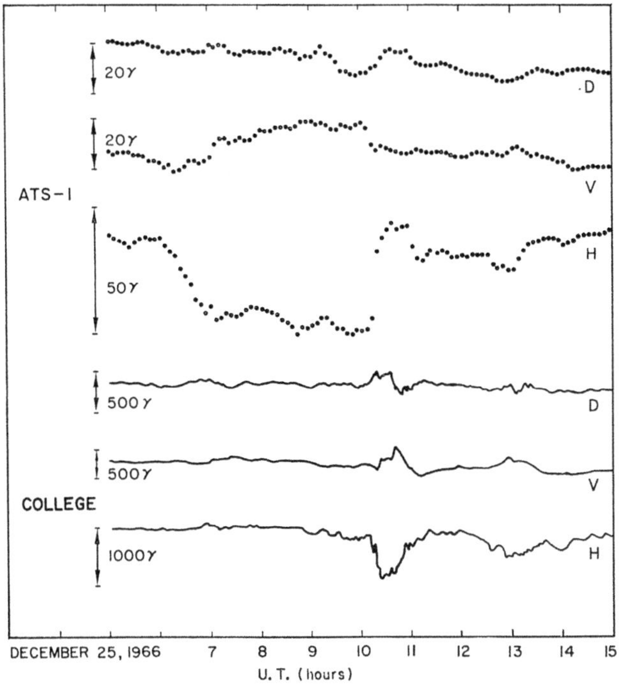

Fig. 3. Comparison of *D*, *H*, and *V* at ATS 1 with those at College, Alaska, during the 10 hr interval of December 25, 1966, centered about midnight at ATS 1.

expansion phase of a geomagnetic substorm occurred at College just after 1000 UT (2400 LT) and 1300 UT (0300 LT) as shown also in Figure 3. The records of *H* from College, ATS 1 and Honolulu, for the same day are plotted on the same scale in Figure 4. The data in the Honolulu and College records in this figure are 5 min samples of the respective magnetograms. There are gaps in the College data where the value was too low to appear on the plot.

This example shows several features that have been found to be typical of substorm related field fluctuations at ATS 1. First, a substorm (auroral) expansion that occurs when ATS 1 is within a few hours of local midnight is nearly always preceded by a decrease in H below its quiet day values and accompanied by a nearly simultaneous recovery of H toward its quiet day value. Second, the deviation in H from

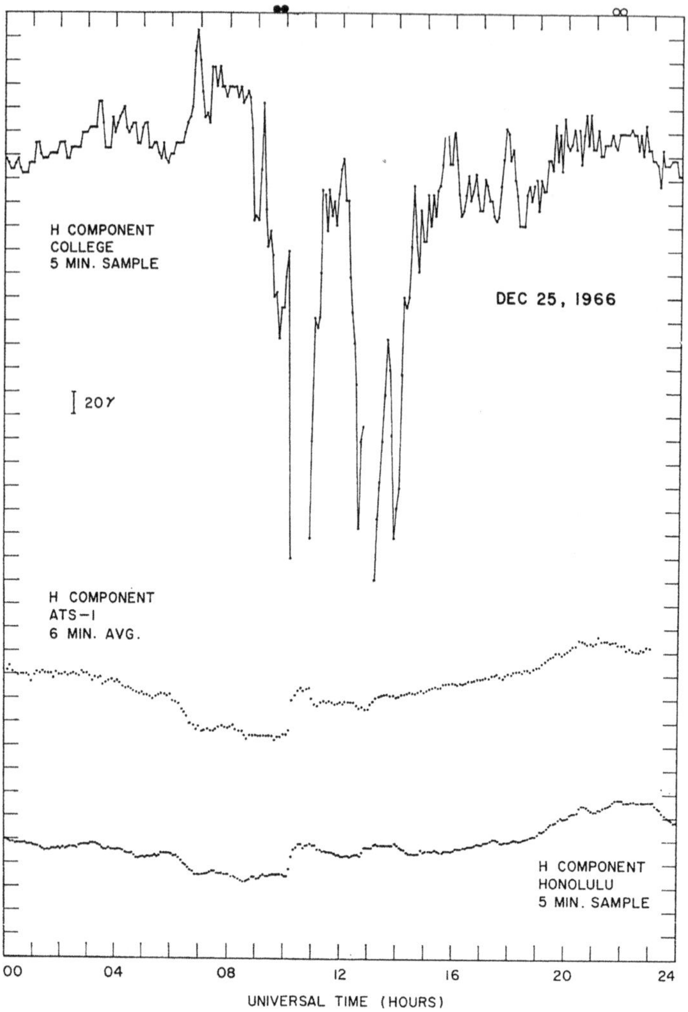

Fig. 4. Comparison of H at College, ATS 1 and Honolulu, all plotted on the same scale, for December 25, 1966.

its quiet day behavior is accompanied by deviations in D and V. Third, H at Honolulu shows a strong positive correlation with H at ATS 1.

Figure 5 shows superimposed ATS 1 magnetograms for five substorm days, two in December 1966 and three in January 1967. The plots of D, H, and V may be compared to those for the quiet days of January 1967 in Section 4. This figure shows the tendency of *H* to decrease and the tendency of *V* to increase around local midnight during intervals marked by substorms. It also shows that *D* tended to increase before midnight and decrease after midnight. In comparison to the quiet day field, the field during periods of substorm activity exhibits a greater variability at all local

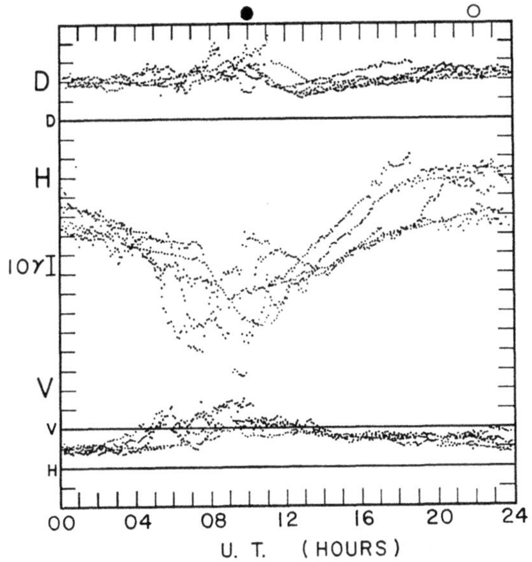

Fig. 5. A superposition of ATS 1 magnetograms for five substorm days, December 24 and 25, 1966, and January 6, 16, and 21, 1967.

times and a tendency for *H* to increase on the average over the daylight hemisphere. Further, the field in the night sector was more active before local midnight than after on these five days.

The data shown in Figure 5 were taken within about a month of the (northern) winter solstice. Subsequent observations indicate that there is a seasonal variation in the changes in *D* and *V* that is obviously substorm related, i.e., those that accompany decreased *H*. The signs of the substorm disturbances in *D* and *V* reverse from winter to summer. At the equinoxes they are comparatively weaker during intervals of comparable depression in *H* and may well average zero. The seasonal reversals in the *D* and *V* components of the disturbance field are due apparently to the fact that on the nightside of the earth ATS 1 is south of the 'plane of symmetry' of the geomagnetic cavity during the winter and north of it during the summer.

January 21, 1967, was another day of substorm related activity. The records of *H* from College, ATS 1, and Honolulu are shown in Figure 6.

In Figure 7 the ATS 1 data for January 21 are displayed again, this time in a manner that exhibits the orientation of the equatorial plane (*DV*) component of the disturbance field. Figure 7 shows the disturbance vectors in the geographical equatorial plane which is also the *DV* plane of our *DHV* system. The quiet day values of *D* and

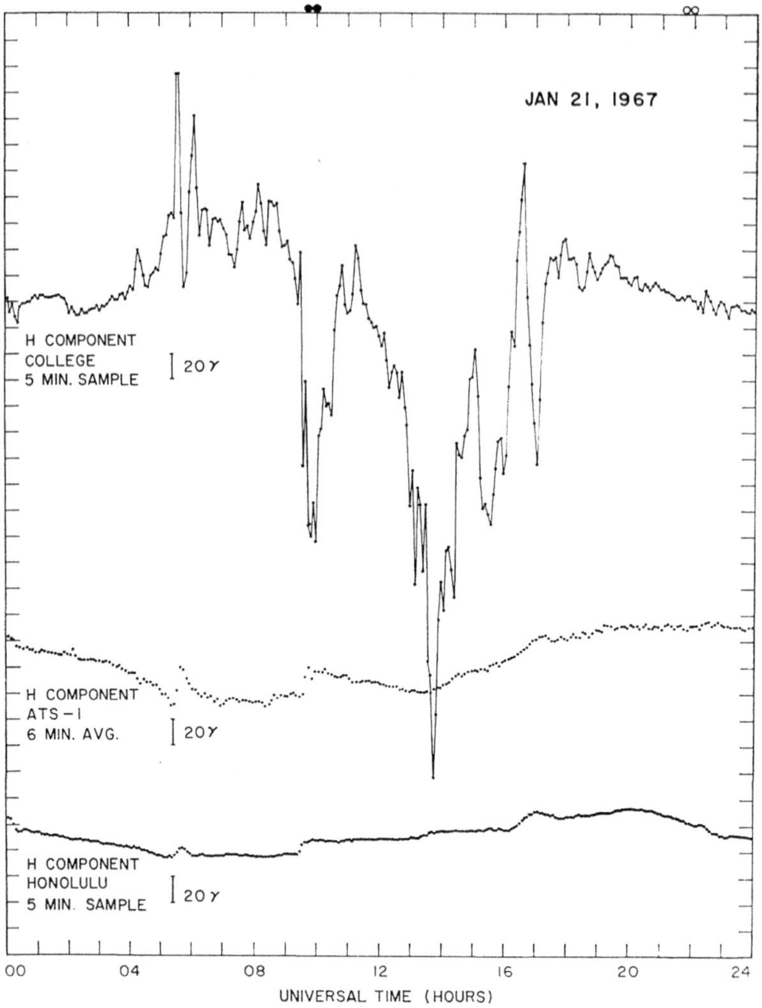

Fig. 6. Same as Figure 4 for January 21, 1967.

V were subtracted from the values measured during this period to obtain the plotted components of the vectors. Every other one of the 6 min averages was used.

It is clear from Figure 7 that when *H* was depressed in the dusk-to-midnight quadrant (0400 to 1000 UT), it was also inclined radially outward. This behavior is particularly evident prior to the recovery at about 0530 UT and during the depression that followed this recovery. A disturbance field of this type could have been produced by a ring current mostly beyond the orbit of ATS 1 with its plane of symmetry north of ATS 1. This position for the plane of symmetry in the night sector is most likely in the winter as shown in Figure 8. Cummings, Barfield, and Coleman (1968) have speculated that this behavior of *V* associated with the decrease in *H* indicates that a partial ring current was flowing in the dusk-to-midnight sector. Of course, it is likely

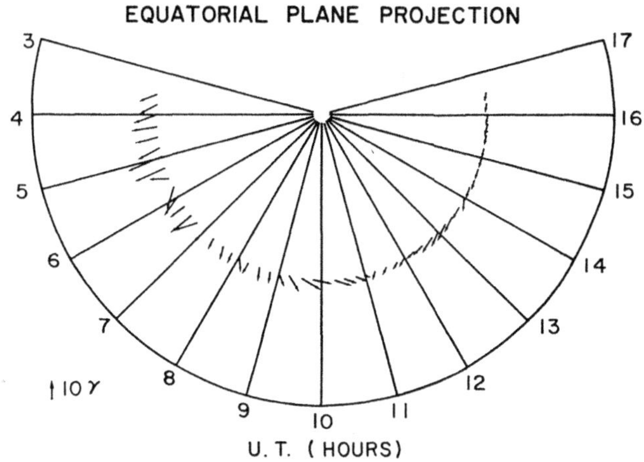

Fig. 7. The *DV* component of the *disturbance* field in the night sector for January 21, 1967.

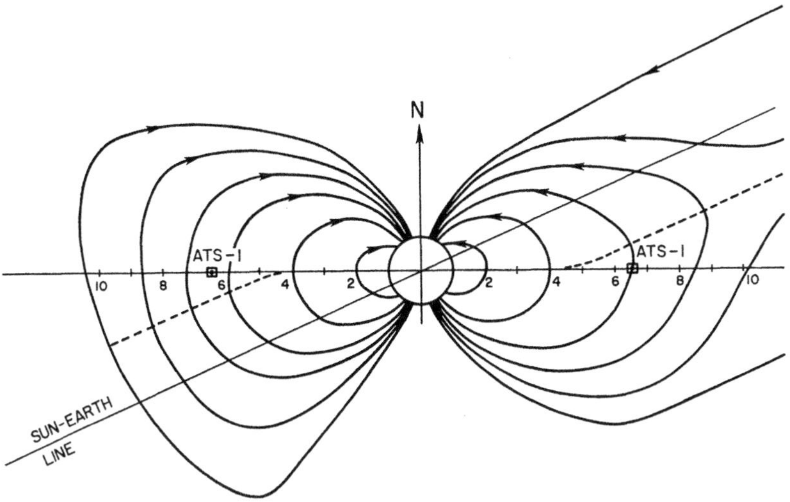

Fig. 8. A sketch of the magnetic field in the noon-midnight meridian at the winter solstice. Note that on the nightside ATS 1 is below the ecliptic plane and thus it is probably below the plane of the surface of symmetry for any ring current or tail sheet current.

that the ring current was more nearly closed. Meng and Akasofu (1969) have found no evidence for a simultaneous depression in *H* at the mid-latitude stations around the world while *H* was depressed in the dusk-to-midnight sector. Notice that the depression in *H* that occurred at ATS 1 near 1300 UT, i.e., in the morning sector was not accompanied by a radial (*V*) perturbation field in the *DV* plane, although there was an appreciable azimuthal disturbance field in the $-D$ direction at the time that *H* reached this local minimum.

A prominent feature of the disturbance field during the 2 hr interval centered at local midnight (1000 UT) was a strong positive D component. This is difficult to account for using a partial ring current or a neutral sheet current that is symmetrical about the aberrated sun–earth line. A current with a major component in the V direction or field aligned could produce the observed effect in D.

To determine the local time dependence of the substorm related disturbance in H at ATS 1, the data from March 1967 were analyzed. A month nearest the equinox was selected because the geomagnetic cavity should be in its simplest configuration during such an interval.

Field variations recorded on the ground during three auroral disturbances that occurred on March 3, 1967, are shown in Figure 9. These curves are 24 hr tracings

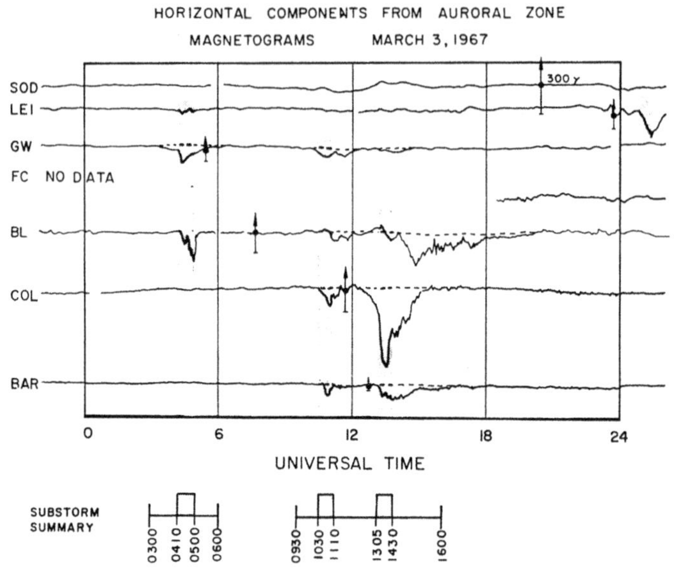

Fig. 9. Comparison of H at several high latitude stations on March 3, 1967, and the substorm summary for the day.

of H at seven high latitude stations. Typical tracings for a quiet day are shown by the dashed curves. The vertical scale at each station is shown by an arrow of length 300γ. The heavy black dot at the center of the arrow is magnetic midnight for each station. The seven stations span 180° in longitude from Finland west to Alaska.

The three shaded sections correspond to the expansion phases of three successive substorms. The beginnings of the growth phases and the endings of the recoveries are roughly defined as the beginning and ending of the deviations from quiet traces, as indicated in the substorm summary at the bottom of the figure.

The 6 min average ATS 1 magnetogram for this same day is shown in Figure 10. The occurrence of polar substorms, as summarized at the bottom of the previous figure, is also shown at the top. The dashed line shows the typical quiet day field at ATS 1.

The effects of a substorm at ATS 1 clearly are evident for the two substorms near local midnight. A 40γ depression in H exists for several hours prior to the expansion phase of the second substorm. A partial recovery in H accompanies the expansion. This event is followed by another depression and then a full recovery, again corre-

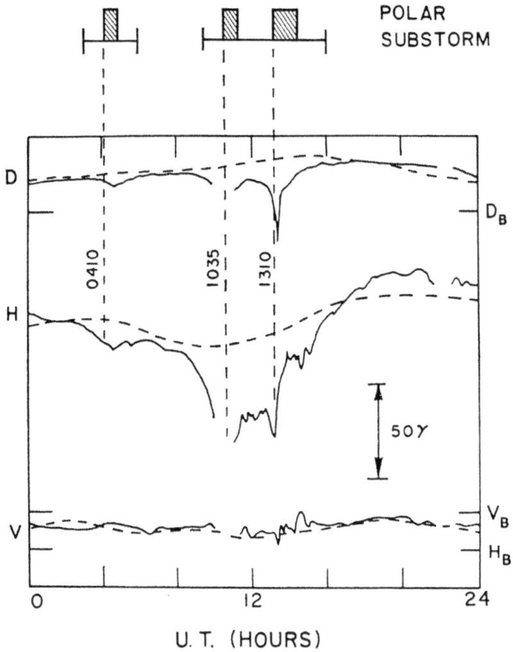

Fig. 10. ATS 1 magnetogram for March 3, 1967, and the substorm summary for the day.

sponding to the expansion phase of the third substorm. The first substorm of the day occurred when ATS 1 was at the dusk meridian (0400 UT). A depression in H preceded this event also, but the start of the recovery was delayed relative to the midnight expansion.

During these particular events interesting variations in the D component also were observed. Note just after the start of the expansion phase of the third substorm a negative pulse with a magnitude of about 30γ occurred in D. There is a suggestion that the earlier substorm also had such a 'D spike', although data are missing because of the satellite eclipse.

For all the days covered by ATS 1 data in March 1967, records from these same seven stations have been similarly examined for substorms. ATS 1 data were available for 15 days in March. During these days 40 substorms were identified. Since the substorms were not selected on a basis of strength, there is a considerable bias in their universal time distribution. The bias occurs because no ground magnetometer data were available from Russia, so that only strong substorms could be detected when Russia was in the night sector. In contrast, very weak substorms could be identified when North America was in the night sector.

Despite these difficulties rather clear results emerged. These are summarized in Figure 11. In this figure the date on which a particular substorm occurred is recorded in a column corresponding to the 3 hr interval of local time at ATS during which the substorm occurred. Substorms in which changes obviously related to substorms

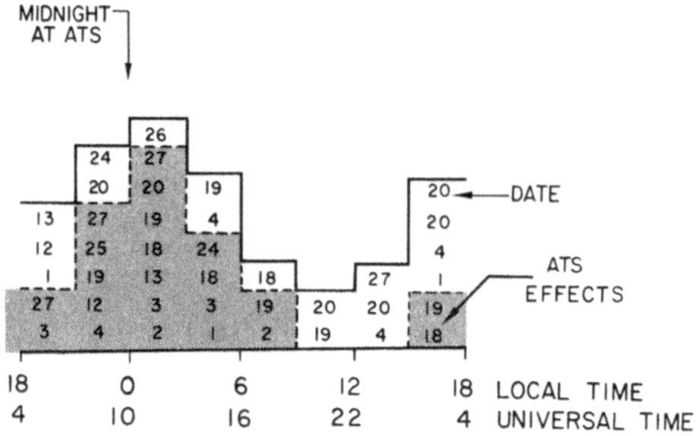

Fig. 11. Local time distribution of those substorms in March 1967 for which ATS 1 data are available. A shaded block indicates that the substorm was detected at ATS 1.

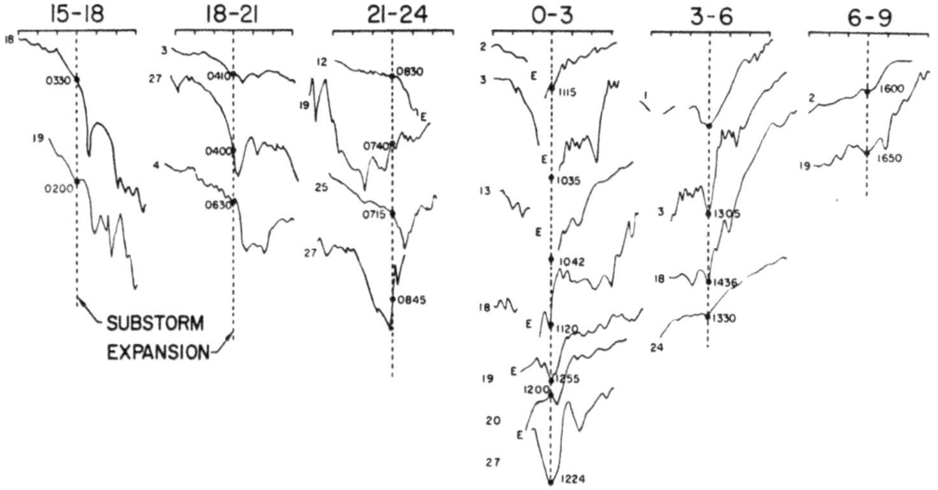

Fig. 12. Substorm effects in H at ATS 1 during March 1967.

occurred at ATS are indicated by shading. Note the almost complete absence of effects in the day sector. In contrast, most of the substorms that occurred while ATS 1 was in the night sector produced significant effects at the spacecraft. The substorms that did not affect the field at ATS 1 were almost all very weak.

In order to obtain a more comprehensive view of the substorm effects on *H* at ATS 1, the *H* record from ATS 1 for each event in March is shown in Figure 12. All events are organized according to the 3 hr interval of local time at ATS in which they occurred. The start time of the associated expansion is shown by a vertical dashed line.

When ATS 1 was located in the afternoon and evening sectors, the recovery of *H* was usually delayed relative to the occurrence of the corresponding substorm expansion as indicated by the time of the negative bay at local midnight. When ATS 1 was near midnight there was usually little delay between the midnight expansion and the recovery in *H* at ATS 1. In fact, delay was usually very small when ATS 1 was anywhere in the sector between midnight and early morning local time. However, this conclusion concerning the delay in the pre- and post-dawn hours is somewhat tentative because of the uncertainty in the times of the substorm expansions that occurred near midnight in the Soviet Union.

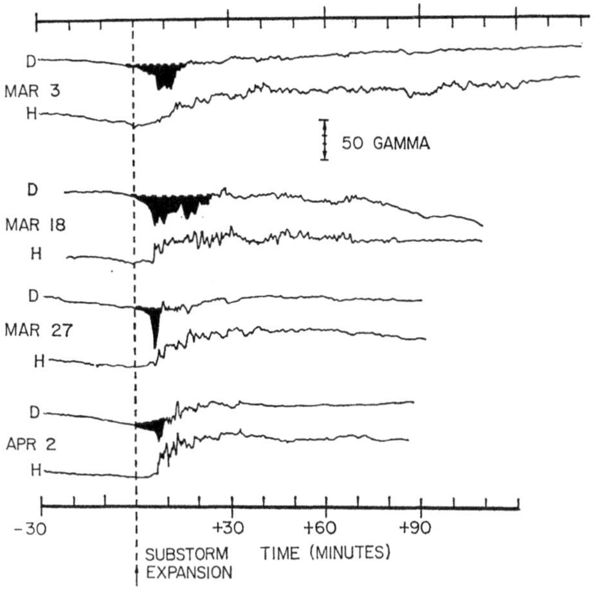

Fig. 13. Some *D* spike events recorded during March and April 1967. The substorm data are given in Table II.

Several of the March substorms exhibited the previously mentioned large amplitude '*D* spikes'. These occurred just after the beginning of the recovery of the *H* component. Records for all of the *D* spike events observed in March and one detected in April are summarized in Figure 13. The vertical dashed line is the start of the expansion phase of a polar substorm. Higher time resolution is used here to illustrate the details of the effects at ATS 1. From this figure it is apparent that the large negative deviations

TABLE II

D spike events at ATS 1 near Spring Equinox 1967

Date	Start time of expansion of polar substorm	Time of *D* spike minimum	Duration (Minutes)
March 3	1305 UT	1318 UT	4
March 18	1120 UT	1126 UT	3.5
March 27	1200 UT	1236 UT	3
April 2	1134 UT	1136 UT	2.5

in the *D* component began at the same time as the recovery in *H*. The times of the substorm expansions are listed in Table II.

6. Substorm Associated Pulsations

Irregular magnetic pulsations are observed on the ground during an auroral breakup. They occur abruptly at the start of the expansion of the substorm.

The power spectrum of d*H*/d*t* usually falls off sharply with increasing frequency. There are no significant spectral peaks. The cross correlation between this signal and its orthogonal component in the horizontal plane, the time derivative of *D*, shows no significant coherence between the two variables.

Similar magnetic pulsations have been observed at ATS 1 during substorms. Figure 14 is an expanded plot of the ATS 1 data taken during the second substorm

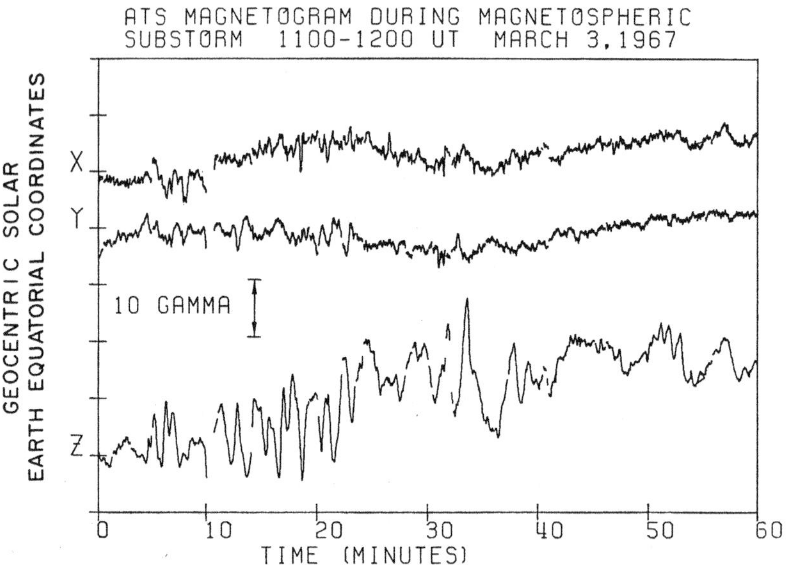

Fig. 14. Pulsations recorded at ATS 1 during the second substorm in Figures 9 and 10.

on March 3 (see Figure 9). This section of the record is from the interval marked by the recovery of the *H* component during the expansion phase of the substorm. Note the large amplitude fluctuations in the *Z* (or *H*) component. These fluctuations are 10 to 20γ peak-to-peak amplitude, superimposed on an 80γ average field. Note that the fluctuations in *Z* are much stronger than those in *X* and *Y*. The irregular nature of the pulsations is quite evident.

The end of this event and the beginning of the depression prior to the third substorm of March 3 is shown in Figure 15. The pulsations stop rather abruptly at about

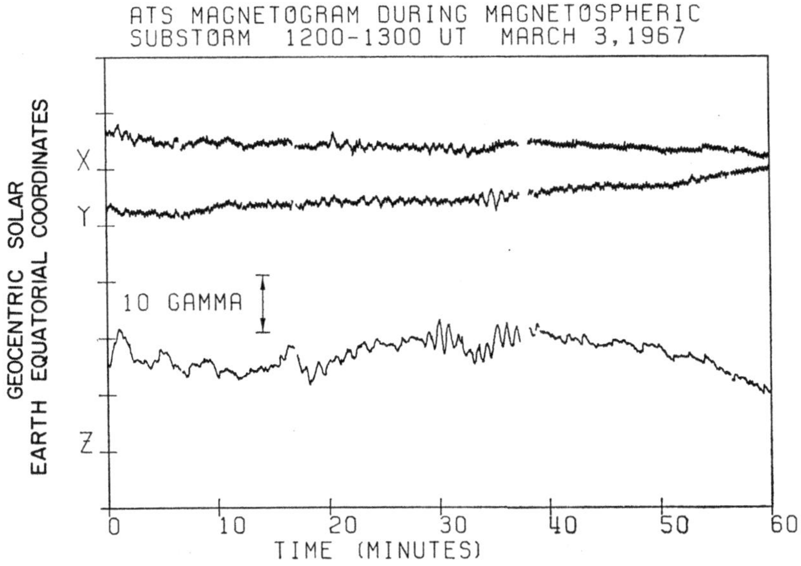

Fig. 15. Continuation of the record of Figure 14.

1240 UT, and the depression in *H* develops in an otherwise relatively quiet field. The expansion phase of the third substorm of March 3 is shown in Figure 16. The beginning of the recovery in the *Z* (or *H*) component is well defined. This is followed within 8 min by a relatively clear *D* spike. The trailing edge of the *D* spike appears to mark an enhancement of the large amplitude irregular pulsations in *Z*. (The spikes in *X* and *Y* are just the *X* and *Y* components of the *D* spike, i.e., the spike in the azimuthal field. Recall that *D* is parallel to *X* at 1600 UT (0600 LT) and to *Y* at 2200 UT (1200 LT).)

To obtain a better idea of the characteristics of these pulsations a spectral analysis was performed. The spectra for the first of the pulsation events shown in the preceding three figures are shown in Figure 17. Note the steep spectra, approximately $1/f^2$ overall, and the inflection point at 10^{-1} Hz in the spectrum for *Z* which indicates a maximum in the pulsation power near this frequency. Note also an order of magnitude greater power in *Z* than in the transverse components.

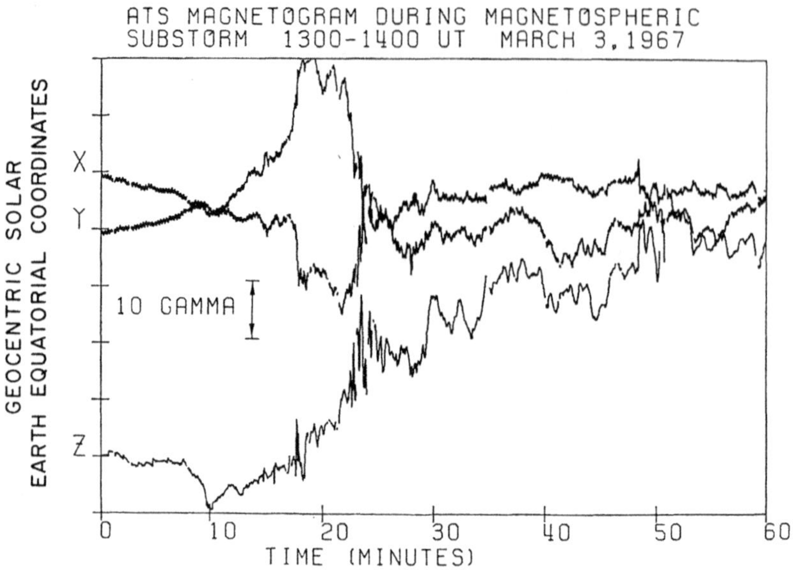

Fig. 16. Continuation of the record of Figure 15. Note the recovery in H associated with the third substorm indicated in Figures 9 and 10.

The cross spectra between pairs of the three vector components show that there is very little coherence between any of the components of the magnetic field. Similar results were obtained for the later event.

Of the 40 substorms in March for which ATS 1 data were available about half were accompanied by pulsation events. The local time distribution of these events is shown in Figure 18. As before, the date of the substorm occurrence is shown as a function of local time. Substorms during which pulsations were detected at ATS 1 are shown by shading. There does not seem to be any dramatic local time dependence in their occurrence.

A preliminary search for local time dependence in the other properties of these pulsations has yielded some positive but tentative results. For example, pulsations in the pre-midnight sector have a tendency to begin gradually, and generally earlier than the associated recovery in H. (The beginning of the pulsations is generally delayed relative to the expansion at midnight even in these cases.) The spectral properties of the pulsations in the pre-midnight section have not been determined yet. However, from their time series they appear to be similar to those associated with midnight events and illustrated in the foregoing. There is some evidence that they are somewhat smaller in amplitude.

Band limited pulsations similar to those observed at surface stations also have been detected at ATS 1. The band limited pulsations in the time derivative of the magnetic field at the earth's surface usually are confined to the spectral region between a 50 sec period and a 5 sec period. The spectral peak occurs at a period of roughly 10 to 15 sec.

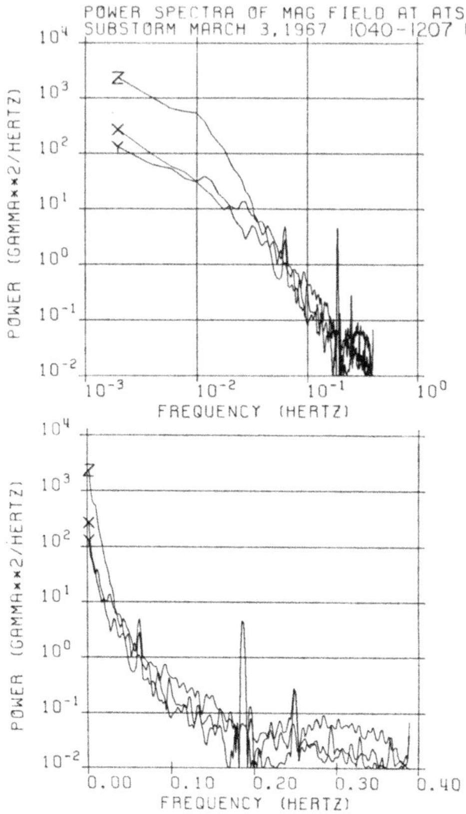

Fig. 17. Auto spectra of the vector components of the field at ATS 1 for the fluctuations illustrated in Figure 14.

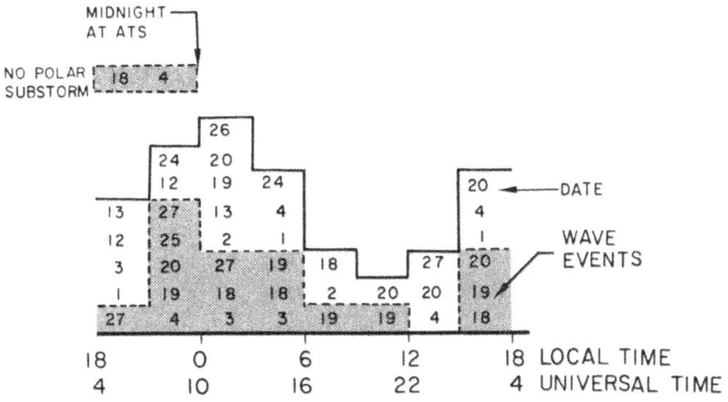

Fig. 18. Local time distribution of those substorms in March 1967 for which ATS 1 data are available. A shaded block indicates that pulsations were recorded at ATS 1 during the substorm.

These pulsations are accompanied by pulsating aurora and modulated precipitation of energetic electrons. The auto spectra of these fluctuations in auroral intensity and electron precipitation are similar to that of the band limited pulsations in the field, but no significant cross correlations have been detected.

Nevertheless, the simultaneous occurrence of fluctuations with comparable spectra in all three quantities led to the prediction that the band limited pulsations in the field would occur at 7 R_E (McPherron, 1968) during the expansion and recovery phases of geomagnetic substorms. In our preliminary examination of the records from the ATS 1 magnetometer we discovered that such pulsations, in fact, do occur in the distant geomagnetic cavity. Two events have been found in the short section of the record examined to date.

Figure 19 shows a section of the record from the ATS 1 magnetometer during one

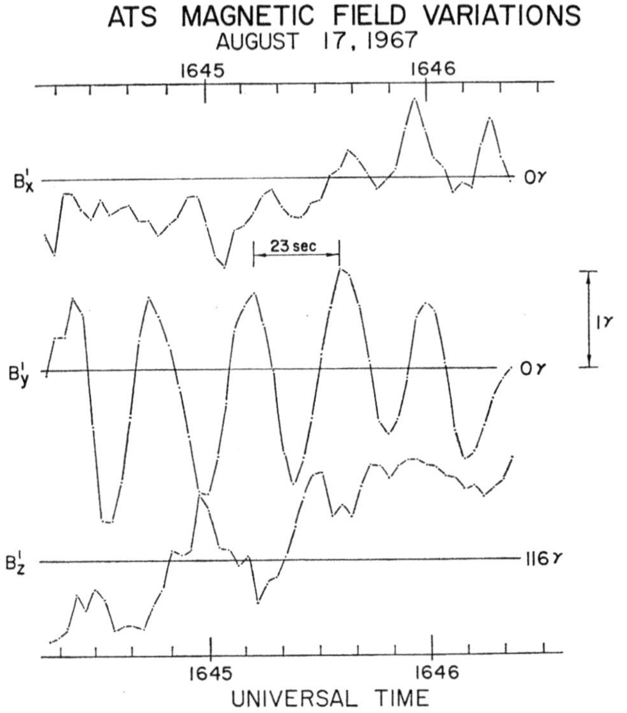

Fig. 19. Field pulsations at ATS 1 in the frequency range of band limited pulsations at the surface.

of these events. The data are 2.56 sec averages. The vector components **X′Y′Z′** are defined as follows: **Z′** is in the direction of the average field $\mathbf{B}_0$ for the interval during which the plotted data were taken; **X′** is in the direction of $\mathbf{B}_0 \times \hat{\mathbf{H}}$ and **Y′** completes the right handed system. Thus, **Y′** is in the plane defined by $\mathbf{B}_0$ and $\hat{\mathbf{Z}}$, the direction of the earth's spin axis. For a dipolar field $\mathbf{B}_0$ is in the *HD* plane and inclined from *H* by 11.5° in the direction of *D*.

It is apparent from Figure 19 that the field fluctuations were rather well confined to the **Y**′ direction. Thus, they were transverse to $\mathbf{B}_0$ and in the plane defined by the undistorted field and the mean distorted field $\mathbf{B}_0$. The fluctuations in **Y**′ are nearly monochromatic with a period near 20 sec. The amplitude is roughly 1γ.

The event occurred at 1600 UT (0600 LT) on August 17, 1967, during the recovery phase of a rather weak geomagnetic substorm. This substorm was preceded by a very strong substorm during which tail like fields were observed in the dusk-to-dawn sector. The $\mathbf{B}_0$ field was still considerably distorted when the pulsations were recorded.

Modulated electron intensities were detected at ATS 1 and pulsating electron precipitation was recorded at College during this event. Both variations exhibited power spectra with peaks at periods of 7, 11, and 21 sec (Parks, 1969). The second event was detected about 1700 UT on June 8, 1968, with the UCLA magnetometer on board OGO 5 as well as at ATS 1. OGO 5 was at 6.3 R_E near 0500 LT inbound in the equatorial plane and ATS 1 was at 6.6 R_E at 0700 LT. This event also occurred during the recovery of a substorm.

The period of the pulsations was again nearly constant at about 20 sec and the amplitude was about 1γ. The fluctuations were transverse to the mean field, but this time the mean field was relatively undistorted, i.e., dipolar. The field fluctuations were azimuthal, i.e., along *D*.

The record from the TRW VLF electric field detector onboard OGO 5 for the interval of this second event indicates that VLF chorus modulated coherently with the pulsations in the magnetic field (Kennel, 1969).

7. Summary

The results of our studies of the substorm disturbance field at ATS 1 may be summarized as follows:

(a) Relatively abrupt recoveries of the *H* component at ATS 1, from a magnitude less than the quiet day average toward the quiet day average, were often recorded when the spacecraft was in the night sector. Each such recovery was accompanied by a negative bay at local midnight.

However, the inverse was not true, i.e., negative bays occurred near local midnight without an accompanying recovery in *H* at ATS 1. In fact, three different types of behavior were observed in *H* at ATS 1 during negative bays at midnight. The first is a recovery in *H* as just mentioned. The second is a decrease in *H*, and the third is no change in *H*. The second type of behavior was observed during the storm in December 1966 as shown in Figure 4. (See Figure 4 of Cummings *et al.* (1968).) As was pointed out in that paper, the depression in *H* at ATS 1 began about 0600 UT (2000 LT) on December 25, 1967, and there was a negative bay at local midnight at nearly the same time, although none was recorded at College. However, a relatively slow decrease in *H* at ATS 1 preceded this 0600 UT event. The decrease probably began about 0330 to 0400 UT as shown in Figure 10. The onset of this disturbance is more obvious in *D* and *V* than in *H*.

It is suggested therefore that the behavior in *H* at ATS 1 during the event at 0600 UT can be considered as the combined effect of two different but nearly simultaneous phenomena. One is a recovery of *H* from the depressed state. This recovery is presumed to have occurred because of occurrence at about this time of a negative bay in the midnight sector. The other is a rapid decrease in *H* which was of sufficient magnitude to completely cancel out the bay associated recovery and produce a net decrease in *H*.

Assuming that this superpositional model is correct, the question remains as to whether the two effects normally accompany a substorm even though only the recovery is usually detected at 6.6 R_E and at mid-latitudes on the surface, or whether they occur simultaneously or only at random. We presently favor the latter alternative. However, since relatively few events of this type were contained in the records analyzed to date, this opinion is unsupported by statistics. However, in the December 25 event, for example, a weak depression probably began about 0330 to 0400 UT as shown in Figure 3. The onset of the disturbance is more obvious in *D* and *V* than in *H*. We would argue therefore that the negative bay about 0600 UT was actually associated with a recovery from this decrease, but the recovery was more than cancelled out by a greatly enhanced and nearly simultaneous decrease.

A case in which the field at ATS 1 remained nearly constant during a negative bay at College is shown in Figure 6 of Cummings *et al.* (1968). The figure includes the *H* records from College, ATS 1, and Honolulu for January 28, 1967. In this case, a rather weak negative bay occurred at College about 1200 UT (0200 LT) and the field at ATS 1 changed very little if at all. Further, there was no obvious effect in the field at Honolulu.

We cannot rule out the possibility that the absence of disturbances at ATS 1 and Honolulu was in this instance simply a consequence of the weakness of the activity. Nevertheless, we suggest that the behavior of the field at both locations was related instead to the absence of depressions in *H*. Of course, this possibility is suggested by the fact that the most frequently observed bay associated effect in *H* at ATS 1 is a recovery toward the quiet day magnitude since a recovery can hardly occur without a prior depression.

(b) During December 1966 and January 1967 decreases in *H* at ATS 1 were accompanied by increases of somewhat lesser magnitudes in *V* there. This behavior is apparent in Figure 5. We interpret this behavior as evidence that the disturbance field is produced by a current with a ring-current-like geometry rather than by a current in the tail with a simple sheet-like geometry (cf., Cummings *et al.*, 1968) or a more nearly completed ring current, or a sheet current with the near earth edge curved to form a '*U*' as viewed from above or below the plane of the sheet.

(c) When ATS 1 was located in the late afternoon and early evening sectors, the recovery in *H* was delayed relative to the negative bay at local midnight. The delays at dusk were something like 30 min. At locations nearer midnight the delay was shorter and, in many cases, it was too short to be measurable by our methods. From midnight to dawn the delays were usually negligible, although the accuracy of our measurements of the onset times of the substorm expansions at midnight was reduced by the lack of data from high latitude stations in the Soviet Union.

(d) Negative 'spikes' were recorded in the *D* component at ATS 1 during several of the March substorms. These spikes occurred at the beginning of the recovery in *H*. Although we have not yet studied the relationships between the occurrence of this phenomenon with those of other substorm phenomena, we would mention here the possibility that the spikes may be associated with field aligned currents. For currents flowing along lines of force outside the orbit of ATS 1, a negative *D* spike would correspond to a pulse of northward current. For currents on lines of force inside ATS 1 it would correspond to a southward pulse. Alternatively, if the field aligned current flows on two *L* shells, parallel to **B** on one shell and antiparallel on the other, then the E–W magnetic field would only be observed when the satellite is between the two shells. According to this interpretation, the near edge of the plasma sheet must have been inside the synchronous orbit prior to the onset of the expansive phase, and it must have moved outward across the synchronous orbit during onset (Cummings *et al.*, 1969b).

(e) The pulsations recorded at 6.6 R_E near midnight and in the morning sector follow the onset of the substorm-associated recovery in *H* with delays of the order of 10 min. As mentioned previously, in this local time sector, the recoveries in *H* are almost simultaneous with the substorm expansion. The pulsations recorded at locations duskward of midnight also followed the midnight expansion after a relatively short delay despite the fact that the recoveries in *H* were usually much more delayed relative to these expansions.

(f) Band limited pulsations similar to those observed at surface stations have been detected also at ATS 1. Two occurrences have been found in the data, although as yet no systematic search of the records has been made. On both occasions the waves were observed between 0500 and 0700 LT during recovery phases of substorms.

Acknowledgments

This work was supported in part by the National Aeronautics and Space Administration under Research Grant NGR 05-007-004.

We wish to thank G. K. Parks and C. F. Kennel for calling our attention to the two events that we have compared to band limited pulsations.

References

Barry, J. D. and Snare, R. C.: 1966, *IEEE Trans. Nuc. Sci.* **NS-13**, 326.
Cummings, W. D. and Coleman, P. J., Jr.: 1968, *J. Geophys. Res.* **73**, 5699.
Cummings, W. D., Barfield, J. N., and Coleman, P. J., Jr.: 1968, *J. Geophys. Res.* **73**, 6687.
Cummings, W. D., Coleman, P. J., Jr., and Siscoe, G. L.: 1969a, *The Quiet Day Magnetic Field at ATS 1*, Preprint, Publication No. 752, Institute of Geophysics and Planetary Physics, University of California, Los Angeles.

Cummings, W. D., Lewis, R. R., and Coleman, P. J., Jr.: 1969b, *Trans. Amer. Geophys. Union* **50**, 280.
Hundhausen, A. J., Gilbert, H. E., and Asbridge, J. R.: 1968, *Trans. Amer. Geophys. Union* **49**, 267.
Kahle, A. B.: 1967, *Harmonic Analysis for a Spheroidal Earth*, Rep. P-3684, Rand Corporation, Santa Monica, California.
Kennel, C. F.: 1969, private communication.
McPherron, R. L.: 1968, Relation of Auroral Zone Micropulsations to Magnetospheric Substorms. Dissertation, University of California, Berkeley.
Meng, C.-I. and Akasofu, S.-I.: 1969, *J. Geophys. Res.* **74**, 4035.
Olson, W. P. and Cummings, W. D.: 1968, *Trans. Amer. Geophys. Union* **49**, 742.
Parks, G. K.: 1969, private communication.
Roederer, J. G.: 1969, *Rev. Geophys.* **7**, 77.
Roederer, J. G., Coleman, P. J., Jr., Cummings, W. D., and Robbins, M. F.: 1968, *Trans. Amer. Geophys. Union* **49**, 227.
Walters, G. K.: 1964, *J. Geophys. Res.* **69**, 1769.

AC MAGNETIC FIELDS

CHRISTOPHER T. RUSSELL and ROBERT E. HOLZER
University of California, Institute of Geophysics and Planetary Physics, Los Angeles, Calif., U.S.A.

Abstract. Magnetic fluctuations have been observed by ground observatories from the Pc5 band to the VLF band. Not all these fluctuations have been observed also in space however. The precision of satellite borne instruments, data rates and orbital characteristics have restricted observations in the past, so that only the lowest and the highest frequencies have been examined extensively. Preliminary results from recent experiments are beginning to fill this gap. Some of these new results are presented together with a review of past work.

These latest results suggest first that the large amplitude 100 to 300 sec period waves are field line resonances driven by waves generated by the motion of the magnetopause and second that ELF hiss and chorus are the result of a plasma instability in the equatorial region of the magnetosphere.

1. Introduction

Variations in the earth's magnetic field have been studied for many years. These variations can be both periodic or quasi-periodic and nonperiodic. Examples of nonperiodic variations are sudden commencements, the main phase of geomagnetic storms, and the signature of the expansion phase of magnetospheric substorms. In this paper, however, only variations of an oscillatory nature will be reviewed. Furthermore, we shall discuss extensively only satellite observations. Nevertheless, the long history of ground observations will be used to order the magnetospheric measurements.

Oscillations observed on the ground have been classified according to their properties and their frequency band. This classification is illustrated in Figure 1. At the low fre-

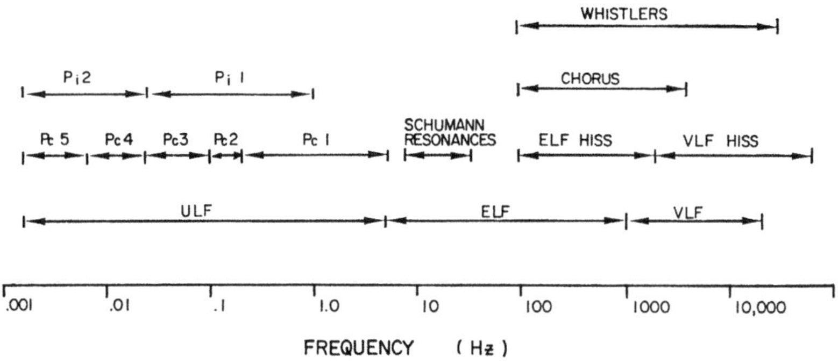

Fig. 1. The classification of ac magnetic fields according to frequency.

quency end of the chart below about 5 Hz are the Pc and Pi pulsations (Jacobs *et al.*, 1964). The Pc pulsations are continuous trains of waves whereas the Pi's are irregular pulsations.

From 5 to 1000 Hz is the ELF band. Defining the upper limit as 1000 Hz is somewhat arbitrary. Some authors consider oscillations up to 3000 Hz to be ELF. In any event the observed phenomena at these frequencies pay little heed to these definitions.

At the low frequency end of the ELF band are the Schumann resonances (Madden and Thompson, 1965). These waves are relatively well understood. The energy source is lightning discharges. The earth ionospheric cavity acts as a spherical waveguide which resonates at several well defined frequencies. Measurements in the magnetosphere as would be expected do not exhibit the Schumann resonances.

Hiss refers to noise that is unstructured. This noise can be band limited. The hiss observed in the ELF region is usually restricted to a narrow band of frequencies. Chorus refers to band limited noise with quasi-periodic structure. Whistlers are the result of the dispersive propagation through the magnetosphere of noise generated by a lightning stroke.

All of the phenomena observed in the high frequency end of the ELF band are observed also in the low frequency end of the VLF band which extends from 1000 to 30000 Hz. An additional feature of the VLF band is VLF or auroral hiss. This noise is a common feature of the auroral zone and appears to be Cherenkov radiation emitted by particle precipitation (Jørgensen, 1968). This emission extends from about 2000 Hz well into the LF band above 30000 Hz.

The low frequency extent of ELF hiss, chorus, and whistlers of Figure 1 does not correspond to the minimum frequency observed on the ground. Whistlers, chorus, and hiss can be observed at much lower frequencies in space than on the ground. But first, before examining the magnetospheric observations, we shall examine how these measurements are made and discuss the constraints of the various techniques and the constraints imposed by satellite orbits.

2. The Measurement of Fields in Space

Three types of instruments have been flown on spacecraft to measure magnetic fields: total field instruments, vector field instruments, and search coils or loops. It should be noted that these instruments measure only the magnetic component of plasma waves and therefore do not reveal the presence of electrostatic waves such as discussed by Kennel (1970). Furthermore, the total field devices do not register transverse oscillations which severely limit their usefulness. In addition, they cannot be used to measure polarizations.

In order to survey a sizable region of the magnetosphere vector field instruments require a combination of large dynamic range plus precision. This in the past has not been achieved. Some instruments have sacrificed dynamic range for precision (Ness *et al.*, 1964). Others have sacrificed precision for more dynamic range (Cahill and Amazeen, 1963). Only recently have attempts been made to obtain large dynamic range and adequate precision with a triaxial fluxgate magnetometer (Snare and Benjamin, 1966). The search coil magnetometer does not measure the static magnetic field and can easily achieve adequate precision since its dynamic range need not be as great. However, if the spacecraft spins the search coil cannot measure the fluctuation spectrum in the spin plane in high fields unless the spin frequency is not in the band-pass of the instrument or unless the spin frequency is very low. A disadvantage of a search coil magnetometer for low frequency measurements is that the sensitivity is proportional to frequency.

Figure 2 compares the minimum detectable amplitude of a typical precise fluxgate

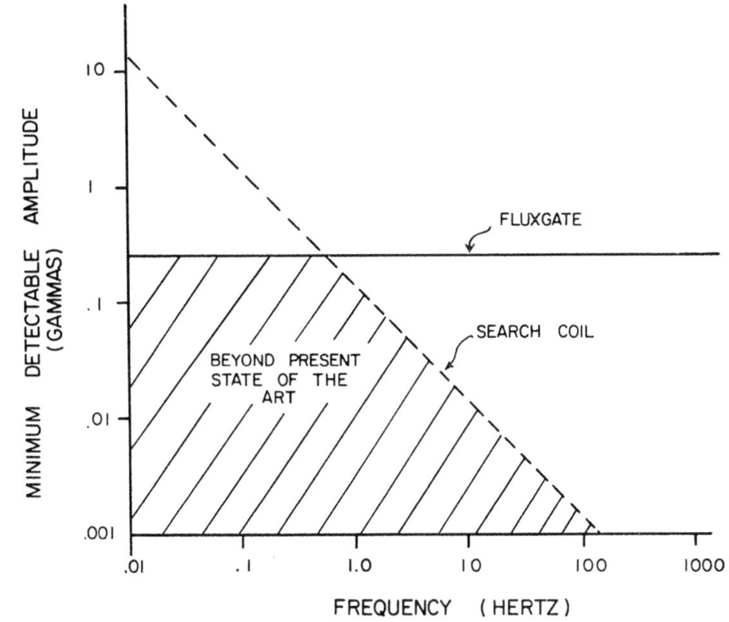

Fig. 2. The relative sensitivity of a typical precise fluxgate magnetometer and a search coil magnetometer such as used on the OGO spacecraft.

magnetometer with a search coil magnetometersuch as used on the OGO satellites (Russell *et al.*, 1969b). The fluxgate is more precise at frequencies below about 1 Hz, and thus is optimum for studies of the Pc bands. The search coil is best for the ELF and VLF band.

The most straightforward way of measuring magnetic field oscillations is to telemeter periodic samples of the waveform of the oscillation. It is obvious that one can only reconstruct the oscillations up to one-half the sampling rate, the Nyquist frequency, since two points are required to define an oscillation. It is not as obvious that if the instrument permits oscillations above the Nyquist frequency to reach the telemetry system, the oscillations below the Nyquist frequency cannot be reconstructed.

There is another means besides periodic sampling to reconstruct the waveform of the signals. Some experiments have used FM systems to reconstruct the waveform up to 1 kHz and above (Gurnett and O'Brien, 1964; Smith *et al.*, 1969). This is far above the Nyquist frequencies of the magnetometers on IMP 1 of 0.0061 Hz, and on Explorer 12 of 1.5 Hz or the maximum Nyquist frequency of the search coil magnetometer on OGO 1 and 3 of 139 Hz.

Another technique that is used when the FM method is not available and information is desired above the Nyquist frequency or to supplement the FM data is to filter the data into different frequency bands and monitor the power in each band. This has been done on the OGO search coil magnetometers. This method, however, destroys the phase information contained in the original signal.

The character of the orbit determines also the suitability of a spacecraft for magnetospheric observations. The highly eccentric orbiters such as OGO 1, 3, and 5 and the IMP satellites spend only a small fraction of their orbital period within the magnetosphere. Furthermore, they have radial velocities of from 2 to 5 km/sec within the magnetosphere. In addition such spacecraft require one full year of operation to entirely survey local time space. Polar orbiters near 1000 km such as OGO 2, 4, and 6 require only about 6 mo to cover all local times. However, their orbital velocity is much higher, about 7 km/sec, and they traverse regions of interest very rapidly. For instance, they cross the auroral oval in the order of 1 min. Clearly, the low polar orbiters do not provide a good platform for measurements of temporal variations with periods of more than several seconds. On the other hand, if we expect low period waves to be coherent over about $\frac{1}{10}$ R_E, the eccentric orbiter can provide measurements of waves with periods shorter than about 300 sec.

The synchronous orbit is excellent for magnetospheric observations. It provides complete local time coverage each day and the satellite moves about 3 km/sec in a direction that is perpendicular to the magnetospheric gradients, not parallel to them as with the eccentric orbiters and polar orbiters. The major disadvantage is that only one radial distance and a small range of magnetic latitudes is accessible to this orbit. One of the largest sets of experimental data available in the magnetosphere has come from the ATS 1, which is in such an orbit.

3. Low Frequency Observations

To this date there have been few published reports of low frequency magnetospheric waves. The existence of hydromagnetic waves was confirmed on one of the first space probes Pioneer 1 by Sonett *et al.* (1962). Similar observations were made by Coleman *et al.* (1960) on Pioneer V but of a qualitative nature. Later, Judge and Coleman (1962) observed 100 to 500 sec period waves on Explorer VI. They found both a transverse elliptically polarized mode and a compressional mode. These experiments, however, measured only two of the three components of the magnetic field as they were using a spinning search coil magnetometer. Thus they could make no unambiguous statement about direction of propagation relative to the field nor polarization.

Patel (1965), using a three axis fluxgate magnetometer, was able to measure the polarization of waves near the equator with periods longer than 60 sec. However, he could not resolve waves with amplitudes less than 4γ.

In this data he found both compressional and transverse waves with periods from 100 to 200 sec. He also found that the polarization of the transverse waves was right handed with respect to the field after 1045 LT and left handed before 1045. This observation agreed with those of Nagata *et al.* (1963) for giant pulsations on the ground. The amplitudes ranged from 4 to 11γ and the waves were observed from 7 to 10 R_E. However, the giant pulsations have amplitudes of up to 200γ on the ground and thus appear to be amplified as they propagate to the earth.

Cummings *et al.* (1969) with the larger set of ATS data at 6.6 R_E found waves with periods ranging from 50 to 300 sec and amplitudes ranging from 2 to 20γ. Figure 3 shows one component of the fluctuations measured by Judge and Coleman

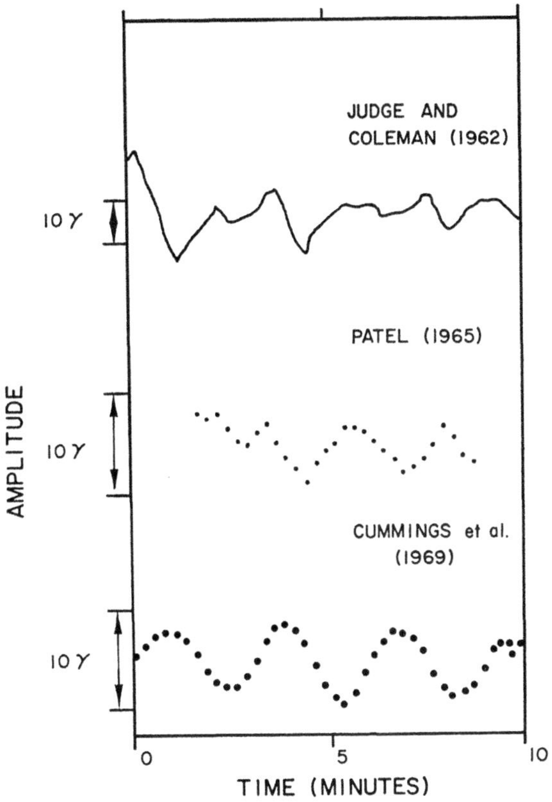

Fig. 3. One component of the low frequency fluctuation observed by Judge and Coleman (1962), Patel (1965) and Cummings *et al.* (1969) plotted on the same time scale.

(1962), Patel (1965), and Cummings *et al.* on the same time scale. The waves measured on ATS 1 are more sinusoidal and last much longer than those measured by the other two spacecraft. This may be due to the fact that ATS is not moving radially outwards through the wave field and because ATS has finer amplitude resolution. Unfortunately, the polarization of the ATS data was not examined and we do not know whether these waves have the same characteristics as determined by Patel except that they are both elliptically polarized.

Cummings *et al.* interpreted these waves as standing Alfvén waves. In order that these waves not have zero amplitude at the equator the observed waves must be even harmonics. Furthermore, the analysis of Cummings *et al.* showed that the amplitude at the surface of the earth should be about ten times that in space. This ratio appears to be larger than that observed, but no data on simultaneous ATS and ground measurements were presented. Simple tests of this interpretation would be first to examine simultaneous rapid run magnetograms on the ground and second, to see if the simultaneously measured plasma density at 6.6 R_E agrees with that predicted with the theory.

The latest paper on low frequency magnetospheric waves is by Sonnerup *et al.* (1969). Again, the wave period is about 300 sec. This one event is reported to have a polarization opposite to the giant pulsations. Simultaneous with the magnetic field oscillations measurements of protons from 100 to 350 keV on the same spacecraft

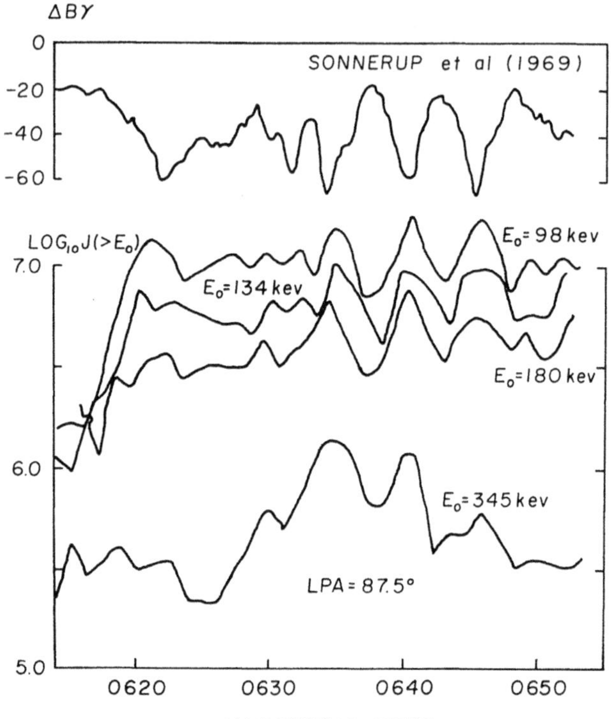

Fig. 4. Correlated particle and field measurements made by Sonnerup *et al.* (1969). The magnetic field amplitude is the depression below the Jensen–Cain reference field and the particle flux is measured in protons cm^{-2} sec^{-1} sr^{-1} at a local pitch angle of 87.5°.

reveal similar oscillations out of phase with the field oscillations (Figure 4). These oscillations were predominantly at large pitch angles. Sonnerup *et al.* interpret these oscillations as a particle population sloshing back and forth along field lines in the acoustic mode.

This measurement was made during a geomagnetic storm. Of the three examples presented by Judge and Coleman (1962) one also was during a geomagnetic storm. Similar to Sonnerup *et al.* they found that the particles, in this case electrons with $E > 200$ keV, oscillated out of phase with the waves. They also attributed these waves to the acoustic mode. However, during the two events at geomagnetically quiet times, the waves and particles oscillated in phase. It would appear thus that at these frequencies there are at least two separate phenomena which can occur.

The existence of these large amplitude long period waves seems to require a large energy source and since these waves appear at least 8% of the time from 0900 to 1200 LT during quiet conditions and are only slightly less common at other local times (Cummings *et al.*, 1969), they require a source of energy that is normally present in the magnetosphere.

Smith and Davis (1969) suggest that the 2 to 3γ waves with approximately 300 sec periods observed on Mariner 5 inside the magnetosphere but close to the magnetopause are waves produced by magnetopause motions. These waves propagating inwards could supply energy to the resonant modes of the magnetosphere. Cummings *et al.* report long durations for these waves which in turn could be the result of a large Q-value, i.e., a small loss of energy per unit cycle. Thus the input to the resonant modes need not be large to maintain the observed large amplitudes. The appearance of waves at ATS would occur when the input spectrum of waves produced by the boundary motions matched a resonant frequency at the ATS orbit.

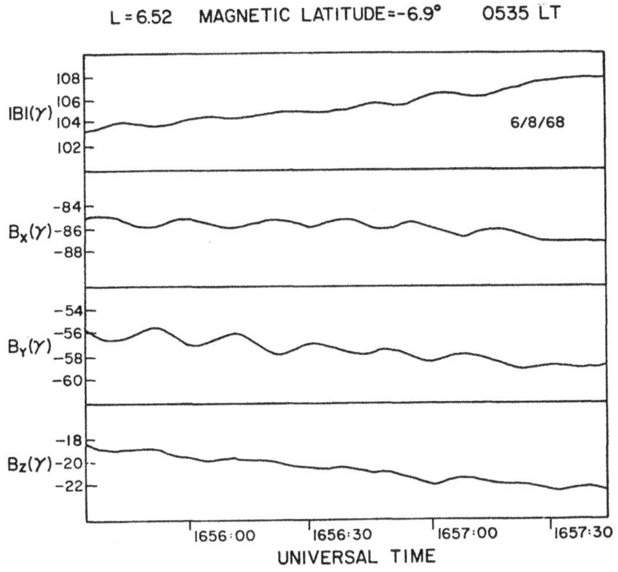

Fig. 5. Three components of the magnetic field in the spacecraft coordinate system plus the field magnitude measured by the UCLA OGO 5 fluxgate magnetometer during a magnetospheric substorm.

Waves at slightly higher frequencies are also present in the magnetosphere and are presently being studied with the UCLA magnetometers on OGO 5 and ATS 1. Figure 5 shows a 20 sec period wave with a 1γ amplitude occurring during a substorm at $L=6.52$ as measured by OGO 5 (McPherron, 1969). This oscillation is highly elliptically polarized with the major axis of the perturbation ellipse perpendicular to the main field and in the azimuthal direction. The normal to the perturbation ellipse is nearly radial and the perturbation magnetic field is left handed with respect to the main field. These characteristics identify the wave as an Alfvén wave propagating almost radially outwards or inwards nearly normal to the main magnetic field.

Figure 6 shows waves at even higher frequencies occurring during a mainly com-

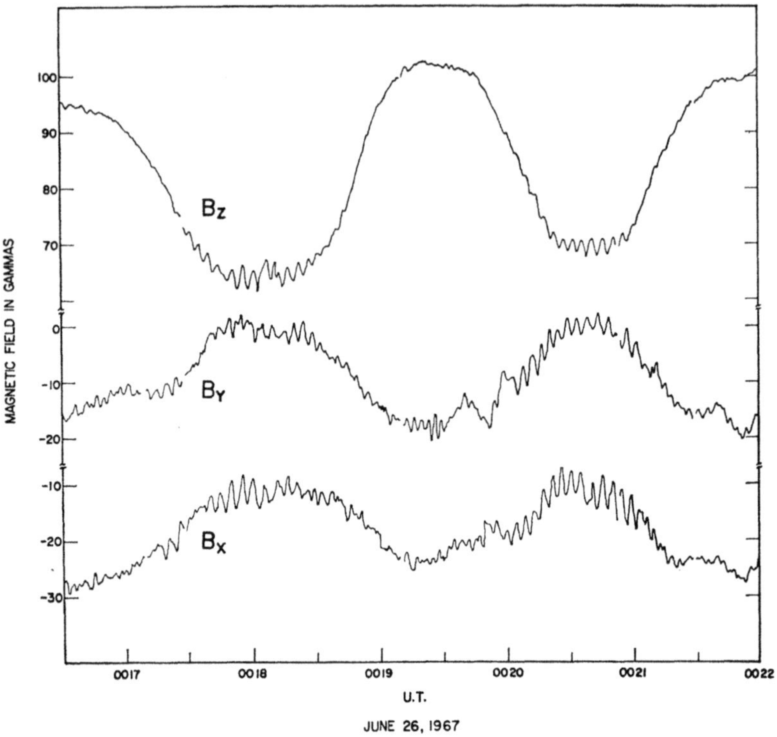

Fig. 6. Three components of the magnetic field in the geocentric solar magnetic coordinate system measured by the UCLA ATS 1 fluxgate magnetometer, showing waves with periods near 5 sec occurring in conjunction with a long period compressional wave.

pressional wave with a period of 150 sec as observed by the ATS 1 satellite (Barfield, 1969). These 5 sec period waves have greatest amplitudes in the troughs of the compressional waves, but high frequency waves exist even at the peak of the compressional wave. In addition to the decrease in amplitude at the peak, the period decreases and the relative amplitudes also change indicating a change in the source of the waves. What may be occurring here is a modulation of the growth rate of waves as the pitch angle distribution of resonant particles is modulated by the compressional wave.

Studies of both the OGO 5 and ATS 1 data are continuing.

4. The Intermediate Frequencies

In the band between Pc1 and the frequencies at which chorus, hiss, and whistlers are observed only the Schumann resonances are observed on the ground. The Schumann resonances are not observed in the magnetosphere. However, there are waves in the outer plasmasphere at these frequencies. They are linearly polarized waves with their magnetic perturbation aligned with the field. This in turn implies that they are propagating perpendicular to the field. These waves are confined within about 2° of the equator and therefore have a field aligned gradient in amplitude. They also exist only

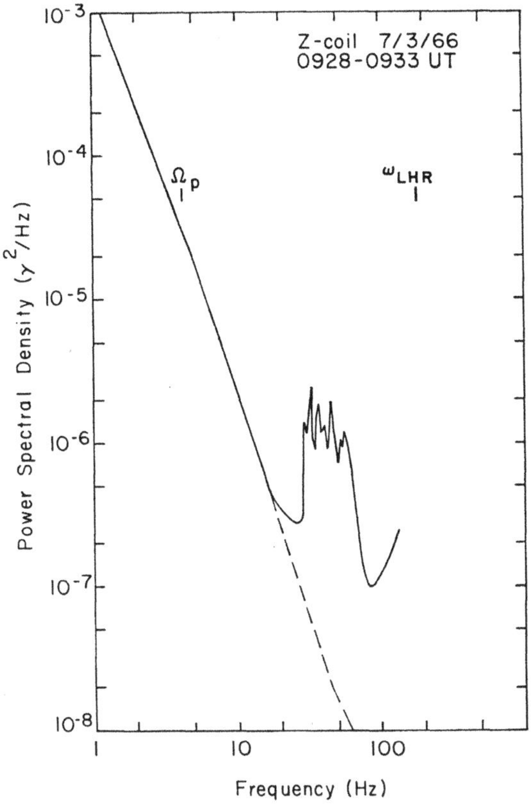

Fig. 7. Power spectrum of waves occurring at the equator in the outer plasmasphere measured by the UCLA OGO 3 search coil magnetometer. Ω_p is its local proton gyrofrequency and ω_{LHR} is the local lower hybrid resonance frequency.

between about twice the proton gyrofrequency and half the lower hybrid frequency. Figure 7 shows a power spectrum of these waves.

Waves with these characteristics can cause violation of the second adiabatic invariant for energetic electrons since the field aligned waves will resonate with harmonisc of the bounce frequency acting as a periodic force on the magnetic moment of the electrons.

These waves do not exist above the sensitivity of the OGO search coils in the inner plasmasphere nor outside the plasmasphere. However, in the outer 2 R_E of the plasmasphere their typical amplitude of 10mγ is well above the instrument's threshold. It

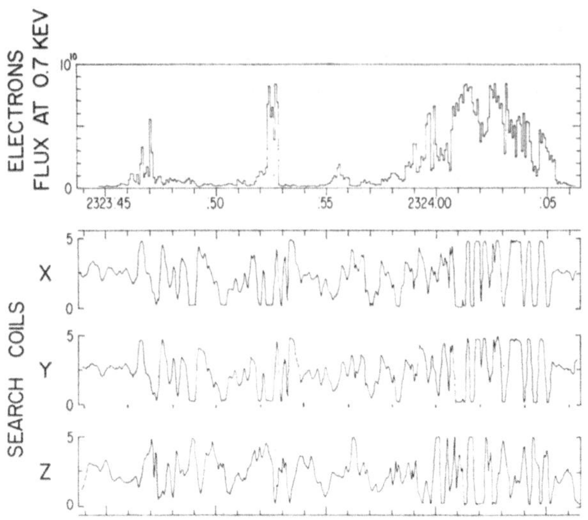

Fig. 8. The correlation between the electron flux greater than 0.7 keV as measured by R. A. Hoffman of GSFC onboard the OGO 4 satellite with the simultaneously measured field derivative measured by the UCLA search coil magnetometer on the same spacecraft. The flux is measured in electrons cm^{-2} sec^{-1} keV^{-1} sr^{-1} and the magnetic field derivative is in arbitrary units. The field fluctuations here are greater than 50γ and are transverse to the field in good agreement with the observed field aligned currents.

should be noted that it has not been established that these waves occur at all local times but they have been observed from late afternoon until midnight. It should be remarked also that they provide a pitch angle scattering mechanism that is not present (in exactly this way) in other regions of the magnetosphere.

Because there is only one report of waves at these frequencies in the magnetosphere does not imply that other waves do not exist. Experimentalists tend to look at the largest signals in their data first and these are the largest waves in this frequency band in the magnetosphere.

Transverse fluctuations of the magnetic field have been detected by polar orbiters in the auroral zone (Zmuda *et al.*, 1966; Burton *et al.*, 1969). The period of these fluctuations range from about 1 to 10 sec and have amplitudes up to 300γ. These fluctuations are not due to propagating waves but are caused by local field aligned currents. The apparent fluctuations are caused by the temporal and spatial variations of these currents. Figure 8 shows measurement of the field derivative by the OGO 4 search coil and the simultaneous measurements of electron fluxes $>$0.7 keV (Burton and Hoffman, 1969). The correlation between these two measurements however is complicated by the fact that the search coil measures both the temporal and spatial gradient of the field created by all nearby currents whereas the particle detector measures the temporal and spatial variations of the currents only at the satellite.

5. The High Frequencies

Measurements in the highest three decades of Figure 1 have been made principally by two techniques. The first method is to frequency modulate a carrier with the wave form of the observed signals. This has been done on the polar orbiters OGO 2 and 4 and the eccentric orbiters OGO 1 and 3 up to 12.5 kHz (Burtis and Helliwell, 1969), and on the polar orbiters Injun 3 and 5 up to 9 kHz (Gurnett and O'Brien, 1964; Gurnett *et al.*, 1969). These experiments, however, measured only one component of the magnetic fluctuation. The eccentric orbiter OGO 5 search coil experiment (Smith *et al.*, 1969) measured all three components but only up to 1000 Hz.

The second technique measures the power in different frequency bands and transmits this information to the earth. This, however, destroys all phase information in the original signal and inherently contains less frequency resolution than the FM technique because the transmission rate is governed by the relatively slow digital telemetry system of the spacecraft. All the above experiments had some such system along with their FM measurements.

Besides measuring the three components another important difference between the OGO search coil magnetometer and the OGO and Injun 'loops' is the use of an AGC circuit by the 'loop' experiments while the search coil has a completely linear system. The noticeable effect of such a nonlinear system is the variation in the 'darkness' of the noise background of the frequency time diagrams obtained by these groups, while the signal strength of the predominant noise remains the same 'darkness'. Changes in signal strength of the noise observed by the OGO 5 search coil on the other hand appear as changes in the 'darkness' of the signal and not the background in such a display.

Whistlers have been detected by all the ELF and VLF experiments. The analysis of whistlers has become quite developed and different manifestations of the whistler have received varied names. On the ground only two special names arose. 'Nose' whistlers were those which had frequency components above the frequency of minimum time delay. 'Knee' whistlers were a set of nose whistlers traveling in different field aligned ducts of enhanced ionization which revealed a density discontinuity or 'knee' in the magnetospheric plasma. Whistlers were also observed to echo in these ducts. The number of hops could be determined and thus whistlers were 1-hop, 2-hop, etc.

In space whistlers, called magnetospherically reflected whistlers, were observed propagating outside of ducts and furthermore were observed before and after crossing the equator. In analogy to the 'hop' terminology they were classified as N_+ whistlers if they had made N crossings of the equator and were returning again and N_- whistlers if they had crossed the equator N times and were proceeding away from the equator. Polar orbiters also measured 'subprotonospheric' whistlers which echoed back and forth between the ionosphere and a reflection level near 1000 km and both 'proton' and 'helium' whistlers which resonate with the proton and helium gyrofrequency, respectively. These depend on the presence of a plasma constituent with higher mass number. Furthermore, ion cutoff whistlers were observed in which the downgoing whistler is reflected locally and just below the satellite at the left handed cutoff frequency. A more detailed review of these observations is given by Helliwell (1969).

Whistlers, however, are not the most common signals observed by the ELF and VLF experiments in space. Hiss and chorus are observed much more frequently and in some regions of the magnetosphere occur almost continually. Hiss is a broadband signal with almost no discrete structure whereas chorus usually occurs in a narrower band and has discrete quasi-periodic structure. Figure 9 shows an example of

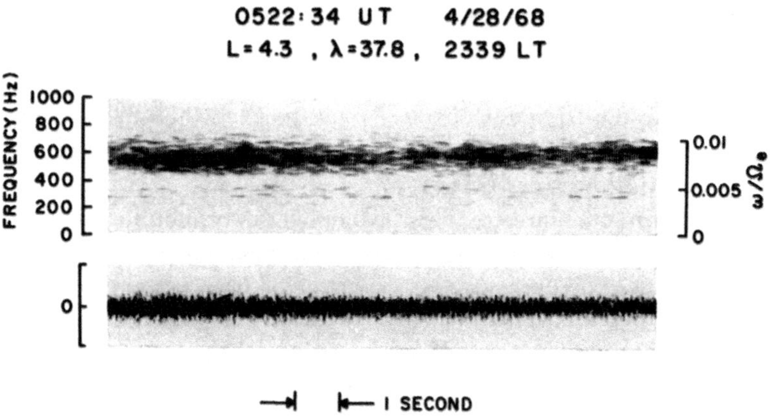

Fig. 9. Plasmaspheric hiss as observed by the UCLA OGO 5 search coil magnetometer. The upper half of the Figure is a frequency time display in which the darkness is proportional to the intensity. The left hand scale is frequency in hertz and the right hand scale is frequency measured in terms of the local electron gyro frequency. The lower half of the figure shows the waveform from which the frequency time display was derived. The vertical lines occurring 20 times per sec in the f-t display originate in the analysis.

hiss inside the plasmasphere at $L=4.3$. Figure 10 shows an example of chorus at $L=8.7$.

Both polar orbiters and eccentric orbiters can and do observe such signals. However, measurements made near the equator are most relevant to those studying electron precipitation because it is here that the pitch angle scattering of electrons occurs as a result of these waves (Kennel and Petschek, 1966).

Polar orbiters do not provide the information required because the amplitudes at the feet of field lines are not the same as at the equator. As waves propagate down field lines they undergo a competition between damping and an amplification caused by field line convergence and increased index of refraction (Russell *et al.*, 1969a). Furthermore, ELF waves can be reflected at the lower hybrid resonance frequency (Thorne and Kennel, 1967) and the left handed cutoff frequency (Gurnett and Burns, 1968). Thus, the polar orbiters do not measure the full bandwidth of the signals present at the equator. For instance, Figure 11 shows a strong band from 200 to 250 Hz whereas the lower cutoff observed by Gurnett and Burns (1968) is typically 400 Hz at 1000 km.

Fortunately solar-lunar perturbations change the inclination of eccentric orbiters

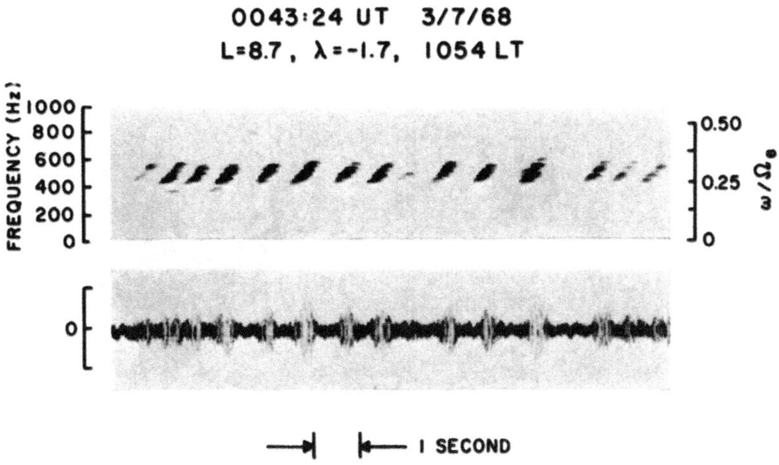

Fig. 10. An example of chorus selected because of its regular well spaced elements. The comments of Figure 9 apply here.

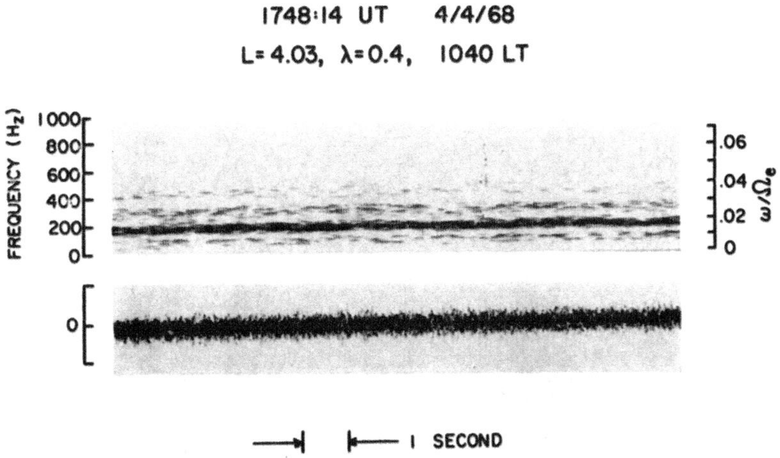

Fig. 11. An intense band of hiss near 200 Hz in the plasmasphere occurring in the middle of a weaker but broader hiss band. The absence of noise near 400 Hz is caused by a sharp notch filter in the instrument. The comments of Figure 9 apply here.

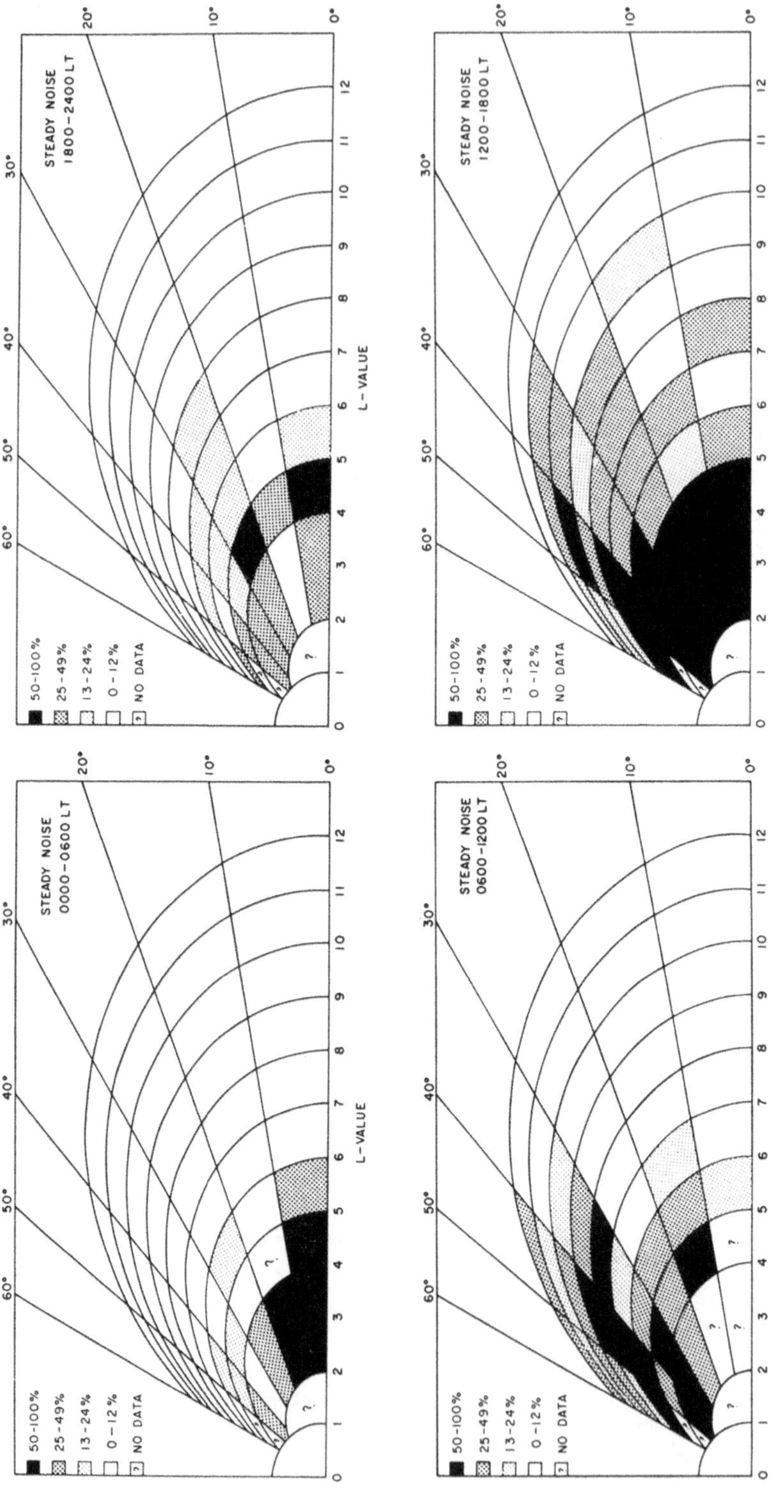

Fig. 12. The distribution of steady noise in the frequency range from 100–1000 Hz with amplitudes greater than about 3 mγ in the magnetosphere as determined by the OGO 3 search coil magnetometer in 4 LT quadrants. The *L*-values and magnetic latitudes are derived from a multipole expansion of the earth's main field.

slowly so that given a sufficiently long period of observation they can map out almost the entire magnetosphere. Figure 12 shows the distribution of hiss in the frequency range from 100 to 1000 Hz as determined from the OGO 3 spectrum analyzers for signals above about 3mγ. Almost two complete years of data have been used. The data have been divided into 4 LT quadrants. Hiss is seen to occur within the plasmasphere at all local times. It is most common over the sunlit hemisphere and especially in the afternoon. Within the plasmasphere the noise is not strongly latitude dependent except in the pre-dawn quadrant. Outside the plasmasphere on the dayside of the earth hiss exists at high L values at latitudes near 45°. This noise appears to be a different phenomenon than the plasmaspheric hiss.

Even with these distributions it is difficult to unambiguously identify the source of the hiss. In the absence of damping we expect the hiss to be amplified by field line convergence and the increase in the index of refraction as the energy proceeds away from the equator. Since the distribution is either fairly constant with magnetic latitude inside the plasmasphere or peaked near the equator it argues for near equatorial generation. Furthermore, Mosier and Gurnett (1969) have shown ELF hiss to be down going except near the left hand cutoff frequency at low polar altitudes. We feel, thus, that the plasmaspheric hiss is generated somewhere near the equator.

The occurrence rate of the daytime high latitude, high L value hiss outside the plasmapause decreases both to higher and lower magnetic latitudes. It appears thus that this noise is generated at high latitudes.

Figure 13 shows the occurrence of bursts which may be either whistlers or chorus in the same four quadrants. The whistler occurrence rate is known to decrease markedly outside the plasmasphere (Carpenter *et al.*, 1968) whereas chorus has not been observed within the plasmasphere.

The pre-midnight quadrant is very quiet. In the outer magnetosphere there is very little chorus and the noise that does occur is confined to the equator. The activity picks up in the pre-dawn quadrant but in the outer magnetosphere is still confined to the equator. If the analysis of Kennel and Thorne (1967) applies here, the confinement to the equator implies the presence of soft electron fluxes. This is in agreement with observations by Vasyliunas (1968).

In the morning quadrant the activity increases even more. The bursts are no longer confined solely to the equatorial region in the outer magnetosphere. A harder electron spectrum at these local times would result in such a broadening of the growth region.

In the afternoon quadrant the overall activity diminishes but shows little decrease in activity with increasing magnetic latitude until past about 30°. It appears from Figure 13 that the equatorial region is the source of chorus, since in the absence of damping the chorus amplitude and hence the rate of occurrence on our plots would increase as magnetic latitude increased. This is due to the convergence of the field lines plus the increase in the index of refraction and this increase is not observed.

There is further evidence for the source being at the equator. First, Burtis and Helliwell (1969) have shown that the chorus tends to occur at some fraction of the equatorial

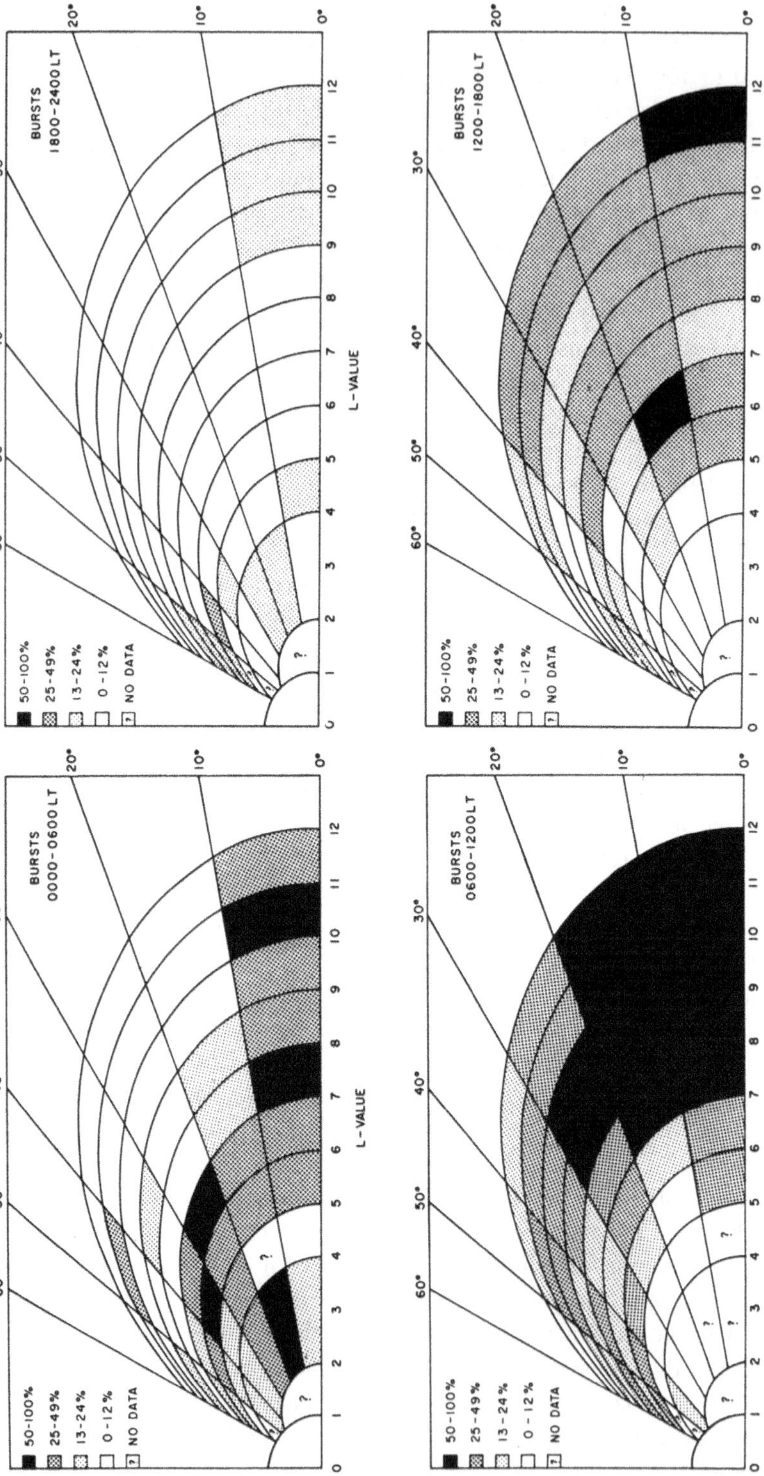

Fig. 13. The same distributions as Figure 12 except for bursts of noise rather than steady noise. Both chorus and whistlers are included as bursts in this analysis.

gyrofrequency rather than any local plasma frequency. Furthermore, Gurnett *et al.*'s (1969) Injun 5 data have shown at least one example of chorus to be down going at polar orbiter altitudes.

The only energy source available for the growth of these waves is the magnetospheric plasma. Kennel and Petschek (1966) have shown that the existence of a loss cone maintains an anisotropy in the pitch angle distribution which results in wave growth of whistler mode waves in cyclotron resonance with electrons. The observed frequencies are right for resonance with electrons of energy near 40 keV, but their analysis was not meant for a time varying instability as the chorus appears to be. However, their analysis does seem to apply to the ELF hiss in the plasmasphere.

In order to talk about this instability which obviously exists and which generates whistler mode signals because of the presence of the loss cone we have proposed the term whistler loss cone instability (Russell *et al.*, 1969a). The details of this wave particle interaction still remain to be resolved.

For those who do attempt to explain the details of this interaction, preliminary OGO 5 observations show that these chorus waves are nearly circularly polarized and elements can have amplitudes as high as 100mγ. Furthermore, the frequency of these waves measured at the equator in the outer magnetosphere can be much less than the half the electron gyrofrequency ratio that Burtis and Helliwell (1969) have measured just outside the plasmasphere. In fact they are sometimes less than one quarter of the gyrofrequency.

Further demonstration of the fact that the chorus is a result of an instability that causes pitch angle scattering is the fact that the region of 40 keV electron microbursts coincides with the region of observed chorus (Oliven and Gurnett, 1968). Furthermore, the duration and repetition rate of microbursts and the elements of chorus are similar. One difficulty in finding the chorus element responsible for an observed microburst results from the fact that the wave and the particles in this interaction travel in opposite directions.

6. Summary

Observations of waves in the magnetosphere reveal waves with periods of from hundreds of seconds to frequencies of several kilohertz. Many of these waves appear to be the result of local plasma instabilities near the equator and a few are the result of propagation up from the atmosphere. However different explanations appear likely for the lowest frequency and highest frequency waves. The lowest frequency waves appear to be field line resonances possibly driven by the motions of the magnetospheric boundary while the highest frequency waves, the auroral or VLF hiss, are probably Cherenkov radiation from precipitating electrons.

The waves cover the frequency range from the hydromagnetic regime to the whistler mode and some measurements have permitted positive identification of wave modes. Work is continuing in this field and since some of the most promising experiments are also the most recent new results should be soon forthcoming.

Acknowledgments

We gratefully acknowledge J. Barfield's permission to show in advance of publication the ATS data and Dr. R. L. McPherron's permission to show the OGO 5 results. The new research reported upon here was supported by National Aeronautics and Space Administration grants NGR 05-007-004 for ATS, NGR 05-007-235 for the OGO 3 and OGO 5 search coil measurements and NAS 5-9098 for the OGO 5 fluxgate magnetometer.

References

Barfield, J. N.: 1969, private communication.
Burtis, W. J. and Helliwell, R. A.: 1969, *J. Geophys. Res.* **74**, 3002.
Burton, R. K. and Hoffman, R. A.: 1969, private communication.
Burton, R. K., Holzer, R. E., and Smith, E. J.: 1969, *Trans. Amer. Geophys. Union* **50**, 292.
Cahill, L. J. and Amazeen, P. G.: 1963, *J. Geophys. Res.* **68**, 1835.
Carpenter, D. L., Walter, F., Barrington, R. E., and McEwen, D. J.: 1968, *J. Geophys. Res.* **73**, 2929.
Coleman, P. J., Jr., Sonett, C. P., Judge, D. L., and Smith, E. J.: 1960, *J. Geophys. Res.* **65**, 1856.
Cummings, W. D., O'Sullivan, R. J., and Coleman, P. J., Jr.: 1969, *J. Geophys. Res.* **74**, 778.
Gurnett, D. A. and O'Brien, B. J.: 1964, *J. Geophys. Res.* **69**, 65.
Gurnett, D. A. and Burns, T. B.: 1968, *J. Geophys. Res.* **73**, 7437.
Gurnett, D. A., Pfeiffer, G. W., Anderson, R. R., Mosier, S. R., and Cauffman, D. P.: 1969, University of Iowa Report 69-19.
Helliwell, R. A.: 1969, *Rev. Geophys.* **7**, 281.
Jacobs, J. A., Kato, Y., Matsushita, S. and Troitskaya, V. A.: 1964, *J. Geophys. Res.* **69**, 180.
Jørgensen, T. S.: 1968, *J. Geophys. Res.* **73**, 1055.
Judge, D. L. and Coleman, P. J., Jr.: 1962, *J. Geophys. Res.* **67**, 5071.
Kennel, C. F.: 1970, this volume, p. 257.
Kennel, C. F. and Petschek, H. E.: 1966, *J. Geophys. Res.* **71**, 1.
Kennel, C. F. and Thorne, R. M.: 1967, *J. Geophys. Res.* **72**, 871.
Madden, T. and Thompson, W.: 1965, *Rev. Geophys.* **3**, 211.
McPherron, R. L.: 1969, private communication.
Mosier, S. R. and Gurnett, D. A.: 1969, University of Iowa Report 69-27.
Nagata, T., Kokuban, S. and Iijima, T.: 1963, *J. Geophys. Res.* **68**, 4621.
Ness, N. F., Scearce, C. S. and Seek, J. B.: 1964, *J. Geophys. Res.* **69**, 3531.
Oliven, M. H. and Gurnett, D. A.: 1968, *J. Geophys. Res.* **73**, 2355.
Patel, V. L.: 1965, *Planet. Space Sci.* **13**, 485.
Russell, C. T., Holzer, R. E., and Smith, E. J.: 1969a, *J. Geophys. Res.* **74**, 755.
Russell, C. T., Holzer, R. E., and Smith, E. J.: 1969b, *J. Geophys. Res.*, in press.
Smith, E. J. and Davis, L., Jr.: 1969, *J. Geophys. Res.* **74**, in press.
Smith, E. J., Holzer, R. E., and Russell, C. T.: 1969, *J. Geophys. Res.* **74**, 3027.
Snare, R. C. and Benjamin, C. R.: 1966, *IEEE Trans. Nuc. Sci.* **NS-13**, 333.
Sonett, C. P., Sims, A. R., and Abrams, I. J.: 1962, *J. Geophys. Res.* **67**, 1191.
Sonnerup, B. U. Ö., Cahill, L. J., and Davis, L. R.: 1969, *J. Geophys. Res.* **74**, 2276.
Thorne, R. M. and Kennel, C. F.: 1967, *J. Geophys. Res.* **72**, 857.
Vasyliunas, V. M.: 1968, *J. Geophys. Res.* **73**, 7519.
Zmuda, A. J., Martin, J. H., and Heuring, F. T.: 1966, *J. Geophys. Res.* **71**, 5033.

ELECTRIC FIELDS IN THE IONOSPHERE AND MAGNETOSPHERE

G. HAERENDEL and R. LÜST

Max-Planck-Institut für Physik und Astrophysik, Institut für extraterrestrische Physik, Garching b. München, Germany

1. Introduction

First indications of magnetospheric convection were derived by relating the magnetic perturbations on the ground to an overhead ionospheric Hall current (Axford and Hines, 1961). The direct measurement of ionospheric drifts with the Ba ion cloud technique (Haerendel *et al.*, 1967; Föppl *et al.*, 1968; Haerendel and Lüst, 1968; Wescott *et al.*, 1969) and of the electrostatic field by probes flown with rockets and satellites (Mozer and Bruston, 1967; Aggson, 1969; Heppner, 1969; Gurnett, 1970) and balloons (Mozer and Serlin, 1969) was a step forward.

The extrapolation of the observed ionospheric motions or fields to the magnetosphere usually involves the assumptions of infinite conductivity and of the magnetic field remaining undistorted under the convective motions of the magnetospheric plasma. Consequently, magnetic lines of force are not only regarded as frozen in, but also as equipotential lines of the electrostatic field with the drift motions being directed along equipotential surfaces. The equation relating the drift motion $\mathbf{v}_\perp$, transverse to $\mathbf{B}$ and the transverse component of the electrostatic field, $\mathbf{E}_\perp$, is assumed to hold approximately everywhere above the E layer of the ionosphere:

$$\mathbf{E} + \mathbf{v} \times \mathbf{B} = 0. \tag{1}$$

Given a stationary model of the magnetic field the mapping of ionospheric motions into the magnetosphere is a straightforward procedure under these assumptions. Temporal changes of the convective motions are normally regarded as being quasi-static. The time of propagation through the magnetosphere has to be short compared with the time scales involved for this assumption to be applicable. Information on changes in the plasma motion is propagated with Alfvén velocity along magnetic field lines. Therefore, the travel time of an Alfvén wave between conjugate points of typically a few minutes sets the minimum time scale for changes to be interpreted in the quasi-static approximation.

Deviations from this simple concept arise where the electric conductivity can no longer be regarded as being infinite. There can be two reasons for that. In the lower ionosphere the finite transverse Pedersen conductivity tends to short out transverse electric fields of scales less than a few kilometers. The occurrence of an anomalous resistivity far above the ionosphere in regions of low plasma density in the presence of strong magnetic field aligned currents would be more serious, as for instance predicted by Swift (1965).

B. M. McCormac (ed.), Particles and Fields in the Magnetosphere.

In the first part of this report a brief synopsis of ion cloud experiments in the auroral zone from the point of view of magnetospheric convection is given. In interpreting the results the above discussed limitations of the simple mapping procedures should be kept in mind. In the second part very preliminary results of the first ion cloud experiment in the distant magnetosphere will be discussed.

2. Electric Fields in the Auroral Zone

A. SURVEY OF THE EXPERIMENTS

A survey of the drift motions of 17 Ba ion clouds in the auroral zone will be presented. The experiments were carried out in rocket launchings from Kiruna, Andenes, and Ft. Churchill by the Institut für extraterrestrische Physik and Goddard Space Flight Center (Table I). A W in front of the experiment number indicates experiments by

TABLE I

List of Ba ion cloud experiments

No.	Date	Launching range	Time of cloud appearance (UT)	Altitude (km)	K_p
22.1	Aug. 17, 66	Fort Churchill	3:44:42	258	10
22.2	Aug. 17, 66	Fort Churchill	3:46:26	376–230	
26	Apr. 7, 67	Kiruna	20:01	231	2 +
27	Apr. 8, 67	Kiruna	20:07	237	1 +
28	Apr. 9, 67	Kiruna	01:04	243	30
29	Apr. 10, 67	Kiruna	20:17:59	230	2 +
30	Apr. 11, 67	Kiruna	20:25	237	1 +
W1.1	Aug. 31, 67	Andenes	22:08	200–300	2 +
W2.1	Sept. 2, 67	Andenes	20:33	200–300	2 −
W3.1	Sept. 12, 67	Andenes	20:59	200–300	2 −
W3.3	Sept. 12, 67	Andenes	21:02	200–300	
33	Oct. 23, 67	Kiruna	16:08:52	217	1 −
34	March 20, 68	Kiruna	18:26:52	215	20
35	March 23, 68	Kiruna	18:38:53	200–210	30
42	March 15, 69	Kiruna	17:58:48	211	30
43	March 17, 69	Kiruna	18:13:51	255	4 +
45	April 17, 69	Andenes	21:55:29	250	3 +

W = Exp. of Wescott, Stolarik, and Heppner, 1969.

Wescott *et al.* (1969). In several flights two or more clouds were generated. Wherever their paths were rather similar only one has been chosen for presentation in this context.

A wide variety of geomagnetic conditions were met; K_p ranges between 1 − and 4 +. Some of the clouds were right in the middle of a region of auroral activity, sometimes even along the same field lines with auroral arcs; some of them were north and some south of the auroral oval. In one case (exp. 30) the cloud was placed at field lines confined just inside of the plasmapause. In Figure 1 the *X*-component of the magnetic

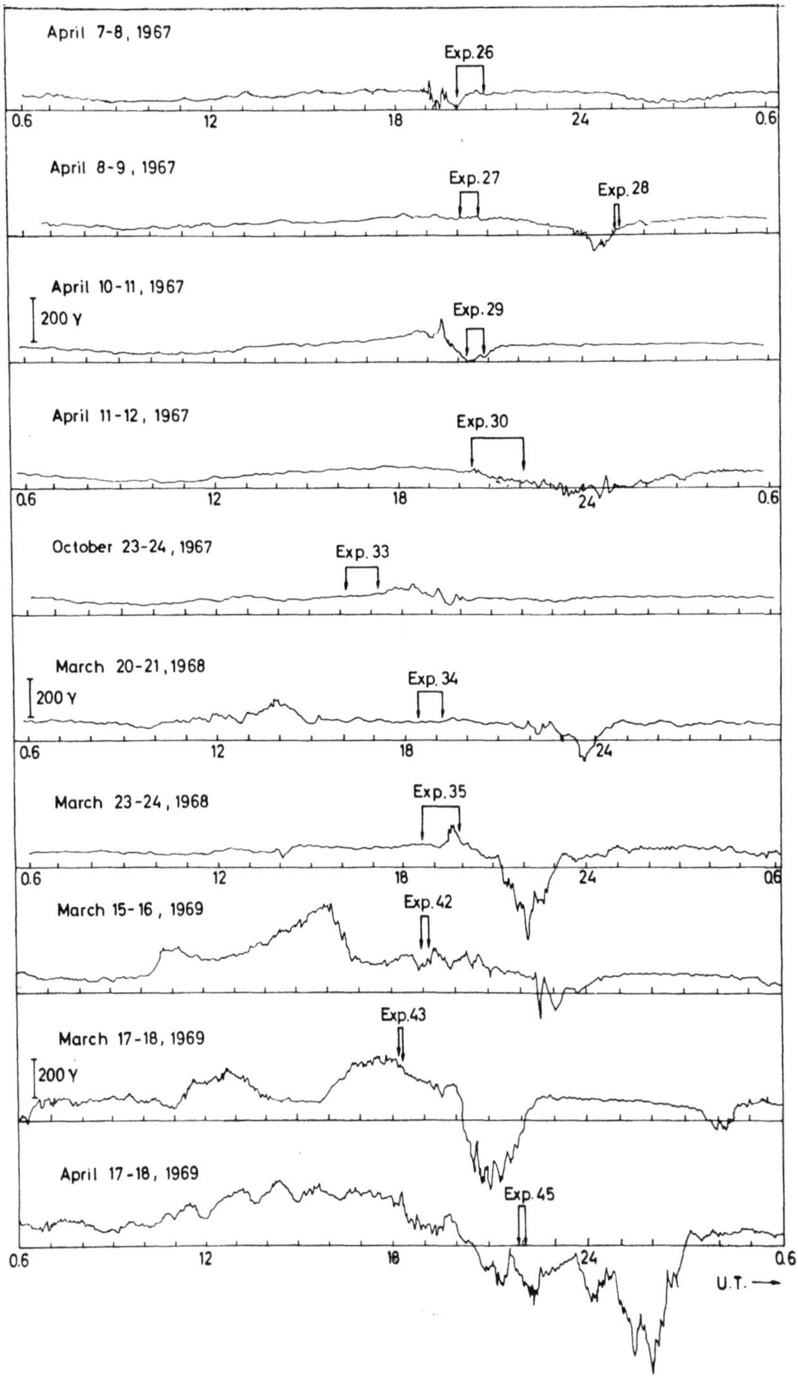

Fig. 1. X-component of the magnetic field in Kiruna during the days of the Ba ion cloud experiments.

field in Kiruna is shown for the days of our experiments in Kiruna and Andenes. During exp. 22 in Fort Churchill the magnetic field was very quiet $\Delta H=20$ to 30γ. The magnetograms of exp. W1, 2, 3 can be found in Wescott *et al.* (1969).

Although all these experiments were carried out during twilight, the variation in latitude and season results in a coverage of a rather wide range in local time. At present it is not known whether the twilight conditions introduce any systematic effects that do not apply to the same local times in a different season.

B. MAGNETIC PERTURBATIONS AND DRIFT MOTIONS OF ION CLOUDS

Although only a rather limited number of experiments has been carried out so far, we shall try to extract some systematic features from these seventeen cases. As a cloud can change or even reverse its sense of motion, the sums in the following tables need not always be seventeen.

One of the most prominent features is the close correlation of positive perturbations of the horizontal magnetic field with westward directed drifts and of negative perturbations with eastward directed drifts (Table II). This correlation is known to exist also

TABLE II

Magnetic perturbation	Drift in rotating frame		Drift in stationary frame	
H	West	East	Dusk	Dawn
positive	10	2[a]	8	5
negative	1[b]	6	0	7

[a] One case (W3) is irregular, another (34) weak.
[b] Inside the plasmasphere (36).

for auroral motions (Harang and Tröim, 1961) and is expected on the basis of the significance of the Hall current for the magnetic perturbation fields. In one of the exceptional cases (W3.3) where an eastward motion was observed during positive ΔX, the electric field, as judged from the motion of three simultaneous clouds, was rather irregular (Wescott *et al.*, 1969); in the other case (exp. 34) the drifts were slow and ΔX very small. A westward drift during a negative bay (exp. 30) was found for a cloud which was located at field lines just inside the plasmapause according to OGO 2 measurements (Carpenter, 1969).

Although strong electric fields tend to occur during magnetically disturbed periods there is no simple proportionality between the magnitudes of the electric field and the magnetic perturbation. This observation suggests that in the magnetic substorm the ionospheric conductivity is playing a key role (Haerendel *et al.*, 1969). If we switch from the aspect of an observer on the earth to a nonrotating frame of reference in the sense that we plot the trajectory of the clouds in ML and MLT, we get Figure 2. Consecutive dots are separated by 10 min, open circles only by 5 min. By subtracting the corotation some of the rather slowly westward drifting clouds reverse their di-

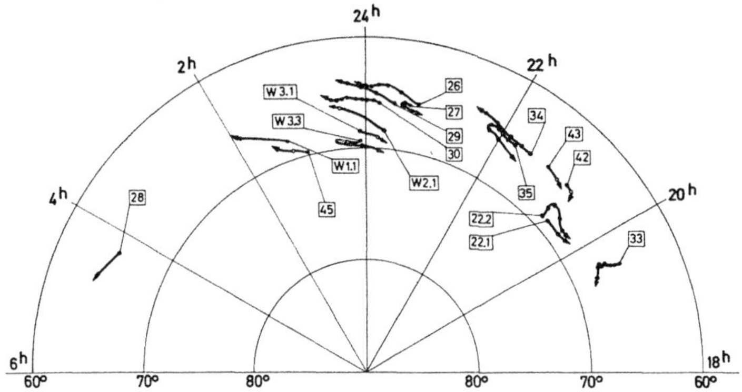

Fig. 2. Trajectory of the ion clouds in ML and MLT in a nonrotating frame of reference.

rection of motion. This appears in the last two columns of Table II. By presenting it in this way a very pronounced correlation appears between velocity components directed towards dawn (eastward) and southward components and between those directed towards dusk (westward) and northward components (Table III). In exp. 33

TABLE III

Drift in stationary frame	Magnetic meridional drift		Local magnetic time	
	S	N	before 2300 hours	after
Dusk	0	7	5	3
Dawn	11	0	2[a]	9

[a] Both cases are nearly corotating.

no clear E–W component could be assigned, in 22.1 no clear N–S component. This figure demonstrates furthermore that the motions to the east occur mostly after 2300 MLT and the westward motion predominantly before 2300 MLT. The number of cases is given in Table III. A similar result was derived from magnetic perturbations by Heppner (1969).

C. MOTIONS IN THE MAGNETOSPHERE

Slight N–S motions can correspond to rather large radial motions in the magnetosphere. In order to give an impression of how the motion of the plasma might look like in the magnetosphere, the magnetic latitude and altitude of the clouds' positions for a given time have been converted into a corresponding L value by using the simple dipole formula (Figure 3). For $L<7$ this should give an approximate picture of the drifts in the magnetic equatorial plane. This method of projection, however, becomes rather questionable in the neighborhood of the tail, when field lines are strongly distorted and it is sometimes difficult to decide whether they are already tail field lines or not. Furthermore, the magnetic configuration can be subject to significant changes

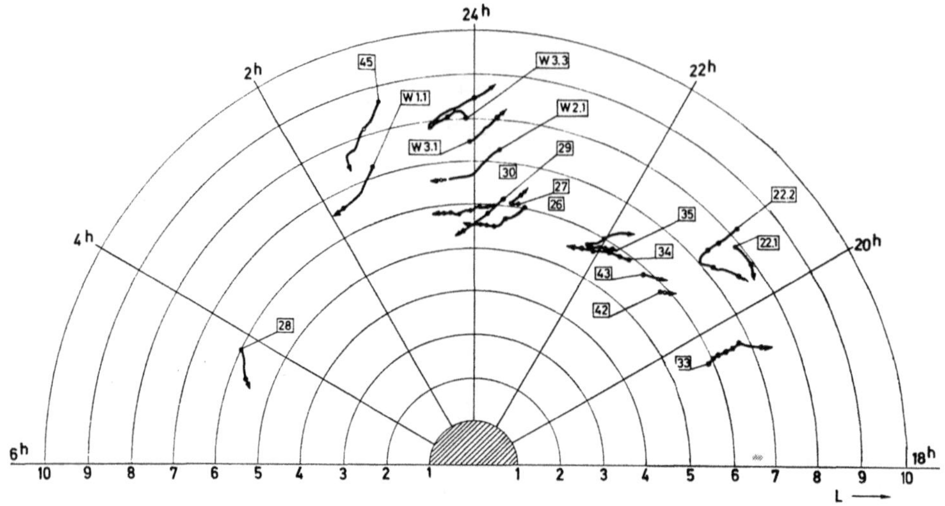

Fig. 3. Trajectory of the ion clouds L-coordinates and MLT in a nonrotating frame of reference.

during substorms. For this reason we refrained from applying a more realistic model of the earth's magnetic field. Although these observations originated in entirely independent events it is tempting to understand the main characteristics of the motions as collected in Tables II and III and Figures 2 and 3 as applicable to the average substorm. Positive and negative phases seem to be clearly distinguished by nearly opposite directions of the drifts. At the plasmasphere there seems to exist a sharp boundary in the convection pattern. So far the data are not sufficient to derive definite conclusions about the overall topology of auroral drifts and currents.

Our results on the general drift directions are in good qualitative agreement with electric field data from balloon measurements of Mozer and Serlin (1969).

The foregoing results and especially the observed antisymmetry between late evening and early morning hours in the N–S components of the drifts or E–W components of the electric fields might be related to other asymmetries observed. (1) One of the consequences is that particles with energies on the order of a few to a few tens of keV will be accelerated in the morning and decelerated in the evening sector. The predominance of X-ray and CNA events in the post-midnight hours and the general hardening of the primary electrons with increasing local time may result there from. (2) As the auroral oval is proceeding to northern latitudes after midnight, the observed motions in this sector have an appreciable component normal to the boundary of the oval. On the evening side this normal component seems to be less significant. According to Wescott *et al.* (1969) auroral arcs tend to be aligned along the direction of drift motions; this may be the reason for the rather short and patchy auroral forms found in the noon to morning quadrant as opposed to the long arcs typical for evening hours. (3) The diffuseness of the inner boundary of the plasma sheet on the morning side might be attributed to the relatively strong normal component of the drift velocity. (4) The typical recession of the plasmapause towards smaller L with increasing local time during night hours may also be related to the convection pattern indicated in Figures 2 and 3.

It is interesting to note that from the interpretation of changes in the electron fluxes at the geostationary orbit (Winckler, 1970) derives similar conclusions about the E–W component of the electric field before and after midnight.

3. Barium Ion Cloud Experiments in the Distant Magnetosphere

A. PURPOSE OF THE EXPERIMENT

A Ba ion cloud release experiment in the distant magnetosphere, as in the ionosphere, will provide a tracer for the movement of the ambient plasma. However, the physical conditions in the magnetosphere are so different from those in the ionosphere that the experiment will have a number of additional features which distinguish it strongly from an ionospheric experiment. One of the main differences is that collisions between the particles are not of importance at all and that the artificial plasma density is much higher than the ambient plasma density for an appreciable period. The result is a prolonged time scale necessary for destroying any initial difference in the velocities between the natural and artificial plasmas. Consequently, the emphasis in such an experiment will lie to a considerable extent on how the ambient plasma velocity is actually imparted to the cloud. This aspect is closely related to the physics of comet tails, which was originally the starting point for the Ba cloud experiments.

Secondly, there are the aspects introduced by the rather low magnetic field strength. During the initial phase of the expansion the kinetic pressure of the cloud plasma exceeds the magnetic pressure; therefore, a magnetic cavity is formed and later collapses. The initial shape of the Ba cloud will be determined by this phase and the surrounding magnetic field will be affected for a considerable period. Instabilities may give rise to certain structures.

B. PARAMETERS OF THE BARIUM ION CLOUD

On March 18, 1969, at 07.20.52 UT a Ba cloud was released from a capsule carried by the ESRO satellite HEOS I, which was launched from Cape Kennedy on December 5, 1968, into an eccentric earth orbit. The time of release was determined by the satellite's orbit and the conditions on the observability of the cloud from the main observation sites. The coordinates of the release are: distance = 12.5 R_E, geogr. lat. = 31°.5N, geogr. long. = 47°.5W, magnetic lat. = 42°, and MLT = 0430. The capsule filled with 2.4 kg of a mixture of Ba and CuO was ejected from the satellite on ground command about 10^4 sec prior to release, which occurred at a distance of slightly more than 30 km from the satellite. The cloud could be observed for 25 min from the observation sites of the Institut für extraterrestrische Physik located at Kitt Peak, Arizona, and La Silla, Chile. Several others participated in the observations using the 48″ Schmidt telescope at Mt. Palomar, television cameras at Fairbanks, Alaska, and Goddard Space Flight Center, Baker-Nunn Cameras on Mt. Hopkins (Arizona), Arequipa (Peru), and Natal (Brazil), and other instruments.

Fig. 4. Ion cloud at 12 R_E distance from the surface of the earth (photographed with the 48″ Schmidt telescope on Mt. Palomar by Braccessi and Münch).

Figure 4 shows the ion cloud after 7 min as photographed by the 48″ Schmidt telescope. Adopting an efficiency of a few per cent for the production of Ba atoms the ion cloud consisted of about 100 g or 4.4×10^{23} Ba ions. At the end of the observation time the cloud had a length of about 5000 km and a width of about 100 km. The elongation of the cloud along magnetic field lines was used to determine the direction of the magnetic field. In agreement with magnetometer data from HEOS I (Hedgecock and Sear, 1969) the inclination of the field turned out to be 65° and the declination decreased from 55 to 45° during the observation period. This was most likely a temporal rather than a local change. The direction of the magnetic field is consistent with the model field of Fairfield (1968) which has been used in Figures 5 and 6 showing the cloud's position and the derived **B** vector in the magnetic meridian and equatorial planes. The magnetic field is pointing roughly towards the sun. Most likely the cloud was located above the plasma sheet at the beginning of the geomagnetic tail.

The cloud showed a sharp edge on one side and a short diffuse tail on the other indicating the interaction with the ambient plasma flow. Magnetic field aligned density modulations cause the appearance of a few striations.

The magnetic field strength B at the release point as measured on the satellite was 49γ corresponding to an energy density $E_m = B^2/8\pi = 10^{-8}$ ergs/cm^3. Estimates of the ambient plasma density n_a are rather uncertain. Assuming $n_a \approx 0.01$ cm^{-3} or $n_a \approx 1$ cm^{-3}

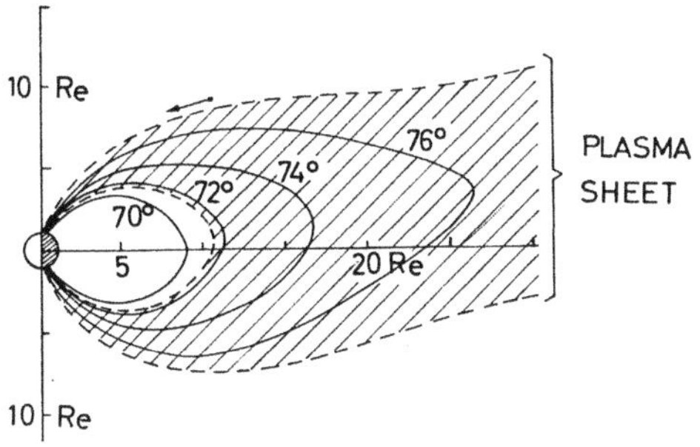

Fig. 5. Ion cloud position in the meridian plane with the model field of Fairfield (1968).

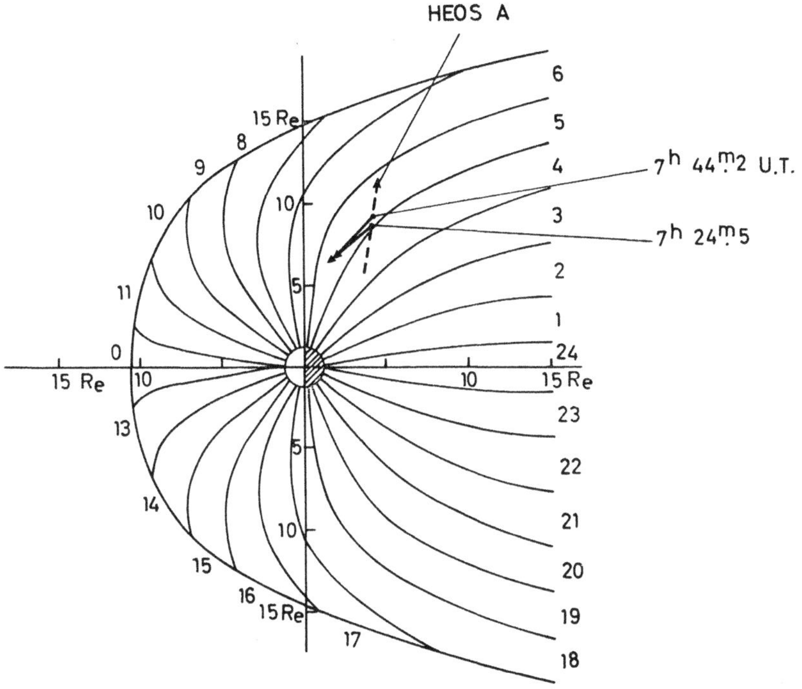

Fig. 6. Ion cloud position at two different times in the equatorial plane with the model field of Fairfield (1968). (The dashed arrow indicates the trajectory of the HEOS I satellite.)

one obtains Alfvén velocities $v_A \approx 1.1 \times 10^9$ cm/sec and $v_A \approx 1.1 \times 10^8$ cm/sec, respectively. The temperature of the ambient plasma might have been of the order of a few keV.

C. THE INITIAL EXPANSION OF THE ION CLOUD

During the first second the neutral Ba gas typically assumes a velocity distribution forming a nearly spherical shell with a mean expansion velocity of about 1.2 km/sec, corresponding to kinetic energy per ion of 1 eV$=1.6 \times 10^{-12}$ ergs. The Ba atoms are photoionized by solar UV radiation with a time scale of 20 sec. In this process the electrons gain an average thermal energy of the order of 0.5 eV corresponding to a temperature of about 5000 K, while the ions remain relatively cool with a temperature of about 1000 K.

The expansion of ion clouds in the magnetosphere has been investigated in detail by Pilipp (1969). Initially the kinetic energy density of the ions and the thermal energy density of the electrons is much larger than the energy density of the ambient magnetic field. Since the electrical conductivity is very high, the radially expanding plasma will carry the magnetic field with it forming a magnetic cavity in the interior of the cloud and deforming it somewhat at the outside. This effect was observed by the

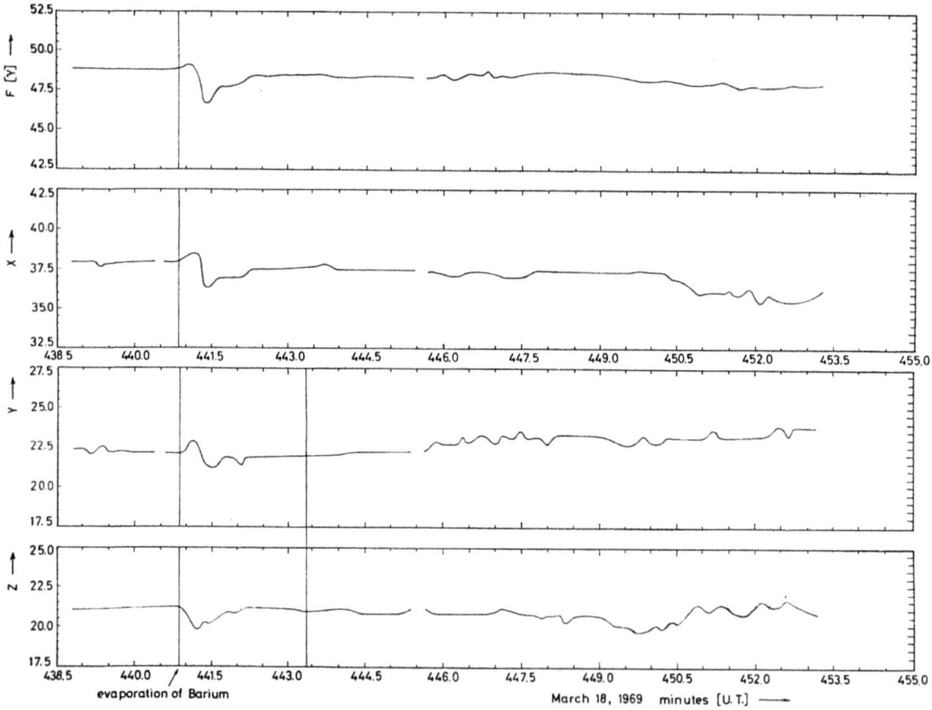

Fig. 7. The total field strength F and the X, Y, and Z components of the earth's magnetic field in 10^{-5} G as measured by the magnetometer onboard of the HEOS I satellite during the time of the ion cloud experiment (Hedgecock and Sear, 1969).

magnetometer onboard the satellite (Figure 7). The radial motion of the already ionized Ba atoms is stopped eventually when the ion and electron pressures are balanced by the magnetic pressure and stresses. This stage is reached after about 17 sec when the cloud has obtained a radius of about 20 km. Because of the high electron pressure the Ba plasma has the tendency to expand faster than the neutrals. However, when the initially created plasma is stopped by the magnetic field the neutrals will overshoot. As the photoionization of the remaining Ba atoms is still continuing, the central part of the cloud, which is strongly diamagnetic, will be surrounded by a region of quickly decreasing plasma density and increasing magnetic field strength. After 23 sec the neutrals apparently reached the satellite, since at this time the onset of a small negative perturbation is observed, which should be due to the Ba plasma created here (Figure 7).

The striations were created during the first minute after release before we were able to separate optically the neutral and ionized clouds. A possible origin of these striations may be a break-up of the initially created density distribution due to flute instabilities (Pilipp, 1969). A rough estimate of the time constant τ for this instability is given by $\tau \approx v_{th}/g$, where v_{th} is the thermal velocity of the ions and g the deceleration. We estimate g by v_{exp}^2/R_m, where v_{exp} is the initial velocity of radial expansion of the ions ($v_{exp} \approx 1.7$ km/sec) and R_m the maximum radius of the diamagnetic cloud. With these and the above given numbers we get $\tau \approx 2.5$ sec.

D. THE ACCELERATION OF THE BARIUM IONS

Initially the Ba plasma has a bulk velocity equal to that of the satellite, which at the instant of release was 2.3 km/sec with a strong radial component. The momentum exchange between ambient and injected plasma was studied by Scholer (1969) in a simple model. The dominant effect is the propagation of the electric field related to the motion of the injected plasma with Alfvén velocity along the undisturbed magnetic field. By moving in a different direction the artificial plasma carries magnetic flux with it, thereby distorting the field and forcing the background to participate in its motion. When the front of this distortion (traveling with Alfvén speed) has passed over a mass comparable to that of the injected plasma, i.e., after a time of the order of:

$$\tau = M/2Fv_A\varrho_a \qquad (2)$$

the momentum exchange is essentially accomplished (M = mass of injected plasma, F = cross section of magnetic flux tube occupied, and ϱ_a = ambient mass density). Taking the total mass of Ba atoms released and dividing by the cloud's cross section (mean radius 50 km) we get $\tau \approx 10$ hr for $n_a = 10^{-2}$ cm^{-3} and $\tau \approx 1$ hr for $n_a = 1$ cm^{-3}.

This time scale will be shortened by several effects. One of them is that the electric field inside the cloud is also acting on neighboring flux tubes. More important is the inhomogeneity of the magnetospheric plasma and the existence of the ionosphere. The increase in Alfvén velocity towards the earth will result in a faster growth of the mass that is coupled to the cloud. The ohmic dissipation in the ionosphere tends to slow down the cloud further. However it turns out that the electric field, which can

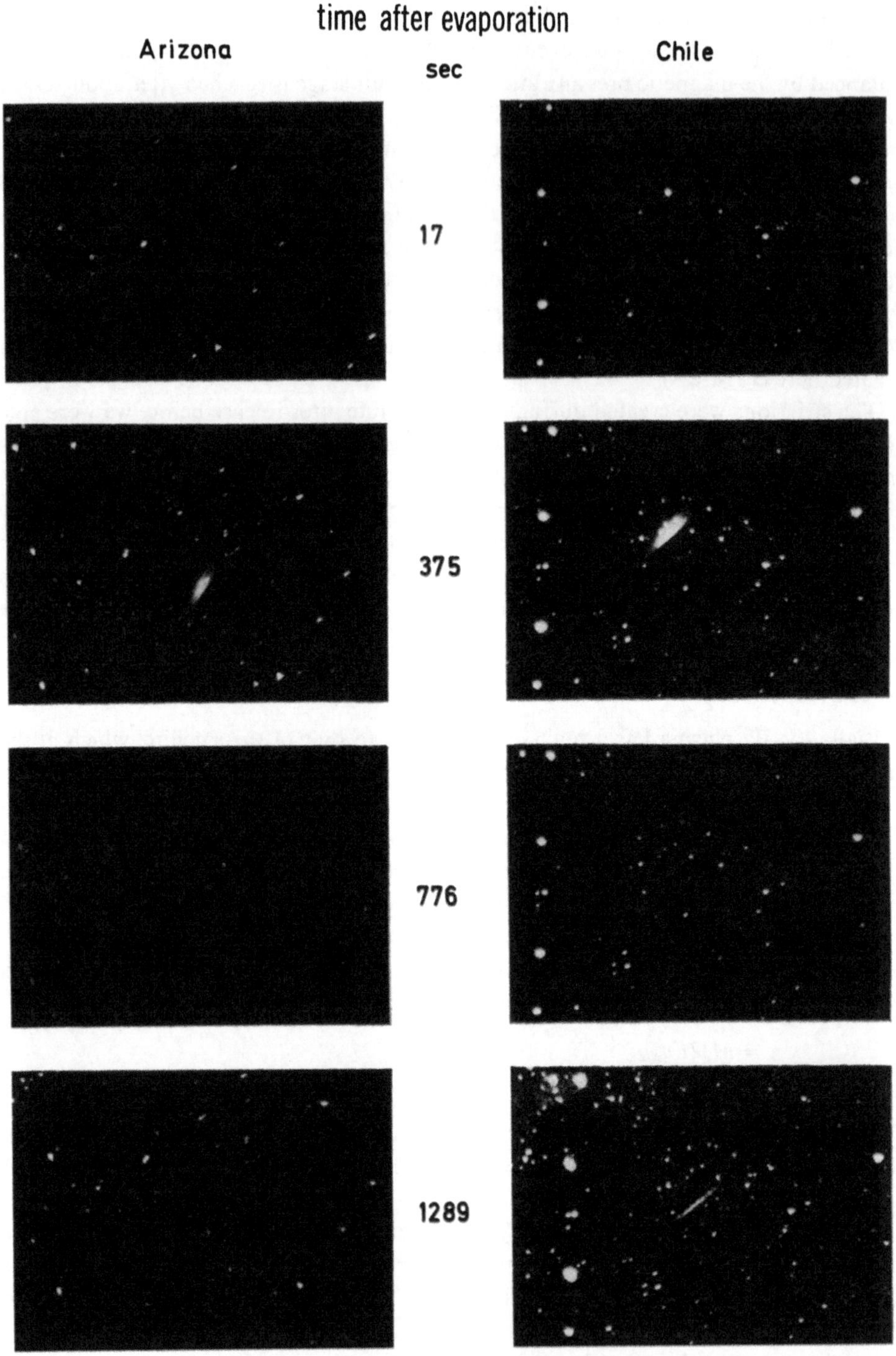

Fig. 8. Ion cloud at different times as observed from Kitt Peak, Arizona and La Silla, Chile.

be applied to the ionosphere, may be much smaller than the original field above the ionosphere because of the polarizing effect of the Pedersen current. Therefore, part of the electric field will be reflected. However, in several subsequent reflections between cloud and ionosphere the ionospheric electric field can build up. Altogether, the momentum exchange between cloud and magnetosphere is a very complex phenomenon. Certainly, however, Equation (2) gives an upper limit to the acceleration time scale for a non-diamagnetic cloud.

E. INTERPRETATION OF THE 'TAIL'

The most interesting feature of the Ba ion cloud is the existence of a sharp edge on one side and of a short diffuse tail on the other. This can be attributed to a strong variation of the acceleration time scales, decreasing from the cloud's interior to the outskirts. The dilute outer parts follow the motion of the ambient plasma fairly quickly while the inner parts lag behind. (This is much different from the behavior at low altitudes where the whole cloud is accelerated in a fraction of 1 sec.) The resulting dilution of Ba plasma leads to the appearance of a short tail close to the cloud's core where the final velocity is not reached yet. The history of the cloud is demonstrated by a sequence of photographs in Figure 8.

The spatial variation of the Ba ion density is only one reason for the resulting variation of time scales, the other being the reduction of the magnetic flux intersecting the cloud by the diamagnetic effect. The values of M/F used in Equation (2) for the inner core should be much larger than the mean values of the Ba cloud. Consequently, it appears quite understandable that until shortly before the end of the period of observation the core did not deviate noticeably from a trajectory close to that of the satellite. However, after about 1200 sec, near the end of the observational period, a deviation of the cloud's trajectory from the unperturbed path is indicated (Figure 9). For times t that are short as compared with the time scale of interaction τ this deviation Δs will be

$$\Delta s \cong \frac{t^2}{2\tau} v_0 \tag{3}$$

where v_0 is the relative velocity of the ambient plasma. The observed upper limit of 200 km for Δs after $t=20$ min leads to the following estimate for the total acceleration: $v_0/\tau < 0.59$ km sec^{-1}/hr. Using a time scale of 10 hr we get $v_0 < 5.9$ km sec^{-1}. However, much larger values of τ may be more realistic for the diamagnetic core of the cloud.

These considerations so far do not take into account the leakage of magnetic flux into the cloud. As the outer parts are carried away by the ambient plasma flow, the boundary layer between cloud and magnetospheric plasma will be steadily removed. Such an interaction may account for the rather sudden disappearance of the remaining thin core of the ion cloud.

The preliminary conclusion from this experiment is that unlike in the ionosphere the electric field cannot be deduced simply from the bulk motion of a plasma cloud

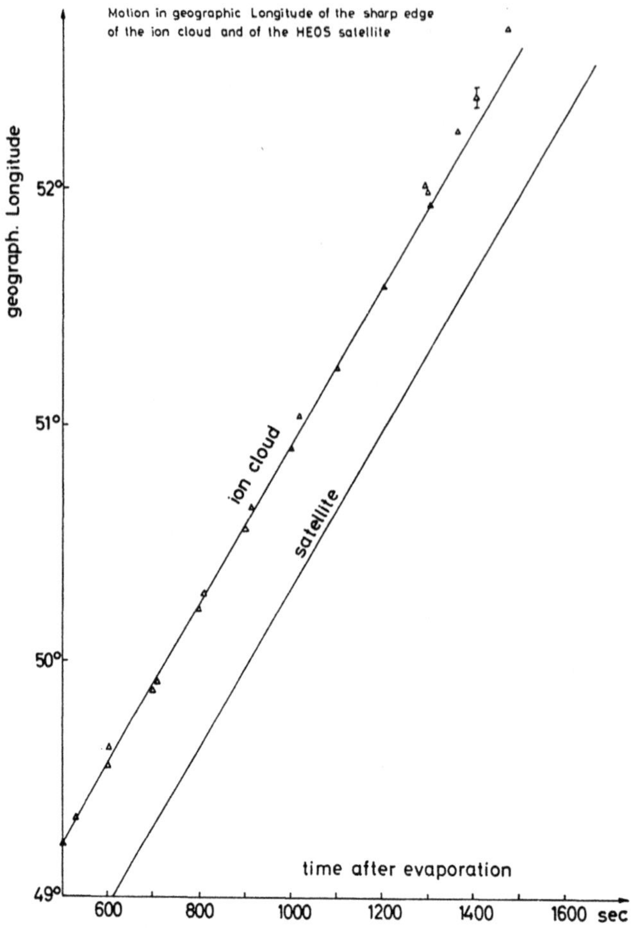

Fig. 9. The position (geographic longitude) of the ion cloud as a function of time.

in a weak magnetic field; one has to analyze carefully the acceleration of the cloud to the ambient plasma velocity. The main difference with comet tails lies in the ratios of drift velocities to Alfvén velocity which are much smaller than unity inside the magnetosphere. It is not clear yet to what extent the ambient plasma drift is expressed in observed motions of structural elements of the cloud.

The direction of the 'tail' could be determined during the separation of one of the striations from the central one, assuming that this motion was in the direction of the tail. The accuracy of this determination is not very high. The tail pointed roughly azimuthally towards midnight. Any radial component so far could not be definitely established. In the neighborhood of the core, we expect the tail to point into the direction of the plasma velocity vector as seen by an observer moving with the satellite. If v_0 is large compared with the satellite's velocity, the aberration angle is negligible and the tail indicates nearly the direction of plasma flow.

The experiment took place right in the middle of a well developed negative bay. In Figure 10 some of the magnetic perturbation vectors are indicated. The field lines intersecting the cloud meet the ionosphere close to Godhavn (Greenland). If the projection of the plasma motions between ionosphere and magnetosphere can be

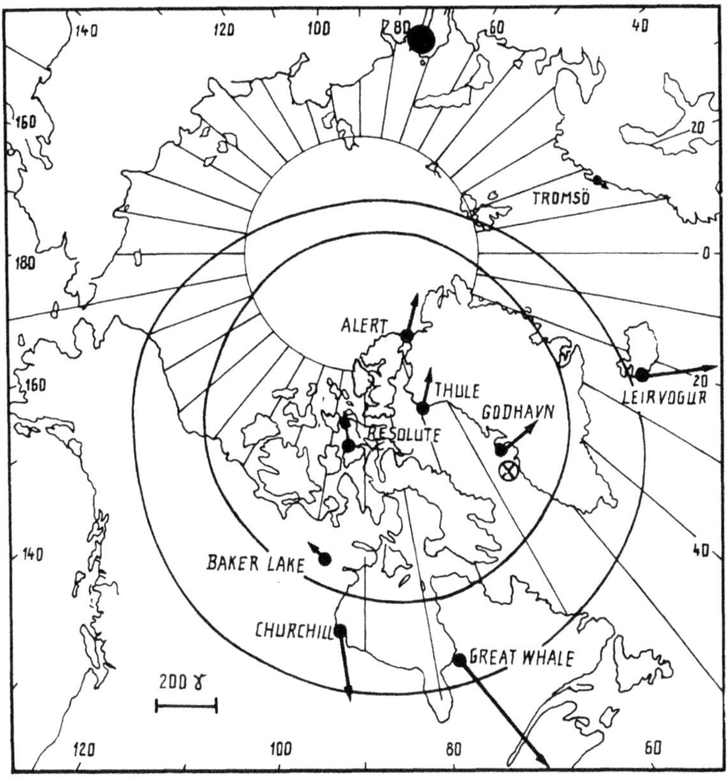

Fig. 10. Magnetic perturbation vectors near the foot point of the magnetic lines of force (cross) intersecting the ion cloud.

carried out as discussed in the introduction, we expect an ionospheric plasma drift directed towards magnetic west near Godhavn, possibly with a certain component to the north because we are not sure about the smallness of the aberration. In any case such an ionospheric drift would not be consistent with the magnetic perturbation in Godhavn being caused by an overhead horizontal ionospheric current, neither with overwhelming Hall nor Pedersen conductivity.

This finding is opposite to all experiences with ion clouds in the auroral zone (Section 2B). Wescott (1970) reports disagreements similar to this one found with Ba releases in the ionosphere north of the auroral oval. One of the possible interpretations of this result is that the magnetic perturbations on the polar cap are to a large extent determined by field aligned currents. If we assume as in the model of Akasofu (1970) that strong field aligned currents enter the auroral oval south of Godhavn, we can resolve the discrepancy. In fact, the distribution of magnetic perturbations (Figure 9) is strongly reminiscent of the fields from Akasofu's simple model.

The last conclusions were of course very tentative since there may be other ways out of the discrepancy. More measurements of the electric field on the polar cap are necessary in order to draw more definite conclusions. However, it is quite apparent that experiments like this one can help in establishing the convection pattern in the distant magnetosphere including the tail.

Acknowledgments

The HEOS I ion cloud experiment was possible only by the large amount of work of many coworkers of our Institute and by the close collaboration with many staff members of the European Space Research Organization (ESRO) and of the Junkers Flugzeug- und Motorenwerke GmbH. In particular we want to thank our collegues H. Föppl, H. Göllnitz, J. Loidl, F. Melzner, G. Metzner, B. Meyer, K.-W. Michel, H. Neuss, E. Rieger, J. Stöcker and H. Völk.

References

Aggson, T. L.: 1969, in *Atmospheric Emissions* (ed. by B. M. McCormac and A. Omholt), Van Nostrand Reinhold, Company, New York, p. 305.
Akasofu, S-I.: 1970, this volume, p. 34.
Axford, W. I. and Hines, C. O.: 1961, *Can. J. Phys.* **39**, 1433.
Carpenter, D. L.: 1969, private communication.
Fairfield, D. H.: 1968, *J. Geophys. Res.* **73**, 7329.
Föppl, H., Haerendel, G., Haser, L., Lüst, R., Melzner, F., Meyer, B., Neuss, H., Rabben, H., Rieger, E., Stöcker, J., and Stoffregen, W.: 1968, *J. Geophys. Res.* **73**, 21.
Gurnett, D.: 1970, this volume, p. 239.
Haerendel, G., Lüst, R., and Rieger, E.: 1967, *Planetary Space Sci.* **15**, 1.
Haerendel, G. and Lüst, R.: 1968, in *Earth's Particles and Fields* (ed. by B. M. McCormac), Reinhold Publishing Corporation, New York, p. 271.
Haerendel, G., Lüst, R., Rieger, E., and Völk, H.: 1969, in *Atmospheric Emissions* (ed. by B. M. McCormac and A. Omholt), Van Nostrand Reinhold Company, New York, p. 293.
Harang, L. and Tröim, J.: 1961, *Planetary Space Sci.* **5**, 33.
Hedgecock, P. C. and Sear, J.: 1969, to be published.
Heppner, J. P.: 1969, in *Atmospheric Emissions* (ed. by B. M. McCormac and A. Omholt), Van Nostrand Reinhold Company, New York, p. 243.
Mozer, F. S. and Bruston, P.: 1967, *J. Geophys. Res.* **72**, 1109.
Mozer, F. S. and Serlin, R.: 1969, *J. Geophys. Res.* **74**, 4739.
Pilipp, W.: 1969, Ph.D. Thesis, Technische Hochschule München, to be published.
Scholer, H.: 1969, Ph.D. Thesis, Technische Hochschule München, to be published.
Swift, D. W.: 1965, *J. Geophys. Res.* **70**, 3061.
Wescott, E. M.: 1970, this volume, p. 229.
Wescott, E. M., Stolarik, J. D., and Heppner, J. P.: 1969 *J. Geophys. Res.* **74**, 3469.
Winckler, J. R.: 1970, this volume, p. 332.

AURORAL AND POLAR CAP ELECTRIC FIELDS FROM BARIUM RELEASES

EUGENE M. WESCOTT*, JOHN D. STOLARIK, and JAMES P. HEPPNER
NASA Goddard Space Flight Center, Greenbelt, Md., U.S.A.

1. Introduction

High latitude magnetic disturbances and accompanying auroral displays are obvious evidence of the release of large amounts of energy, presumably from the magnetosphere and magnetotail regions. Until recently experimental efforts towards understanding these phenomena were directed toward measurement of magnetic fields, energetic particles, optical emissions, and the morphology of the observables. The importance of ionospheric and magnetospheric electric fields was recognized in theoretical studies, but actual measurements were not available due to the lack of suitable techniques. In the past several years valid electric field measurements have been made from sounding rockets and satellites with long antennas (Aggson, 1969; Maynard and Heppner, 1970). Another method more suited for making E field observations simultaneously at several points and for extended times at a given location was developed at the Max Planck Institut (Föppl *et al.*, 1967). Barium vapor released from a sounding rocket above about 150 km partially ionizes and produces a visible ion cloud during twilight which can be tracked photographically. The ion cloud drifts under the influence of the electric and magnetic fields, and it can be shown that the velocity of a small cloud above 200 km altitude is given essentially by $\mathbf{v}=\mathbf{E}\times\mathbf{B}/B^2$. Inverting this equation to $\mathbf{E}=-\mathbf{v}\times\mathbf{B}$, the measurement of $\mathbf{v}$ in a known magnetic field $\mathbf{B}$ gives $\mathbf{E}$.

The GSFC-NASA Ba release experiments have produced high latitude observations of 23 ion clouds in the auroral zone and 12 in the polar cap region. From these observations we can draw conclusions about the nature of the ionospheric electric fields and the tensor conductivity elements that are most effective in producing ionospheric currents. The measurements also test the usual assumption that the surface magnetic disturbance is explained by ionospheric currents. From ground based magnetometers a typical equivalent ionospheric current pattern for the polar cap and auroral zone disturbance can be inferred. Figure 1 shows a diagram illustrating the essential features. Our E field investigations have sampled four regions of interest: (a) in the westward electrojet or negative magnetic bay region, (b) in the eastward electrojet region, (c) in the transition region where the westward electrojet passes polewards of the eastward jet, and (d) in the polar cap region. Visible auroral displays were observed in the Ba release region or close by, in all flights except those in the polar cap where auroras were seen only near the southern horizon.

* NAS Resident Research Associate, NASA Goddard Space Flight Center, Greenbelt, Maryland 20771, now at Geophysical Institute, University of Alaska, 99701.

B. M. McCormac (ed.), Particles and Fields in the Magnetosphere.

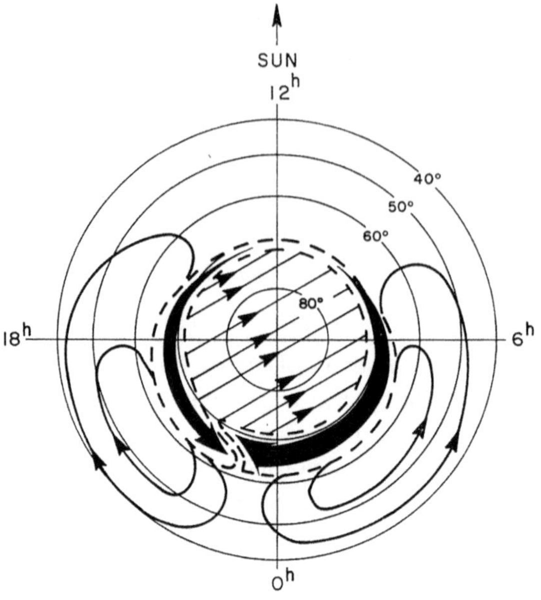

Fig. 1. Idealized equivalent current pattern for the auroral and polar cap substorm as viewed from above the N magnetic pole in magnetic latitudes and magnetic time.

2. Westward Electrojet Region

On the evenings of August 31 and September 2, 1967, rocket flights from Andøya, Norway, released 8 Ba clouds in or slightly north of the westward electrojet (Wescott *et al.*, 1969). The ion cloud speeds differed on the two occasions, but the direction of motion with respect to the magnetic disturbance vector was the same. The August 31, 1967, flight illustrates the salient features of the electric field in the westward electrojet region with auroral forms present. At the time of the Ba releases the nearby ground magnetograms at Andøya and Tromsø, Norway, and Kiruna, Sweden, showed a negative bay in H of -150γ which continued during the observation period to -340γ. Typical post-breakup auroral arcs of varying brightness and activity existed in the release region during the observation and a quiet faint arc remained nearly stationary about 100 km to the south.

Figure 2 illustrates the ion cloud positions, projected down the magnetic field lines to 100 km altitude vs time. All motions were eastward with speeds most typically between 700 and 1000 m/sec. The corresponding electric fields were 35 to 50 mV/m directed southward and perpendicular to the velocity. Figure 3 shows the velocity vectors and the positions of the auroral arcs at four times during the observations. The ion clouds were observed to move closely parallel to the arc alignments. A comparison of the electric field direction with the magnetic perturbation vector revealed agreement within 5°. This implies that the current was nearly perpendicular to the E field in the Hall current direction.

All results of the September 2 flight agreed with those of August 31 except that higher E fields ≈ 130 mV/m during a smaller negative bay were observed. This fact illustrates that the ionospheric conductivity varies considerably and that one cannot predict the magnitude of E solely from the ground magnetograms. The surface disturbance, however, does appear to be reliable in determining the direction of E as would be expected for Hall currents.

3. Transition and Eastward Electrojet Region

Two flights on September 12, 1967, and September 20, 1968, from Andøya, Norway, were planned to measure the electric field in the region of the eastward electrojet. The launch conditions were auroral activity in the release area and a positive magnetic bay in H at the nearest magnetic stations, Andøya and Tromsø. In each case three

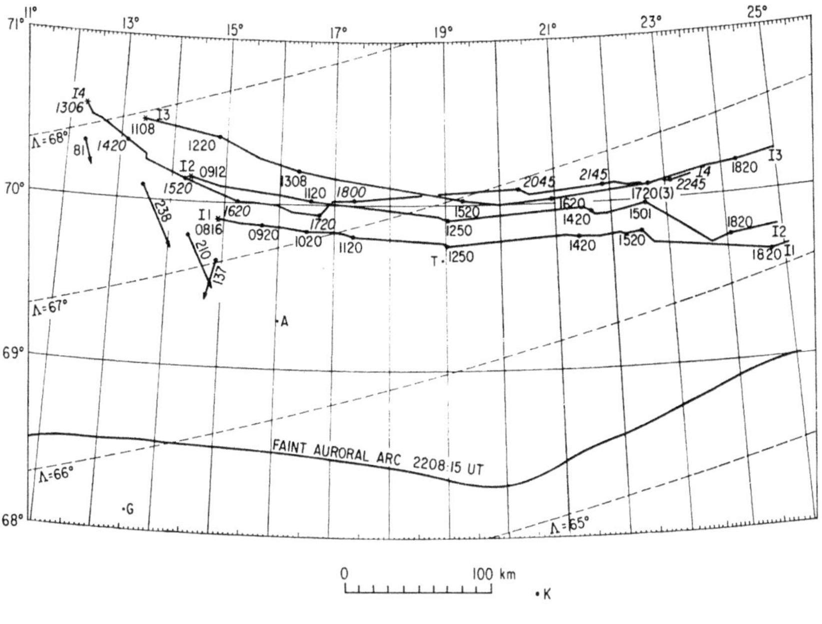

Fig. 2. Tracks of 4 ionized Ba clouds on August 31, 1967, projected along magnetic field lines to 100 km altitudes on geographic coordinates. Numbers along the tracks are minutes and seconds of 2200 UT. Dashed grid lines are invariant latitudes and arrows indicate average neutral cloud velocities in m/sec.

useful Ba clouds were formed, the most southernly two were obviously in the eastward current region. On September 12, 1967, the northernmost cloud, however, apparently spanned the transition region and temporarily moved eastward in accordance with the electric field of the westward electrojet before moving westward at the time of an enhancement in the positive bay disturbance. This flight has been discussed previously (Wescott *et al.*, 1969).

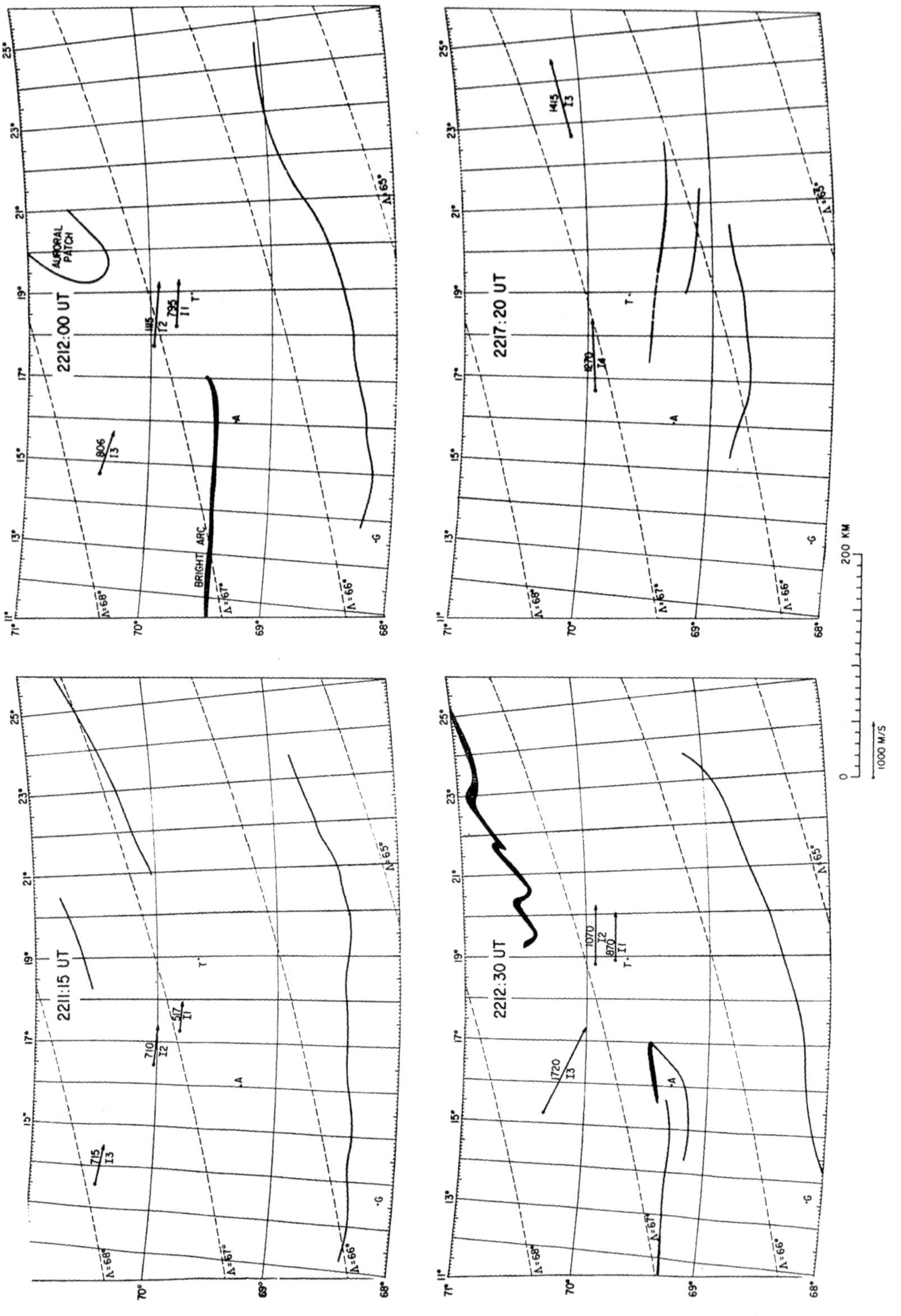

... flight of August 21, 1967. All cloud positions

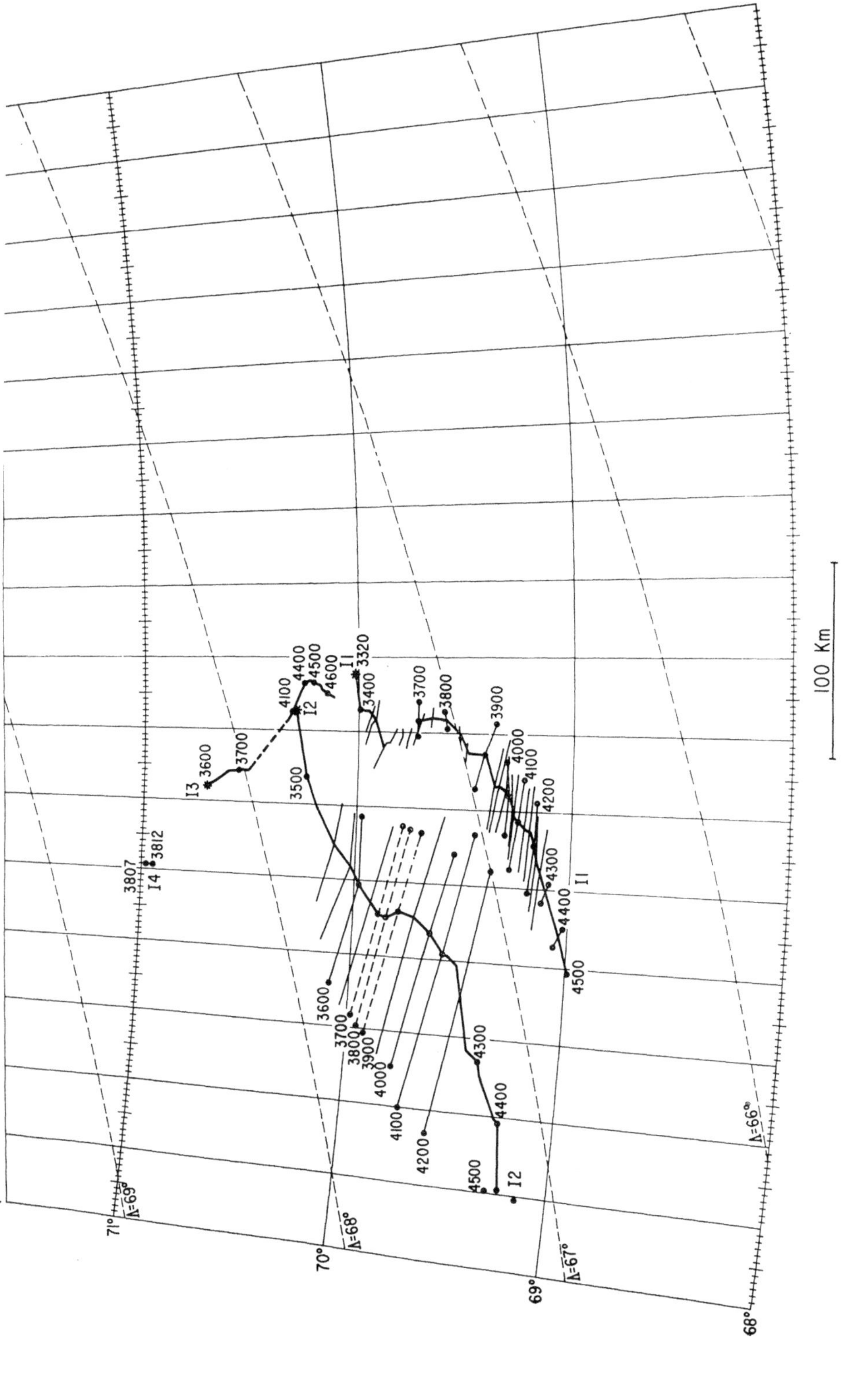

Fig. 4. Tracks of 3 ionized Ba clouds on September 20, 1968, projected to 100 km altitude down magnetic field lines and plotted on geographic coordinates. Numbers along the tracks are minutes and seconds of 1900 UT. Dashed grid lines are invariant latitudes. Clouds 1 and 2 became elongated E–W, the triangulated end and middle points are shown. Note the velocity shear between clouds 3 (Southeast) and 1, 2 (west-southwest).

The results of the flight of September 20, 1968, are more complex but are more interesting because of more complete auroral information. During the observation interval the magnetogram shows a ΔH of about $+30\gamma$ and a west component of 20γ. The 100 km projected ion cloud positions vs time are shown in Figure 4. Both clouds 1 and 2 became elongated E–W while moving generally west or southwest. This motion is in reasonable agreement with the magnetic perturbation at Andøya (*A* on the map) if the current is predominantly Hall current. The most striking feature of the tracks in Figure 4 is the motion of the third cloud which is to the southeast.

A fairly active auroral arc at launch was nearly overhead at Andøya, the launch site. After the Ba releases the arc moved polewards and passed first through the projected position of cloud 1 and then of cloud 2 and eventually remained in the cloud 3 position for most of the observation period. The ion cloud motion did not correspond to the poleward motion of the arc, which implies that the cause of the poleward auroral motion was not directly associated with an outward plasma motion in the magnetosphere.

The ion cloud velocity was low at the time the Ba clouds and auroras were in the

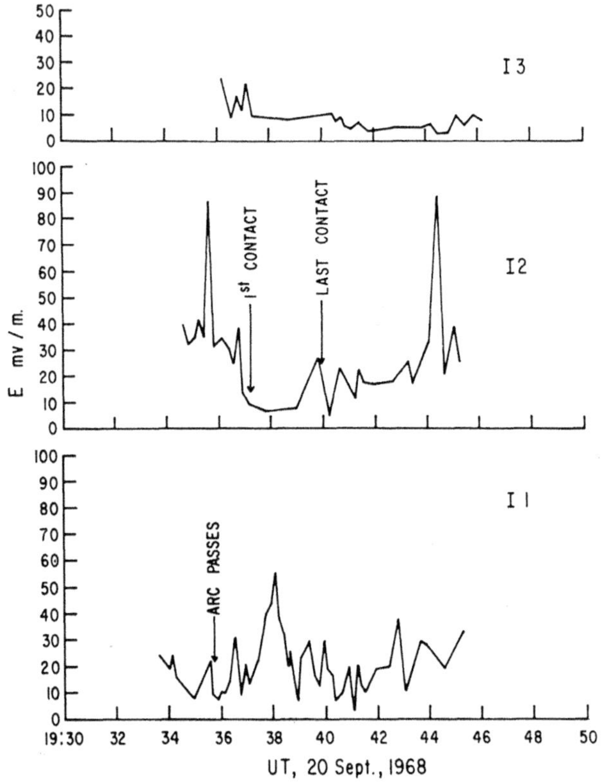

Fig. 5. Magnitude of E vs. time for 3 clouds released on September 20, 1968. Note the variability of E in time and position. The low E is apparent at times when the visible auroral arc is on the same magnetic field lines as the various clouds.

same shell. Figure 5 illustrates the calculated E field magnitude vs time for the three clouds. At the time of auroral crossings the E field dropped to less than 10 mV/m. This is particularly evident in the cloud 2 data where the contact extended over 3 min. These observations confirm probe measurements which showed a low E field inside auroral forms (Aggson, 1969).

4. Conclusions on Electrojet Region

(a) The E field direction is consistent with observed magnetic perturbations for an ionospheric current which is predominantly Hall current.

(b) $|E|$ ranges from less than 10 mV/m to greater than 100 mV/m.

(c) Observations show that $|E|$ is low inside auroral forms, which argues against having a polarization field drive Cowling currents to explain the electrojets.

(d) Assuming that magnetospheric convection is closely related to the ion drifts observed, the auroral substorm convection is primarily E–W or W–E with dominant inward motions confined to the auroral breakup region where E reverses.

(e) $|E|$ cannot be simply related to the magnitude of ΔB hence the conductivity distribution is needed to predict $|E|$ from $|\Delta B|$.

(f) The polewards motion of an auroral arc does not imply any direct connection with the E field or outward plasma drift in the magnetosphere. Explanation in terms of a change in the energy of precipitation particles appears more promising.

(g) The strong northward and southward directed E fields, respectively, in the eastward and westward electrojet regions and the behavior of Ba^+ clouds in the transition between the two electrojets is in excellent agreement with the two cell configuration (Figure 1) and contrary to treatments of the eastward electrojet as a mere return current from the westward electrojet.

(h) E is not uniform in space or time and there can be large shears over a short distance.

(i) The velocity of the auroral zone ionosphere is often supersonic with respect to the neutral atmosphere.

(j) Striations aligned along B are a common feature of the auroral zone Ba^+ clouds. Their existence, growth, and size does not appear to be related to altitude above 200 km or the time after release. Some form of $E \times B$ instability is suggested.

(k) Abnormal vertical motions of the Ba^+ clouds that might suggest a strong E field parallel to **B** have not been observed.

5. Polar Cap Electric Fields

Referring to Figure 1, the essential characteristic of magnetic disturbance in the polar cap region is the wide spread uniform perturbation, such that the equivalent current lines run parallel across the cap. To determine the E field in the polar cap 3 rocket flights, releasing a total of 12 Ba ion clouds were conducted in March 1969 from Cape Parry, N.W.T. (74.7 °INL). The flight of March 8, 1969 – an evening shot – has been analyzed. The magnetic perturbation vectors in the polar cap and auroral oval (Figure 6) indicate that a classic type polar substorm (Figure 1) was in progress. The polar cap vectors from Mould Bay, Resolute Bay, and Alert show a very uniform 50 to 70γ disturbance. The vectors at Cape Parry and Cape Young (immediately to

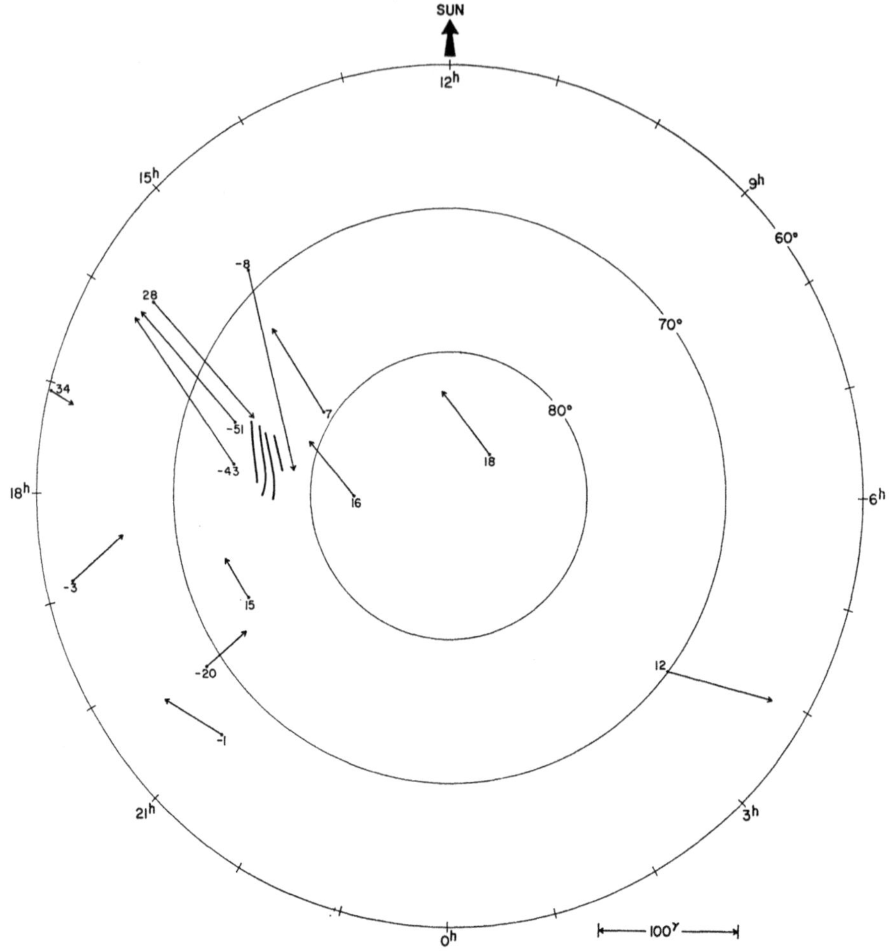

Fig. 6. North polar plot in magnetic latitude and magnetic time of the horizontal magnetic disturbance vectors at the time of release of the fourth Ba cloud at 03:25 UT on March 8, 1969. Numbers with each vector give the vertical component of the magnetic disturbance. Ba^+ cloud tracks projected to 100 km altitude also are shown.

the south of the 4 tracks) are parallel to those at the higher latitude stations but have a slightly greater magnitude. Vectors at Point Barrow and College, Alaska approximately 10° to the southwest show a $+\Delta H$ disturbance that is in good agreement with the location of an auroral arc which was visible on the south horizon from Cape Parry and Cape Young. Figure 7 shows the tracks of the 4 clouds. All motions were nearly

Fig. 7. Tracks (projected to 100 km altitude) for March 8, 1969, plotted on geographic coordinates (solid lines) with invariant latitude (lines dashed).

parallel over a wide area and the velocity was reasonably smooth also. The corresponding E field was 30 to 40 mV/m. The remarkable result however is that the direction of E is such that only a component of the magnetic perturbation could be caused by an overhead Hall current. The disagreement between the normal to the cloud motion and the disturbance vector would be even greater if a Pedersen current was present. If the overhead current was all Hall current the disagreement is between 55 to 65°.

This serious disagreement clearly shows that an appreciable component of the polar cap disturbance is *not* due to horizontal ionospheric current. The nonionospheric disturbance component remains to be explained. Aside from the ever popular field aligned currents we can suggest another possible mechanism, namely, that in addition to the Hall current produced perturbation there is a perturbation vector directed towards the sun; this arises from the combination of solar wind compression of the magnetosphere, the tension on the magnetic field lines extending into the distant tail, and the nightside inflation of the near magnetosphere. These factors all produce a sunward directed vector in the polar cap and a self-consistent model is needed to see how large a sunward vector can be produced from these distant effects. Roughly 50γ are needed under the $K_p = 3$ conditions of this example. It is believed also that the study of variations in the sunward component of the polar cap disturbance as a function of magnetospheric parameters simultaneously measured by satellites will provide important tests.

6. Polar Cap Conclusions

(a) E appeared to be uniform over a large area and averaged 30 to 40 mV/m.

(b) The direction of E is such that only a component of the magnetic disturbance could be caused by Hall currents.

(c) Pedersen currents cannot provide an explanation for the magnetic disturbance and inclusion of a Pedersen current component would only increase the discrepancy between the observed disturbance and a Hall current component.

(d) The results imply a nonionospheric component at least comparable in magnitude to the magnitude of a Hall current component.

References

Aggson, T. L.: 1969, in *Atmospheric Emissions* (ed. by B. M. McCormac and A. Omholt), Van Nostrand Reinhold Company, New York, p. 305.

Föppl, H., Haerendel, G., Loidl, J., Lüst, R., Melzner, F., Meyer, B., Neuss, H., and Rieger, E.: 1967, *Planetary Space Sci.* **15**, 357.

Maynard, N. C. and Heppner, J. P.: 1970, this volume, p. 247.

Wescott, E. M., Stolarik, J. D., and Heppner, J. P.: 1969, *J. Geophys. Res.* **74**, 3469.

SATELLITE MEASUREMENTS OF DC ELECTRIC FIELDS IN THE IONOSPHERE

DONALD A. GURNETT

Dept. of Physics and Astronomy, The University of Iowa, Iowa City, Iowa, U.S.A.

Abstract. This paper presents initial results from the dc electric field experiment on the low altitude polar orbiting Injun 5 satellite. At low latitudes, where the ionospheric plasma is expected to corotate with the earth, the only observed electric field is the $\mathbf{V} \times \mathbf{B}$ electric field due to the motion of the satellite through the ionosphere. Comparisons of the computed $\mathbf{V} \times \mathbf{B}$ field agree with the measured electric field within about ± 10 mV/m. At the plasmapause boundary small changes (~ 5 to 15 mV/m) in the dc electric field are observed which are believed to be due to magnetospheric convection phenomena at the plasmapause boundary. At high latitudes large amplitude (~ 100 mV/m) electric field irregularities and disturbances, believed to be due to ionospheric electric fields, are observed on virtually every passage over the polar regions. Long period electric field oscillations with periods on the order of 15 sec also are commonly observed at high latitudes.

1. Introduction

The importance of magnetospheric electric fields and the associated convection of plasma in the magnetosphere has been recognized for a number of years (Axford and Hines, 1961; Dungey, 1961; Piddington, 1962a, 1962b; Boström, 1966; Axford, 1969); however, only recently have techniques been developed for the measurement of magnetospheric electric fields and the study of magnetospheric convection. Of the variety of techniques from which electric fields can be directly or indirectly deduced (see Axford, 1969) probably the most extensive and sensitive measurements have been obtained from artificial Ba cloud releases (Haerendel *et al.*, 1967, 1969; Föppl *et al.*, 1968; Wescott *et al.*, 1969). Relatively few direct probe measurements of dc electric field measurements have been reported. Mozer and Bruston (1967), Fahleson *et al.* (1968), and Aggson (1969) have used the double probe technique on sounding rockets to measure ionospheric electric fields. Heppner (1968) has reported on results from the electric field experiment on the OV1 10 satellite which detected electric field disturbances with periods less than about 60 sec. This paper presents initial results from the dc electric field experiment on the low altitude (677 to 2528 km) polar orbiting Injun 5 satellite. These measurements are among the first reported satellite measurements of *dc* electric fields using the double probe technique.

2. Instrumentation

The electric field experiment on Injun 5 is of the double probe type described by Fahleson (1967), Aggson (1969), and others. The probes used consist of two conducting spheres 20.3 cm in diameter mounted on booms with a center-to-center separation of 2.85 m as shown in Figure 1.

If the two spheres have identical characteristics and if wake effects and other

B. M. McCormac (ed.), Particles and Fields in the Magnetosphere.

spacecraft related perturbations can be ignored, then the difference between the floating potential and the plasma potential will be the same for each sphere and the electric field in the plasma can be determined directly from the difference in the floating potential and the separation distance of the two spheres. Possible errors in

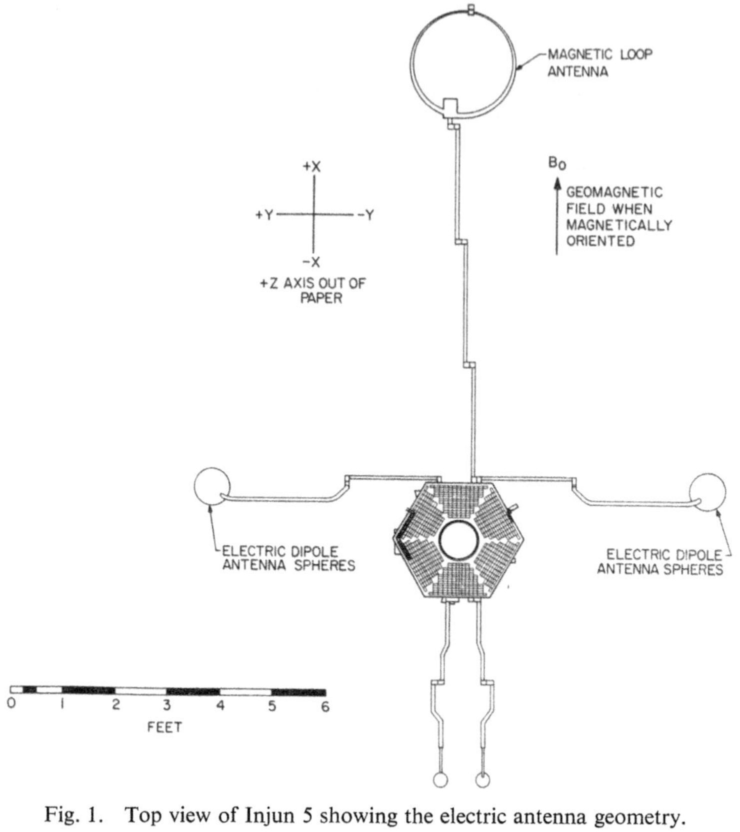

Fig. 1. Top view of Injun 5 showing the electric antenna geometry.

the double probe technique due to differences in the contact potential of the spheres, unequal photoelectron emission, and various other effects are discussed by Fahleson (1967).

The Injun 5 spacecraft is magnetically oriented by a bar magnet within the spacecraft such that when properly aligned the x-axis of the spacecraft is parallel to the geomagnetic field with the positive x-axis pointing downward in the Northern Hemisphere. Typical maximum alignment errors between the x-axis and the geomagnetic field are about 10 to 15°. As shown in Figure 1, when the spacecraft is oriented magnetically, the axis through the two spheres is perpendicular to the geomagnetic field. The dc electric field experiment, therefore, is primarily sensitive to the electric field $\mathbf{E}_\perp$ perpendicular to the geomagnetic field.

The potential difference between the spheres is obtained from a high input impedance (20 MΩ) differential amplifier located in the main spacecraft electronics. The differential amplifier has a dynamic range of ±1.0 V and an RC time constant of 0.4 sec. The output from the differential amplifier is sampled by the digital data system once every 4 sec with 8 bit (256 step) accuracy. The minimum resolvable electric field strength increment is approximately 2.75 mV/m. In order to make a quantitative estimate of the error due to the finite voltmeter impedance, the ac impedance of the spheres is measured every 30 sec by differentially driving the spheres with a constant amplitude ac current source and measuring the resulting ac potential difference between the spheres. Further details of the Injun 5 dc electric field experiment are given by Gurnett *et al.* (1969).

3. Comparison with V × B

The potential difference between the spheres observed for a typical mid-latitude pass is shown in Figure 2. The systematic sinusoidal variation evident in the potential difference, with a period of about 20 min, is due to the **V** × **B** electric field arising from the satellite motion through the ionosphere. The sinusoidal modulation of the **V** × **B** potential is caused by the slow rotation of the satellite and the electric antenna axis around the geomagnetic field with a period of about 20 min. The dashed lines in Figure 2 are the potential limits of the **V** × **B** electric field as computed from the satellite orbit. The satellite velocity **V** was computed relative to a coordinate system corotating with the earth. At middle and low latitudes, where the ionospheric plasma is expected to corotate with the earth (Axford, 1967), only the **V** × **B** electric field should be observed. In order to fit the computed **V** × **B** potential limits to the observed maxima and minima of the sphere potential difference (when the electric antenna axis

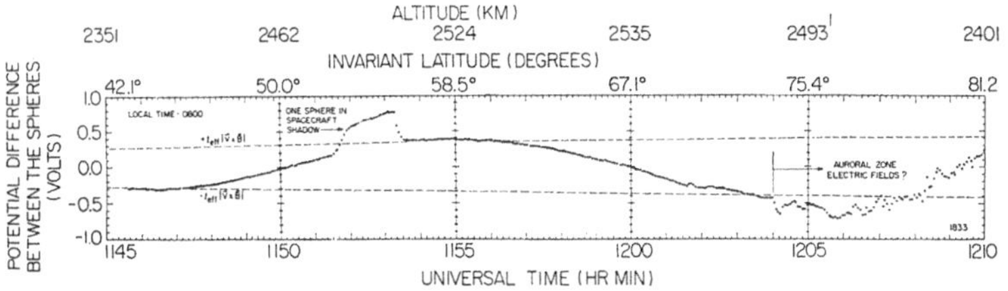

Fig. 2. DC potential difference between the spheres compared with the **V** × **B** potential limits (dotted lines).

becomes aligned and anti-aligned respectively with **V** × **B**), it was found to be necessary to subtract a constant offset voltage of about 40 mV from the measured sphere potential difference. This offset voltage is believed to be due to differences in the surface properties of the two spheres. As can be seen from Figure 2, when this offset voltage is taken into account the maximum and minimum sphere potential differences agree very well with the limits for the **V** × **B** potential. This excellent agreement demonstrates that the double probe electric field experiment on Injun 5 can measure accurately electric fields in the ionosphere. From comparisons between the measured electric field and $\mathbf{V} \times \mathbf{B}$ under various conditions, the error limits for the absolute dc electric field measurements have been estimated to be about ± 10 mV/m.

The abrupt 0.3 V increase in the sphere potential difference from 1152 to 1153 UT in Figure 2 is due to the change in the photoelectron emission of one of the spheres as it passes through the shadow of the spacecraft body. Similar optical shadowing effects are commonly observed in the dc electric field data and are easily identified. Small (10 mV/m) errors due to unequal shadowing of the spheres by the booms also are observed for some orientations. No wake effects have been observed. Sheath resistances observed in orbit generally are considerably less than the input impedance of the differential amplifier so that errors due to the 'loading' of the antenna by the differential amplifier are usually negligible.

4. High Latitude Electric Field Irregularities

Starting at approximately 1204 UT and 73° INL during the pass illustrated in Figure 2, large amplitude (0.3 V) irregular perturbations from the $\mathbf{V} \times \mathbf{B}$ potential are observed. These irregular perturbations in the sphere potential difference, observed on virtually every pass over the polar regions, are believed to be due to spatial irregularities in the ionospheric electric field of the same type reported by Heppner *et al.* (1968).

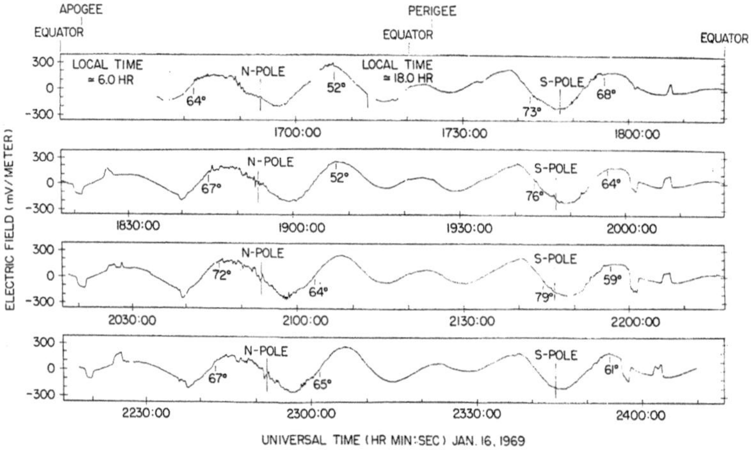

Fig. 3. DC electric field for a sequence of four consecutive orbits.

To illustrate the general character of these electric field irregularities, Figure 3 shows the dc electric field data for a sequence of four entire orbits during a period of relatively little magnetic activity ($K_p \leqslant 3$). The time scales for these orbits have been aligned to permit direct comparison of successive passes over each polar region. As can be seen electric field irregularities are evident on every pass over the polar region. The lowest latitude at which the electric field irregularities are observed is quite variable, from as low as 52° to as high as 79° INL. Generally the largest amplitudes are observed at INL > 70°. Although the general region of the irregularities is similar for successive passes over a given polar cap, the detailed electric field structure shows little similarity from one orbit to the next. Typical peak amplitudes for these electric field irregularities during a given orbit are on the order of 30 to 100 mV/m.

Preliminary investigations of electron density data obtained from the AFCRL electron density experiment on Injun 5 have shown that large amplitude (5 to 25%) irregular electron density fluctuations are observed often in the same general region that the electric field irregularities are observed. A similar correlation also has been observed with the OGO 6 satellite (Hanson, 1969; Heppner, 1969).

5. The Plasmapause Boundary

Since various current magnetospheric models picture the plasmapause boundary as being the inner boundary of magnetospheric convection (Nishida, 1966; Brice, 1967), several plasmapause crossings have been investigated to determine if any net disturbance from the **V** × **B** field could be detected as the satellite crossed the plasmapause boundary.

Investigation of these plasmapause crossings has shown that the electric field generally varies smoothly near the plasmapause boundary and that a small (5 to 15 mV/m) but measurable change in the dc electric field is often evident at the plasma-

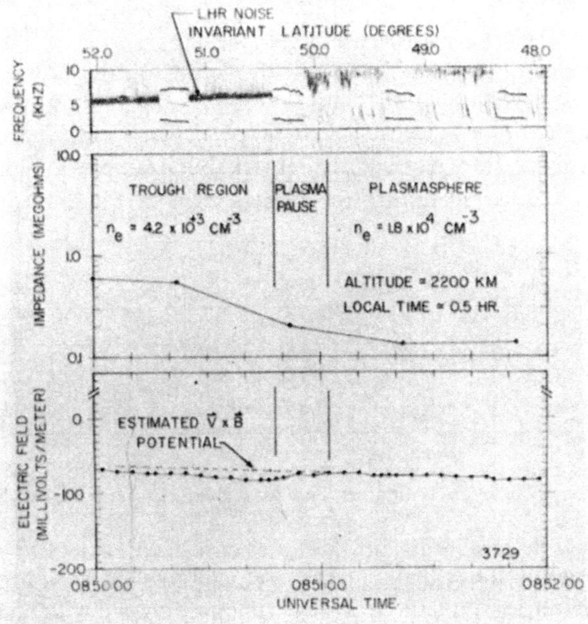

Fig. 4. DC electric field variations at a typical plasmapause crossing.

pause boundary. The electric field irregularities described in the previous section are normally encountered at latitudes somewhat higher than the plasmapause.

A typical plasmapause crossing is shown in Figure 4 at 0851:20 UT. The plasmapause location in this case was identified from the characteristic 'lower hybrid resonance (LHR) breakup' effect commonly found in the VLF electric field data at the plasmapause boundary (Carpenter *et al.*, 1968), the characteristic increase in the sheath resistance at the plasmapause boundary (Gurnett *et al.*, 1969), and the change in the electron density as measured directly by the AFCRL electron density measurement on Injun 5.

The dc electric field is seen to vary smoothly as the satellite crosses the plasmapause, with no evidence of the large amplitude electric field irregularities of the type typically found at higher latitudes. By extrapolating the $\mathbf{V} \times \mathbf{B}$ potential from the low latitude side of the plasmapause (dotted line in Figure 4), where we assume that there is no ionospheric electric field, a small shift in the dc electric field on the order of 15 mV/m is evident upon crossing the plasmapause boundary. Estimates of errors (Fahleson, 1967) due to unequal sunlight and ram ion shadowing of the two spheres cannot account for the observed change in the sphere potential difference. This shift in the dc electric field, therefore, is attributed to a convection electric field on the order of 15 mV/m, corresponding to a convection velocity on the order of 0.4 km/sec. The sense and magnitude of the change in the electric field component measured (the spheres were aligned approximately in the N–S direction) is consistent with a transition from corotation inside the plasmasphere to nearly noncorotation outside the plasmasphere.

6. Long Period Electric Field Oscillations

Large amplitude electric field oscillations with periods on the order of 15 to 20 sec are frequently found in the Injun 5 electric field data, particularly in the latitude range from about 70 to 80° INL. Two particularly remarkable examples of long period electric field oscillations occurred during the orbit shown in Figure 5 on the low

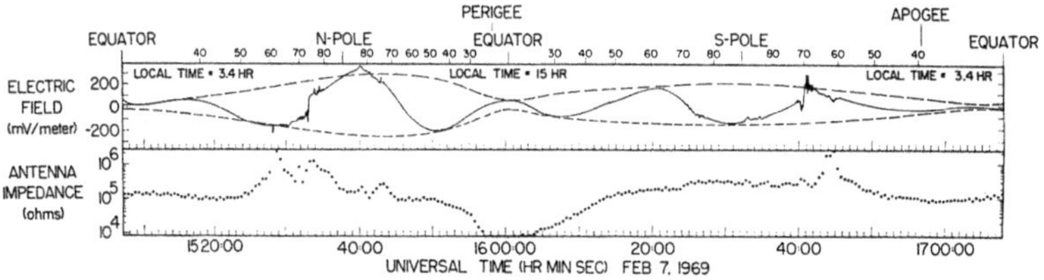

Fig. 5. Conjugate electric field discontinuities and associated electric field oscillations.

latitude side of the discontinuities in the dc electric field evident at about 1534:00 and 1643:00 UT. The electric field data near these times are shown with an expanded scale in Figure 6. These electric field discontinuities and associated electric field oscillations are remarkable in that at the times of the electric field discontinuities the satellite was at very nearly conjugate points in the geomagnetic field. The invariant latitude at these times differed by only 3° (73° vs. 70°, respectively) and the magnetic local time differed by only 0.14 hr (3.27 hr vs. 3.41 hr, respectively). A detailed examination of these two events strongly suggest a close magnetically conjugate relationship. The N–S electric field component (which was the approximate orientation of the electric antenna axis in both cases) decreases toward higher latitudes in both hemispheres and the magnitude of the electric field discontinuity is very nearly the same in both hemispheres. The periods of the electric field oscillations observed in both hemispheres are also very nearly the same – about 17 sec. This close magnetically

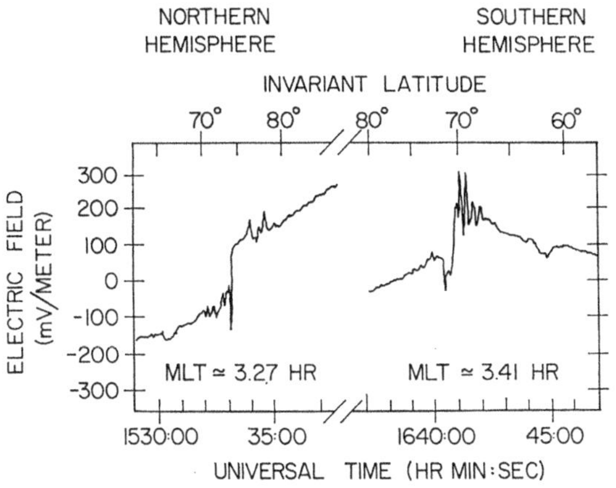

Fig. 6. Expanded illustrations of the discontinuities and electric field oscillations shown in Figure 5.

conjugate relationship, particularly for the electric field discontinuities, is very suggestive that the electric field and associated convection are mapped from one hemisphere to the other along the geomagnetic field line, as is often presumed for the 'frozen field' model of magnetospheric convection.

At the present time the detailed nature of these long period electric field oscillations is largely unknown. It has not yet been established whether the observed oscillations are basically a temporal variation or a long wavelength (100 km) spatial structure converted to a temporal variation by the satellite motion through the ionosphere. Since very similar fields were observed at magnetically conjugate points, it is reasonably certain that the oscillations are a magnetospheric phenomena occurring over a relatively wide range of L shells. If these electric field oscillations are due to hydromagnetic waves in the magnetosphere, then corresponding large amplitude magnetic field oscillations (possibly corresponding to pc 2 micropulsations) also should be observed. No comparable magnetic field oscillation could be found from ground magnetometer records investigated for the two cases illustrated in Figures 5 and 6, although in neither case was the subsatellite point closer than 1000 km to the nearest ground magnetometer.

Acknowledgments

The author wishes to express his thanks to Mr. David P. Cauffman for his effort in analyzing this data and to Dr. S.-I. Akasofu for his effort in providing ground magnetometer records and Dr. R. Sagalyn for her effort in providing electron density data.

This research was supported by the National Aeronautics and Space Administration under Contracts NAS5-10625, NAS1-8141, NAS1-8144(f), NAS1-8150(f), and NGR-16-001-043 and by the Office of Naval Research under Contract Nonr 1509(06).

References

Aggson, T. L.: 1969, AGU Conference on Electric Fields in the Magnetosphere, March 10–13, Houston, Texas.

Axford, W. I.: 1967, in *Physics of Geomagnetic Phenomena* (ed. by S. Matsushita and W. H. Campbell), Academic Press, New York, p. 1243.

Axford, W. I.: 1969, *Rev. Geophys.* **7**, 421.

Axford, W. I. and Hines, C. O.: 1961, *Can. J. Phys.* **39**, 1433.

Boström, R.: 1966, in *Aurora and Airglow* (ed. by B. M. McCormac), Reinhold Publishing Corporation, New York, p. 293.

Brice, N. M.: 1967, *J. Geophys. Res.* **72**, 5193.

Carpenter, D. L., Walter, F., Barrington, R. E., and McEwen, D. J.: 1968, *J. Geophys. Res.* **73**, 2929.

Dungey, J. W.: 1961, *Phys. Rev. Letters* **6**, 47.

Fahleson, U. V.: 1967, *Space Sci. Rev.* **7**, 238.

Fahleson, U. V., Kelley, M. C., and Mozer, F. S.: 1968, University of California, preprint.

Föppl, H., Haerendel, G., Haser, L., Lüst, R., Melzner, F., Meyer, B., Neuss, H., Rabben, H., Rieger, E., Stocker, H., and Stoffregen, W.: 1968, *J. Geophys. Res.* **73**, 21.

Gurnett, D. A., Pfeiffer, G. W., Anderson, R. R., Mosier, S. R., and Cauffman, D. P.: 1969, *J. Geophys. Res.*, **74**, 4631.

Haerendel, G., Lüst, R., and Reiger, E.: 1967, *Planet. Space Sci.* **15**, 1.

Haerendel, G., Lüst, R., Rieger, E., and Völk, H.: 1969, in *Atmospheric Emissions* (ed. by B. M. McCormac and A. Omholt), Van Nostrand and Reinhold Company, p. 293.

Hanson, W. B.: 1969, private communication.

Heppner, J. P.: 1968, NASA-Goddard Space Flight Center, preprint X-612-68-331.

Heppner, J. P.: 1969, private communication.

Heppner, J. P., Aggson, T. L., and Maynard, N. C.: 1968, Symposium on the Physics of the Magnetosphere, September, Washington, D.C.

Mozer, F. X. and Bruston, P.: 1967, *J. Geophys. Res.* **72**, 1109.

Nishida, A.: 1966, *J. Geophys. Res.* **71**, 5669.

Piddington, J. H.: 1962a, *J. Geophys.* **7**, 183.

Piddington, J. H.: 1962b, *Planet. Space Sci.* **9**, 947.

Wescott, E. M., Stolarik, J. D., and Heppner, J. P.: 1969, *J. Geophys. Res.* **74**, 3469.

VARIATIONS IN ELECTRIC FIELDS FROM POLAR ORBITING SATELLITES

N. C. MAYNARD and J. P. HEPPNER
NASA Goddard Space Flight Center, Greenbelt, Md., U.S.A.

1. Introduction

The advent of techniques to measure electric fields in the ionosphere and above has put new emphasis on the questions of what electric fields exist in the ionosphere and magnetosphere and how they vary. The probe technique using long antennas on rockets and satellites (Aggson, 1969) is useful in studying the spatial variations of electric fields, complimenting the study of temporal variations using the motions of Ba ion clouds (Föppl *et al.*, 1967; Wescott *et al.*, 1969; Haerendel and Lüst, 1970; Wescott *et al.*, 1970).

Electric fields in the magnetosphere have been postulated theoretically as convection patterns derived from the distributions of ionospheric currents (Axford and Hines, 1961; Axford, 1969). These assume that the electric field is related to the convective velocity by:

$$\mathbf{E} = - \mathbf{v} \times \mathbf{B}$$

and that the conductivity along magnetic field lines is infinite allowing them to be equipotentials. While these patterns are useful in estimating gross behavior, it has been noted (Heppner, 1969) that the smaller scale irregularity structure must be understood to explain many questions regarding auroral forms and morphology, and it was suggested that this structure is related to the presence of larger scale dc electric fields. It is the purpose here to further define the characteristics of the electric fields inferred from OV1 10 measurements using preliminary data from OGO 6 (launched in June 1969) which substantiates the OV1 10 data on spatial irregularities and waves in the ELF-VLF regime. OGO 6 also is providing highly accurate measurement of large scale dc electric fields, but presentation of these results (other than figures regarding typical magnitudes) will have to await the merging of measurements with orbit and vehicle orientation data.

2. The Experiments

The probe technique (Aggson, 1969) using long cylindrical antennas has been used to measure electric fields from dc to low VLF frequencies on both OV1 10 and OGO 6. Briefly, the experiment monitors the potential difference between two axial antennas using a high input impedance voltmeter. This potential difference ϕ is related to the electric field by

$$\phi = (\mathbf{E} + \mathbf{v} \times \mathbf{B}) \cdot \mathbf{d},$$

B. M. McCormac (ed.), Particles and Fields in the Magnetosphere.

where **v** is the velocity of the vehicle and **d** is the vector distance between the antenna mid-points.

Since the $\mathbf{v} \times \mathbf{B}$ field tends to mask the desired ambient fields, it has to be subtracted to observe the ambient field. As **v** and **B** are well known and/or measured the accuracy of the subtraction is primarily dependent on knowing the vehicle orientation. The principal effect of the $\mathbf{v} \times \mathbf{B}$ term is the limitation it places on the sensitivity of the dc measurement in that the telemetry scale must accommodate a $\mathbf{v} \times \mathbf{B}$ range of plus and minus several hundred millivolts per meter in addition to the ambient field strength. This becomes a severe handicap in looking for variations of less than several millivolts per meter. Thus, for observing small scale irregularities a capacitor coupled logarithmic amplifier (log channel) with a 60 sec time constant is used. For even greater sensitivity at higher frequencies, the rms level of signals in the region between 3 Hz and 4 kHz was measured by separating the range into three bands in the case of OV1 10 (3 to 30 Hz, 30 to 300 Hz, and 300 to 3000 Hz) and five bands in OGO 6 (4 to 16 Hz, 16 to 64 Hz, 64 to 256 Hz, 256 to 1024 Hz, and 1 to 4 kHz).

OV1 10, launched in December 1966 (648 by 776 km orbit with a 93.4° inclination), had 51 ft antennas aligned along the local vertical for use in the gravity gradient stabilization of the satellite. Each antenna was insulated over the inner 19 ft to move the active element away from spacecraft interference fields. Unfortunately, a short on one antenna prevented dc measurements from being made; however, the ac measurements have proved very useful. Since the configuration of the booms was not normal due to the short, and the gravity gradient stabilization was weak, the direction of the component measured is ambiguous.

OGO 6 was launched in June 1969 into a more elliptical polar orbit (400 to 1600 km with an inclination of 82°). The 30 ft antennas extended from each solar array and were insulated on the inner 15 ft (to move the active element away from spacecraft fields such that the baseline **d** was 20 m). Stabilization is such that the antennas are always perpendicular to the normal to the earth and vary in orientation from N–S in the dawn-dusk plane to E–W in the noon-midnight plane. Results presented here are all from the dawn-dusk plane, hence the component of field being measured is essentially N–S.

3. Observations

A. LOW FREQUENCY IRREGULARITIES

The data from OV1 10 indicated a usually distinct boundary as one passed from the equator toward the pole where the 3 to 30 Hz channel rose sharply above the noise (Heppner, 1969; Heppner *et al.*, 1968). It was also noted that the activity in the log channel increased in this region, usually between 60 and 75° INL, indicating the possible presence of strong spatially varying dc electric fields. The 3 to 30 Hz signal, in the majority of cases, reached peak intensity at slightly higher latitudes than the region of maximum auroral occurrence and decreased in the central region of the polar cap.

Since the satellite is moving in the vicinity of 8 km/sec, the most probable con-

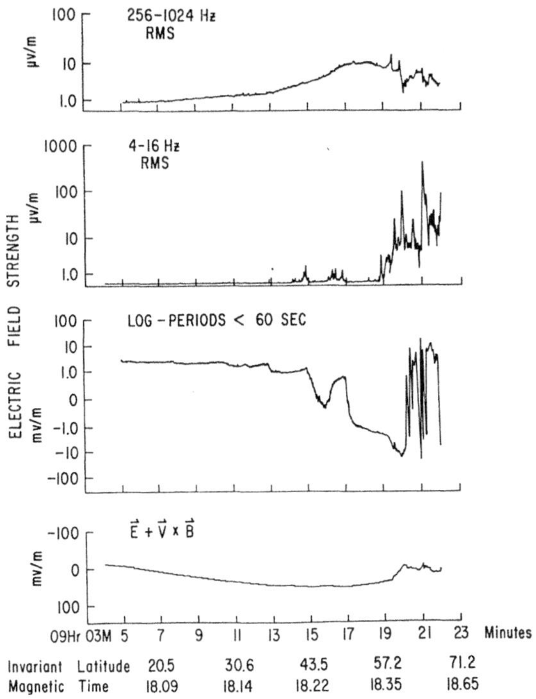

Fig. 1. OGO 6 electric field data (June 12, 1969) depicting the latitude variation from near the equator to the auroral region at dusk in the Northern Hemisphere.

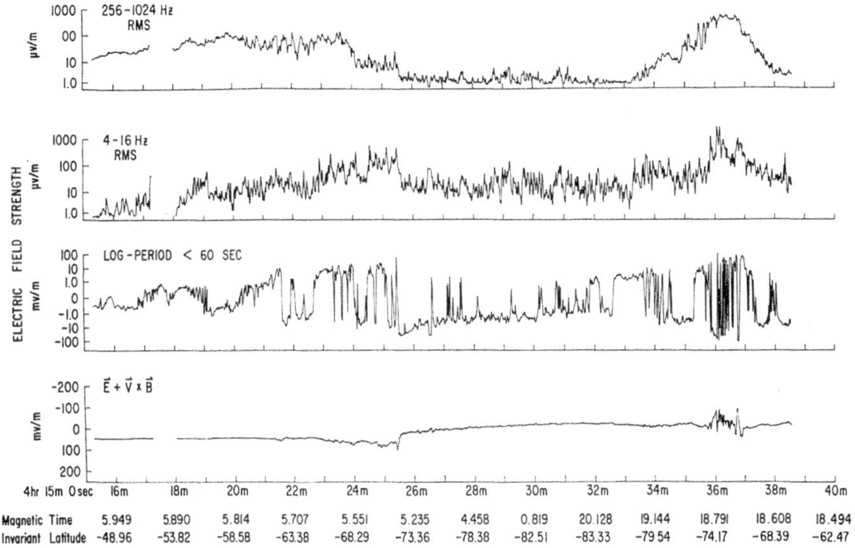

Fig. 2. OGO 6 electric field data (June 13, 1969) showing the latitude variation in a dawn-dusk pass across the southern auroral and polar cap regions.

clusion as to the source of the signal was that the satellite was moving through magnetic field aligned irregularity structures typically 0.2 to 2 km in width. In interpreting the OV1 10 data there was the possibility that the detected signal could have resulted from electron density irregularities as a consequence of the probe asymmetry caused by the short circuit to one of the axial probes. Thus, the measurements might not have truly represented electric field irregularities. It was assumed, however, that such an irregularity structure in the electron density would have to be accompanied by a nearly identical electric field structure to maintain current continuity in the *E* region – subject, of course, to the assumption that the irregularity structure extended downward to the *E* region. Additional confidence in the assumption that the irregularities did extend to the *E* region was later gained from the observation of highly similar and simultaneous striation structure in Ba^+ clouds above 200 km and aurora near 100 km (Wescott *et al.*, 1969). Another factor which indicated that OV1 10 was measuring electric field irregularities was the observation of conjugate agreement in the minimum latitude of occurrence, noted later in this paper. The OGO 6 results provide strong evidence that the OV1 10 assumptions were correct.

The same characteristic response is seen in preliminary data from OGO 6. Figure 1 depicts data from two spectrum channels (4 to 16 Hz and 256 to 1024 Hz) and from the log and dc channels in the Northern Hemisphere from near the equator to the auroral regions. The onset of activity in the 4 to 16 Hz channel occurs near and slightly before the obvious activity in the log and dc channels. The level quickly rises from the background to 10 to 30 μV/m rms with spikes up to several hundred μV/m. The log channel recorded variations of the order of 15 mV/m. Of particular interest is the essentially complete absence of the irregularity signal at low latitudes. The signal level is below 0.4 μV/m. Figure 2 shows a more active pass across the southern auroral regions from dawn to dusk. In the vicinity of 65° INL, the 4 to 16 Hz channel increased from around 10 μV/m to several hundred μV/m near the onset of activity in the dc channel. The activity is depressed over the polar cap and returns again on the high latitude side of the evening auroral region. The peaks of the 4 to 16 Hz signal typically are several hundred μV/m and occasionally go as high as several mV/m.

The overall distribution of this irregularity signal can be seen from the OV1 10 data. Figure 3 shows the locations in latitude and magnetic time of the peak intensities encountered on each polar pass. K_p values at these times are noted for three levels of activity by the symbols used. Although a shift to lower latitudes with increasing K_p is apparent in peak intensities, it appears most clearly in plots of the average minimum latitude of occurrence (Heppner, 1969).

As a consequence of the nearly infinite conductivity along magnetic field lines above 600 km, electric field irregularities should appear magnetically conjugate in both hemispheres. Figure 4 represents an attempt to look at the conjugacy of the lowest latitude of occurrence. Obviously, as sampling of opposite hemispheres involves time separations of 25 to 40 or 55 to 70 min and the orbit is tilted relative to the earth's magnetic axis, good comparison cannot be made for most of the orbits. However, by restricting comparisons to those cases where the magnetic local times in opposite

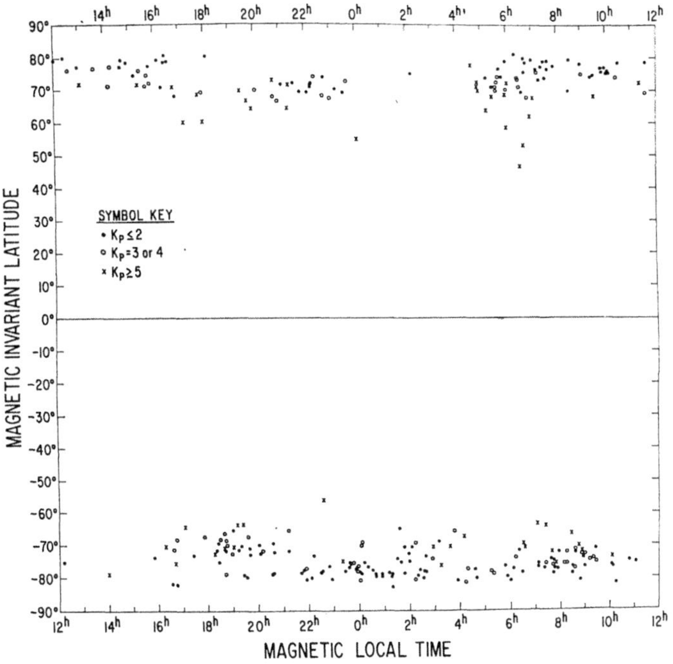

Fig. 3. The location of the maxima in the 3 to 30 Hz irregularity signal from OV1 10 data as a function of INL and MLT. Variation with K_p is denoted by the different symbols.

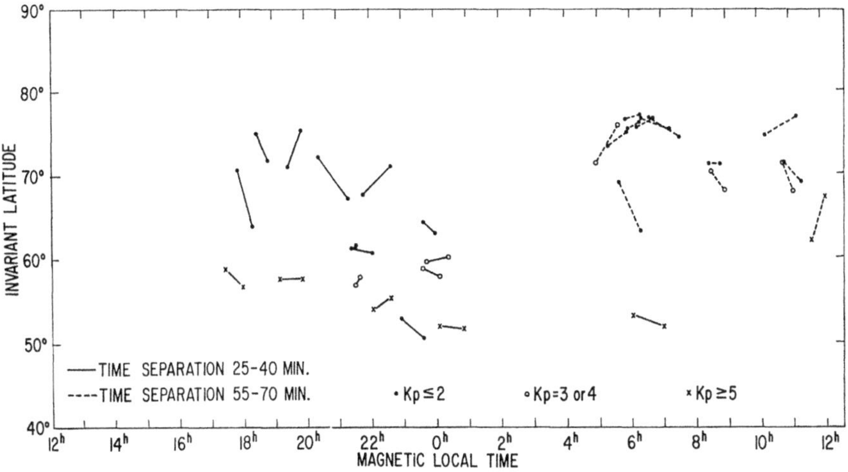

Fig. 4. The degree of conjugacy of the minimum latitude of occurrence of the 3 to 30 Hz irregularity signal from OV1 10 data in INL-MLT coordinates.

hemispheres are within 1 hr, it becomes quite clear that conjugacy exists. Figure 4 shows all the OV1 10 cases fulfilling the 1 hr criteria where good data were available from both hemispheres on the same orbit.

The electric field picture presented here is one of a small scale irregularity structure superimposed on the irregular dc field seen at auroral latitudes. While this structure is only a few percent or less in magnitude of the larger variations, its importance can be inferred by it being characteristic of the high latitude ionosphere, always being present with dc fields. Interpreting this low frequency signal as a spatial variation, one notes that the length scale is similar to that of auroral rays and the striations observed in Ba clouds (Wescott *et al.*, 1969). It has been suggested by them that an **E** × **B** plasma instability (e.g., Simon, 1963) could be the cause of the observed irregularities.

B. DC AND SLOW VARIATIONS

As final orbit and vehicle attitude information was not available the **v** × **B** variations

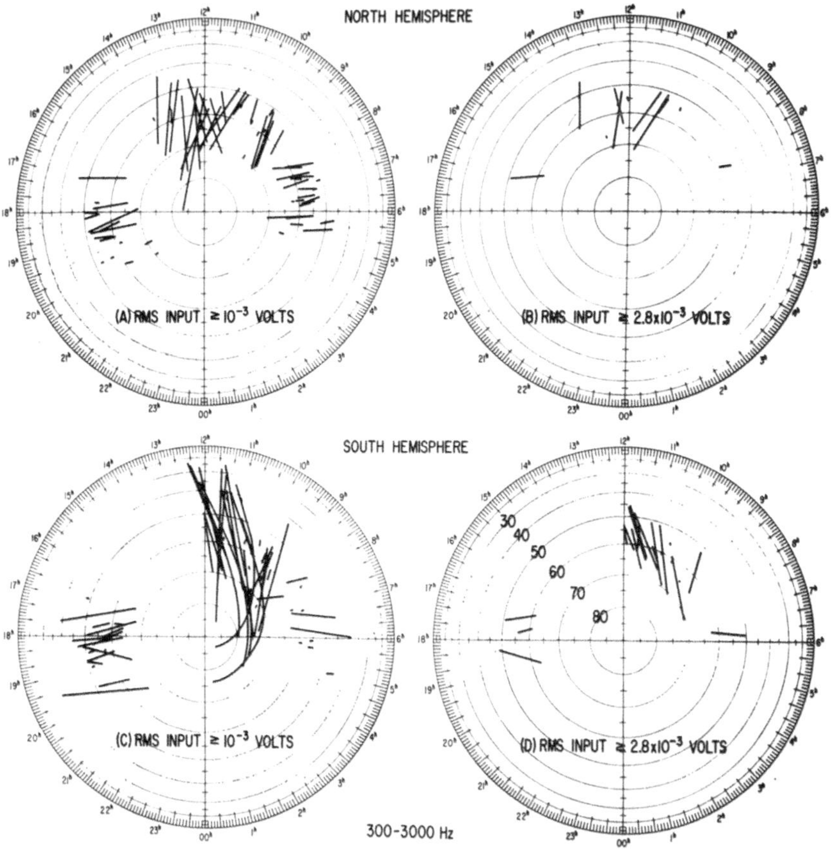

Fig. 5. The distribution of regions where the rms input to the 300 to 3000 Hz channel on OV1 10 exceeded 10^{-3} V (approximately 66 μV/m) and 2.8×10^{-3} V (approximately 180 μV/m) in the Northern and Southern Hemispheres.

could not be removed from the dc channel of Figures 1 and 2. However, this variation is a slowly varying smooth curve; hence, rapid variations seen in the 60 to 80° INL regions can be interpreted as varying dc fields. The magnitude of these N–S directed fields is in the 10 to 100 mV/m range and is in good agreement with the rocket results from both Ba clouds (Wescott *et al.*, 1969) and probes (Aggson, 1969). One notes that the largest dc fields occur at the time of maximum activity in the 4 to 16 Hz channel, with the start of small variations commencing near the onset of the 4 to 16 Hz irregularity signal. DC fields have not been seen without the presence of the 4 to 16 Hz signal.

C. ELF ELECTRIC FIELDS

The two higher spectrometer channels on OGO 6 tend to follow each other relatively closely. The 256 to 1024 Hz variations plotted in Figures 1 and 2 exemplify the general behavior in that they usually peak before the main onset of the 4 to 16 Hz signal. The poleward decrease in the signal occurs before the decrease in the 4 to 16 Hz signal, usually in the regions of strong dc fields. Typically, field strengths vary between 50 and 100 μV/m, occasionally reaching several hundred μV/m.

This is in good agreement with signal levels observed by OV1 10. Figure 5 shows the distribution of orbital segments where the signal intensity in the 300–3000 Hz band exceeded 1.0 or 2.8 mV. Taking 15 m as the effective antenna length, this corresponds to rms values of 66 and 180 μV/m, respectively. Although the data gaps between 200 to 400 hr and 1400 to 1600 hr limit complete comparison with Taylor and Gurnett's (1968) plots of the diurnal distribution of these signals observed by means of the magnetic component, it is clear from both the spatial diurnal distribution and the magnitudes that these signals are propagating in the whistler mode. Thus, there is not any reason to view them as electrostatic waves.

References

Aggson, T. L.: 1969, in *Atmospheric Emissions* (ed. by B. M. McCormac and A. Omholt), Van Nostrand Reinhold Publishing Corporation, New York, p. 305.

Axford, W. I.: 1969, in *Magnetospheric Physics* (ed. by D. J. Williams and G. D. Mead), American Geophysical Union, Washington, D.C., p. 421.

Axford, W. I. and Hines, C. O.: 1961, *Can. J. Phys.* **39**, 1433.

Föppl, H., Haerendel, G., Haser, L., Loidl, J., Lütjens, P., Lüst, R., Nielzner, F., Meyer, B., Neuss, H. and Rieger, E.: 1967, *Planet. Space Sci.* **15**, 357.

Haerendel, G. and Lüst, R.: 1970, this volume, p. 213.

Heppner, J. P.: 1969, in *Atmospheric Emissions* (ed. by B. M. McCormac and A. Omholt), Van Nostrand Reinhold Publishing Corporation, New York, p. 251.

Heppner, J. P., Aggson, T. L., and Maynard, N. C.: 1968, Symposium on the Physics of the Magnetosphere, Washington, D.C., Sept. 1968, Abstract.

Simon, A.: 1963, *Phys. Fluids* **6**, 382.

Taylor, W. L. and Gurnett, D. A.: 1968, *J. Geophys. Res.* **73**, 5615.

Wescott, E. M., Stolarik, J. D., and Heppner, J. P.: 1969, *J. Geophys. Res.* **74**, 3469.

Wescott, E. M., Heppner, J. P., Stolarik, J. D., Halliman, T., Davis, T. N., and Romick, G. J.: 1969, Rapid Ray Motions Through Barium Ion Clouds and Associated Auroras, AGU Annual Meeting Washington, D.C., April, Abstract.

Wescott, E. M., Stolarik, J. D., and Heppner, J. P.: 1970, this volume, p. 229.

PART V

WAVE-PARTICLE INTERACTIONS

HIGH FREQUENCY ELECTROSTATIC WAVES IN THE MAGNETOSPHERE

C. F. KENNEL

Dept. of Physics, University of California, Los Angeles, Calif., U.S.A.

and

R. W. FREDRICKS and F. L. SCARF

Space Sciences Lab., TRW Systems Group, Redondo Beach, Calif., U.S.A.

Abstract. Several theories of high frequency electrostatic instabilities are reviewed, and examples of relevant OGO 5 observations are presented.

1. Introduction

While electrostatic waves dominate the phenomenology of controlled thermonuclear fusion and other laboratory plasma experiments, they have received relatively little attention from the space plasma physics community. One reason for this seems to be historical. Because electromagnetic waves penetrate to the earth's surface, much was already known about them prior to the satellite era; thus, it was natural to investigate first the well formulated questions relating to electromagnetic wave phenomenology. Whatever the reason, electrostatic waves have only recently been discovered in deep space. This dearth of experimental knowledge makes electrostatic waves one of the last remaining frontiers of magnetospheric physics.

A brief review will be given of several instabilities in which high frequency electrostatic waves are generated by electrons. While there are many hot ion electrostatic instabilities (some of which will undoubtedly be important for space physics) we restrict ourselves to electron instabilities, primarily because present electrostatic detectors are not yet sensitive at the low frequencies relevant to most ion instabilities. Finally, examples of several magnetospheric electrostatic emissions observed on OGO 5 will be discussed.

2. General Electrostatic Dispersion Relation

The general dispersion relation for electrostatic waves in a uniform plasma and straight magnetic field has been given by Rosenbluth (1965),

$$K^2 = \sum_{+,-} \omega_{p\pm}^2 \sum_n \int \mathrm{d}^3 v \frac{\left(K_\parallel \dfrac{\partial F^\pm}{\partial v_\parallel} + \dfrac{n\Omega_\pm}{v_\perp} \dfrac{\partial F^\pm}{\partial v_\perp}\right) J_n^2}{\nu + n\Omega_\pm + K_\parallel v_\parallel} \tag{1}$$

where $\nu = \omega + i\gamma$ is the wave frequency, $K_\parallel$ and $K_\perp$, the wave vector components parallel and perpendicular to the magnetic field, and $K^2 = K_\parallel^2 + K_\perp^2$ ($\gamma < 0$ denotes

B. M. McCormac (ed.), Particles and Fields in the Magnetosphere.

instability). $\sum_{+,-}$ denotes a summation over ions (+) and electrons (−), and $\sum_n$ denotes summation over cyclotron harmonics, where n is an integer. $\omega_{p\pm}$ is the ion or electron plasma frequency, and $\Omega_\pm$ the gyrofrequency, defined by

$$\omega_{p\pm}^2 = \frac{4\pi Ne^2}{M_\pm}; \quad \Omega_\pm = \pm\frac{eB_0}{M_\pm c} \tag{2}$$

where B_0 is the magnetic field strength, N is the equilibrium number density, e is the electronic charge, $M_\pm$ the ion or electron mass, and c the speed of light. Gaussian units will be used throughout. The equilibrium velocity distribution functions $F^\pm(V_\perp, V_\parallel)$ are normalized to unity. $V_\parallel$ and $V_\perp$ are the parallel and perpendicular velocities, respectively. J_n denotes a Bessel function of integer order n and argument $K_\perp V_\perp/\Omega_\pm$. This argument will be dropped wherever the meaning is clear. Special solutions to Equation (1) will be derived in the following sections.

3. Thermal Anisotropy Instabilities

When $F^\pm$ are bi-Maxwellians, with parallel and perpendicular temperatures $T_\parallel$ and $T_\perp$ respectively, Hall and Heckrotte (1964) and Shima and Hall (1965) have shown that unstable roots are possible only when

$$n + \tfrac{1}{2} < |\omega/\Omega_\pm| < n + \frac{T_\perp - T_\parallel}{T_\perp} \tag{3}$$

and

$$|\omega/\Omega_\pm| < \frac{T_\perp - T_\parallel}{T_\parallel}.$$

While instability remains to be proven, the above theorem enables us to draw the following conclusions for bi-Maxwellian distributions:

(a) There are no firehose-type ($T_\parallel > T_\perp$) electrostatic instabilities.

(b) Instabilities are only possible when $T_\perp > 2T_\parallel$.

(c) When the thermal anisotropy is relatively small, any unstable roots must lie near half-harmonics of the cyclotron frequency; for instance when $T_\perp/T_\parallel = 3$, the unstable waves can lie only in the range $\frac{1}{2} < |\omega/\Omega_\pm| < \frac{2}{3}$ and $\frac{3}{2} < |\omega/\Omega_\pm| < \frac{5}{3}$.

4. Electrostatic Electron Cyclotron Waves

Hall *et al.* (1965) found that small numbers of cold particles can be crucial in producing instabilities near but above cyclotron harmonics. Let the number density of cold and hot electrons be qN and ΔN, respectively, so that $q + \Delta = 1$. For $\omega \gg \omega_{p+}$ ion motions may be neglected altogether. If the perpendicular wavelength is longer than the cold electron gyroradius, the Rosenbluth dispersion Equation (1) reduces to

$$\frac{K^2}{\omega_p^2} = q\left\{\frac{K_\parallel^2}{\nu^2} + \frac{K_\perp^2}{2\Omega^2}\left(\frac{\Omega}{\nu-\Omega} - \frac{\Omega}{\nu+\Omega}\right)\right\} + \Delta \sum_n \int d^3v \frac{\left(K_\parallel \frac{\partial F}{\partial v_\parallel} + \frac{n\Omega}{v_\perp}\frac{\partial F}{\partial v_\perp}\right) J_n^2}{\nu + n\Omega + K_\parallel v_\parallel} \tag{4}$$

where $F(V_\perp, V_\parallel)$ is now the hot electron distribution function, and the $\pm$ subscript has been dropped since all quantities refer to electrons. If $|(\nu-\Omega)/\Omega| \ll 1$, we may drop all but the $n=-1$ terms in Equation (4). Dropping higher n terms requires $K_\parallel a/\Omega \ll 1$, where a is a typical hot electron speed; consequently,

$$\frac{K^2}{\omega_p^2} = \frac{qK_\perp^2}{2\Omega(\nu-\Omega)} + \Delta \int d^3v \frac{\left(K_\parallel \frac{\partial F}{\partial v_\parallel} - \frac{\Omega}{v_\perp}\frac{\partial F}{\partial v_\perp}\right) J_1^2}{\nu - \Omega + K_\parallel v_\parallel}. \tag{5}$$

We search for instabilities with $|(\nu-\Omega)/K_\parallel a| \ll 1$. Furthermore, if $|\gamma/\omega| \ll 1$ the resonant denominator in Equation (5) may be expanded using the DIRAC relation:

$$\frac{1}{\nu - \Omega + K_\parallel v_\parallel} \approx \frac{1}{K_\parallel v_\parallel}\left(1 - \frac{\omega-\Omega}{K_\parallel v_\parallel} + \cdots\right) + \frac{i\pi}{|K_\parallel|}\delta\left(\frac{\omega-\Omega}{K_\parallel} + v_\parallel\right), \tag{6}$$

whereupon, if F is even in $V_\parallel$,

$$\frac{K^2}{\omega_p^2} = \frac{qK_\perp^2}{2\Omega(\nu-\Omega)} - \frac{\Delta}{a^2} + \frac{i\pi\Delta}{|K_\parallel|}\int d^3v J_1^2\left[K_\parallel \frac{\partial F}{\partial v_\parallel} - \frac{\Omega}{v_\perp}\frac{\partial F}{\partial v_\perp}\right]\delta\left(\frac{\omega-\Omega}{K_\parallel} + v_\parallel\right) \tag{7}$$

where a is defined more precisely by

$$\frac{1}{a^2} = -\int d^3v \frac{1}{v_\parallel}\frac{\partial F}{\partial v_\parallel}. \tag{8}$$

For $F(V_\perp, V_\parallel) = f(V_\perp)\, g(V_\parallel)$, where g is Maxwellian, a is the parallel thermal speed. Then since $(\nu-\Omega)/K_\parallel a \ll 1$ we may drop the $\partial F/\partial v_\parallel$ term in Equation (7), and since $|\gamma/\omega| \ll 1$, Equation (7) reduces approximately to

$$\frac{\omega-\Omega}{\Omega} + \frac{i\gamma}{\Omega} = \frac{q}{2}\frac{\omega_p^2}{\Omega^2}\frac{K_\perp^2}{K^2 + K_H^2} - i\Delta\sqrt{\frac{\pi}{2}}\frac{\omega_p^2}{K_2 + K_H^2}\frac{\omega-\Omega}{K_\parallel a}R_1 \tag{9}$$

where K_H is the Debye wave number of the hot electron gas, $K_H^2 \equiv \Delta\omega_p^2/a^2$, and the sign of R_1 determines instability:

$$R_1 \equiv 2\pi \int_0^\infty dv_\perp J_1^2 \frac{\partial F}{\partial v_\perp} = 2\pi \int df\, J_1^2. \tag{10}$$

Since $(\omega-\Omega)/\Omega > 0$, $R_1 > 0$ implies instability; however, this necessarily requires

$\partial f/\partial v_{\perp} > 0$ somewhere. This condition is satisfied by loss cone distributions, two humped energy distributions, and so on. Let U denote the location of the maximum positive $\partial f/\partial v_{\perp}$. Then, R_1 is most likely to be positive if $K_{\perp}$ is chosen so that J_1^2 has its first maximum near U, i.e., $K_{\perp} \approx \Omega/U$.

The above calculation will be self consistent if $(\omega - \Omega)/K_{\parallel} a \ll 1$ for $K_{\perp} \approx \Omega/U$ and $K_{\parallel} \ll \Omega/a$. For two humped distributions, $U \gtrsim a$, so that $K_{\parallel} \ll K_{\perp}$. Then, using Equation (9), we see that with $K \approx K_{\perp} \approx \Omega/\omega$,

$$\frac{\omega - \Omega}{K_{\parallel} a} \approx \frac{q}{2} \frac{\Omega}{K_{\parallel} a} \left[\frac{\Omega^2}{\omega_p^2} + \Delta \frac{U^2}{a^2} \right]^{-1}. \tag{11}$$

Thus, we require $q/2[\Omega^2/\omega_p^2 + \Delta U^2/a^2]^{-1} \ll K_{\parallel} a/\Omega \ll 1$. Therefore, if q is sufficiently small, or Ω^2/ω_p^2 or U^2/a^2 sufficiently large, instabilities near but above the electron cyclotron frequency can be generated.

5. Upper Hybrid Flute Instability of Pearlstein et al. (1966)

For 'flute' modes ($K_{\parallel} = 0$), Equation (4) reduces to

$$\begin{aligned} \frac{K^2}{\omega_p^2} \left\{ \frac{\nu^2 - \omega_H^2}{\nu^2 - \Omega^2} \right\} &= \Delta \int \frac{d^3 v}{v_{\perp}} \frac{\partial F}{\partial v_{\perp}} \sum_n \frac{J_n^2}{n + \nu/\Omega} \\ &= \Delta \int \frac{d^3 v}{v_{\perp}} \frac{\partial F}{\partial v_{\perp}} \left(1 - \frac{\nu}{\Omega} \sum_n \frac{J_n^2}{n + \nu/\Omega} \right) \end{aligned} \tag{12}$$

where $\omega_H = (q\omega_p^2 + \Omega^2)^{1/2}$, the cold plasma upper hybrid frequency. $\sum_n J_n^2 = 1$ was used in arriving at the second form of Equation (12). For $Ka/\Omega \gg 1$, we may replace J_n^2 by its average $\Omega/\pi K v_{\perp}$. Then $\sum_n (J_n^2/(n + \nu/\Omega))$ reduces to $(\Omega/K v_{\perp}) \cot(\pi v/\Omega)$, so that Equation (12) becomes

$$\frac{\nu^2 - \omega_H^2}{\nu^2 - \Omega^2} = \Delta \frac{\omega_p^2}{K^2} \int \frac{d^3 v}{v_{\perp}} \frac{\partial F}{\partial v_{\perp}} \left(1 - \frac{\nu}{K v_{\perp}} \cot \frac{\pi \nu}{\Omega} \right). \tag{13}$$

We consider cases where both $|\omega/\Omega| \gg 1$ and $|\gamma/\Omega| \gg 1$. Then, since $\cot \pi\nu/\Omega \to +i$ when $\gamma/\Omega < -1$,

$$\frac{\nu^2 - \omega_H^2}{\nu^2 - \Omega^2} = \frac{\Delta \omega_p^2}{K^2} \int \frac{d^3 v}{v_{\perp}} \frac{\partial F}{\partial v_{\perp}} \left(1 - i \frac{\nu}{K v_{\perp}} \right). \tag{14}$$

For the 'loss-cone' distributions typical of mirror confined plasmas, $F(V_{\perp} = 0) = 0$, which also implies $\partial F/\partial V_{\perp} > 0$ somewhere. Then, if $|\gamma/\omega| < 1$, Equation (14) has the approximate solutions:

$$\omega^2 \simeq \omega_H^2; \quad \gamma = -\frac{\Delta \omega_p^4}{2K^3} \int \frac{d^3 v}{v_{\perp}} \frac{\partial F}{\partial v_{\perp}}. \tag{15}$$

Thus, if $\int (d^3 v/v_{\perp})(\partial F/\partial v_{\perp}) > 0$, instability will occur at the hybrid frequency, which by assumption is well above the cyclotron frequency. Perkins (1968) has generalized this instability to $K_{\parallel} \neq 0$, where it becomes convective, and has applied it to the mono-energetic electron fluxes occasionally observed in the auroral ionosphere.

6. Anti-Loss Cone Instabilities

Berk and Galeev (1967) found that guiding center drifts preferentially carry $V_{\parallel} \approx 0$ particles to the walls of toroidal confinement devices. By contrast with mirror devices, this creates an 'anti-loss cone' with a deficit of small $V_{\parallel}$ particles. A similar situation exists in the magnetosphere. Roederer (1968) has shown that on the earth's nightside, $V_{\parallel} \approx 0$ particles drift to the magnetopause (presumably to be lost), while $V_{\parallel} \neq 0$ particles remain trapped on drift orbits closed within the magnetosphere. On auroral lines of force, the anti-loss cone is larger, in fact, than the loss cone.

A hot electron distribution function with anti-loss cone properties is

$$F(v_{\perp}, v_{\parallel}) = \frac{f_{\mathrm{M}}(T_{\perp})}{\sqrt{2\pi}} \left\{ \frac{e^{-v_{\parallel}^2/2a_1^2} - \varrho\, e^{-v_{\parallel}^2/2a_2^2}}{a_1 - \varrho a_2} \right\} \tag{16}$$

where $f_{\mathrm{M}}(T_{\perp})$ is a Maxwellian distribution of perpendicular velocities, a_1 is the thermal speed of trapped electrons, and $a_2 < a_1$ is the thermal speed of the missing particles. $\varrho\,(0 \leqslant \varrho \leqslant 1)$ is a depth parameter describing the strength of the anti-loss cone distribution. When $\varrho = 1$, $F(V_{\perp}, 0) = 0$. In the magnetosphere, where acceleration occurs in time scales comparable with the drift time scale, it is unlikely that ϱ is often near one. Let us consider the ion acoustic mode with $K_{\perp} = 0$; then Equation (1) reduces to

$$K^2 = \sum_{+,-} \omega_{p\pm}^2 \int \frac{\partial F^{\pm}/\partial v_{\parallel}}{(v/K) + v_{\parallel}}. \tag{17}$$

For phase velocities above the ion thermal speed and below the two electron thermal speeds, the resonant denominator may be expanded, assuming $|\gamma/\omega| \ll 1$, as follows:

for electrons,

$$\frac{1}{(v/K) + v_{\parallel}} \simeq \frac{1}{v_{\parallel}} \left(1 - \frac{\omega}{K v_{\parallel}} + \cdots \right) + i\pi \operatorname{sign} K \delta \left(\frac{\omega}{K} + v_{\parallel} \right) \tag{18a}$$

and for ions,

$$\frac{1}{(v/K) + v_{\parallel}} \simeq \frac{K}{\omega} \left(1 - \frac{K v_{\parallel}}{\omega} + \cdots \right) + i\pi \operatorname{sign} K \delta \left(\frac{\omega}{K} + v_{\parallel} \right). \tag{18b}$$

Carrying out the analysis neglecting the exponentially small resonant ion term it is concluded that:

$$\omega^2 = \frac{K^2 \tilde{c}_s^2}{\dfrac{a_2 - \varrho a_1}{a_1 - \varrho a_2} + K^2 \tilde{\lambda}_{\mathrm{D}}^2}; \quad \frac{\gamma}{\omega} = -\sqrt{\frac{\pi}{3}} \frac{\omega}{|K|(a_1 - \varrho a_2)} \frac{a_1}{a_2} \left[\varrho - \frac{a_2^2}{a_1^2} \right] \tag{19}$$

where we have defined an ion acoustic speed and Debye length based upon the geometric mean of the two electron temperatures:

$$\tilde{c}_s^2 \equiv \frac{M_-}{M_+} a_1 a_2 = \frac{\sqrt{T_1 T_2}}{M_+}; \quad \tilde{\lambda}_D^2 = \tilde{c}_s^2/\omega_{p+}^2. \tag{20}$$

For $\varrho > a_2/a_1$ and $K\tilde{\lambda}_D \ll 1$ there are non-resonant instabilities. For $(a_2/a_1)^2 < \varrho < a_2/a_1$ the mode propagates, but is unstable to resonant electrons. $\varrho < (a_2/a_1)^2$ is clearly damped. For $K\tilde{\lambda}_D \gtrsim 1$, $\omega \to \omega_{p+}$.

Only energetic electrons, whose motions are dominated by guiding center drifts, probably have an anti-loss cone in the magnetosphere. Lower energy electrons have a different dynamics and presumably no anti-loss cone. For simplicity, these low energy electrons are treated as cold. Then, performing the analysis as before we find

$$\omega^2 = \frac{K^2 \tilde{c}_s^2 (1 + qM_+/M_-)}{\Delta \dfrac{a_2 - \varrho a_1}{a_1 - \varrho a_2} + K^2 \tilde{\lambda}_D^2}; \quad \frac{\gamma}{\omega} = -\Delta \sqrt{\frac{\pi}{3}} \frac{\omega}{|K|(a_1 - \varrho a_2)} \frac{a_1}{a_2} \left[\varrho - \left(\frac{a_2}{a_1}\right)^2\right] \tag{21}$$

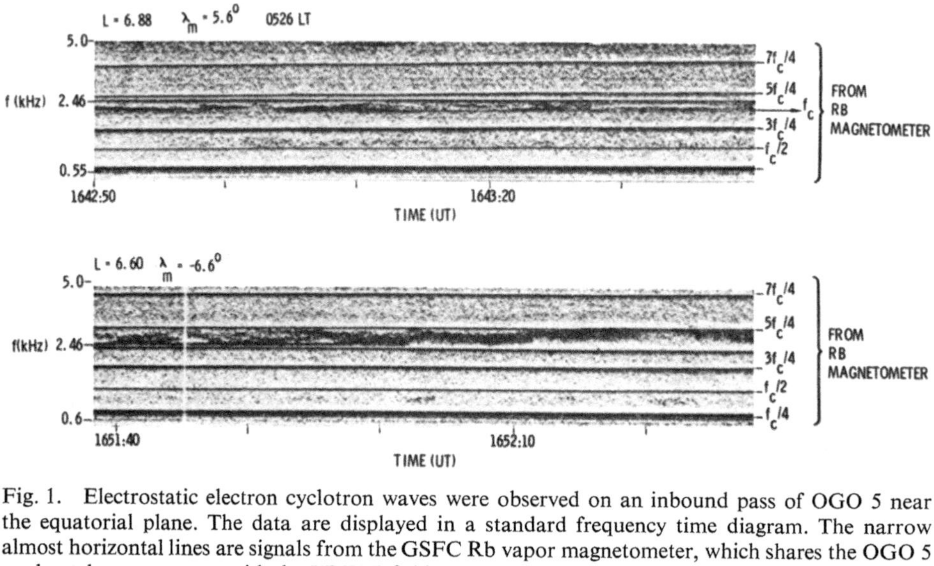

Fig. 1. Electrostatic electron cyclotron waves were observed on an inbound pass of OGO 5 near the equatorial plane. The data are displayed in a standard frequency time diagram. The narrow almost horizontal lines are signals from the GSFC Rb vapor magnetometer, which shares the OGO 5 analog telemetry system with the TRW E field sensor output. These signals which are at the quarter electron cyclotron frequency and its harmonics enable us to identify the local cyclotron frequency. The cyclotron wave is the fuzzy smudge riding above the cyclotron frequency line. Two pieces of the same event ten minutes apart are shown, indicating that as the satellite moved inbound, the local cyclotron frequency increased, and the emission continued to ride above the cyclotron frequency.

where q and Δ are defined as before and where ion and cold electron damping has been neglected. A very small fraction of cold electrons, $q > M_-/M_+$, increases the phase velocity of the ion acoustic wave. When $K\lambda_D \geqslant 1$, $\omega \to q\omega_{p-}$. Since $\gamma \propto \omega^2$, the growth rate tends to increase as well. In order that $\omega/Ka_2 < 1$, the system parameters must satisfy the constraint

$$1 + q\frac{M_+}{M_-} < \frac{a_2}{a_1}\frac{M_+}{M_-}\left[\Delta\frac{a_2 - \varrho a_1}{a_1 - \varrho a_2} + K^2\tilde{\lambda}_{\mathrm{D}}^2\right]. \tag{22}$$

Thus, without cold electrons the anti-loss cone distribution generates waves near ω_{p+}, and with cold electrons near $q\omega_{p-}$.

7. Discussion

Since the above analyses are clearly limited in generality, rigor, and applicability to realistic magnetospheric conditions, they can only illustrate what might occur. However, it is interesting to note that intense, presumably electrostatic emissions have been

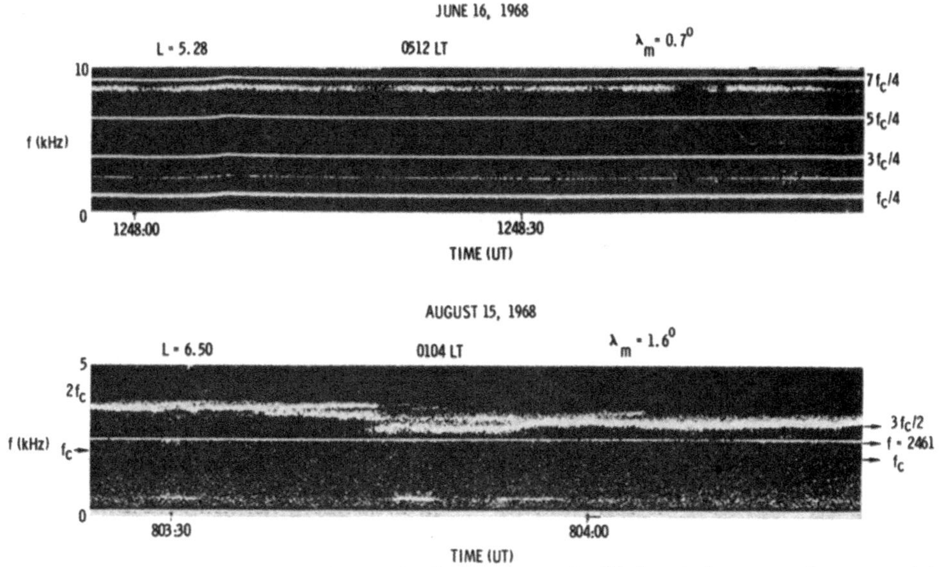

Fig. 2. 3/2 Cyclotron waves. A similar f-t display shows $f \approx 3/2 f_c$ emissions near the equatorial plane. The local cyclotron frequency was obtained either with the help of the GSFC Rb magnetometer, or with the UCLA fluxgate magnetometer (Coleman and Russell, 1969).

detected on OGO 5 on nightside auroral lines of force (Kennel, *et al.*, 1969) in the frequency ranges mentioned above. Figure 1 shows an electron cyclotron emission, Figure 2 shows two examples of $\frac{3}{2}$ cyclotron emissions, and Figure 3 shows two examples of 6 to 9 kHz emissions which are well above the electron cyclotron frequency but could be near ω_{p-} or ω_H if $N \approx (0.3\text{–}1)/\mathrm{cm}^3$.

It is premature, both theoretically and experimentally, to identify these observations with the above simple theories. However, our theoretical discussion does indicate that several difficult experimental questions must be resolved before a definitive understanding of magnetospheric electrostatic emissions is possible.

(a) How common is a second peak in the electron energy distribution? This involves high time resolution measurements of 1 to 100 keV electrons, including especially the 10 to 40 keV range, which is still poorly documented.

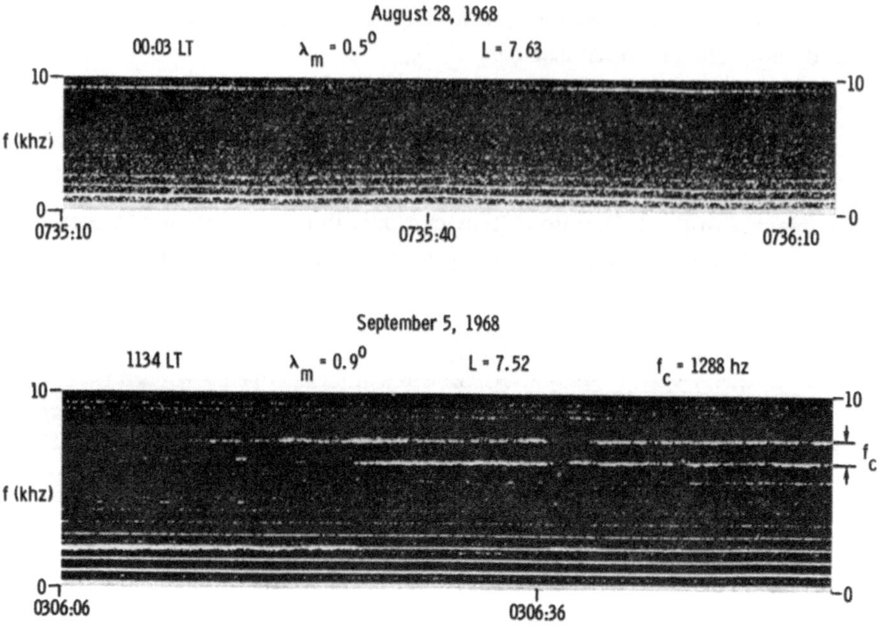

Fig. 3. High frequency waves. In both these cases, the Rb traces lie at low frequencies, indicating the emission is well above the electron cyclotron frequency. The identification of f_c was checked using the UCLA fluxgate magnetometer as well. These emissions could be near the plasma or hybrid frequency if the electron density were $(0.3 - 1)/cm^3$.

(b) How prevalent are anti-loss cone distributions?

(c) What is the ratio of hot to cold electron densities? Even very small cold electron densities can significantly alter the nature of the possible instabilities.

A similar list applies to ion instabilities.

Acknowledgments

The theoretical work was supported at UCLA by National Aeronautics and Space Administration-NGR-05-007-190, and the experimental work at TRW Systems, by National Aeronautics and Space Administration-5-9278. It is a pleasure to thank our colleagues in the UCLA fluxgate magnetometer group, P. J. Coleman and C. T. Russell, for their unstinting generosity in providing magnetic field data to us. We also wish to thank J. P. Heppner for allowing us to display and draw conclusions from the Rb magnetometer traces. J. M. Cornwall and C. S. Liu made useful comments on the manuscript.

References

Berk, H. L. and Galeev, A. A.: 1967, *Phys. Fluids* **10**, 441.
Coleman, P. J. and Russell, C. T.: 1969, private communication.
Hall, L. S. and Heckrotte, W.: 1964, *Phys. Rev.* **134**, A1474.
Hall, L. S., Heckrotte, W., and Kammash, T.: 1965, *Phys. Rev.* **139**, A1117.
Kennel, C. F., Fredricks, R. W., and Scarf, F. L.: 1969, *Trans. Amer. Geophys. Union* **50**, 291.
Pearlstein, L. D., Rosenbluth, M. N., and Chang, D. B.: 1966, *Phys. Fluids* **9**, 953.
Perkins, F. W.: 1968, *J. Geophys. Res.* **73**, 6631.
Roederer, J.: 1968, in *Earth's Particles and Fields* (ed. by B. M. McCormac) Reinhold Publishing Corporation, New York, p. 193.
Rosenbluth, M. N.: 1965, *Plasma Physics*, IAEA, Vienna, 499.
Shima, Y. and Hall, L. S.: 1965, *Phys. Rev.* **139**, A1115.

MUTUALLY INTERACTING INSTABILITIES IN THE MAGNETOSPHERE

JOHN M. CORNWALL*
Dept. of Physics, University of California, Los Angeles, Calif., U.S.A., and Aerospace Corporation, El Segundo, Calif., U.S.A.

1. Introduction

It may happen that a magnetospheric plasma can admit two or more instabilities of different types, for example, a high frequency instability (electrostatic or cyclotron) and a low frequency instability (hydromagnetic or drift wave). In many such cases, the plasma wave dynamics cannot even approximately be described by a simple linear superposition of the various instabilities which are possible, because slight changes in the plasma produced by one instability may cause major changes in the growth rate of another. It is always possible to describe this interaction *mathematically* by saving enough terms in the Vlasov-Maxwell equations (e.g., mode-mode coupling). To some extent a physically concise and useful description is supplied by following the analysis of Dupree (1966, 1967, 1968), certain aspects of which were rediscovered independently by the author in the course of the present work. In this analysis, one takes into account the influence of the instability wave fields on particle orbits which are used to calculate the dispersion relation. In this fashion, growth rates taken from the dispersion relation depend explicitly on wave intensities, in contrast to the usual linear stability calculations. In certain cases, this explicit dependence on wave intensities is essential for description of equilibrium, since the equilibrium conditions of quasi-linear theory (Kennel and Engelmann, 1966) cannot be applied.

Even with a simplified physical picture, it is extremely difficult to carry out a self-consistent calculation of the mutual interaction of two or more instabilities, and we shall not attempt this. Sometimes it is possible to give a simple verbal description. For example, a drift wave instability may give rise to strong radial diffusion (Cornwall, 1968). Particles which diffuse inward conserving the first invariant will increase their perpendicular energy and pitch angle anisotropy, which will promote the growth of high frequency instabilities. In turn, these high frequency instabilities will (usually) tend to stabilize the drift wave instability. The present work is primarily concerned with this second point.

In Section 2 some general aspects of the equilibrium problem are described briefly. Further sections concentrate on one aspect of a specific example: an Alfvén drift mode in an auroral plasma which is subject to an unspecified high frequency instability that causes pitch angle diffusion of the plasma. We calculate the influence of the high frequency instability of the drift wave growth rate, in terms of the pitch angle

* Alfred P. Sloan Foundation Fellow.

diffusion coefficient (which depends on the high frequency wave intensity). It turns out to be easy to bring the drift mode to equilibrium long before ordinary quasi-linear equilibrium is reached. As mentioned above, on a time scale long compared to drift wave periods, one effect of the drift wave is radial diffusion which can serve to enhance the high frequency instabilities. On the drift wave time scale there are other interesting phenomena: the pitch angle anisotropy will be modulated at the drift wave frequency, which results in pronounced modulation of the intensity of cyclotron instabilities driven by pitch angle anisotropy (Coroniti and Kennel, 1969).

2. The Equilibrium Problem

It would be possible to give an extensive mathematical description of what happens when a plasma subject to two or more instabilities tries to relax toward equilibrium by continuing the quasi-linear analysis of Kennel and Engelmann (1966). Such an analysis in full generality would be of little interest to space physicists. Instead, let us be more specific and consider the magnetosphere to be subject to two kinds of instabilities: low frequency drift waves (the instability is driven, in part, by spatial gradients) and high frequency electrostatic or cyclotron instabilities (driven, in part, by pitch angle anisotropy). We label the instabilities with an index $i=1$ and 2 using 1 for low and 2 for high frequency. (Extra labels to distinguish various wave numbers in a given instability will be suppressed.)

In the usual linear analysis, each instability has a growth rate given by

$$\gamma_i = D_i f \tag{1}$$

where f is the distribution function and D_i a linear operator which depends on phase-space coordinates $\mathbf{x}$, $\mathbf{v}$. At the same time, the plasma diffuses according to

$$\frac{\partial f}{\partial t} = \sum (E_i D_i)^* M_{ij} (E_j D_j) f + S - \lambda f \tag{2}$$

where the sum is over resonant wave-particle interactions, E_i is a wave amplitude, and M_{ij} a positive definite matrix. S is a source of plasma and λ a loss rate; both such terms are present in the magnetosphere (e.g., convective source, atmospheric loss).

In the absence of mode-mode coupling or other relaxation mechanisms, the equation for wave intensity is

$$\frac{\partial}{\partial t} |E_i|^2 = 2\gamma_i |E_i|^2 . \tag{3}$$

The high frequency mode ($i=2$) has a loss term which should be added to Equation (3) simply from transport out a flux tube. The question is, what relaxes the low frequency drift mode? Ordinary quasi-linear theory would suggest that f diffuses in such a way that

$$D_1 f = 0 . \tag{4}$$

But this occurs over such a long time scale that other physical effects must be important (e.g., Krall, 1967). We propose a simple model in which the growth rate of mode 1 is modified by terms which describe the influence of mode 2 upon it, so that we have (as will be described in the following sections)

$$\gamma_1 = D_1 f + |E_2|^2 Q_{12} f \tag{5}$$

where Q_{12} is another linear operator. It is possible then for γ_1 to be zero with a finite drift wave amplitude, even though Equation (4) is not satisfied.

The problem does not close on itself here. The amplitude E_2 depends on the spatial diffusion terms in Equation (2), because such spatial diffusion drives the plasma to higher pitch angle anisotropy. The spatial diffusion in Equation (2) depends on $|E_1|^2$, and it becomes clear that neither mode can in any way be described independently of the other. It is clear also that a self-consistent solution of the entire problem is extremely difficult.

In the rest of this paper, we carry out a very modest program, in which we look at Equation (5) for a specific drift mode, and ask whether this mode can be brought to equilibrium by a high frequency instability of unspecified type, which causes pitch angle diffusion at a rate consistent with experimental facts of the magnetosphere. The next section introduces the necessary concepts, as elucidated by Dupree and discovered independently by the author.

3. Influence of Wave Fields on Particle Orbits

To derive a dispersion relation for a wave in a plasma one usually begins with a description of an equilibrium plasma with no waves present. All that is needed is knowledge of the single particle orbits in the unperturbed magnetic field. This will be inappropriate if the waves actually distort the orbits to such an extent that they deviate appreciably from unperturbed orbits in a few wave periods. We have in mind a situation where a fairly broadband incoherent high frequency wave field causes random disturbances in the particles' trajectories much as collisions would. These disturbances must be taken into account when the effective collision frequency is comparable to the frequency of the mode whose dispersion relations we are trying to derive.

Other situations can also be envisaged in which a reasonably monochromatic and coherent wave field perturbs the particle orbits. For example, Ivanov *et al.* (1968) discuss stabilization of an instability by application of a high frequency monochromatic wave; Coroniti and Kennel (1969) discuss the modulation of instability growth rates by application of low frequency fields.

The standard method of deriving a dispersion relation is as follows: the distribution function and perturbing fields are expanded in a Fourier series:

$$F = f + \sum_k \delta f_k(\mathbf{v}, t)\, e^{i\mathbf{k}\cdot\mathbf{x}} \tag{6}$$

$$\mathbf{E} = \sum_k \mathbf{E}_k\, e^{i(\mathbf{k}\cdot\mathbf{x} - \omega t)} \tag{7}$$

where f is the equilibrium function with no waves present. To first order in the wave fields the perturbed distribution function is

$$\delta f_k = (\text{I.V.}) + \int_0^t \mathrm{d}\tau U_k(t-\tau)\, P_k(\tau) f \,. \tag{8}$$

In Equation (8), (I.V.) means an initial value term which does not contribute to the dispersion relation, and

$$P_k(\tau) = -\frac{e}{m} e^{-i\omega\tau} \left(\mathbf{E}_k + \frac{1}{c}\mathbf{v}\times\mathbf{B}_k\right)\cdot\frac{\partial}{\partial\mathbf{v}} . \tag{9}$$

The evolution function U_k is given by

$$U_k(t) = e^{-i\mathbf{k}\cdot\mathbf{x}(t)} \tag{10}$$

where $\mathbf{x}(t)$ are the particle orbits in the unperturbed magnetic field with initial conditions $\mathbf{x}(0)=\mathbf{0}$. According to Krall and Rosenbluth (1963) these orbits are:

$$x = \frac{v}{\Omega}\sin(\Omega t + \phi) - \frac{v}{\Omega}\sin\phi ,$$

$$y = -\frac{v}{\Omega}\cos(\Omega t + \phi) + \frac{v}{\Omega}\cos\phi + v_{\mathrm{D}}t ,$$

$$z = ut . \tag{11}$$

Here V_D is the usual drift velocity; in a curl free field,

$$v_{\mathrm{D}} = \frac{1}{2\Omega B}\frac{\partial B}{\partial s}(v^2 + 2u^2) \tag{12}$$

where Ω is the (appropriately signed) cyclotron frequency, B the field strength, s an element of length along $\mathbf{B}$, v the velocity perpendicular to $\mathbf{B}$, and u the velocity along $\mathbf{B}$. In the standard way, substitution of Equation (10) and (11) into Equation (8) gives a series of resonance denominators of the type $\omega - k_{\|}u - k_{\perp}v_{\mathrm{D}} - n\Omega$; n is an integer.

In fact, the orbit Equations (11) are not correct because they omit the influence of the wave field itself on the orbits. Dupree (1966, 1967, 1968) has discussed the appropriate modifications in detail, and a simple example is given to illustrate the principles. Imagine a plasma with no magnetic field and perturbing wave vectors in the z direction only. The evolution function $U_k(t)$ satisfies the unperturbed Boltzmann equation:

$$\left(\frac{\partial}{\partial t} + iku\right) U_k(t) = 0 \qquad U_k(0) = 1 . \tag{13}$$

Obviously the solution is just $\exp(-ikut)$. Now suppose that we consider the 'unperturbed' Boltzmann equation to represent velocity space diffusion driven by the unstable wave field. Let D_v be the diffusion coefficient, a pure constant for simplicity;

then Equation (13) should be replaced by

$$\left(\frac{\partial}{\partial t} + iku - D_v \frac{\partial^2}{\partial u^2}\right) U_k(t) = 0 \qquad U_k(0) = 1 \tag{14}$$

whose solution is

$$U_k(t) = \exp\{-\, ikut - \tfrac{1}{3}D_v k^2 t^2\}. \tag{15}$$

We substitute Equation (15) in Equation (8), and encounter integrals of the type (as $t \to \infty$)

$$I = \int_0^\infty \mathrm{d}\tau \exp\{+\, i\tau(\omega - ku) - \tfrac{1}{3}D_v k^2 \tau^3\}. \tag{16}$$

This integral is evaluated simply in two limits:

$$\begin{aligned} &\omega - ku \to \infty \qquad I \to \frac{i}{\omega - ku} \\ &\omega - ku \to 0 \qquad I \to (k^2 D_v)^{-1/3}\, 3^{-2/3}\,\Gamma(1/3) \equiv \gamma^{-1} \simeq (k^2 D_v)^{-1/3}. \end{aligned} \tag{17}$$

The first equation in (17) is derived by the method of steepest descents and the second is trivial. It should be an adequate approximation to take

$$I \simeq \frac{i}{\omega - ku + i\gamma}. \tag{18}$$

In other words, the frequency ω is replaced by $\omega + i\gamma$ and diffusion has introduced an imaginary part to the frequency as if there actually were collisions.

Another approach, also discussed by Dupree, is to use Equation (10) for the evolution function, but to replace it by its average over the ensemble of particle orbits, which are diffusing because of the instability. For example, the orbit equation for z in Equation (11) is replaced by $z = (u + \delta u)\, t$, where δu is a Gaussian distributed fluctuating component. Suppose that there is diffusion in pitch angle only; then we have for Gaussian processes,

$$\langle(\delta u)^2\rangle = 2v^2 Dt \qquad \langle(\delta u)^{2m}\rangle = \frac{(2m)!}{2^m m!}\langle(\delta u)^2\rangle^m \tag{19}$$

where D is the pitch angle diffusion coefficient. Then we find

$$\langle U_k(t)\rangle = \exp\{-\, ikut - k^2 v^2 D t^3\}. \tag{20}$$

As long as the inequality

$$D \ll \Omega(\Omega/k_{\|} u)^2 \tag{21}$$

holds, we need not consider the fluctuations in the perpendicular velocity. Equation (20) does not agree perfectly with Equation (15), because we should not write $\delta z = t\delta u$ but rather $\delta z = \int u \mathrm{d}t$. The numerical error in our final expressions is small.

Our result then is: In the dispersion relations, replace all resonant denominators $\lambda - n\Omega$ by $\lambda + i\gamma - n\Omega$, where

$$\gamma = [k_{\|}^2 v^2 D]^{1/3} \qquad \lambda = \omega - k_{\|} u - k_{\perp} v_{\mathrm{D}} \tag{22}$$

and D is the pitch angle diffusion coefficient. Of course, we could derive similar results for diffusion in energy or combinations of the two cases.

4. Application to Drift Waves in the Magnetosphere

Several authors (e.g., Chamberlain, 1963; D'Angelo, 1969) have speculated that drift wave instabilities might be important in modulating auroral phenomena, leading to pulsations, for example. These instabilities may also play a role in radial diffusion leading to what is known phenomenologically as Bohm diffusion (Cornwall, 1968). Drift waves can occur in plasmas with spatial gradients and are characterized by low frequencies ($\lesssim 1$ Hz) and long growth times (up to 10 min in the magnetosphere). Because the growth times are long, ordinary quasi-linear diffusion may be insufficient to bring drift wave instabilities to equilibrium. Our calculations will show that high frequency instabilities can furnish the requisite damping mechanism. Ideally, one would then turn the problem around to see what effect the drift wave has on the high frequency instability (for instance, a drift wave which promotes radial diffusion can encourage anisotropic velocity-space distributions, as the plasma diffuses inward with $E_{\perp}/B \sim$ constant). This is too difficult for us, and we content ourselves with asking whether a high frequency instability which causes pitch angle diffusion of experimentally reasonable size can stabilize unstable drift waves.

There are a number of potentially unstable drift waves in the magnetosphere; we concentrate on a specific example. The conclusions should be more general than the particular example studied. Our study centers on the mode whose importance for magnetospheric physics was first pointed out by Chamberlain (1963). However, in his analysis, Chamberlain assumed infinite wavelength parallel to the earth's magnetic field. In fact, we must have $k_{\|} L \gtrsim 1$, where L is the length of the field line.

Let us begin with some simple numerology. In the magnetosphere, $k_{\|}$ might be of the order of 10^{-9} and a typical velocity for electrons might be 7×10^9 (all numbers in this section are in CGS units). If D, the pitch angle diffusion coefficient is $\gtrsim 3 \times 10^{-4}$ (corresponding to 1 hr lifetimes) during auroral activity, then from Equation (22) $\gamma \sim 0.25$. We expect typical drift wave frequencies to be of the order of 1, and growth rates less than this by 1 or 2 orders of magnitude. Therefore, γ is quite comparable to typical growth rates, and we can expect a substantial effect coming from the high frequency instability.

We take Maxwellian velocity distributions for auroral ions and electrons, and replace the gradient operators which occur in the dispersion relation by an inverse plasma scale length ε_{p}. In these equations, therefore, we set

$$f_0 \to (\alpha/\pi)^{3/2} \exp\{-\alpha(u^2 + v^2)\}. \tag{23}$$

The plasma scale length $\varepsilon_{\mathrm{p}}^{-1}$ is doubtless much shorter than the magnetic field scale length which turns out to mean that ω is somewhat greater than $k_\perp v_D$. We will ignore, therefore, $k_\perp v_D$ in the resonant denominators, although this is not highly accurate for magnetospheric parameters, especially for the ions.

The mode of interest to us has been studied for finite $k_\parallel$ by Mikhailovskii and Rudakov (1963); it is a socalled high β mode (β is the ratio of particle energy density to magnetic field energy density), but high only means $\beta > m/M$, which is easily fulfilled in the magnetosphere. For smaller β only electrostatic modes exist. In the limit $k_\parallel = 0$, the mode discussed by Chamberlain is recovered. We derive the electric field from a scalar potential plus a vector potential in the z direction:

$$E_y = - ik_\perp \phi , \qquad E_z = - ik_\parallel \phi + i \frac{\omega}{c} A . \tag{24}$$

The dispersion relations come from two of Maxwell's equations:

$$[k_\perp^2 - (\omega^2/c^2)]\, A = \frac{4\pi}{c} J_z - \frac{k_\parallel \omega}{c} \phi , \tag{25}$$

and

$$k^2 \phi = 4\pi \varrho + \frac{k_\parallel \omega}{c} A , \tag{26}$$

We can ignore ω^2/c^2 compared to k^2 for these low frequency modes, and also the $k_\parallel$ terms. After straightforward analysis the charge and current densities are:

$$4\pi J_z = \sum 2\alpha\omega_p^2 (\omega - \omega_0)\, e^{-b} I_0(b) \left[- \frac{A(\omega + i\gamma)}{ck_\parallel^2} W(Z) + \frac{\phi}{k_\parallel} W(Z) \right] \tag{27}$$

$$4\pi\varrho = \sum 2\alpha\omega_p^2 \left\{ e^{-b} I_0(b)\, (\omega - \omega_0) \times \left[- \frac{AW(Z)}{ck_\parallel} + \frac{\phi}{\omega + i\gamma} [W(Z) + 1] \right] - \phi \right\} \tag{28}$$

where,

$$W(Z) = \frac{1}{\sqrt{\pi}} \int_{-\infty}^{\infty} \mathrm{d}t \frac{te^{-t^2}}{z - t}, \qquad Z = \frac{(\omega + i\gamma)\,\alpha^{1/2}}{k_\parallel}$$

$$W(Z) \underset{Z \to 0}{\to} - 1 + 2Z^2 - iZ\sqrt{\pi}$$

$$W(Z) \underset{Z \to \infty}{\to} \frac{1}{2Z^2} - i\sqrt{\pi} Z e^{-Z^2}$$

$$I_0(b) = J_0(ib), \qquad b = k_\perp^2 / 2\alpha\Omega^2$$

$$\omega_0 = \frac{3}{2} \frac{k_\perp \varepsilon_p}{\alpha\Omega} \quad \left(\text{setting } 1 + \frac{2u^2}{v^2} \simeq 3 \right) . \tag{29}$$

The summation in Equations (27) and (28) is over charge species; when it becomes necessary to distinguish them labels e or i will be used. In the $k_\parallel = 0$ limit, the dispersion relations split into an electrostatic mode, and the mode discussed by Chamberlain. In fact, something like the following holds: $k_\parallel u_i \ll \omega \ll k_\parallel u_e$. It is important, therefore, to save $k_\parallel u$ for the electrons.

It seems likely that only the electrons will be seriously affected by the high frequency instabilities (Section 3), so we set $\gamma_i = 0$. Define

$$X = \sum 2\alpha\omega_p^2 [(\omega - \omega_0)/\omega]\, e^{-b} I_0(b)\, W(Z) \tag{30}$$

$$Y = \sum 2\alpha(\omega_p^2/\omega)\, [(\omega - \omega_0)\, e^{-b} J_0(b) - (\omega + i\gamma)]\,. \tag{31}$$

The approximate dispersion relation based on Equation (25) through (29) is:

$$k_\perp^2 \left[1 + \frac{Y}{X}\right] + \frac{\omega(\omega + i\gamma)\, Y}{c^2 k_\parallel^2} = 0\,. \tag{32}$$

We have dropped terms in ω^2/c^2k^2 or $k^2/2\alpha\omega_p^2$ compared to 1. For the conditions $k_\parallel u_i \ll \omega \ll k_\parallel u_e$, and equal electron and ion temperatures ($\alpha_i m = \alpha_e M$) we approximate X and Y by:

$$X \simeq 2\alpha\omega_p^2 \left[-\left(\frac{\omega + \omega_{0i}}{\omega}\right)(1 + i\sqrt{\pi}Z_e) + \left(\frac{\omega - \omega_{0i}}{\omega}\right) \frac{k_\parallel^2}{2\alpha_i(\omega + i\gamma)^2}\right] \tag{33}$$

$$Y \simeq \frac{2\alpha\omega_p^2}{\omega} [b_i(\omega_{0i} - \omega) - i\gamma] \tag{34}$$

where $Z_e = (\omega\alpha_e^{1/2})/k_\parallel$, and we have used $-\omega_{0e} = +\omega_{0i} \equiv \omega_0$, $b_e \ll 1$, $1 - e^{-b_i} I_0(b_i) \simeq b_i \equiv b$. For our purposes, the $k_\parallel^2$ term in X can be dropped; it is important for the quasi-electrostatic mode whose dispersion relation is approximately $X + k^2 = 0$.

Set $\omega = \omega_1 + i\omega_2$ with $\omega_2 \ll \omega_1$. The approximate solution of the dispersion relation (Equation (32)) is (v_A is the Alfvén velocity)

$$\omega_1 = \tfrac{1}{2}[\omega_0 \pm (\omega_0^2 + 4k_\parallel^2 v_A^2)^{1/2}] \tag{35}$$

$$\omega_2 = \frac{k_\parallel^2 v_A^2}{2\omega_1 - \omega_0} \times \left\{ \frac{b(\omega_0 - \omega_1)}{\omega_0 + \omega_1} \sqrt{\pi} Z_e + \gamma \left[-\frac{\omega_0}{\omega_1(\omega_0 + \omega_1)} + \frac{1}{b(\omega_0 - \omega_1)}\right]\right\}. \tag{36}$$

Both solutions are unstable ($\omega_2 > 0$) for $Z_e > 0$, $\omega_0 < 0$, and $k_\parallel^2 v_A^2 < 2\omega_0^2$ if the terms in γ are omitted. These terms always tend to stabilize the modes as may be easily checked. As a consequence, the drift mode can come to equilibrium ($\omega_2 = 0$) with a finite drift wave intensity and a finite high frequency instability intensity (γ is proportional to this intensity).

As for numbers, with $k_\perp \sim 10^{-7}$ and $\varepsilon_p \sim 10^{-8}$ we can get $|\omega_0| = 1$ for auroral particles high in the magnetosphere. The parameter b will be of the order of 1. (So in Equation (36), b should really be replaced by $1 - e^{-b} I_0(b)$.) The plasma temperature will be of the order of kilovolts. Under these circumstances $\omega_2 \simeq 0$ for γ in the range 10^{-1}–10^{-3} which covers a broad range of pitch angle diffusion coefficients ($D \sim 10^{-3}$–10^{-9} sec^{-1}). Although no direct measurements of pitch angle lifetimes have been made for electrons of less than about 40 keV, Parks (1969) has made measurements on electrons > 50 keV which are consistent with this broad range of values for D.

5. Summary

When two instabilities clash they may compete with one another for the plasma's free energy in such a way that ordinary quasi-linear equilibrium is inappropriate. This can be an especially important consideration for drift waves, because the time scale for quasi-linear equilibrium can be so long that many other physical processes come into play to establish equilibrium (e.g., Krall, 1967).

For the auroral magnetosphere high frequency electron instabilities may well be the most important stabilizing effect on drift waves, as the calculations of Section 4 indicate. There are two possibilities: (a) the damping effect of the high frequency instability is so large that the drift mode never appears; and (b) there is a reasonable balance between drift mode growth and high frequency damping so that there can be a finite amplitude drift wave in the magnetosphere. These possibilities can be resolved only experimentally as the calculations are hardly accurate enough. (It is intriguing to speculate that some low frequency magnetic activity with periods of seconds or more may be manifestations of drift wave instabilities.) If case (b) holds, we can imagine that drift wave instabilities occurring near the auroral precipitation boundary cause radial diffusion, which in turn creates enhanced anisotropy for the diffusing electrons. The resulting high frequency instability stabilized the drift mode at some finite amplitude.

The present work is a very preliminary attempt at a quantitative calculation of such effects. One of the most important remaining problems is to estimate the amplitude of equilibrium drift waves, and thereby estimate the consequent radial diffusion coefficient. Simple Bohm diffusion estimates show that it could be large.

Acknowledgments

I thank Professor C.-G. Fälthammar for his hospitality at the Royal Institute of Technology during March 1969 when part of this paper was written. Thanks also to Professor C. F. Kennel and other colleagues at UCLA for valuable discussions.

This work was supported in part by National Aeronautics and Space Administration contract NGR-05-007-190 and in part by the Air Force under contract F04701-68-C-0200.

References

Chamberlain, J. W.: 1963, *J. Geophys. Res.* **68**, 5667.
Cornwall, J. M.: 1968, *Radio Sci.* **3**, 740.
Coroniti, F. V. and Kennel, C. F.: 1969, *J. Geophys. Res.*, to be published.
D'Angelo, N.: 1969, *J. Geophys. Res.* **74**, 909.
Dupree, T. H.: 1966, *Phys. Fluids* **9**, 1773.
Dupree, T. H.: 1967, *Phys. Fluids* **10**, 1049.
Dupree, T. H.: 1968, *Phys. Fluids* **11**, 2680.
Ivanov, A. A., Rudakov, L. I., and Teikhmann, I.: 1968, *Soviet Phys.-JETP* **27**, 739.
Kennel, C. F. and Engelmann, F.: 1966, *Phys. Fluids* **9**, 2377.
Krall, N. A.: 1967, *Phys. Fluids* **10**, 2263.
Krall, N. A. and Rosenbluth, M. N.: 1963, *Phys. Fluids* **6**, 254.
Mikhailovskii, A. B. and Rudakov, L. I.: 1963, *Soviet Phys.-JETP* **17**, 621.
Parks, G. K.: 1969, *J. Geophys. Res.*, to be published.

AC FIELDS AND WAVE PARTICLE INTERACTIONS

F. L. SCARF, C. F. KENNEL*, R. W. FREDRICKS,
I. M. GREEN, and G. M. CROOK
Space Sciences Laboratory, TRW Systems Group, Redondo Beach, Calif., U.S.A.

Abstract. Several new experiments provide detailed information about electrostatic and electromagnetic plasma waves in the outer magnetosphere, magnetosheath, and solar wind. Dipole antennas now commonly detect the electric components of whistler mode chorus and triggered emissions near 20 to 50% of the local or equatorial gyrofrequency, and purely electrostatic bursts are sometimes found in the same frequency range. Non whistler mode electrostatic noise with $f > f_c$(local) is very frequently detected on the nightside auroral lines of force, and strong emissions that appear to be related to the Bernstein modes are found beyond the plasmapause very near the magnetic equator. These electrostatic signals, as well as the chorus bursts, represent strong sources of turbulence that can produce significant pitch angle scattering, particle precipitation, and acceleration. On Pioneer 8 and 9 low frequency electrostatic waves are found near the nightside magnetopause boundary; these signals may be associated with fast diffusion and with the 'viscous' interaction that produces the geomagnetic tail. In the magnetosheath and solar wind other types of wave particle interactions are observed. Magnetic 'null' regions at filament boundaries generally involve excitation of intense high frequency electrostatic noise and locally trapped electromagnetic waves are sometimes detected in density troughs. The modern comprehensive payloads allow some wave modes to be identified completely and in-orbit verification of the *E* field amplitude measurements is obtained.

1. Introduction

The account of the Freising meeting on *Earth's Particles and Fields* includes several papers concerned with the general area of magnetospheric and interplanetary wave particle interactions, but in 1967 little direct information could be summarized. The discussions were based largely on theoretical concepts, extrapolation of incomplete low altitude measurements to distant regions, or generalizations from observations made using magnetic antennas alone. In the last few years the situation has improved considerably and direct measurements of **E** and **B** wave components over extended frequency ranges are now available throughout much of the magnetosphere, magnetosheath, and solar wind. Here we select a few topics for examination, but the present discussion should be regarded as an interim progress report, rather than a comprehensive review.

Our OGO 5 and Pioneer 8 and 9 measurements already have been used to demonstrate that ac electric fields play an important role in providing collisionless shock dissipation (Scarf *et al.*, 1970) and it will become apparent that many other dynamical phenomena in the magnetosphere and solar wind are dominated by or strongly influenced by time varying electric fields. However, in this difficult experimental area it is not useful to separate completely a discussion of the observations from an evaluation

* Permanent address: Department of Physics, University of California, Los Angeles, California 90024.

. *M. McCormac (ed.), Particles and Fields in the Magnetosphere.*

of the techniques. For this reason we now stress those measurements that shed light on the in-orbit operation of the electric antennas.

We show below that in several cases where known electromagnetic waves are detected using both the magnetic and the electric sensors of OGO 5 the measured E field amplitudes are in excellent agreement with computed values (the computations are carried out using plasma parameters obtained from on-board diagnostics). Several other indirect tests demonstrate the self and mutual consistency of these observations and give us confidence that accurate amplitude values are obtained for electrostatic as well as electromagnetic waves beyond the plasmapause.

2. Capacitive Coupling for Dilute Plasmas

The TRW experimenters have always tried to achieve capacitive coupling for their electric field antennas. As described in reports on the OV3 3 experiment (Scarf, 1968; Scarf *et al.*, 1968b, 1969a), the Javelin 8.45 antenna (Scarf *et al.*, 1969a), the Pioneer 8 detector (Scarf *et al.*, 1968a) and the OGO 5 instrumentation (Crook *et al.*, 1969), this criterion requires a small sensor ($r_0 \ll L_D$, where L_D is the Debye length and $L_D^2 = 4\pi Ne^2/\kappa T$) and a small current collecting area A. The dynamic sheath resistance R_A is approximately given by $R_A = \kappa T/(ej_+ A)$ where j_+ is the positive ion current density (typically, $j_+ \simeq NeU_+$ or $j_+ \simeq j_+$ (photo)) and for $r_0 \ll L_D$ the antenna capacitance C_A is nearly equal to its free space value. When $\omega R_A C_A \gg 1$, we say that the sensor is capacitively coupled and in this case the antenna response is essentially independent of sheath characteristics.

Many arguments favor this design approach for use in low density regions beyond the plasmapause. If $(\omega R_A C_A) \gg 1$ is not satisfied, then the experimenter generally must place great reliance on his ability to describe and interpret the operation of a Langmuir probe in a space environment dominated by photoelectrons, secondaries, solar panel voltages, wake effects, etc. We know that this type of resistive probe can certainly be used in the ionosphere. However, several comparisons of simultaneous data from ion traps, retarding potential analyzers, and Langmuir probes strongly suggest that serious ambiguities in interpretation can arise in low density plasmas.

Much of the previous discussion of capacitive coupling was related to analysis of the OV3 3 data (Scarf, 1968; Scarf *et al.*, 1968b, 1969a), but this experiment did not provide an adequate test of the technique; in a low polar orbit many observations are made in high density regions where the two conditions $\omega R_A C_A \gg 1$ and $r_0 \ll L_D$ are not satisfied (Scarf *et al.*, 1968b). The OV3 3 experiment had several other serious flaws (strong spacecraft interference, use of an unbalanced dipole, limited telemetry), and it appears that the first genuine test of the capacitive method at low altitudes came with the flight of the University of Iowa Javelin 8.45 in September 1967. Shawhan and Gurnett (1968) presented a comprehensive description of the payload and results, and the observations were discussed further by Scarf *et al.* (1969a). As shown in Figure 2 of the latter reference, the high frequency ($f \geqslant 650$ Hz) output from the short TRW wire grid dipole approached the value expected using $\omega R_A C_A \gg 1$, $C_A \rightarrow C_A$

(free space) as the altitude increased beyond about 400 km, but for low frequencies and low altitudes the electric amplitudes obtained using the capacitive formalism are in error by significant factors (Scarf *et al.*, 1968b, 1969a).

OGO 5 offers a vastly superior ride for a VLF E field experiment. The spacecraft is never tracked at very low altitudes (by the summer of 1969 the perigee height was over 9000 km and rising) and virtually all of the data are obtained beyond the plasmapause, so that the OV3 3 type of amplitude uncertainty is essentially absent. Interference problems are not serious (balanced dipoles are mounted far from the spacecraft), excellent telemetry is available and simultaneous E and B measurements are made in the 560 Hz and 70 kHz channels.

On OGO 5 we directly verify the fact that for the main wire grid antennas the electric field amplitudes are independent of sheath properties. As demonstrated by Crook *et al.* (1969), simultaneous measurements from spatially separated sensors mounted in shadow or sunlight in the wake or in the free stream, respectively, show that changes in sheath properties do not produce variations in antenna response. We regard this as an indirect impedance measurement that validates the design approach for the grid antennas (see Crook *et al.* (1969) for details).

3. In-Flight Calibration Using EM Wave Observations

The verification of sheath independent antenna response is gratifying and reassuring. However, the most basic and reliable method for investigating the accuracy of the E field measurement involves the analysis of a known electromagnetic wave. We tacitly assume that all search coil amplitude measurements are correct; we identify an electromagnetic wave mode, compute $E=cB/n$ (where n is the index of refraction), and compare the computed value of E with the measured one. If the two agree then the measurement of the electric field can be considered to be accurate. On OGO 5 this program has been carried out using 70 kHz and 560 Hz wave observations with the TRW grid antenna and search coils. At 200 Hz the output of the JPL-UCLA search coils is compared with the response of our 200 Hz foil dipoles (Crook *et al.*, 1969).

Figure 1 shows some simultaneous interplanetary 70 kHz E and B measurements that have been analyzed in detail by Scarf *et al.* (1969b). For each (E, B) pair the index of refraction was computed and the n values were compared with those predicted by the full Appleton-Hartree equations. We used $f_{ce} \simeq 500$ Hz (Coleman, 1969) and varied f_{pe} over the range 66 to 72 kHz. A subset of the observations could be clearly identified with the mode having

$$c^2B^2/E^2 = n^2 \simeq 1 - f_{pe}^2/f^2, \tag{1}$$

$(f_{pe}^2 = Ne^2/\pi m)$, and for these readings the observed E/B ratio was used to evaluate f_{pe} and N, the electron density. In the lower part of Figure 1 the results are shown and compared with simultaneous direct measurements of the electron density. It is clear that the measured E/B ratio, the measured value for N, and the measured wave frequency lead to a set of self and mutually consistent interpretations.

The JPL probe provides complete electron density measurements in the solar wind; however, we cannot calibrate our 560 Hz channels using measurements of interplanetary whistler components because the electron cyclotron frequency in the solar wind exceeds 600 to 800 Hz only under extremely disturbed and nonuniform storm con-

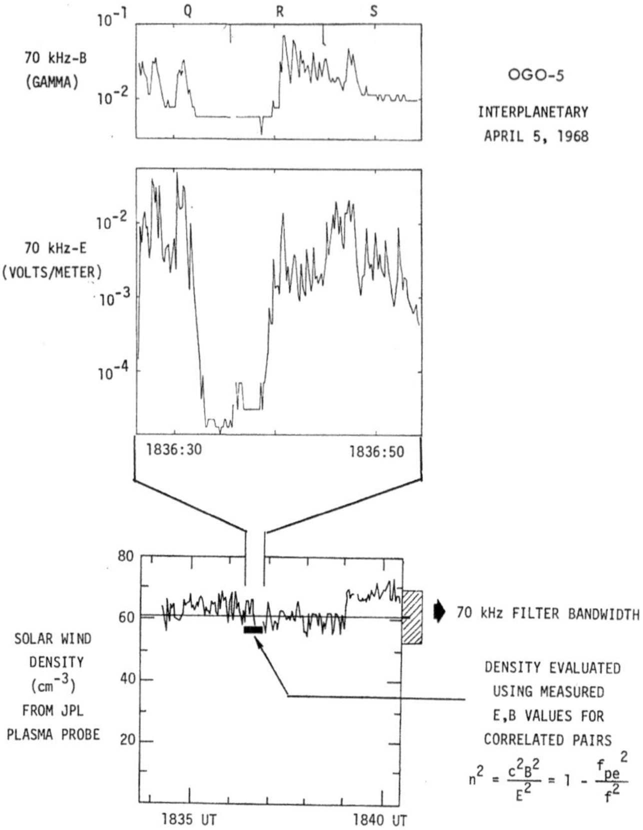

Fig. 1. Detection of quasi-trapped high frequency electromagnetic waves on OGO 5. The exceptionally high densities were encountered several hours after a flare associated sudden commencement was detected. The lower box shows how the measured wave amplitudes are used to evaluate the local density, leading to an absolute calibration for the E field dipole in the plasma.

ditions. In the magnetosheath we do occasionally check the 560 Hz calibration using our E and B measurements of whistler emission amplitudes, the UCLA fluxgate value for f_{ce} and an inferred value for N (N(sheath) is calculated using shock jump relations and interplanetary electron densities from the JPL probe). In all cases examined to date we find agreement with

$$c^2B^2/E^2 = n^2 = 1 + f_{pe}^2/f\,(f_{ce} - f) \tag{2}$$

to within a factor of 2 for these 560 Hz magnetosheath whistlers, but even these signals are detected under such disturbed conditions that it is not really clear that Equation (2) is applicable.

Equation (2) can be used with more confidence when magnetospheric whistler mode signals are analyzed. Figure 2 shows an event that is fairly typical for auroral L shells

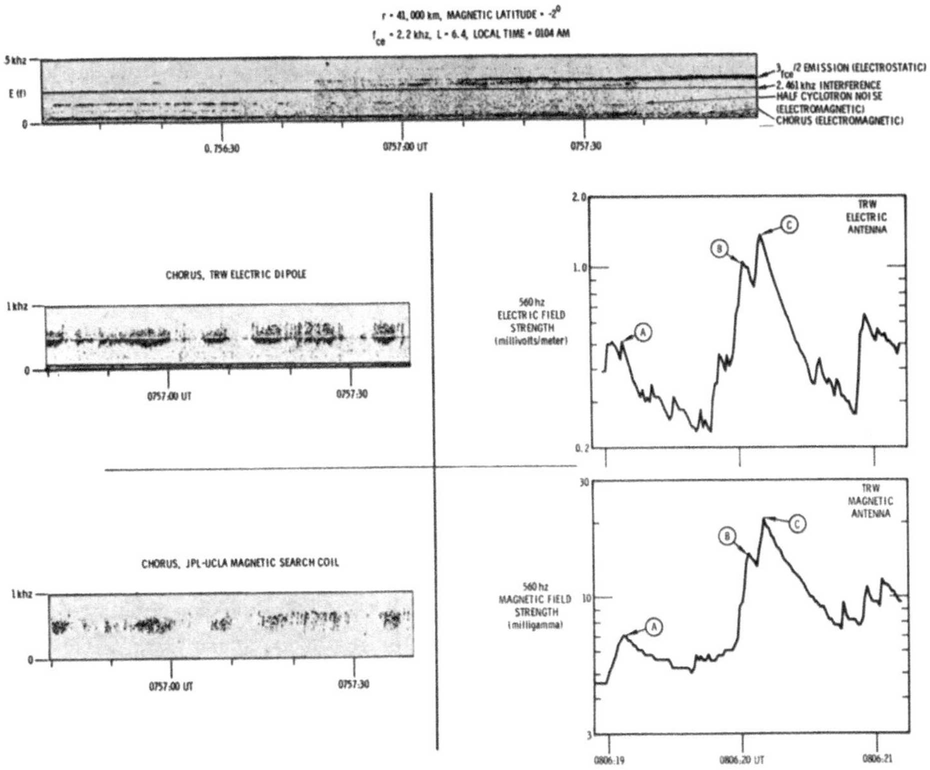

Fig. 2. Detection of electromagnetic waves (chorus and half cyclotron frequency noise) and an electrostatic emission with $f > f_{ce}$(local) on OGO 5. The measured E and B amplitudes for chorus are used to evaluate N, assuming that we have whistler mode propagation along **B**. The points labeled A, B, C all yield $N \simeq 0.2$ electrons/cm³.

and nightside equatorial crossings. At 0800 on August 28, 1968, OGO 5 was inbound with $r \simeq 41\,000$ km, $\lambda_M \simeq -1°$, $L = 6.4$, and LT $\simeq 0104$. The electron cyclotron frequency was 2.2 kHz and since OGO was essentially at the magnetic equator the local value for f_{ce} was equal to the minimum value along the L shell. The upper f-t diagram in Figure 2 shows an electrostatic noise signal having $f \simeq 3f_{ce}/2$, along with some short bursts of low frequency ($f \simeq 0.25 f_{ce}$) noise. The left hand side of the lower part of the figure contains expanded f-t diagrams for these low frequency bursts. Here the E sonogram comes from the TRW dipole and the B sonogram has the simultaneous broadband output of the JPL-UCLA search coil (courtesy of E. J. Smith, R. Holzer and C. Russell). This electromagnetic chorus signal has $f \simeq 560$ Hz ($\pm 15\%$), and we may examine simultaneous amplitude values using the identical TRW 560 Hz magnetic and electric bandpass channels. The right hand side of Figure 2 shows a short sequence of these measurements.

For points labeled A, B, and C the E/B ratios from Figure 2 have been inserted in Equation (2). We find $n(\mathrm{A})=4.2$, $n(\mathrm{B})=4.3$, and $n(\mathrm{C})=4.45$. For the mean index of refraction, Equation (2) yields $f_{pe}=4$ kHz or $N\simeq 0.2$ electrons/cm^3. Since August 15, 1968, was a day that had some bay activity, this low density at $L\simeq 6.4$ is in good agreement with results of the OGO 1 and OGO 3 electron surveys (Vasyliunas, 1968); it can also be seen that an electron density of 0.2 cm^{-3} at $L=6.4$ fits very well onto any of the curves given in Figure 3 of the introductory lecture by Walt (1970).

Our 200 Hz electric field sensor elements are foil cylinders wrapped on a fiberglass mast (Crook *et al.*, 1969; Figure 2), and for these antennas we do not satisfy $\omega R_A C_A \gg 1$ [$f(200)=0.36 f(560)$, $A(200)=5A(560)$, and $C_A(200)=0.3\ C_A(560)$ so that $\omega R_A C_A(200)\simeq 0.02\ \omega R_A C_A(560)$]. The 200 Hz antennas operate as resistively coupled sensors with the dynamic resistance governed by photoemission currents. J. Olson has kindly furnished data from the 220 Hz channels of the JPL-UCLA search coils and we have used measurements of 200 Hz whistlers in the magnetosheath to verify the resistive interpretation.

4. Pioneer-OGO Comparisons

Late on December 14, 1967, Pioneer 8 entered the magnetosheath at a distance of about 45 $\mathrm{R_E}$ and near 0200 LT. For several hours the probe traversed a 'diffusive layer' in which low nonthermal fluxes of positive ions with rapidly varying angular distributions were detected (Wolfe *et al.*, 1968). In this region of space the electric field experiment measured enhanced broadband ($f \gtrsim 100$ Hz) noise, and beyond the magnetopause sporadic large amplitude bursts were detected in the 400 Hz channel. Pioneer 9, launched on November 8, 1968, passed through a similar diffusive particle layer near the magnetopause at $r\simeq 15$ to 17 $\mathrm{R_E}$, LT$\simeq$1800. Once again an enhanced low frequency noise was detected within the magnetopause and sporadic 400 Hz enhancements were found in the inner magnetosheath.

The Pioneer 8 and 9 electric field experiments are quite limited because the available telemetry is restricted, only two quantitative bandpass channels are used, and the antenna is an unbalanced dipole. Fortunately, it is possible to compare OGO 5 and Pioneer 9 data during the magnetopause traversal of November 8, 1968, and this greatly clarifies the Pioneer results. On November 8 OGO 5 was inbound near the evening meridian, and in a rotated (cylindrically symmetric) coordinate system the two spacecraft were within a few earth radii of each other in the vicinity of the magnetopause ($\Delta\mathrm{R}\simeq 2$ to 3 $\mathrm{R_E}$ near 1700 UT). The lower right hand part of Figure 3 shows the spacecraft trajectories, and the magnetopause curve connects the actual Pioneer and OGO crossings. Data taken along the solid segments of the two trajectories (A→D) are used to construct the power spectra displayed at the top of Figure 3. These results suggest that a quasi-stationary low frequency hiss band (lasting in more or less the same form for at least as long as 2 to 3 hr) is to be found in the diffusive layer along the flanks of the magnetopause. In segment **B** the plasma is stationary, while in segment C it is streaming; the spectral broadening is consistent with an interpretation that the noise signals have $(\omega/k) \ll u$, so that Doppler shifts modify **B** into C.

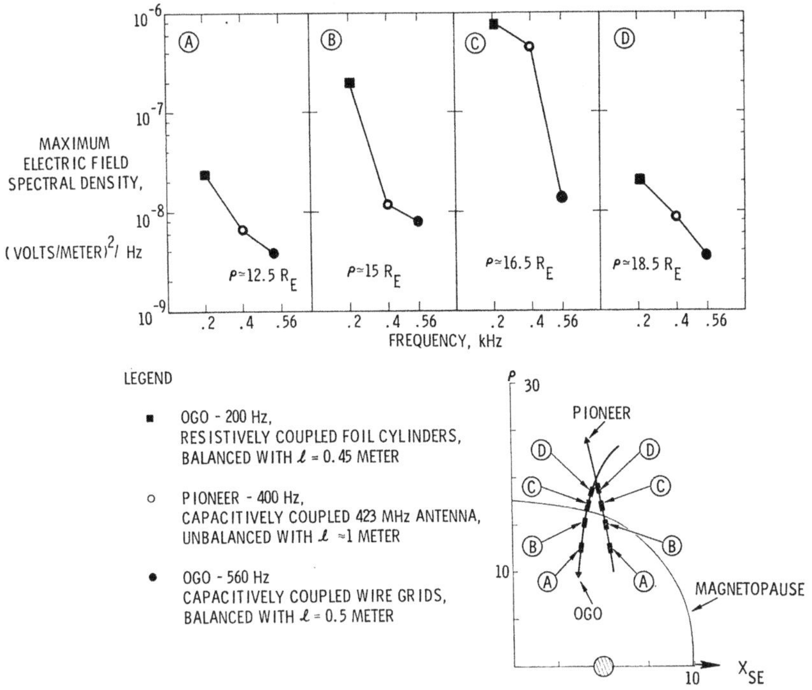

Fig. 3. Comparison of OGO 5 and Pioneer 9 low frequency measurements near the magnetopause on November 8, 1968. For Pioneer 9 we use an effective length of 1 m and for the OGO 5 200 Hz antenna we assume resistive coupling via the photoemission electron current. This interpretation has been verified by comparing E and B components for whistler mode waves.

Excitation of these waves may be related to the viscous drag interaction that produces the geomagnetic tail, and we can also speculate that the waves allow fast particle diffusion across the magnetopause. Finally, the mutual consistency of these observations using three distinct sensors again strongly supports our overall interpretative approach.

5. General Description of VLF Measurements

Many distinct wave particle interactions have been analyzed experimentally to date, and in this concluding section we describe in brief some of the characteristic events that have been studied in some detail. In the solar wind, magnetosheath and bow shock region the most intense signals are generally the low frequency noise bursts detected in close association with steep magnetic field compressions. These waves are observed at the bow shock, at low Mach number interplanetary shock fronts, and at magnetic field 'nulls' in the magnetosheath or solar wind. The 'null' regions appear to be filament boundaries associated with field rotations, and we believe that these oscillations are electrostatic ion sound waves generated by two stream instabilities. Beyond the magnetopause there is evidence that electron plasma oscillations are also excited, and as shown in Figure 1, electromagnetic waves with $f \simeq f_{pe}$ are presumably radiated by superthermal electron streams in the solar wind.

In the magnetosphere the most prominent VLF wave particle interactions are found on the nightside, near auroral L shells, and near the magnetic equator. Figure 2 illustrates how electromagnetic chorus and emissions at half the local electron cyclotron

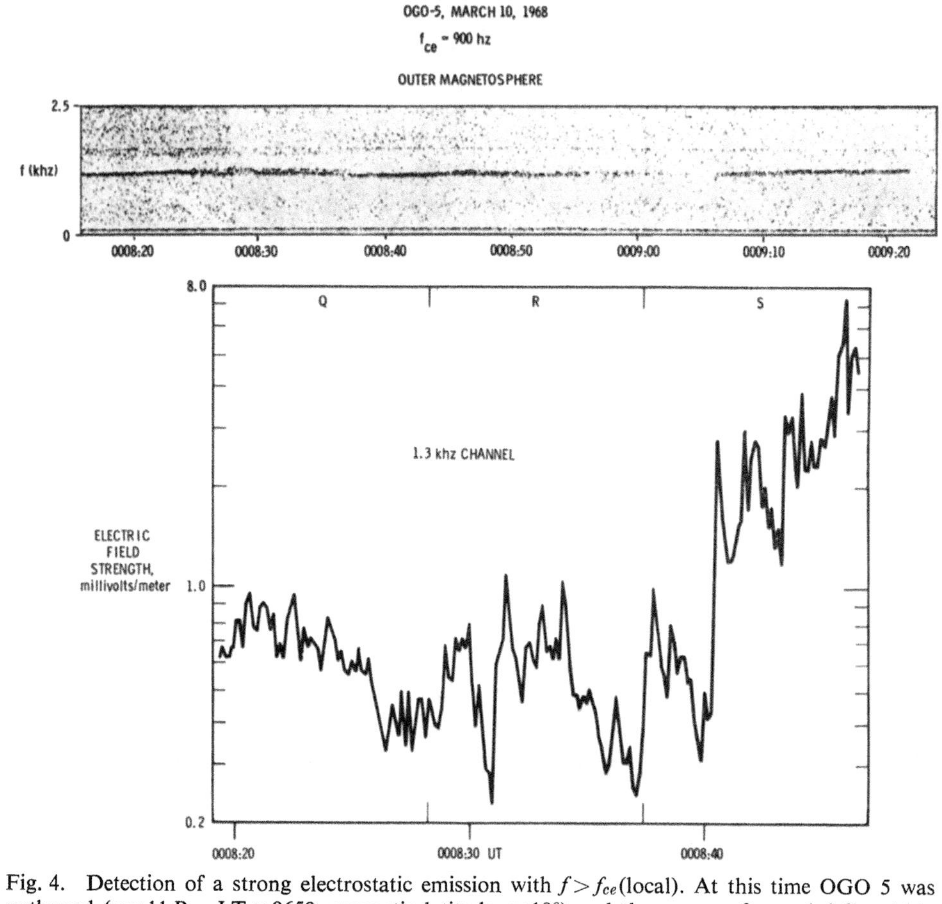

Fig. 4. Detection of a strong electrostatic emission with $f > f_{ce}$(local). At this time OGO 5 was outbound ($r \simeq 11$ R$_E$, LT $\simeq 0650$, magnetic latitude $\simeq 18°$) and the spacecraft was briefly within the outer magnetosphere during a series of multiple magnetopause traversals.

frequency are detected and analyzed (the chorus event has $f \simeq 0.25\, f_{ce}$(min); this is generally the case, and we rarely find f as high as $0.5 f_{ce}$(min)). Figure 2 also shows that a very strong electrostatic wave with $f \simeq 3 f_{ce}/2$ coexists with the chorus. We frequently detect intense electrostatic waves of this type with $f \gtrsim f_{ce}$ or $f \simeq (n+1/2) f_{ce}$.

These electrostatic waves generally are found near the magnetic equator, but they have been observed in other regions as well. Figure 4 shows the characteristics of an intense emission with $f \simeq 3f_{ce}/2$ detected at high latitudes just within the magnetopause on March 10, 1968.

When the diffusion coefficients and scale times for wave particle interactions caused by these oscillations are computed, it is found that the electrostatic waves produce scattering comparable with, or greater than, that produced by the strongest whistler mode chorus events. Details of these interactions and more complete accounts of the observations will be presented elsewhere.

Acknowledgments

We are extremely grateful to those particle and field co-investigators on Pioneer 8 and 9, and OGO 5 who have continuously supported our research by furnishing values for local plasma parameters and by confirming tentative boundary crossing identifications. The OGO 5 analysis of low frequency electromagnetic waves is frequently a cooperative effort with the JPL-UCLA search coil group. We thank E. J. Smith, R. E. Holzer, C. T. Russell, K. I. Brody, and J. V. Olson for much assistance and for permission to display the search coil spectra of Figure 2. This work was supported in part by the National Aeronautics and Space Administration under Contracts NAS 2-4673, NAS 5-9278, and NASW-1589.

References

Coleman, P.: 1969, private communication.

Crook, G. M., Scarf, F. L., Green, I. M., and Lukas, P.: 1969, *IEEE Transactions on Geoscience Electronics* **GE-7**, 120.

Scarf, F. L.: 1968, in *Earth's Particles and Fields* (ed. by B. M. McCormac), Reinhold Book Corporation, New York, p. 247.

Scarf, F. L., Crook, G. M., Green, I. M., and Virobik, P. F.: 1968a, *J. Geophys. Res.* **73**, 6665.

Scarf, F. L., Fredricks, R. W., and Crook, G. M.: 1968b, *J. Geophys. Res.* **73**, 1723.

Scarf, F. L., Crook, G. M., Fredricks, R. W., Green, I. M., and Kennel, C. F.: 1969a, *Plasma Waves in Space and Laboratory* (ed. by J. O. Thomas and B. J. Landmark), Edinburgh University Press, Edinburgh, p. 379.

Scarf, F. L., Green, I. M., Fredricks, R. W., and Neugebauer, M.: 1969b, *Trans. Amer. Geophys. Union* **50**, 303.

Scarf, F. L., Fredricks, R. W., and Kennel, C. F.: 1970, this volume, p. 102.

Shawhan, S. D. and Gurnett, D. A.: 1968, *J. Geophys. Res.* **73**, 5649.

Vasyliunas, V. M.: 1968, *J. Geophys. Res.* **73**, 2839.

Walt, M.: 1970, this volume, p. 410.

Wolfe, J. H., Intriligator, D. S., McKibbin, D. D., Smith, E. K., and Collard, H. R.: 1968, *Book of Summaries and Abstracts, International Symposium on the Physics of the Magnetosphere,* papers 1–9.

HYDROMAGNETIC WAVES AND INSTABILITIES IN THE MAGNETOSPHERE

A. HASEGAWA
Bell Telephone Laboratories, Inc., Murray Hill, N.J., U.S.A.

1. Introduction

The theory of hydromagnetic waves and their instabilities based on kinetic theory ia presented. In Section 2 the behavior of hydromagnetic waves in a uniform plasmr with uniform magnetic field is reviewed by deriving their dispersion relations fos relevant magnetospheric parameters. In Section 3, instabilities associated with velocity space anisotropy are introduced, and in Section 4 an occurrence of such an instability during a storm time of April 18, 1965, is demonstrated.

2. Hydromagnetic Waves

Effects of wave particle interactions have been found to modify the ordinary MHD solutions for the hydromagnetic waves (Kutsenko and Stepanov, 1960; Barnes, 1966; Tajiri, 1967). In this section, these effects are summarized by using the dispersion relation obtained from the Vlasov and the Maxwell's equations for relevant magnetospheric parameters.

For a phase factor of $\exp i(\mathbf{k}\cdot\mathbf{r}-\omega t)$, the Maxwell's equations for electric field $\mathbf{E}$ becomes

$$\mathbf{k}(\mathbf{k}\cdot\mathbf{E}) - k^2\mathbf{E} + \frac{\omega^2}{c^2}\,\boldsymbol{\varepsilon}\cdot\mathbf{E} = 0, \tag{1}$$

where $\boldsymbol{\varepsilon}$ is the plasma dielectric tensor whose value for the hydromagnetic limit, i.e., $\omega/\omega_{ci} \sim k\langle v_i\rangle/\omega_{ci} = \delta \ll 1$, is shown in the Appendix. (ω_{ci} and $\langle v_i\rangle$ are the ion cyclotron frequency and the ion thermal velocity, respectively.)

Satellite observations indicate that from the standpoint of energy, electrons and protons in the magnetosphere can be classified into three groups: high energy ($\sim$ MeV), medium energy ($\sim$ keV), and cold group ($\sim$ eV). The medium energy group contributes the largest energy density to the plasma, while the cold group has the largest number density. Because of the presence of the large density of cold electrons, the ordering of the size of each component of the ε tensor becomes

$$\varepsilon_{xx} \sim \varepsilon_{yy} \sim \varepsilon_{xz} \sim \delta\varepsilon_{yz} \ll \delta^2\varepsilon_{zz}\,.$$

The fact that ε_{zz} is larger than the other components by more than the order of $1/\delta$ means that the parallel (z) electric field is short circuited by the cold electrons. Under these circumstances Equation (1) produces the following three dispersion relations (Hasegawa, 1969):

B. M. McCormac (ed.), Particles and Fields in the Magnetosphere.

1. *Shear (Torsional) Alfvén Wave*

$$\frac{c^2 k_{\|}^2}{\omega^2} - \varepsilon_{xx} = 0\,, \tag{2}$$

(Subscript$_{\|}$ indicates the direction parallel (z) to the dc-magnetic field.)

2. *Compressional Alfvén (Magnetosonic) Wave and Non-Oscillatory (Entropy) Wave*

$$\frac{c^2 k^2}{\omega^2} - \varepsilon_{yy} = 0\,, \quad \text{and} \tag{3}$$

3. *Acoustic Wave*

$$\varepsilon_{zz} = 0\,. \tag{4}$$

Here, for simplicity current parallel to magnetic field was assumed to be zero. In the absence of cold electrons Equations (3) and (4) are not separable and the form derived by Kutsenko and Stepanov (1960),

$$\frac{c^2 k^2}{\omega^2} - \left(\varepsilon_{yy} + \frac{\varepsilon_{yz}^2}{\varepsilon_{zz}}\right) = 0 \tag{5}$$

must be used.

Because ε_{xx} does not involve the integral I, shown in the Appendix, that represents the resonant interaction with particles, the *Shear Alfvén Wave* (Equation (2)) is a wave free of damping, while the other three waves suffer either the transit time

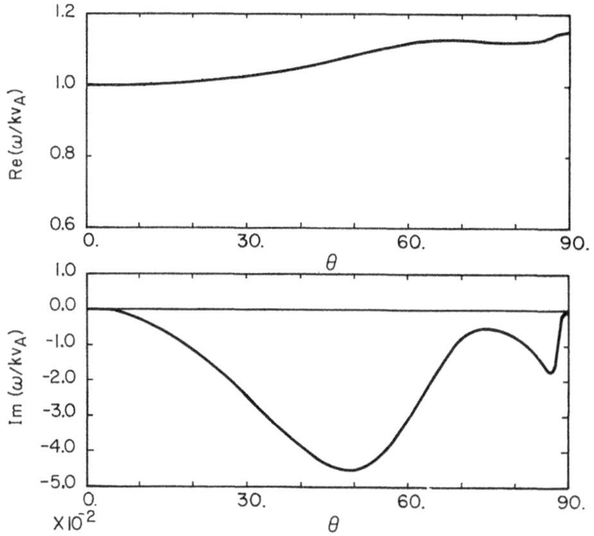

Fig. 1. Numerical solution of Equation (5) for the dispersion relation ω/k vs $\theta = \tan^{-1} k_{\|}/k_{\perp}$, of the Compressional Alfvén Wave. Data used are: ion temperature, 50 keV; electron temperature, 10 keV; cold plasma temperature, 3 eV; hot density, 0.5 cm^{-3} and cold density, 5 cm^{-3}; magnetic field, 200 γ. A root is picked up which is closely located with that of Equation (3).

damping (Equation (3)) (Stix, 1962; Patel, 1967) or the Landau damping (Equation (4)) (Fried and Gould, 1961). For example, Figure 1 shows a numerical solution of the dispersion relation for the *Compressional Alfvén Wave* for a typical magnetospheric parameter. Transit time damping ($\mathrm{Im}\,(\omega/k)<0$) can be due to both ions (near $\theta\sim 48°$) and to electrons (near $\theta\sim 87°$). There exists a heavy damping also in the *Acoustic Wave* owing to the ion Landau damping because the ion temperature is higher than the electron temperature (Fried and Gould, 1961).

It can be concluded that only the *Shear Alfvén Wave* can propagate without any damping in the magnetosphere. The *Compressional Alfvén Wave* can propagate without damping in a limited direction ($\theta\sim 0°$ and $70°$), but the effect of non-uniformity in the magnetic field or plasma density will alter this result.

3. Hydromagnetic Instabilities

Plasma instability occurs owing to its thermodynamically nonequilibrium state. A nonequilibrium state can exist either in coordinate space, in velocity space, or in both.

We treat here, for simplicity, instabilities associated only with velocity space nonequilibrium, such as existence of pressure anisotropy or current parallel to the magnetic field. Consequence of such instabilities appears in diffusions in pitch angle or energy distributions.

A. HOSE INSTABILITY

This instability was found independently by Rosenbluth (1956), Parker (1958), and Vedenov and Sagdeev (1958). The condition of instability is obtained simply by letting the coefficient of $k_{\|}^2/\omega^2$ in Equation (2) or (3) be negative. The condition is satisfied more easily in the presence of a parallel current, which will be discussed later, but even in the absence of such current the instability condition becomes

$$1-\sum_j \frac{\beta_{\| j}-\beta_{\perp j}}{2}<0, \tag{6}$$

where β is the ordinary pressure ratio given by

$$\beta_{\| j}=\frac{2\omega_{pj}^2\langle v_{\|}^2\rangle_j}{\omega_{cj}^2c^2} \quad \text{and} \quad \beta_{\perp j}=\frac{\omega_{pj}^2\langle v_{\perp}^2\rangle_j}{\omega_{cj}^2c^2}$$

and summation is over different species j, ω_p is the plasma frequency, parallel ($\|$) and perpendicular ($\perp$) directions are with respect to the direction of the dc magnetic field. Because Equation (6) is satisfied for large $\beta_{\|}$, such that $\beta_{\|}>\beta_{\perp}\sim 1$, the instability might occur in the magnetospheric tail.

B. MIRROR INSTABILITY

This occurs for large $\beta_{\perp}$, such that $\beta_{\perp}>\beta_{\|}\sim 1$. This instability was found by Rosenbluth (1956), Sagdeev (1958), and Chandrasekhar *et al.* (1958), and is an instability of the nonoscillatory wave represented by Equation (3), which can be obtained by assuming a low frequency limit such that $\omega^2 \ll k^2 v_A^2$ (v_A is the Alfvén speed),

$$\omega = -ik_{\parallel}\langle v_{\parallel}\rangle_i\left(\frac{2}{\pi}\right)^{1/2}\frac{\beta_{\parallel i}}{\beta_{\perp i}^2}\left[\frac{k_{\parallel}^2}{k_{\perp}^2}\left(1+\sum_j\frac{\beta_{\perp j}-\beta_{\parallel j}}{2}\right)+1+\sum_j\beta_{\perp j}\left(1-\frac{\beta_{\perp j}}{\beta_{\parallel j}}\right)\right]. \tag{7}$$

The instability occurs when $\mathrm{Im}(\omega)>0$, i.e., for $k_{\parallel}\to 0$,

$$1+\sum_j\beta_{\perp j}\left(1-\frac{\beta_{\perp j}}{\beta_{\parallel j}}\right)<0. \tag{8}$$

This mode is strictly nonoscillatory, ($\mathrm{Re}(\omega)=0$), unlike the previous derivations where oscillatory solutions were presented for stable cases. Such instability may be possible in the dayside equatorial plane, as will be discussed in Section 4.

C. CURRENT DRIVEN LONGITUDINAL INSTABILITY

This is the well known two stream instability associated with relative motion of electrons with respect to ions in the direction parallel to the magnetic field. Because the electron temperature is lower than that of the ions, such instability occurs only when

$$\langle v_{\parallel}\rangle_e > \sqrt{\langle v_{\parallel}^2\rangle_e - \langle v_{\parallel}\rangle_e^2} \tag{9}$$

which is rather difficult to be met in the magnetosphere.

D. CURRENT DRIVEN TRANSVERSE INSTABILITY

This is an instability of the *Shear Alfvén Wave* driven by current parallel to the magnetic field. Such current generates an azimuthal magnetic field around its path. The total magnetic field then is deformed into a spiral shape which then tries to become straight causing an instability to the current flow.

In the presence of a current ε_{xy} no longer vanishes. If we take **k** in the direction parallel to the local spiral magnetic field, the dispersion relation for the transverse waves is modified to

$$\left(\frac{c^2k_{\parallel}^2}{\omega^2}-\varepsilon_{xx}\right)\left(\frac{c^2k_{\parallel}}{\omega^2}-\varepsilon_{yy}\right)+\varepsilon_{xy}^2=0. \tag{10}$$

Using suitable ε, one can immediately derive the condition of instability from Equation (10) as

$$\frac{\langle v_{\parallel}\rangle}{v_A} > \frac{v_A k_{\parallel}}{\omega_{ci}}. \tag{11}$$

Equation (11) reduces to the famous Kruskal-Shafranov limit for the 'kink' instability if $\langle v_{\parallel}\rangle$ is expressed in terms of the azimuthal magnetic field generated by the current

(Kadomtsev, 1966). This instability is also, as it stands, a nonoscillatory type, *i.e.*, ω is pure imaginary. However it was shown by Hasegawa and Birdsall (1964) and subsequently by Kimura and Matsumoto (1968) that if the plasma is not charge neutral, it can become overstable (oscillatory) with frequency proportional to the excessed charge density.

4. Observation of Hydromagnetic Instability

An example is given of an actual occurrence of a hydromagnetic instability near the equatorial plane in the magnetosphere during a storm time of April 18, 1965. Figure 2 shows the variations of the magnetic field and proton fluxes published by Brown *et al.* (1968). The observation was made by Explorer 26 which was located at 5.11 R_E with

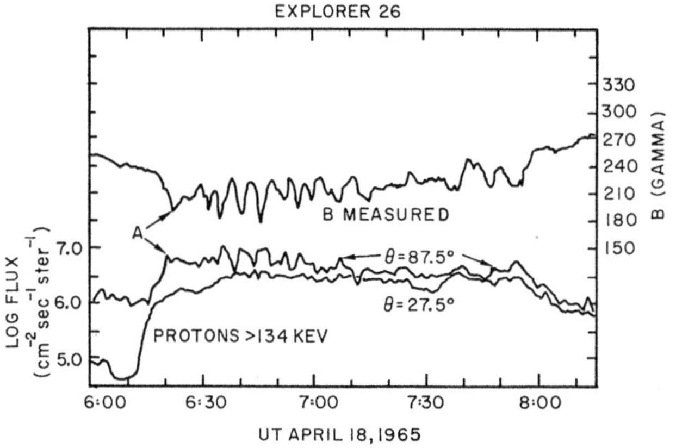

Fig. 2. Observed variation of proton fluxes and geomagnetic field (Brown *et al.*).

1400 LT and latitude of 17°. From this figure can be seen (a) a strong diamagnetic effect (indicating $\beta \sim 1$) starting at 6:15 and suddenly terminating at 6:20 shown by point *A*, (b) a large anisotropy of the proton fluxes ($\beta_\perp > \beta_\parallel$), and (c) start of large out of phase oscillations of the field and the fluxes shortly after the point *A*.

The first two facts strongly indicate an occurrence of mirror instability. When such an instability is set up, it can terminate a further increase and anisotropization of the proton flux and trigger the succeeding oscillations. In fact, by studying the proton data provided by L. Davis of GSFC, we found, corresponding to the point *A* in Figure 2, $\beta_\perp / \beta_\parallel \sim 2$ and $\beta_\perp \sim 1$ satisfying the condition of instability shown in Equation (8) (Hasegawa, 1969; Lanzerotti *et al.*, 1969).

Because the mode causing the mirror instability is nonoscillatory other effects are required to explain the oscillations clearly observed after 6.30. We consider that the oscillation is produced by the coupling with the drift wave created by the ion drift perpendicular to the magnetic field. This will give rise to a real frequency $\omega \sim k_\perp v_d$, where v_d is the proton drift speed and $k_\perp$, the perpendicular wave number, may be chosen to be $\omega_{ci} / \langle v_\perp \rangle_i$, corresponding to its value for the maximum growth rate.

Observed values of average proton energy ($\sim$20 keV) and the magnetic field ($\sim$200γ) give a frequency which is in good agreement with the observed frequency.

The phase relation between the proton flux and the magnetic field is another important point to check. Under the MHD approximation, only the sound wave has the observed out of phase relation. However, as mentioned in Section 2, this mode is a heavily damped mode in the magnetosphere. On the other hand, if the kinetic equation is used allowing anisotropic pressure, one can show that the phase relation between the proton density perturbation $\tilde{n}$ and the magnetic field perturbation $\tilde{B}$ for the compressional modes, Equation (3), becomes

$$\frac{\tilde{n}}{n_0} = \left(1 - \frac{\beta_{\perp i}}{\beta_{\parallel i}}\right) \frac{\tilde{B}}{B_0}. \tag{12}$$

Therefore, if $\beta_\perp > \beta_\parallel$, $\tilde{n}$ and $\tilde{B}$ in fact become out of phase. This is because, when $\beta_\perp > \beta_\parallel$, the effect of $-\nabla \mu B$ force which accelerates plasma into the lower flux region

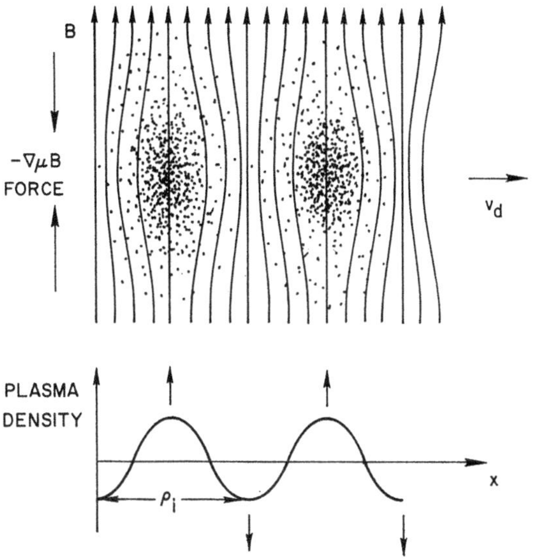

Fig. 3. Schematic diagram of the drift mirror instability. When the instability condition is satisfied, $-\nabla \mu B$ force that accelerates plasma parallel to the field lines into the weaker field region, excludes the fields further by the diamagnetic effect and the process grows. This effect occurs over a region where the instability condition is satisfied. The growth rate becomes maximum when the spatial periodicity is roughly the proton gyroradius ϱ_i ($\sim$ 100 km). In the presence of proton drift across the magnetic field, such periodic bumps of plasma drift also with velocity v_d, hence for a stationary observer an oscillatory proton density, which is out of phase with respect to the field intensity, can be observed with the oscillation frequency given by v_d/ϱ_i.

parallel to $\mathbf{B}_0$ overbalances the effect of $\mathbf{E} \times \mathbf{B}$ drift which moves plasma with the flux density in the perpendicular direction (cf. Figure 3).

As was shown above in an actual magnetospheric problem, we should often consider the effects of non-uniformity because the size of the magnetosphere is not always much larger than the wavelength of hydromagnetic waves.

5. Conclusion

A brief review of hydromagnetic waves and instabilities with an example of actual observation of such an instability in the magnetosphere is presented. Existence of cold electrons in the magnetosphere is pointed out to modify the known dispersion relations of the hydromagnetic waves because they short circuit the parallel electric field.

Among the velocity space instabilities introduced, the mirror instability and the current driven transverse instability seem to be most feasible in the magnetosphere.

The latter instability is worthwhile considering because of its several interesting natures, i.e., frequency of instability being proportional to difference in charge density along the line of force, or the relatively low threshold current.

Finally, saturation of diamagnetization and anisotropization and succeeding oscillations of proton fluxes during a storm time of April 18, 1965, are explained in the context of the mirror instability.

Appendix

Plasma dielectric tensor in the hydromagnetic limit:

$$\frac{k_\perp v_{\perp i}}{\omega_{ci}} \sim \frac{k_\parallel v_{\parallel i}}{\omega_{ci}} \sim \frac{\omega}{\omega_{ci}} \sim \delta \ll 1 .$$

$$\varepsilon_{xx} = \sum_j \frac{\omega_{pj}^2}{\omega_{cj}^2} \left[\left\langle \left(1 - \frac{k_\parallel v_\parallel}{\omega} \right)^2 \right\rangle_j - \frac{k_\parallel^2 \langle v_\perp^2 \rangle_j}{2\omega^2} \right]$$

$$\varepsilon_{xy} = - \varepsilon_{yx} = i \sum_j \frac{\omega_{pj}^2}{\omega \omega_{cj}} \left(1 - \frac{k_\parallel \langle v_\parallel \rangle_j}{\omega} \right)$$

$$\varepsilon_{xz} = \varepsilon_{zx} = \sum_j \frac{\omega_{pj}^2}{\omega_{cj}^2} \frac{k_\perp}{k_\parallel} \left[\frac{k_\parallel^2 \langle v_\perp^2 \rangle_j}{2\omega^2} + \left\langle \frac{k_\parallel v_\parallel}{\omega} \left(1 - \frac{k_\parallel v_\parallel}{\omega} \right) \right\rangle_j \right]$$

$$\varepsilon_{yy} = \varepsilon_{xx} - \sum_j \frac{\omega_{pj}^2}{\omega_{cj}^2} \left(\frac{k_\perp^2 \langle v_\perp^2 \rangle_j}{\omega^2} + \frac{k_\perp^2 \langle v_\perp^4 \rangle_j}{4\omega^2} I_j \right)$$

$$\varepsilon_{yz} = - \varepsilon_{zy} = - i \sum_j \frac{\omega_{pj}^2}{\omega \omega_{cj}} \left(\frac{k_\perp \langle v_\parallel \rangle_j}{\omega} + \frac{k_\perp}{2k_\parallel} \langle v_\perp^2 \rangle_j I_j \right)$$

$$\varepsilon_{zz} = - \sum_j \left[\frac{\omega_{pj}^2}{k_\parallel^2} I_j + \frac{\omega_{pj}^2}{\omega_{cj}^2} \left(\frac{k_\perp^2 \langle v_\perp^2 \rangle_j}{2\omega^2} - \frac{k_\perp^2 \langle v_\parallel^2 \rangle_j}{\omega^2} \right) \right]$$

where

$$I_j = \int_{-\infty}^{\infty} \mathrm{d}v_\parallel \int_0^\infty 2\pi v_\perp \mathrm{d}v_\perp \frac{f_{oj}(v_\perp, v_\parallel)}{(v_\parallel - \omega/k_\parallel)^2},$$

$$\langle v_j^n \rangle = \int_{-\infty}^{\infty} \mathrm{d}v_\parallel \int_0^\infty 2\pi v_\perp \mathrm{d}v_\perp v^n f_{oj}(v_\perp, v_\parallel)$$

dc magnetic field is in z, $\mathbf{k}$ is in $x(\perp)-z(\parallel)$ plane, summation is over species j, ω_p, ω_c, f_0 are plasma frequency, cyclotron frequency (sign included), and velocity distribution function, respectively.

References

Barnes, A.: 1966, *Phys. Fluids* **9**, 1483.

Brown, W. L., Cahill, L. J., Davis, L. R., McIlwain, C. E., and Roberts, C. S.: 1968, *J. Geophys. Res.* **73**, 153.

Chandrasekhar, S., Kaufman, A. N., and Watson, K. M.: 1958, *Proc. Roy. Soc. (London)* **245A**, 435.

Fried, B. D. and Gould, R. W.: 1961, *Phys. Fluids* **4**, 139.

Hasegawa, A.: 1969, *Phys. Fluids* **12**, 2642.

Hasegawa, A. and Birdsall, C. K.: 1964, Electr. Res. Lab. Tech. Rep. No. 64-5, Univ. of Calif., Berkeley.

Kadomtsev, B. B.: 1966, *Review of Plasma Physics*, Vol. 2 (ed. by M. A. Leontovich), Consultants Bureau, New York, p. 190.

Kimura, I. and Matsumoto, H.: 1968, *Radio Sci.* **3**, 333.

Kutsenko, A. B. and Stepanov, K. N.: 1960, *Sov. Phys.-JETP* **11**, 1323.

Lanzerotti, L. J., Hasegawa, A., and Maclennan, C. G.: 1969, *J. Geophys. Res.* **74**, 5565.

Parker, E. N.: 1958, *Phys. Rev.* **109**, 1874.

Patel, V. L.: 1967, *Earth Planetary Sci. Letters* **2**, 36.

Rosenbluth, M. N.: 1956, Los Alamos Sci. Lab. LA-2030.

Sagdeev, R. Z.: 1958, *Review of Plasma Physics*, Vol. 4 (ed. by M. A. Leontovich), Consultants Bureau, New York, p. 32.

Stix, T.: 1962, in *The Theory of Plasma Waves*, McGraw-Hill, New York, p. 196.

Tajiri, M.: 1967, *J. Phys. Soc. Japan* **22**, 1482.

Vedenov, A. A. and Sagdeev, R. Z.: 1958, *Acad. Nauk. SSR* **3**, 278.

INTENSITY OF DISCRETE VLF EMISSIONS

R. A. HELLIWELL

Stanford University, Stanford, Califor., U.S.A.

1. Introduction

Very low frequency whistler mode noise from the magnetosphere frequently appears in the form of narrow band tones of variable frequency called discrete VLF emissions. These tones may appear spontaneously or they may be triggered by transmissions from ground based VLF stations (Helliwell, 1965). They can be observed from within the plasmasphere out to the magnetopause and at frequencies from 300 to 30000 Hz. A phenomenological theory of discrete emissions has been advanced, based on cyclotron resonance between energetic electrons and narrow band whistler mode waves traveling along the static magnetic field (Helliwell, 1967). An important feature of this theory is the maintenance of oscillations through feedback between the waves and the electrons over a path that is long compared with the wavelength.

The feedback mechanism is connected with the period of oscillation of electrons in the magnetic 'potential well' of the wave. This 'trapping' period depends on the wavelength, wave magnetic intensity, and the perpendicular velocity of the resonant electrons. Values of the trapping period are estimated to be of the order of 10^{-1} sec and may be compared with the cyclotron period of the order of 10^{-4} sec and the electron bounce period of the order of 1 sec. Thus we have a new periodicity in the electron motion that falls between the electron gyroperiod and its bounce period. However, the trapping periodicity is important only in the presence of a suitable electromagnetic wave.

It has been shown that electrons entering the 'resonance' region of a coherent wave train would be bunched in phase so that their transverse currents would radiate coherently in the manner of an end fire array of antennas (Brice, 1963). This radiation in turn bunches electrons entering the resonance region at later times so that a feedback loop is established. With sufficient electron flux a self-sustaining oscillation of narrow bandwidth is established. The frequency of oscillation changes with time at a rate that depends on the spatial inhomogeneity of the medium (Helliwell, 1967).

In the present paper the previous theory is extended to include a general electron density model and variable frequency waves. The length of the resonance region is calculated for a representative model of the magnetosphere. This length is then used to estimate the magnetic field intensity of the emission.

2. Interaction Model

Calculation of the resonance length requires knowledge of the response of individual energetic electrons to an applied whistler mode wave. The main features of the inter-

B. M. McCormac (ed.), Particles and Fields in the Magnetosphere.

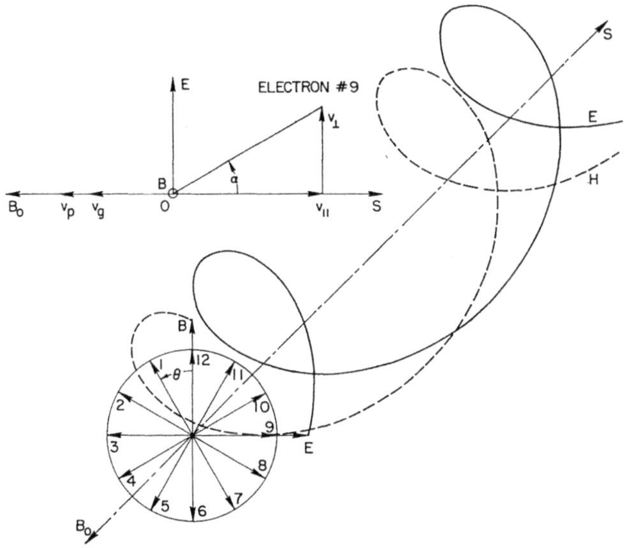

Fig. 1. Schematic diagram showing the initial phase relations between the E and H fields of a right hand circularly polarized wave traveling in the $-S$ direction and a system of 12 test electrons introduced to the wave at $S=0$.

action model are shown in the schematic diagram of Figure 1. A plane whistler mode wave of frequency ω is assumed to be propagating in the direction ($-S$ axis) of the static magnetic field ($\mathbf{B}_0$) at phase velocity v_p. The spatial relation of the wave fields at a particular instant of time ($t=0$) is shown by the vectors **E** and **B**. These vectors are constant in amplitude and rotate counter clockwise at angular frequency ω, to represent a right hand circularly polarized wave as viewed in the direction of propagation. The corresponding variations in direction of the wave fields along the S axis at $t=0$ are shown by the two left hand turning helices.

The propagation of the plane wave is controlled mainly by the thermal component of the plasma, which can be considered cold for the purpose of describing wave propagation. The effects of ions and collisions are neglected. Since we shall be concerned mainly with situations in which the refractive index is much greater than unity, the phase and group velocities and the wave number can be represented respectively by

$$v_p = c\,\frac{\omega^{1/2}(\Omega-\omega)^{1/2}}{\omega_N} \tag{1}$$

$$v_g = 2c\,\frac{\omega^{1/2}(\Omega-\omega)^{3/2}}{\omega_N\Omega} \tag{2}$$

$$k = \frac{\omega_N}{c}\left(\frac{\omega}{\Omega-\omega}\right)^{1/2}, \tag{3}$$

where c = velocity of light, ω = angular wave frequency, Ω = angular gyrofrequency, ω_N = angular plasma frequency = $2\pi\sqrt{81N_e}$, and N_e = electron density.

The spatial growth of the wave is controlled mainly by nonrelativistic energetic electrons near cyclotron resonance. When these electrons interact with the transverse magnetic field of the wave, their orbits are modified in such a way as to produce coherent whistler mode radiation. Resonance occurs when the electron sees a doppler shifted wave frequency equal to its own gyrofrequency. Since the wave frequency must be less than the gyrofrequency for propagation to occur, the parallel velocity $v_{\parallel}$ of the electron must be opposite to the wave phase velocity. The doppler shifted wave frequency is given by

$$\omega' = \omega + kv_{\parallel} \tag{4}$$

and at resonance the electron gyrofrequency is given by

$$\Omega = \omega + kv_{\parallel r}, \tag{5}$$

where $v_{\parallel}$ = parallel velocity of electron, measured in $+S$ direction and $v_{\parallel r}$ = resonance value of $v_{\parallel}$.

Under conditions of exact resonance, as given by Equation (5), the perpendicular velocity $v_{\perp}$ of the electron maintains a constant phase angle θ with respect to the field vectors **E** and **B**. Then a sheet of electrons, such as the 12 test electrons shown in Figure 1, would move through the wave without changing their relative phases and no transverse current would be produced. Actually, however, the wave fields exert forces on the electrons that alter their relative phases, causing a transverse current to be produced.

The total wave force on the electron is $\mathbf{F} = q(\mathbf{E} + \mathbf{v}_{\perp} \times \mathbf{B})$ where q = electronic charge. Although $q|\mathbf{E}|$ is comparable with $q|\mathbf{v}_{\perp} \times \mathbf{B}|$, its effect on the phase is small since it acts in the same direction as the much larger $q\mathbf{v}_{\perp} \times \mathbf{B}_0$ force produced by the earth's static magnetic field. Thus for typical values of the parameters ($B = 10^{-2}\gamma$; $B_0 = 10^3\gamma$) $B/B_0 = 10^{-5}$, while the fractional change of B_0 in one resonance length (discussed later) is of the order of 10^{-3}. This means that spatial inhomogeneity is a much more important factor than the electric field with respect to phase. However, the $qv_{\perp} \times \mathbf{B}$ component, acting longitudinally, causes a significant change in phase. Since the longitudinal force varies like $\sin\theta$, the relative longitudinal motion of the electron is identical to that of the simple pendulum. In Figure 1, each of the 12 test electrons can be thought of as a pendulum responding to a vertical gravitational force equal to $qv_{\perp}B$. When these pendulums reach bottom, the corresponding test electrons are arranged in a helix that is approximately 180° out of phase with the original **B** helix. They then constitute an effective transverse current that is in-phase with the original **B**. The radiated fields from this current are then rotated 90° backwards in phase from the applied fields. The complete problem (not yet solved) is to obtain a consistent description of the transverse currents and the total wave fields. In this paper we shall confine our attention to the variations of the phase angle θ as caused by spatial inhomogeneities only, and to the wave fields required to produce bunching within the resonance region. The important parameter is the 'bunching' time which is defined as one-quarter of the pendulum period for small initial angle (i.e., θ near π in Figure 1), and is given by

$$T_{1/4} = \sqrt{\frac{\pi}{8} \frac{m}{q} \frac{\lambda}{v_{\perp} B}}, \tag{6}$$

where m = electronic mass, q = electronic charge, and λ = wavelength.

In the next section we estimate the length of the resonance region and equate this to the bunching distance $D = v_{\parallel} T_{1/4}$. The corresponding B obtained from Equation (6) is then the minimum field intensity to be expected. By allowing the resonant electrons to bunch more than once within the resonance region (a possibility not considered in the earlier paper) a higher value of B would be obtained.

3. Model of the Medium

Because of the influence of spatial inhomogeneities on the interaction, we must know the variations of the plasma parameters along the S axis. For locations close to the geomagnetic equator of a dipole field, the gyrofrequency is given approximately by

$$\Omega = \Omega_0 \left[1 + \frac{9}{2} \left(\frac{S}{R} \right)^2 \right], \tag{7}$$

where Ω_0 = electron gyrofrequency at the equator, S = distance from the equator ($S \ll R$), and R = distance from the center of the earth.

Electron plasma frequency will be assumed to vary as the power N (called 'plasma

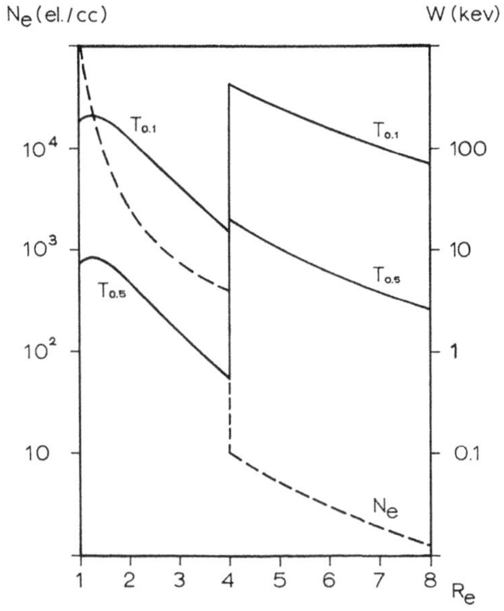

Fig. 2. Electron density model; parallel resonant electron energy for $\omega/\Omega = 0.5$ and 0.1, for tranverse resonance ($\alpha \approx 0°$); plasmapause at $R_E = 4$.

index') of the local gyrofrequency and therefore can be described near the equator by

$$\omega_N = \omega_{N_0}\left[1 + \frac{9}{2}N\left(\frac{S}{R}\right)^2\right], \tag{8}$$

where ω_{N_0} = plasma frequency on the equator.

The variation of equatorial electron density with geocentric distance for a model magnetosphere (Angerami and Carpenter, 1966) is shown in Figure 2. Within the plasmapause $N=0$, corresponding to constant density along a field line. Outside the plasmapause $N=\frac{1}{2}$ which corresponds roughly to the so-called gyrofrequency model, in which $N_e \propto R^{-3}$.

Electron orbit parameters are assumed to be governed by the first adiabatic invariant so that the parallel velocity can be described by

$$v_{\|} = v\left(1 - \Omega\frac{\sin^2\alpha_0}{\Omega_0}\right)^{1/2}, \tag{9}$$

where v = total velocity of electron and α_0 = pitch angle of electron at equator.

With the aid of Equation (7) the parallel velocity can be expressed in terms of the position S by

$$v_{\|} = v_{\|0}\left[1 - \frac{9}{2}\left(\frac{S}{R}\right)^2 \tan^2\alpha_0\right]^{1/2} \tag{10}$$

where $v_{\|0}$ = electron parallel velocity on the equator.

Since we are interested primarily in electrons near resonance it will be convenient to express the parallel resonant velocity in terms of the wave frequency and the parameters of the medium. Thus, if we use Equation (3) to eliminate k in Equation (5), we obtain for the resonant velocity

$$v_{\|r} = \frac{c}{\omega_N}\frac{(\Omega - \omega)^{3/2}}{\omega^{1/2}}. \tag{11}$$

Values of the equatorial parallel resonant energy, calculated from Equation (11) for $\omega/\Omega = 0.5$, 0.1 are shown along with the model magnetosphere in Figure 2.

4. Resonance Conditions

Exchange of energy between electrons and waves depends on the phase angle between the electron perpendicular velocity $v_\perp = v \sin\alpha$ and the electric field **E** of the wave. This phase, measured with respect to the **B** field in Figure 1, in turn depends on the initial phase angle of the perpendicular velocity of the electron and the difference between its gyrofrequency and the doppler shifted wave frequency.

We assume that there are many nearly resonant electrons in each cubic wavelength of space and that their perpendicular velocities are oriented randomly. In each velocity and pitch angle interval the actual population can be represented approximately by a sheet of electrons spaced uniformly in phase before their encounter with the wave. For example, in Figure 1 a sheet of 12 electrons is assumed, in which the perpendicular velocities are equally spaced at 30° intervals.

As each electron moves through the wave train its phase angle *with respect to* **B** changes because of (a) wave induced longitudinal drift (as discussed earlier), and (b) spatial inhomogeneity. In the absence of wave induced drift this phase angle, called the 'unperturbed' phase angle, is given by

$$\theta^u = \theta_0 + \int_0^S \Delta\omega \frac{dS}{v_\parallel}, \tag{12}$$

where θ_0 = phase angle on the equator and $\Delta\omega = \Omega - \omega'$.

An explicit expression for the unperturbed phase angle can be obtained from Equation (12) with the aid of Equations (7), (4), (3), and (10). It will be convenient to express the equatorial parallel velocity by

$$v_{\parallel 0} = v_{\parallel 0r}(1 + A), \tag{13}$$

where $A = \Delta v_{\parallel 0}/v_{\parallel 0r}$, $v_{\parallel 0r}$ = parallel resonant velocity on the equator, and $\Delta v_{\parallel 0}$ = deviation of equatorial parallel velocity from its resonance value.

The form of Equation (12) can be simplified by assuming that terms involving S are each $\ll 1$, and also by using the first two terms of the binomial expansion of each variable factor. Then we find that

$$\frac{\Delta\omega}{v_\parallel} = \frac{\Omega_0}{v_{\parallel 0r}} \left\{ \frac{9}{4}\left(\frac{S}{R}\right)^2 [3 + (1 - \Lambda_0)(\tan^2\alpha_0 - 2N)] - A[1 - \Lambda_0] \right\}, \tag{14}$$

where $\Lambda_0 = \omega_0/\Omega_0$ and ω_0 = frequency on the equator (assumed constant in this case).

Substituting Equation (14) in Equation (12) and integrating we obtain

$$\theta^u = \theta_0 + \frac{\Omega_0}{v_{\parallel 0r}} \left\{ \frac{9}{4}\frac{S^3}{R^2}\left[1 + \left(\frac{1 - \Lambda_0}{3}\right)(\tan^2\alpha_0 - 2N)\right] - AS(1 - \Lambda_0) \right\}. \tag{15}$$

Inspection of Equation (15) shows that for $N = \frac{1}{2}[3/(1 - \Lambda_0) - \tan^2\alpha_0]$ the variation of θ^u is linear with S. If in addition $A = 0$ (i.e., for exact resonance) θ^u is constant. For example, with $\Lambda_0 = \frac{1}{2}$ and small α_0, $N \approx 3$ which represents a relatively rapid increase of ω_N with Ω. However, for low frequencies (i.e., $\Lambda_0 \ll 1$) and small α_0, N drops to 1.5 which is still larger than that for most models (e.g., $N = \frac{1}{2}$ for the 'gyrofrequency' model). Under these special conditions the interaction distance becomes very large so that bunching can occur for low values of wave magnetic field intensity.

The spatial variations of electron gyrofrequency Ω, the doppler shifted wave frequency ω' and the corresponding unperturbed electron phase θ^u for a constant density model are illustrated in Figures 3a and b. In computing these curves $\theta_0 = 0$ in Equation (15) so that the phase angle of each electron with respect to B is zero at the equator.

The variation of θ^u is affected by the electron pitch angle, the plasma frequency,

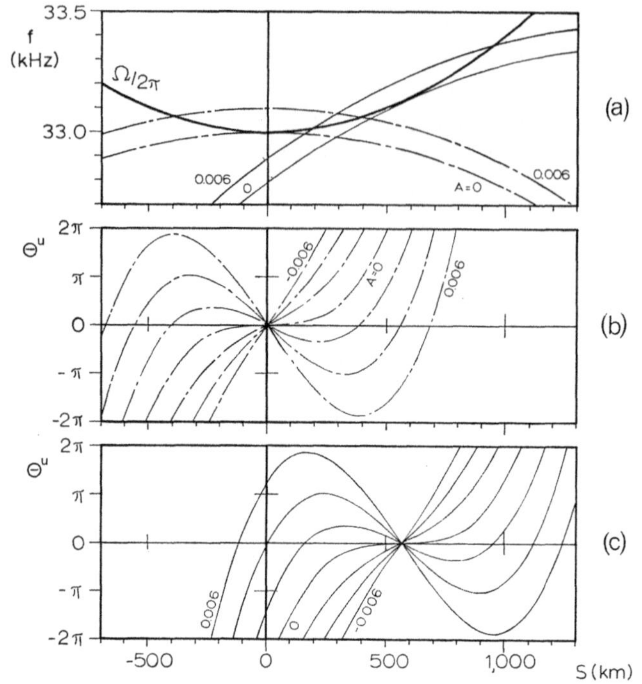

Fig. 3. (a) Heavy solid line shows electron gyrofrequency Ω vs. distance S from the equator. Broken lines show doppler shifted wave frequency ω' for constant wave frequency. Light solid line shows ω' for wave frequency changing linearly with time at equator ($\omega_0 = 16.6$ kHz, $B = 3.7 \times 10^{-7}$ Hz/m. $R_E = 3; \Lambda = 0.5; N = 0; A = 0.006, 0; \alpha_0 = 30°$). – (b) Unperturbed phase shift for constant frequency case; $A = \pm 0.006, \pm 0.004, \pm 0.002, 0$. – (c) Unperturbed phase shift for variable frequency case.

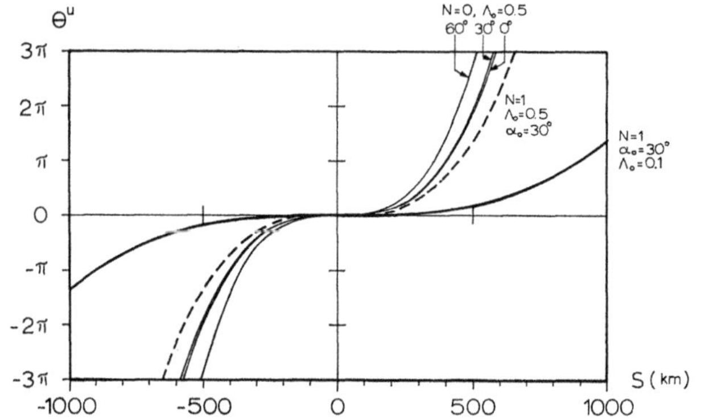

Fig. 4. Effect of normalized frequency Λ_0, equatorial pitch angle α_0, and plasma index N on the unperturbed phase θ^u vs. S.

the plasma index, and normalized frequency as shown in Figure 4, which is drawn for electrons resonant on the equator ($A = 0$) at $R_E = 3$. It is seen that for the smaller pitch angles (0°, 30°) there is little variation of θ^u with pitch angle. For large pitch angles ($\alpha_0 > 60°$) the mirror points approach the equator causing significant increase in phase variation. On the other hand, the phase variation can be reduced by increasing the plasma index N and lowering the normalized frequency Λ_0.

Previously we showed how the perpendicular velocity of an electron is related in direction to the field vectors of a passing wave of fixed frequency. We now consider the case of variable frequency waves. The analysis is the same as before except that ω is now a function of both position S and time t. It is convenient to start with ω vs. t at a fixed position S since this is the condition under which most observations are made. The fundamental problem is to obtain ω' as seen by a test particle. This in turn requires finding $\omega(t)$ in the frame of the particle since k can always be found from ω.

As the particle spirals around a field line it sees a change in ω that depends on $\partial\omega/\partial t$, $v_{\|}$ and v_g. The total change in frequency seen by a test particle traveling in the $+S$ direction can be expressed by

$$\Delta\omega = \frac{\partial\omega}{\partial t}\,\Delta t + \frac{\partial\omega}{\partial s}\,\Delta S \tag{16}$$

where $\Delta t =$ the time the particle moves a distance ΔS.

Since $\Delta S/\Delta t = v_{\|}$ and $\partial\omega/\partial S = (1/v_g)\,(\partial\omega/\partial t)$ Equation (16) becomes

$$\frac{\mathrm{d}\omega}{\mathrm{d}t} = \left(1 + \frac{v_{\|}}{v_g}\right)\frac{\partial\omega}{\partial t} \tag{17}$$

which is the time rate of change of frequency as seen by the moving electron.

For our model calculations we shall start with assumed values of ω_0 and $\partial\omega/\partial t$. Then the frequency seen by a test electron of parallel velocity $v_{\|}$ is given approximately by

$$\omega = \omega_0 + \int_0^t \left(1 + \frac{v_{\|}}{v_g}\right)\frac{\partial\omega}{\partial t}\,\mathrm{d}t\,. \tag{18}$$

Substituting Equations (2) and (9) and replacing $\mathrm{d}t$ by $\mathrm{d}S/v_{\|}$ in Equation (18) we obtain

$$\omega = \omega_0 + \int_0^S \left(1 + \frac{1}{2}\frac{\Omega}{\omega}\right)\left(\frac{\partial\omega}{\partial t}\right)\frac{\mathrm{d}S}{v_{\|}} \tag{19}$$

which is the variation of frequency with distance as seen by the moving electron.

Although quantities Ω, ω, and $\partial\omega/\partial t$ change with S, these changes are usually small so that Equation (19) can be written in terms of equatorial values (subscript 0) as

$$\omega = \omega_0 + BS\,, \tag{20}$$

where

$$B = \frac{1}{v_{\|0}}\left(1 + \frac{1}{2}\frac{\Omega_0}{\omega_0}\right)\left(\frac{\partial\omega}{\partial t}\right)_0 (m - \sec)^{-1}\,.$$

Now the computation of the unperturbed phase angle θ'' can be carried out as before except that Equation (20) is substituted for ω in Equation (4). Neglecting small quantities, the final result is

$$\theta^u = \theta_0 + \frac{\Omega_0}{v_{\parallel 0r}} \left\{ \frac{9S^3}{4R^2} \left[1 + \frac{1 - \Lambda_0}{3} (\tan^2 \alpha_0 - 2N) \right] - AS(1 - \Lambda_0) - \frac{BS^2}{4\omega_0} (1 + 2\Lambda_0) \right\}. \quad (21)$$

Representative curves of doppler shifted wave frequency and unperturbed phase angle θ^u for the variable frequency case are shown in Figures 3a and c. Here the initial value θ_0 has been chosen so that $\theta^u = 0$ at the point of tangency between Ω and ω'. It is seen, with respect to this point of tangency, that the variation of unperturbed phase θ^u is independent of the rate of change of frequency with time.

5. Field Intensity

Using Equation (14) (or curves like those in Figure 3) we can estimate the length of the interaction region. Computer calculations indicate that the significant effects of resonance are confined to a region over which the phase varies by no more than π radians. Applying this criterion to the model of Figure 2 ($N=0$ for $R_E \leqslant 4$ and $N=\frac{1}{2}$ for $R_E > 4$) we calculate the interaction distance D for $\Lambda = 0.5$ as shown in Figure 5. Also shown is the corresponding bunching period $T_{1/4}$ (same as $\frac{1}{4}$ of the trapping period) obtained from $T_{1/4} = D/v_{\parallel}$.

The field intensity corresponding to the bunching period is computed from Equation

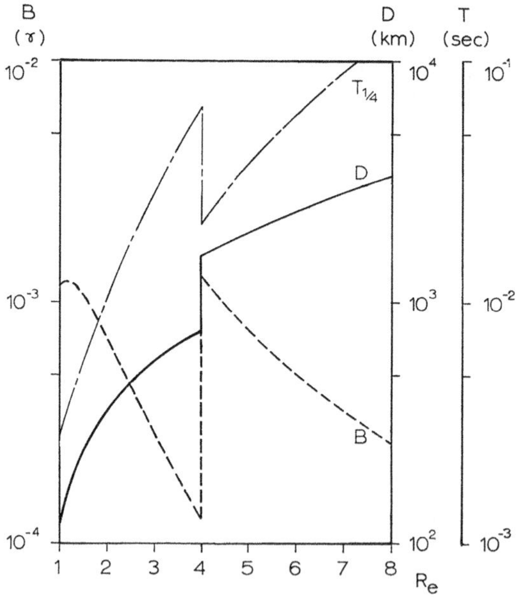

Fig. 5. Predicted minimum field intensity B_w, interaction distance D, and bunching period $T_{1/4}$ vs. R_E for model magnetosphere of Figure 2. $\Lambda = 0.5$, $\alpha_0 = 30°$, $A = 0$.

(6) and is shown by the dashed line on Figure 5. This field intensity would be expected if the particle flux was just sufficient to generate the corresponding wave output power. For large fluxes B_w could increase up to the value at which the electromagnetic power available from the electrons equals the output wave power. Some spatial drift of the interaction region is expected and this would alter the energy balance somewhat. Large fields (producing small bunching times) would also cause appreciable frequency and amplitude modulation of the radiating currents which in turn would tend to destroy the coherence of the generated wave. It is not yet known by what factor the field intensity can exceed the estimated minimum value shown in Figure 5. The marked changes in D, $T_{1/4}$, and B at the plasmapause are controlled by the changes in electron density and in plasma index. Electron density drops about two orders of magnitude which causes the resonant particle energy to increase by the same factor, as shown by Equations (3) and (5) and as illustrated in Figure 2.

The plasma index, on the other hand, increases from $N=0$ to $N=\frac{1}{2}$ as the plasmapause is crossed in the outward direction, causing the resonance length to increase, as shown in Figure 4. However, the increase in D is less than the increase in $v_{\|}$ so that the net effect is a decrease in $T_{1/4}$. Since the changes in λ and $v_{\perp}$ cancel out then B changes inversely as the square of the change in $T_{1/4}$, as shown in Figure 5.

Experimental values of B for discrete emissions have not yet been reported in any quantity. However, measurements of all types of noise including discrete emissions, observed with the OGO 1 VLF experiment, indicate that values for discrete emissions tend to fall in the range of 1 to 100 mγ (Dunckel and Helliwell, 1969) in acceptable agreement with the predictions of the theory.

Acknowledgments

Calculations for figures were made by T. M. Wang. The research was supported under a grant from the Air Force Office of Scientific Research (AFOSR 783-68 a & b).

References

Angerami, J. J. and Carpenter, D. L.: 1966, *J. Geophys. Res.* **71**, 711.
Brice, N.: 1963, *J. Geophys. Res.* **68**, 4626.
Dunckel, N. and Helliwell, R. A.: 1969, *J. Geophys. Res.* **74**, 6371.
Helliwell, R. A.: 1965, in *Whistlers and Related Ionospheric Phenomena*, Stanford University Press, Stanford, Calif.
Helliwell, R. A.: 1967, *J. Geophys. Res.* **72**, 4773.

PART VI

RADIATION BELT OBSERVATIONS

SUMMARY OF PARTICLE POPULATIONS IN THE MAGNETOSPHERE

JAMES I. VETTE

National Space Science Data Center, Goddard Space Flight Center, Greenbelt, Md., U.S.A.

1. Introduction

The purpose of this paper is to provide a succinct description of the charged particle distributions within the magnetosphere which has been made possible by the many satellite observations over the past decade. The main emphasis will be on the time averaged behavior of these particle populations rather than the detailed dynamical charges which occur. Some of the important physical processes which produce these changes are discussed elsewhere in this volume. Furthermore, many of the papers presented later will give new data, new interpretations, and generally new understanding of this complex environment. Consequently this work does not attempt to present a complete review of the previous observational measurements but rather to provide an average baseline from which the transient events depart.

The main constituents from a particle radiation standpoint are electrons and protons. Alpha particles and heavier nuclei play a smaller but significant role in understanding the various processes. Within the magnetosphere electrons range in energy from the thermal region (eV) up to several MeV while proton energies extend from thermal values up to several hundred MeV. The lower energy particles are generally spread throughout the whole cavity although certain natural boundaries can be defined for various classes of particles. The most energetic natural electrons exist around 3 to 4 R_E; the Starfish injected electrons have decayed considerably but still dominate the inner zone for energies above 1 MeV. Energetic protons peak around 1.5 R_E.

Because the particle populations and time behavior are different in many parts of the magnetosphere, it will be convenient to dissect the cavity into three basic regions: the plasma sheet, pseudo-trapping, and stable trapping regions. The average extent of these regions near the earth's geomagnetic plane is shown in Figure 1. During disturbed conditions these boundaries can change by several earth radii. We will now proceed to characterize the fluxes in these regions and to discuss Figure 1 in more detail.

2. Plasma Sheet

The inner boundary of the plasma sheet has been mapped by Vasyliunas (1968) using OGO 1 and 3 electron measurements in the 0.04 to 2 keV energy range. This region is identified with the soft electron band first detected by Gringauz *et al.* (1960a, 1960b). During magnetic bays the boundary of the plasma sheet on the evening side of the magnetosphere moves inward to the stable trapping boundary shown in Figure 1 and

B. M. McCormac (ed.), Particles and Fields in the Magnetosphere.

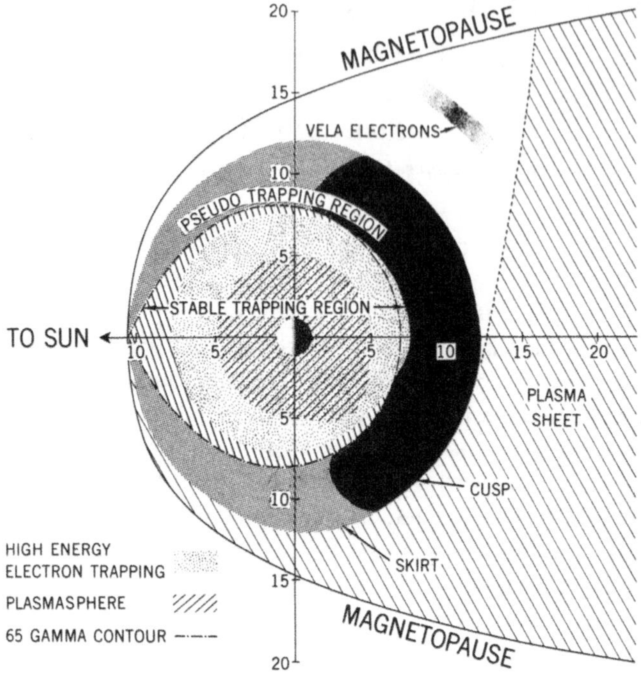

Fig. 1. Regions of the magnetosphere in the equatorial plane. The regions are depicted when magnetic activity is not very high. The 65 γ contour and the magnetopause are taken from Fairfield (1968). The dotted inner boundary of the plasma sheet in the dawn-midnight quadrant indicates this boundary has not been detected yet in this region of space.

sometimes pushes in to 5.5 R_E. However, the plasma sheet has not been observed to penetrate the plasmasphere. The complete radial extent of the sheet is not presently known; it has been studied recently at lunar distances by Nishida *et al.* (1969) where its properties are similar to those at 18 R_E.

The plasma sheet is present all the time and extends across the tail of the magnetosphere. The sheet is 4 to 6 R_E thick in the center and flares out to 8 to 12 R_E in thickness toward the dawn and dusk boundaries. The most extensive spectral properties of this plasma have been measured by the Vela satellites at 17 R_E and have been summarized by Bame (1968). The electron spectrum is quasi-thermal with a high energy tail. The average electron energy is 0.6 keV but varies from 0.1 to 10 keV. Omnidirectional fluxes >100 eV extend to 10^9 cm^{-2}-sec^{-1} and the particle density ranges from 0.1 to 3 cm^{-3} with an average value of 0.5. Vasyliunas (1968) has obtained densities up to 30 cm^{-3} with a lower energy threshold. The energy density of the protons in the plasma sheet exceeds that of electrons by a factor of about 8. The average proton energy is 5 keV and ranges from 1 to 20 keV. Of course, the plasma is neutral so electron and proton densities are the same.

The high energy tail of both these particles distributions has been detected by many different instruments. A radial profile has been obtained using detectors sensitive to electrons >30 to 45 keV with an intensity limit of 10^4 cm^{-4}-sec^{-1} (Anderson, 1965; Murayama, 1966). With these instruments there are many times when no fluxes are observed past ~ 7 R_E. When they are observed, the events are called islands which exhibit fluxes up to 10^7 cm^{-2}-sec^{-1}, which are nearly as intense as those found in the radiation belts. The majority of these events show a fast rise time and a slow decay leading to the suggestion by Anderson that these variations are temporal and not spatial. Murayama has shown the frequency of these events are independent of radial distance (between 15 to 31 R_E) but are strongly correlated with the distance from the magnetic neutral sheet and are confined within the thickness of the plasma sheet. The frequency of events increases with magnetic activity. Konradi (1966) has reported proton islands with intensities $\sim 10^5$ cm^{-2}-sec^{-1} and energies >125 keV. The integral proton energy spectrum can be approximated by a power law with an exponent lying between 5 to 8. Both the high intensity proton and electron island events probably represent an energization within the plasma sheet. Recently Retzler and Simpson (1969) have reported observations of 400 keV electron island events in the plasma sheet. Detailed observations by Montgomery (1968) provide a good picture at 17 to 18 R_E. The frequency of occurrence of electron fluxes >64 keV is shown as a function of solar magnetospheric coordinates in Figure 2. The dawn-dusk asymmetry can

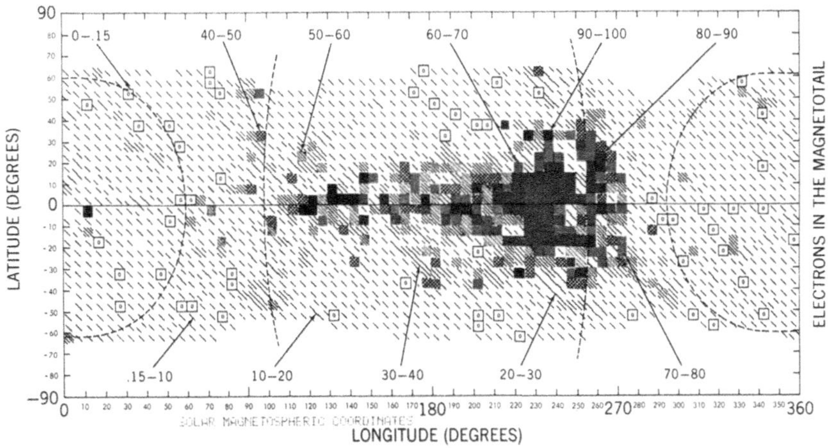

Fig. 2. Frequency of occurrence of >64 keV electron fluxes in the magnetotail. The various shaded blocks represent the frequency of occurrence of fluxes greater than 83 cm^{-2}-sec^{-1}-sr^{-1} in percent for the $5° \times 5°$ solar magnetospheric angular areas. The outer dotted lines represent the average position for the standing shock wave and the inner dotted lines represent the average position of the magnetosphere at 17 R_E (Montgomery, 1968).

clearly be seen (Figure 1). The integral energy spectrum of these electrons can be approximated by a power law with an average exponent ranging between 3 and 4 which shows no systematic change with solar magnetospheric longitude. The extreme excursions of this exponent vary from 1 to 7.

The inner boundary of the plasma sheet has not been reported yet in the midnight-dawn quadrant. We have made a distinction in Figure 1 between the normal plasma sheet and the region where Montgomery nearly always detects a high energy electron tail.

3. Pseudo-Trapping Region

This region has been referred to as the distant radiation zone by Anderson (1965). It was recognized that particles in this region could not execute complete drift motion around the earth but could exhibit complete bounce and partial drift motion. The skirt (Figure 1) was first described by Frank *et al.* (1963) and was subsequently discussed by Anderson *et al.* (1965) and Frank (1965). The cusp, which is confined to low

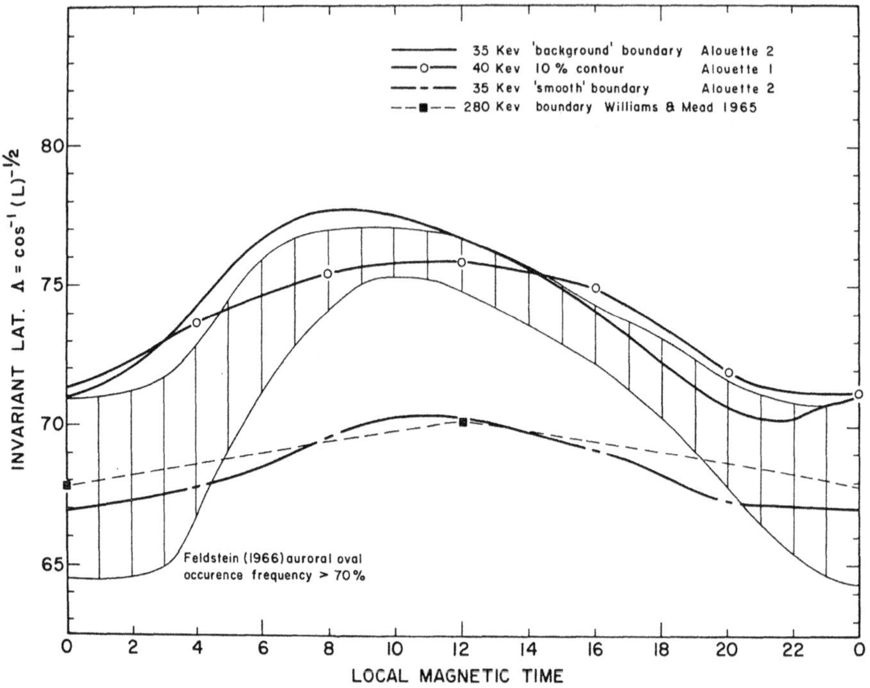

Fig. 3. Comparison of various high latitude boundaries for energetic particle fluxes observed by Alouette 1 and 2 with Feldstein's auroral oval and the trapping boundary of Williams and Mead (McDiarmid and Burrows, 1968).

latitudes $\sim \pm 20°$, was also detected by these investigators. Roederer (1967) calculated the adiabatic motion of particles using the Mead-Williams model (Mead, 1964; Williams and Mead, 1965) and the geometric extent of the pseudo-trapping regions agreed reasonably well with the experimental results for the skirt and the cusp. The outer edge of the pseudo-trapping region shown in Figure 1 was determined by $\sim$40 to 100 keV electrons at low latitudes from results on Explorer 14 by Frank (1965), Explorer 21 (IMP 2) by Rothwell and Lyman (1969) and ERS-17 by Vette

(Peterson *et al.*, 1968). The various particle boundaries also can be detected at high latitude by low altitude satellites. McDiarmid and Burrows (1968) have presented results from Alouette 1 and 2. The various boundaries are shown in Figure 3 along with Feldstein's (1966) auroral oval and the 280 keV boundary given by Williams and Mead (1965). We have projected the Alouette 2 stable trapping boundary and the outer limit of trapping onto the equatorial plane using the average magnetic field configuration given by Fairfield (1968). The stable trapping boundary is shown in Figure 1 as the high energy electron region and is in essential agreement with the low latitude 1.6 MeV electron boundary obtained on IMP 2 by Rothwell and Lyman (1969). On the dusk side of the earth from 1200 to 2400 LT, the Alouette outer trapping boundary follows the low latitude boundary given in Figure 1. However on the dawn side the high latitude boundary maps into the magnetopause over the 1200 to 0600 hr LT interval. For local times between 0200 to 0600 hr, the boundary extends out in the equatorial plane to large distances. These results, along with those of Montgomery (1968), suggest that the clear region shown in Figure 1 is a special one in the magnetosphere bearing further investigation.

The close spatial relationship between the auroral zone and the pseudo-trapping region can readily be seen in Figure 3. Kennel (1969) has recently reviewed the electron and proton discontinuities in the auroral zone. He has made the following conclusions on the basis of experimental evidence:

(a) Soft electrons from the plasma sheet have a sharp inner boundary which corresponds with a sharp outer boundary of harder Van Allen belt electrons;

(b) The auroral electron discontinuity is continuously present but moves in response to geomagnetic activity. During substorms the discontinuity moves closer to the earth;

(c) Electrons up to ~100 keV approach pitch angle isotropy at the discontinuity both inside and near the loss cone suggesting a strong diffusion mechanism;

(d) Electrons are locally accelerated to energies of a few hundred keV at the discontinuity and precipitated particles maximize in this region where isotropy of fluxes are approached; and

(e) A similar proton boundary lies closer to the earth and often coincides with the plasmapause. Precipitating protons also tend toward isotropy.

Recently, Chase (1969) has shown that the flux and spectrum of precipitating electrons in the auroral zone on occasion has the same flux and spectrum as that for electrons in the plasma sheet suggesting that the plasma sheet is the source of these particles.

Frank (1967, 1968) has presented some results of protons and electrons in this region of the magnetosphere covering the energy range ~200 eV to 50 keV in 13 energy bands. These OGO 3 measurements show a rich and varied behavior which is difficult to summarize. Some of Frank's proton results for one pass through the cusp region are shown in Figure 4 in order to present some representative flux values. Electron results on a similar pass are given in Figure 5. The detailed analysis of this experiment will provide a much greater understanding of the pseudo-trapping region

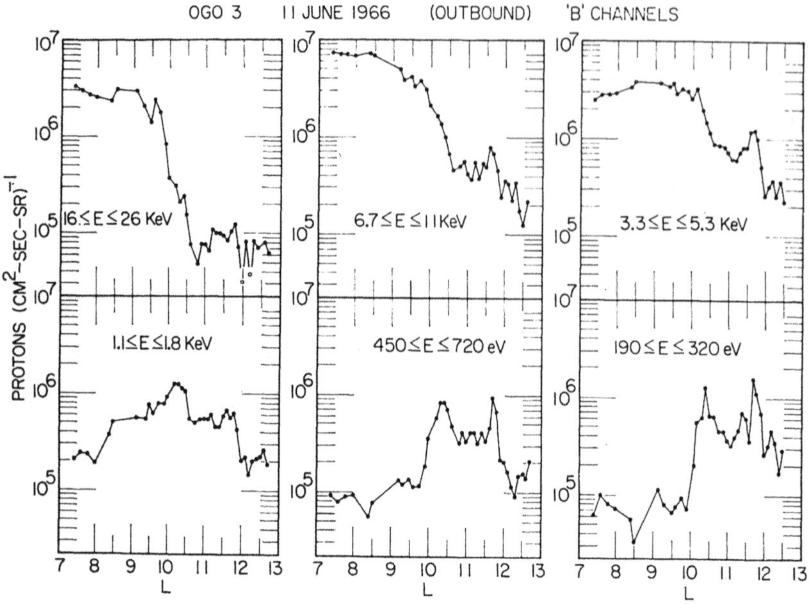

Fig. 4. Directional intensity of protons observed on June 11, 1966. The solar ecliptic longitude varies from 95° at $L = 7$ R_E to 117° at $L = 13$; magnetic latitude lies between 22 and 27° (Frank, 1967).

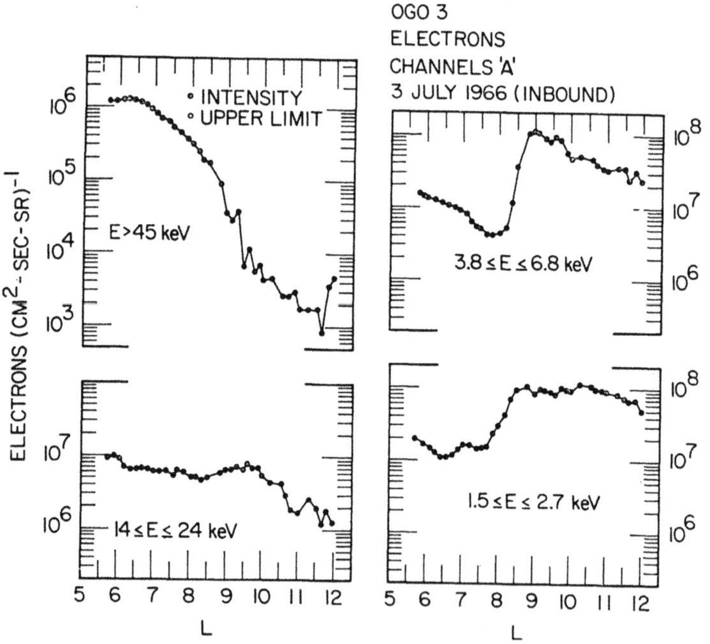

Fig. 5. Directional intensity of electrons observed near the midnight meridian on July 3, 1966. The magnetic latitude was less than 20° for the whole period shown (Frank, 1968).

and will permit average flux values to be obtained in this energy range. Average electron flux values above 40 keV in this spatial region are presented later in Figures 7 and 8. The sharp boundaries which appear on individual satellite passes to distinguish the pseudo-trapping region are not seen in the time averaged radial profile. The pseudo-trapping region seems to be a thick transition layer between the plasma sheet and the Van Allen radiation belt where particles are accelerated to higher energy. Obviously, other accelerations must occur within the stable trapping region in order to account for the even higher energy particles found in the inner and outer radiation zones.

4. Stable Trapping Region

The extent of the stable trapping region is shown in Figure 1 where the 65γ B contour of Fairfield has been used to limit the region on the sunward side. Some of the particles which mirror at high latitude and pass through this region in the equatorial plane are pseudo-trapped so that the omnidirectional equatorial flux is composed of both types of particles. However within the high energy electron boundary all of the particles will normally be trapped. During magnetic storms and substorms the magnetospheric cavity undergoes dramatic changes; particles within the cavity are both adiabatically and nonadiabatically accelerated. The boundaries shown in Figure 1

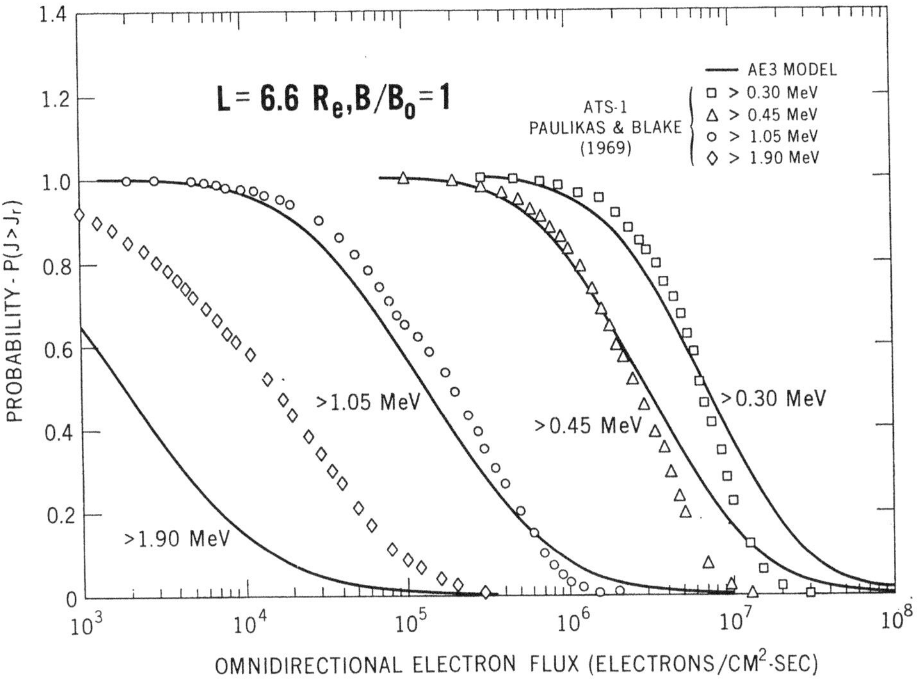

Fig. 6. Comparison of ATS 1 electron fluxes with the electron model environment AE3. The ATS results represent an average from December 6, 1966–March 1, 1968 (Paulikas and Blake, 1969) while AE3 was based on data prior to Jan. 1966.

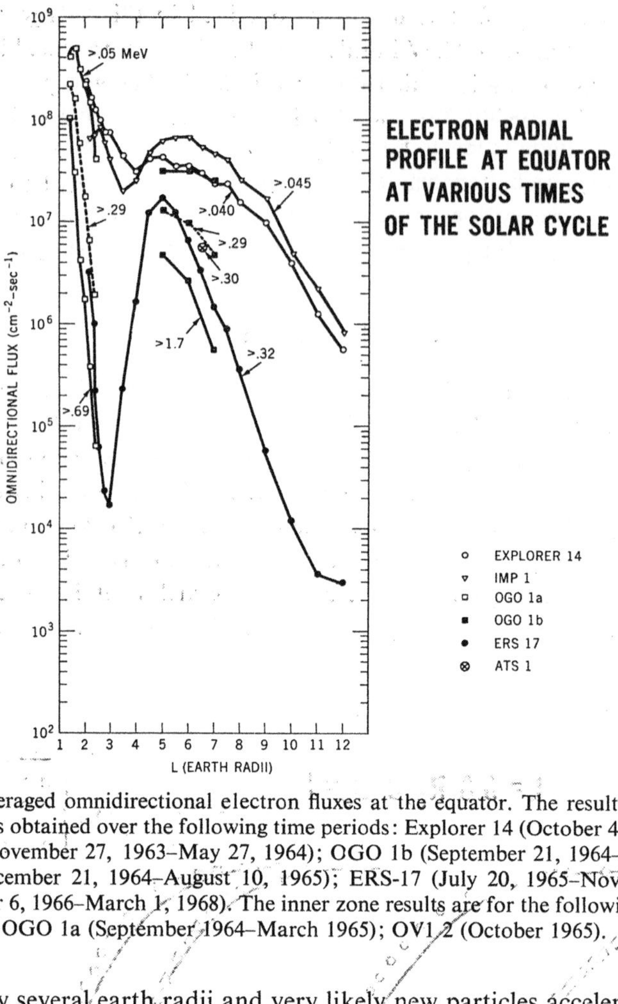

Fig. 7. Time averaged omnidirectional electron fluxes at the equator. The results for $L > 2.2\ R_E$ are average values obtained over the following time periods: Explorer 14 (October 4, 1962–August 5, 1963); IMP 1 (November 27, 1963–May 27, 1964); OGO 1b (September 21, 1964– June 20, 1965); Explorer 26 (December 21, 1964–August 10, 1965); ERS-17 (July 20, 1965–November 3, 1965); ATS 1 (December 6, 1966–March 1, 1968). The inner zone results are for the following time periods: OGO 1a (September 1964–March 1965); OV1 2 (October 1965).

shift inward by several earth radii and very likely new particles accelerated from the plasma sheet in the pseudo-trapping region are fed into the stable trapping region. The main consequence of these complicated events is that the electron and proton fluxes show rather sharp increases or decreases depending on the particle energy, the particular event, and the position. Protons show less change than electrons both in frequency of occurrence and in magnitude. The details of particular storm changes have been discussed by numerous authors.

In view of the one to two order of magnitude change of elections during these events, it is useful to determine the average value of the flux in order to obtain background. Because of the distortion of the geomagnetic cavity and the existence of a tail field, the adiabatic motion of the particles is altered from that in a dipole field. Diurnal variations of the fluxes are evident past 5 R_E but it is difficult to establish the local time behavior with precision using elliptical satellite data. Fortunately, the synchronous satellites provide a platform where these variations can be studied in an

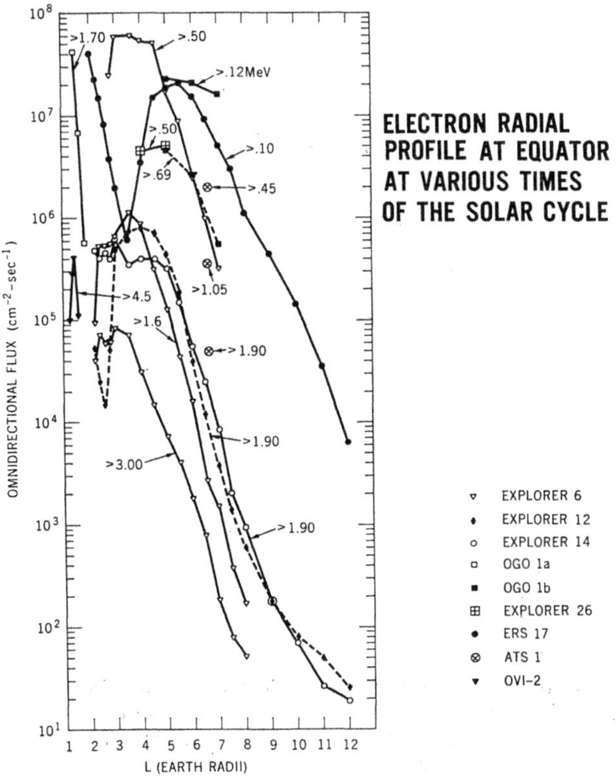

Fig. 8. Time averaged omnidirectional electron fluxes at the equator. The results for $L > 2.2\ R_E$ are average values obtained over the following time periods: Explorer 6 (August 7, 1959–September 10, 1959); Explorer 12 (August 17, 1961–December 5, 1961); all other satellites the same interval as in Figure 7.

important region. Vette and Lucero (1967) showed that fluxes could be approximately represented by a log normal distribution in the outer zone. These results had been based on a small number of points at various L values. We show the recent results of Paulikas and Blake (1969) obtained on ATS 1 in Figure 6 compared to the AE3 model electron environment constructed earlier by Vette and Lucero. For energies below 1 MeV the agreement is excellent except at high flux values. For all energies the log normal distribution with proper parameters seems to be a reasonable description of this data.

The average electron flux near the equator for a selected series of satellite measurements covering the time period August 1959 to March 1968 is shown in Figures 7 and 8 as a function of radial distance. Directional data have been converted to omnidirectional fluxes to facilitate comparison with the other data. The variability of the fluxes about their mean values can be expressed in terms of the standard deviation of the logarithm of the flux. This parameter is shown in Figure 9 as a function of L for some of the various measurements we have processed. The lower energy particles show

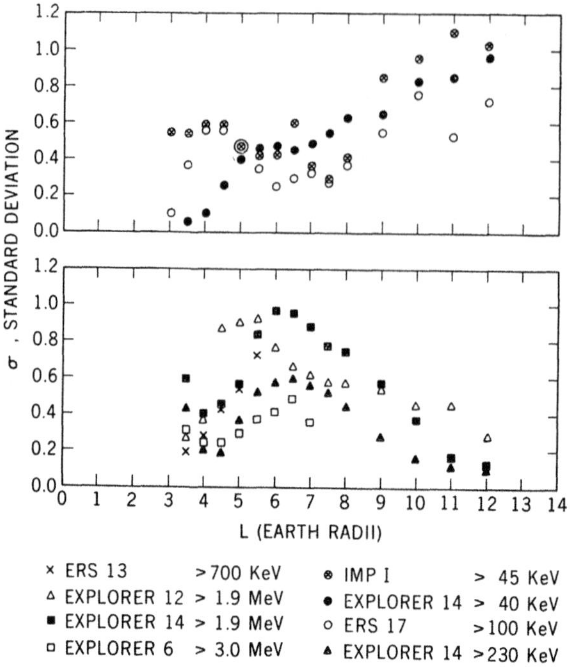

Fig. 9. *L* variation of the standard deviation of the logarithm of the electron flux near the equator. The upper data sets are for energies below 100 keV and the lower sets are for higher energies. Notice the lower energy particles exhibit maximum variability in the pseudo-trapping region whereas the higher energy ones show a peak around 6 to 7 R_E.

a maximum variation at L values corresponding to the pseudo-trapping region whereas the higher energy particles show a peak variation between 6 and 7 R_E. The position of the outer zone flux maximum and slot minimum as a function of the solar cycle has been presented by Vernov *et al.* (1969) and the slot minimum by Frank and Van Allen (1966). It should be pointed out that these parameters are energy dependent and the flux maximum can vary by 1 R_E in a few weeks. The B/B_0 variation for some of the outer zone electron data is shown in Figure 10.

While the lifetime of protons in the inner zone is many years, the effective lifetime of electrons is much less than this. The electrons injected by Starfish have been decaying approximately exponential with time. Walt (1966) has compiled the electron lifetimes obtained by various experiments which show a broad peak of 300 to 400 days between $L=1.25$ and 1.7. There is rapid decline at lower L values caused by the increasing density of the earth's atmosphere; above $L=2.0$ lifetimes drop abruptly to just a few days.

Recently, Bostrom *et al.* (1968) have presented the time behavior of trapped particles in the inner zone from October 1963 through December 1967. Electrons >1.2 MeV have decayed monotonically since the Starfish detonation. For electrons >280 keV the general behavior has been one of exponential decay but there were definite increases during the large magnetic storms in September 1966 and May 1967. Pfitzer and Winckler (1968) have observed these same increases for electrons less than 690 keV.

Based on these latest observations, it appears that large magnetic storms can produce electrons in the inner zone with energies not exceeding about 1 MeV. Thus, all higher energy electrons present are the residue of Starfish. Between 1963 and 1966

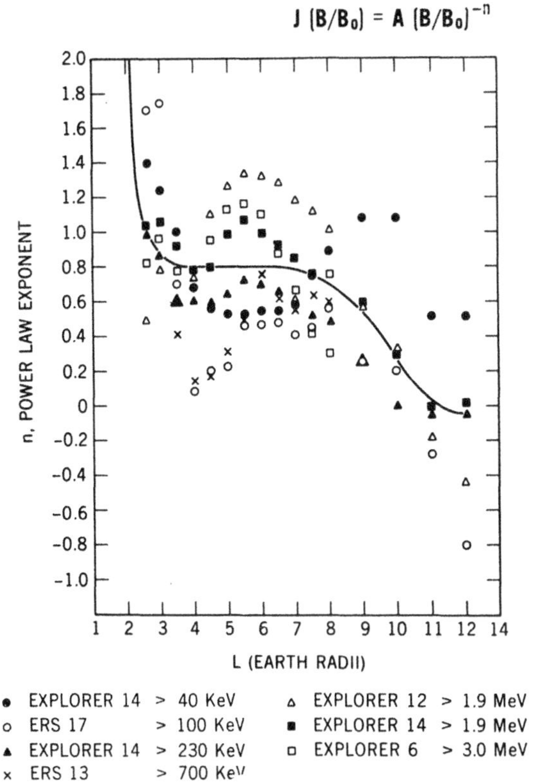

Fig. 10. B/B_0 variation of outer zone average electron fluxes. These data are averaged over time periods given in Figure 7 and 8. The solid line is used to provide a guide for the eye. There seems to be no systematic trend with energy except possibly in the range $4 \leqslant L \leqslant 7$ R_E.

Starfish electrons dominated the whole inner zone. By mid 1966, however, electrons below 1 MeV had decayed to the point where natural effects became apparent. For higher energies the electrons continue to decay.

High energy protons have exhibited great stability in the radiation belts. We show the 40 to 110 MeV results of McIlwain (1963, 1969) in Figure 11 for two different epochs. The second bump around $L=2.2$ had begun to diffuse inward by 1965 and has probably disappeared by the present time. The results of Bostrom *et al.* (1968) at 1100 km showed protons above 8 MeV to be quite stable until the May 1967 event when decreases were seen for L values above 2.6. The 2 MeV protons showed a slight decay over this same time period with both increases and decreases depending on L values seen in the fluxes during the May event. Soraas and Davis (1968) have discussed the

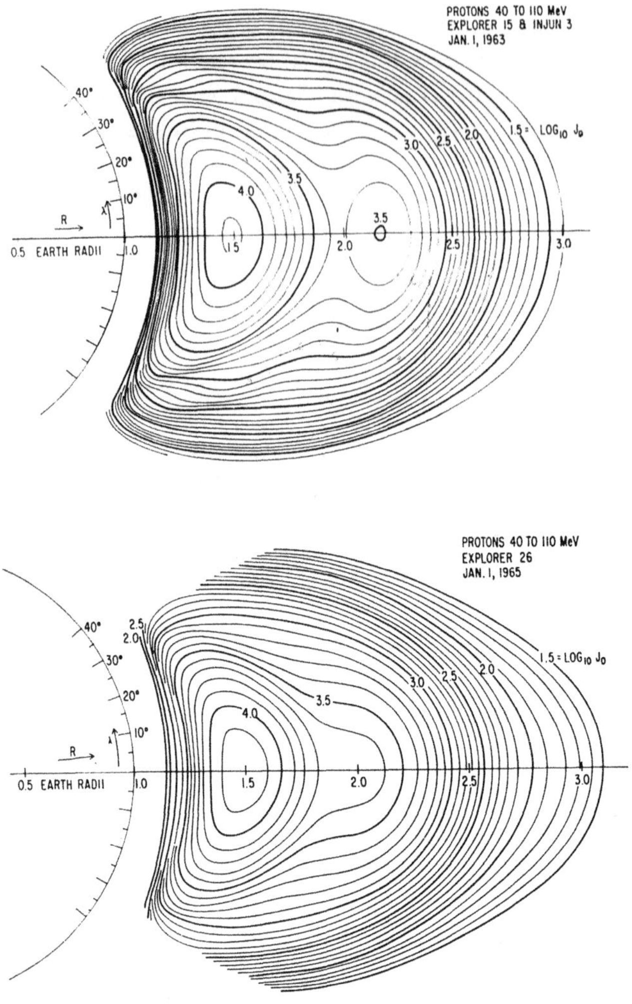

Fig. 11. Comparison of 40–110 MeV protons at two different epochs. The results were obtained on Explorer 15 (McIlwain, 1963) and Explorer 26 (McIlwain, 1969).

increases in protons below 180 KeV and decreases for those above 345 KeV in connection with the April 1965 magnetic storm. They showed that only part of the changes were adiabatic. At low altitudes the solar cycle effect predicted by Blanchard and Hess (1964) finally has become evident. Nakano and Heckman (1968) have reported the 63 MeV protons with minimum altitudes between 220 and 375 km decreased a factor of 2 between mid 1966 and late 1967 after remaining constant for the previous $3\frac{1}{2}$ yr. Measurements of energetic protons with energies above 55, 105, and 170 MeV have been presented recently by Thede (1969) which show that previous estimates near the equator in this range were low by about a factor of 3.

We show in Figure 12 a radial profile of the omnidirectional proton flux at the

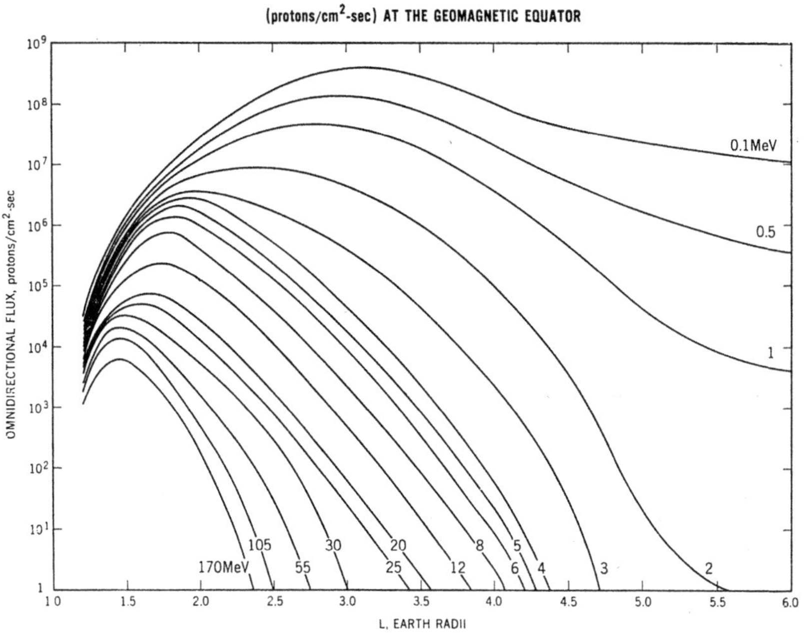

Fig. 12. Radial profile of integral energy omnidirectional proton fluxes at the geomagnetic equator.

equator which incorporates the new high energy data of Thede. Although there are some time variations in the low energy protons, insufficient information has been presented to determine any long term trends near the equator and average values are not yet available. Therefore, at the present time, the static proton models of Vette and co-workers (Vette, 1966; King, 1967; Lavine and Vette, 1969) are all that are justified by the available data.

Finally, we cite that α-particles have been detected in the stable trapping region by Krimigis and Van Allen (1967) with the peak intensity occurring at $L=3.1$ R_E. This peak was enhanced and shifted inward by 0.2 R_E following the April 17, 1965 storm. Trapped α-particle-measurements also have been reported by Paulikas and co-workers (Paulikas *et al.*, 1968; Paulikas and Blake, 1968) who have obtained a power law spectrum over the range 2.8 to 6.2 MeV. The alpha to proton ratio at 0.5 MeV/nucleon varies from 2×10^{-3} to 10^{-4} as L increases from 2.2 to 4.2.

References

Anderson, K. A.: 1965, *J. Geophys. Res.* **70**, 4741.

Anderson, K. A., Harris, H. K., and Paoli, R. J.: 1965, *J. Geophys. Res.* **70**, 1039.

Bame, S. J.: 1968, in *Earth's Particles and Fields* (ed. by B. M. McCormac), Reinhold Publishing Corporation, New York, p. 373.

Blanchard, R. C. and Hess, W. N.: 1964, *J. Geophys. Res.* **69**, 3927.

Bostrom, C. O., Beall, D. S., and Armstrong, J. C.: 1968, Paper presented at International Symposium on the Physics of the Magnetosphere.

Chase, L. M.: 1969, *J. Geophys. Res.* **74**, 348.

Fairfield, D. H.: 1968, *J. Geophys. Res.* **73**, 7329.
Feldstein, Y. I.: 1966, *Planet. Space Sci.* **14**, 121.
Frank, L. A.: 1965, *J. Geophys. Res.* **70**, 1593.
Frank, L. A.: 1967, *J. Geophys. Res.* **72**, 3753.
Frank, L. A.: 1968, in *Physics of the Magnetosphere* (ed. by R. L. Carovillano, J. F. McClay, and H. R. Radoski), D. Reidel Publishing Company, Dordrecht, Holland, p. 271.
Frank, L. A. and Van Allen, J. A.: 1966, *J. Geophys. Res.* **71**, 2697.
Frank, L. A., Van Allen, J. A., and Macagno, E.: 1963, *J. Geophys. Res.* **68**, 3543.
Gringauz, K. I., Bezrukikh, V. V., Ozerov, V. D., and Rybachinsky, R. Ye.: 1960a, *Soviet Phys. Doklady* **5**, 361.
Gringauz, K. I., Kurt, V. G., Moroz, V. I., and Shklousky, I. S.: 1960b, *Soviet Astron. AJ*, **4**, 680.
Kennel, C. F.: 1969, *Rev. Geophys.* **7**, 379.
King, J. H.: 1967, NASA SP-3024 Vol. IV.
Konradi, A.: 1966, *J. Geophys. Res.* **71**, 2317.
Krimigis, S. M. and Van Allen, J. A.: 1967, *J. Geophys. Res.* **72**, 5779.
Lavine, J. P. and Vette, J. I.: 1969, NASA SP-3024 Vol. V.
McDiarmid, I. B. and Burrows, J. R.: 1968, *Can. J. Phys.* **46**, 49.
McIlwain, C. E.: 1963, *Science* **142**, 355.
McIlwain, C. E.: 1969, private communication.
Mead, G. D.: 1964, *J. Geophys. Res.* **69**, 1181.
Montgomery, M. D.: 1968, *J. Geophys.* **73**, 871.
Murayama, T.: 1966, *J. Geophys. Res.* **71**, 5547.
Nakano, G. H. and Heckman, H. H.: 1968, *Phys. Rev. Letters* **20**, 806.
Nishida, A., Lyon, E. F., and Ness, N. F.: 1969, private communication.
Paulikas, G. A. and Blake, J. B.: 1968, Paper presented at International Symposium on the Physics of the Magnetosphere.
Paulikas, G. A. and Blake, J. B.: 1969, private communication.
Paulikas, G. A., Blake, J. B., and Freden, S. C.: 1968, Aerospace Corp. Report No. TR-0200(4260-20)-03.
Peterson, L. E., Matteson, J. L., Huszar, L., and Vette, J. I.: 1968, Univ. of Calif., San Diego Report SP-68-7.
Pfitzer, K. A. and Winckler, J. R.: 1968, *J. Geophys. Res.* **73**, 5792.
Retzler, J. and Simpson, J. A.: 1969, *J. Geophys. Res.* **74**, 2149.
Roederer, J. G.: 1967, *J. Geophys. Res.* **72**, 981.
Rothwell, P. and Lyman, C.: 1969, *Planet. Space Sci.* **17**, 447.
Soraas, F. and Davis, L. R.: 1968, NASA-GSFC X-612-68-328.
Thede, A. L.: 1969, Air Force Weapons Lab. Report AFWL-TR-128.
Vasyliunas, V. M.: 1968, *J. Geophys. Res.* **73**, 2839.
Vernov, S. N., Gorchakov, E. V., Kuznetsov, S. N., Logachev, Y. I., Sosnovets, E. N., and Stolpovsky, V. G.: 1969, *Rev. Geophys.* **7**, 257.
Vette, J. I.: 1966, NASA SP-3024 Vol. I.
Vette, J. I. and Lucero, A. B.: 1967, NASA SP-3024 Vol. III.
Walt, M.: 1966, in *Radiation Trapped in the Earth's Magnetic Field* (ed. by B. M. McCormac), D. Reidel Publishing Company, Dordrecht, Holland, p. 337.
Williams, D. J. and Mead, G. D.: 1965, *J. Geophys. Res.* **70**, 3017.

FURTHER COMMENTS CONCERNING LOW ENERGY CHARGED PARTICLE DISTRIBUTIONS WITHIN THE EARTH'S MAGNETOSPHERE AND ITS ENVIRONS

L. A. FRANK

Department of Physics and Astronomy, The University of Iowa, Iowa City, Iowa, U.S.A.

Abstract. Several recent observations of the distributions of low energy charged particles (100 eV $\lesssim E \lesssim$ 50 keV) within the earth's magnetosphere and its environs are examined for evidences of source (acceleration) mechanisms. Proton (0.5 $\lesssim E \lesssim$ 50 keV) intensities in the outer radiation zone and plasma sheet are concentrated upon since this population forms the largest charged particle energy reservoir in the magnetospheric system. To date no source mechanism for these proton intensities with broad differential energy spectrums (i.e., of the plasma sheet and the storm time and quiescent 'ring currents') has been delineated clearly. It is suggested herein that at least the storm time ring current protons are of solar origin and are convected into the magnetotail under the influence of magnetospheric electrostatic fields or via a diffusion process driven by fluctuating magnetic and/or electric fields.

1. Introduction

The following discussion summarizes recent observational evidences concerning the nature and origin of the vast population of low energy protons (0.5 $\lesssim E \lesssim$ 50 keV) within the earth's outer zone and within the magnetotail. For adequate summaries of the various recent observations of the distributions of low energy electron and proton intensities within the magnetosphere and its environs, the reader is referred to the following literature: the 'storm time' ring current protons (Frank, 1967a); the relationship between the plasmapause and plasma sheet at the magnetic equator (Schield and Frank, 1969); atomic hydrogen densities, 1.5 to 4 R_E altitudes (Swisher and Frank, 1968); the existence of two distinct proton distributions in the earth's magnetic tail (Kanbach and Frank, 1969); asymmetric injection of ring current protons into the outer radiation zone during the early development phase of a magnetic storm (Frank, 1969a); existence of significant proton (5 $\lesssim E \lesssim$ 50 keV) intensities in the interplanetary medium and of solar origin (Frank, 1970); high resolution studies of trapped and precipitated proton and electron intensities at low altitudes over the auroral zone and polar regions (Frank *et al.*, 1969); and comprehensive L value vs time diagrams of low energy proton intensities at the magnetic equator in the outer radiation zone (Frank and Owens, 1969). Our present attention, however, is drawn to the proton (0.5 $\lesssim E \lesssim$ 50 keV) distribution within the distant magnetosphere since this proton population provides the largest single charged particle energy reservoir in the magnetospheric system and by the fact that its presence and behavior are generally not understood.

B. M. McCormac (ed.), Particles and Fields in the Magnetosphere.

2. Overall Morphology of the Extraterrestrial 'Ring Current'

Our brief survey of the character of the proton ($0.5 \lesssim E \lesssim 50$ keV) distributions within the magnetosphere begins with several comments concerning these proton intensities within the outer radiation zone and the earthward edge of the plasma sheet. Omnidirectional differential intensities of protons averaged over the energy range $16 \leqslant E \leqslant 25$ keV as functions of the magnetic shell parameter L and time at the magnetic equator are summarized for the June 10 through July 23, 1966, in Figure 1

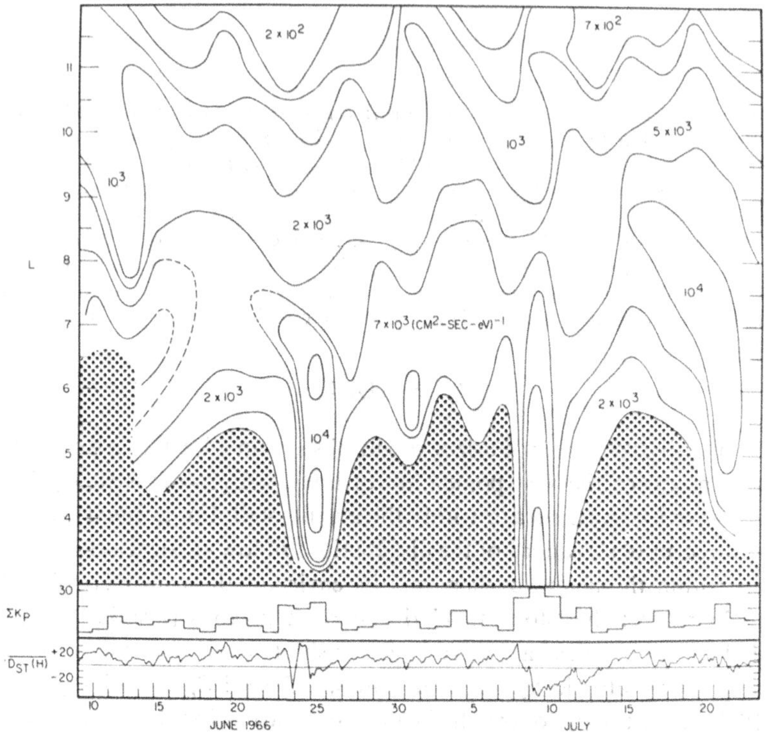

Fig. 1. Contours of constant omnidirectional intensities of protons ($16 \leqslant E \leqslant 25$ keV) as functions of magnetic shell parameter L and time at the magnetic equator for the June 10 through July 23, 1966. The contour intensity increments are 1, 2, 5, and 7×10^n. Units are omnidirectional differential flux, protons$(\text{cm}^2\text{-sec-eV})^{-1}$, averaged over the instrument energy bandpass (Frank and Owens, 1969).

(Frank and Owens, 1969). Two salient features of these proton distributions in the local evening-midnight quadrant of the magnetosphere are of immediate interest: (a) the persistent presence of a 'quiet time ring current' centered at $L \simeq 6.5$ (position of peak proton, $5 \lesssim E \lesssim 50$ keV, energy densities) (cf., Frank, 1967a; Swisher and Frank, 1968) and (b) the penetration of these proton intensities deep into the outer radiation zone during the two moderate magnetic storms on June 23 and July 9. These enhanced proton ($5 \lesssim E \lesssim 50$ keV) distributions are responsible largely for the worldwide decreases of magnetic field intensities (Frank, 1967a) and are known as the storm time 'extraterrestrial ring current'. No clear association among various parameters of the solar wind, such as ion temperatures and velocities and interplanetary magnetic field intensities and direction, and the great enhancement of the extraterrestrial ring current responsible for the main phase geomagnetic storm has been found yet (cf., Snyder *et al.*, 1963; Fairfield and Cahill, 1966; Wilcox *et al.*, 1967; Burlaga and Ogilvie, 1969; Verzariu *et al.*, 1969). The differential energy spectrums of proton intensities at various positions within the storm time ring current should provide several clues as to the mechanism responsible for their origin. Such a series of proton spectrums for July 9 is provided in Figure 2. These spectrums for $L = 4.0$,

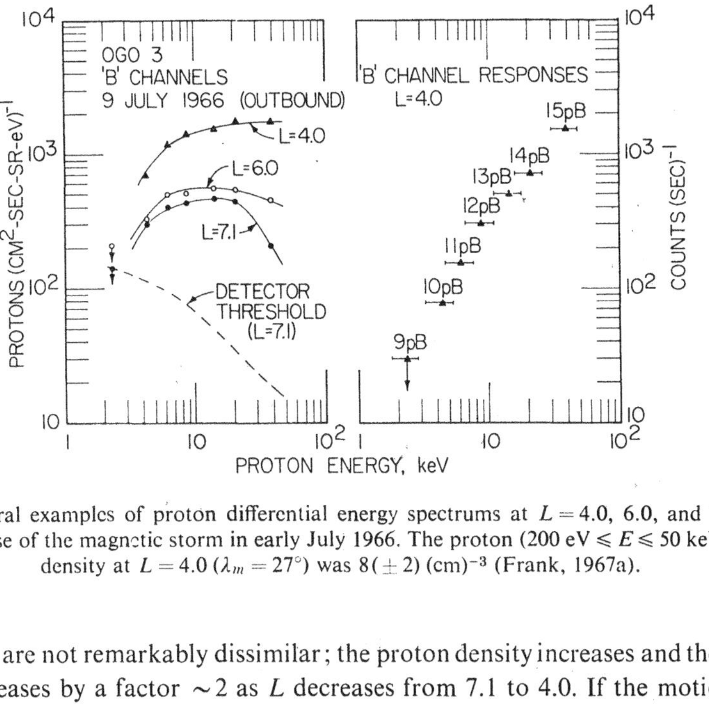

Fig. 2. Several examples of proton differential energy spectrums at $L = 4.0$, 6.0, and 7.1 during the main phase of the magnetic storm in early July 1966. The proton (200 eV $\leqslant E \leqslant$ 50 keV) number density at $L = 4.0$ ($\lambda_m = 27°$) was $8(\pm 2)$ $(\text{cm})^{-3}$ (Frank, 1967a).

6.0, and 7.1 are not remarkably dissimilar; the proton density increases and the average energy increases by a factor ~ 2 as L decreases from 7.1 to 4.0. If the motion of the protons inward proceeded with conservation of the first adiabatic invariant μ in a dipole field the expected increase in the average proton energy is ~ 6. However, the observational fact that this increase is not realized in this series of observations is not surprising due to the high β ($\beta \gtrsim 1$) of the plasma (cf., Frank, 1967a) and associated inflation of the geomagnetic field. The proton spectrums within the 'quiet time' ring current possess similar spectrums when compared to the above storm time distributions, but feature lower number densities (cf., Frank, 1967b) and significant intensities are restricted to larger radial distances $\gtrsim 5.5$ R_E. No clear indication of a region of 'local acceleration', where acceleration of solar wind or magnetosheath ions with kinetic energies approximately several hundred eV to energies typical of the dominant 'ring current' proton distributions is evident in this series of observations.

Even during the earliest developmental phase of a geomagnetic main phase storm, hours before a decisive decrease is evident in the Dst(H) values but during which these proton intensities have penetrated deep into the evening sector of the outer radiation zone, these proton spectrums again are characterized by average energies ~5 to 20 keV and broad width (Frank, 1969a). Direct observations of this asymmetric injection of protons into the outer radiation zone are summarized for four 'snapshots' of proton ($31 \leqslant E \leqslant 49$ keV) intensities as functions of shell parameter L in Figure 3:

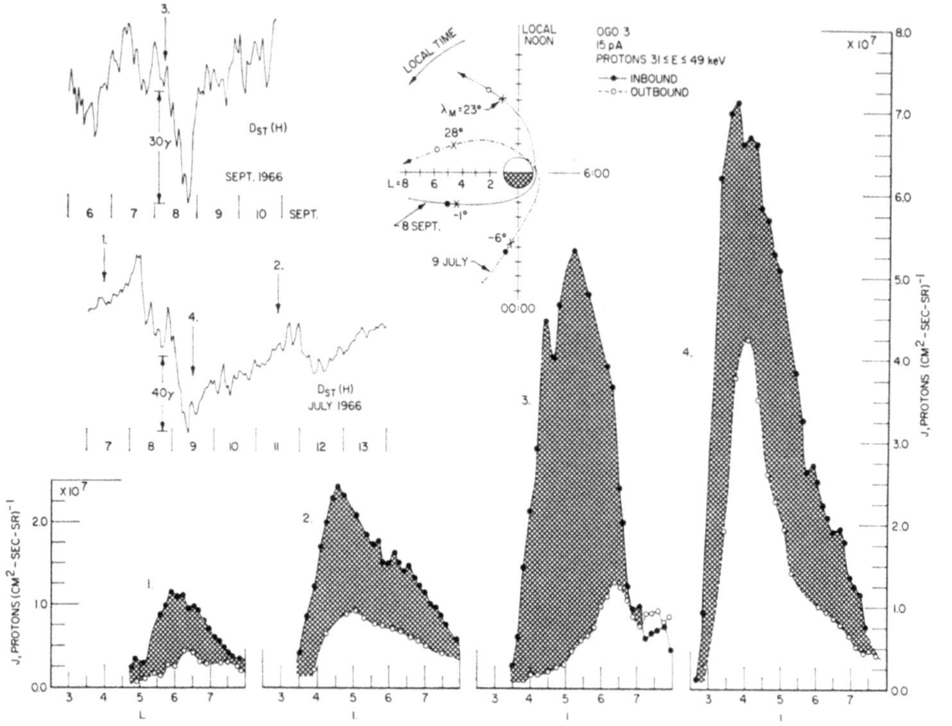

Fig. 3. Proton ($31 \leqslant E \leqslant 49$ keV) directional intensities as functions of L during several phases of two geomagnetic storms on July 9, and September 8, 1966. Hourly values of Dst(H), several useful coordinates for these observations, and the OGO 3 trajectory through the outer radiation zone as functions of local time and shell parameter L have also been included in this graphic summary (Frank, 1969a).

quiescent or pre-storm (snapshot 1); early developmental phase (3); main phase (4); and recovery phase (2). The proton spectrums of snapshot 3 during the early developmental phase in the evening sector of the outer radiation zone (●, local evening; ○, local noon) displayed no evidences of a proton energy spectrum with lower average energy and/or narrower width relative to the quiescent or storm time proton spectrums which perhaps would indicate concurrent acceleration by, for example, a wave particle interaction or a geoelectric potential system. The above overall homogeneity of these proton spectrums with regard to average energy and similar broad widths and the general character of the temporal behavior of the proton distributions over such a large region of the magnetosphere and broad range of geomagnetic activity suggest that bulk transport throughout these regions is an important contributor to the morphology of this proton distribution. A decisive evaluation of the relative importance of transport and acceleration awaits simultaneous measurements of magnetic and electric fields.

3. Comments on the Proton Distributions in the Earth's Magnetic Tail

A discussion of the proton ($0.5 \lesssim E \lesssim 5$ keV) distributions within the magnetosphere is incomplete without at least several comments concerning the character of the continuation of this near earth plasma into the earth's magnetic tail, the so-called 'plasma sheet'. The existence of the earth's magnetic tail and its neutral sheet (Ness, 1965; Ness *et al.*, 1967) is important to a host of magnetospheric models for the acceleration and transport of low energy charged particles, including the auroral

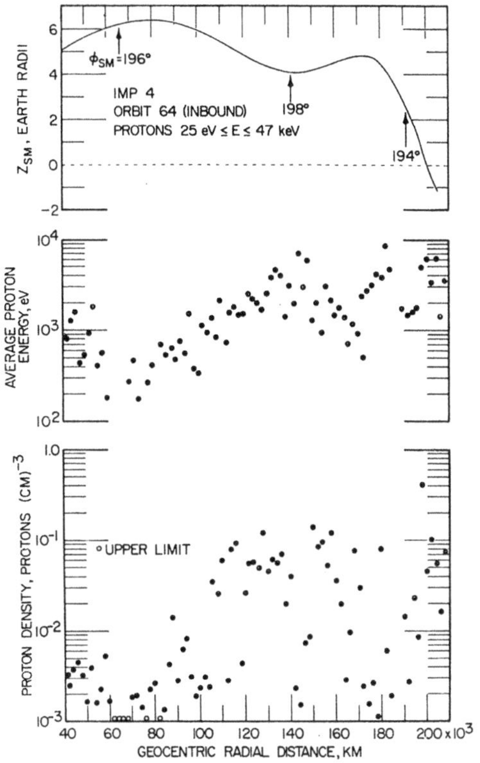

Fig. 4. A typical series of observations of the proton (25 eV $\leqslant E \leqslant$ 47 keV) energy densities and number densities as functions of geocentric radial distance in the earth's magnetic tail. Values for the calculated height of the satellite above the neutral sheet Z_{SM} and for the corresponding earth-centered solar magnetospheric longitudes φ_{SM} are included at the top of the graph (Kanbach and Frank, 1969).

corpuscular radiation, based upon the dynamical behavior of this region (cf., Speiser, 1965, 1967; Taylor and Hones, 1965; Dessler, 1968; Piddington, 1968; Dungey, 1968; Schield, 1969). A typical series of observations of the average energy and number densities for proton (25 eV $\leqslant E \leqslant$ 47 keV) intensities within the magnetic tail as functions of geocentric radial distance are summarized in Figure 4 (Kanbach and Frank, 1969). Proton number densities and average energies range from $<10^{-3}$ to 0.4 protons $(\text{cm})^{-3}$ and $\sim$200 eV to 10 keV, respectively. No obvious order such as a clear increase in average proton energy with increasing (decreasing) radial distance, which would be indicative of an acceleration by geoelectric potential fields, for example, is evident in the large fluctuations of these parameters; however, the average energy and number densities generally do appear to increase with decreasing distance from the neutral sheet (Z_{SM}, top of Figure 4). This ordering arises from the apparent existence of two proton distributions within the magnetotail characterized by the two directional, differential spectrums of proton intensities shown in Figure 5. The energy spectrum with average proton energy approximately several hundred eV

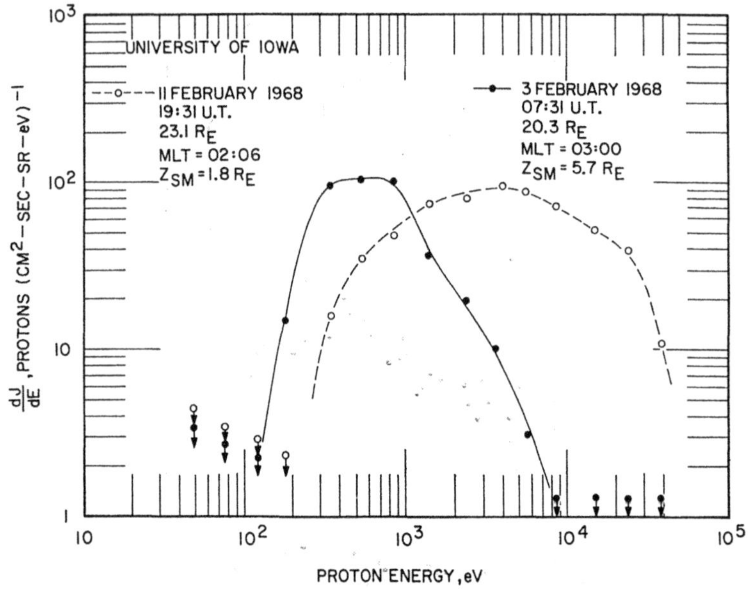

Fig. 5. Two characteristic spectrums of directional, differential proton intensities observed in the earth's magnetic tail (Kanbach and Frank, 1969).

is similar to typical proton spectrums observed within the magnetosheath (Figure 6) but is usually characterized by a lesser number density by factors $\sim 10^{-2}$ to 10^{-3}. In fact these proton intensities within the magnetic tail are similar *in intensities* to the magnetosheath proton intensities directed *normal to the downstream magnetopause surface*; this feature strongly suggests that these proton intensities have access to the magnetic tail. The second proton energy spectrum displayed in Figure 5 has an average energy ~5 keV and is typical of the proton spectrums within the plasma sheet. This second proton population is present, with a large range of number densities and hence energy densities, over a large region of the earth's magnetic tail to geocentric radial distances of a least 35 R_E (see Figure 4, also Kanbach and Frank, 1969).

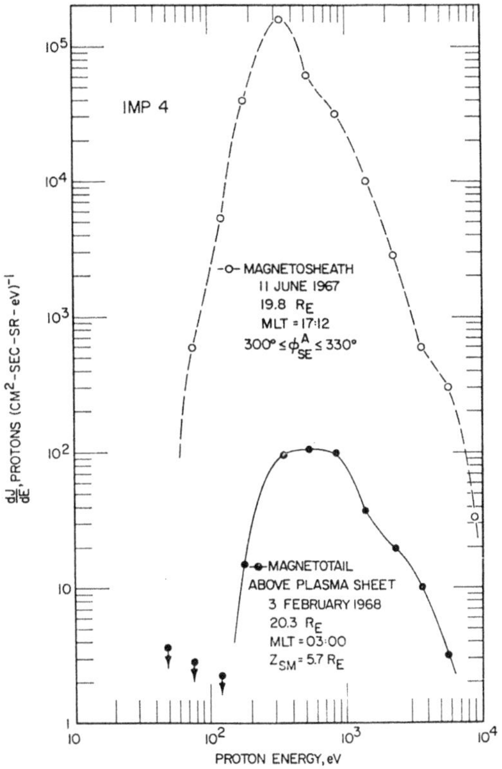

Fig. 6. Comparison of differential spectrums of proton intensities for a typical magnetosheath observation in the direction of bulk flow (○) with the low energy proton spectrum observed in the magnetotail (●, see also Figure 5) (Kanbach and Frank, 1969).

Generally the angular distributions of these proton intensities are closely isotropic (Frank, 1969b). However, detailed studies of these angular distributions in the vicinity of the neutral sheet currently are being undertaken and the results of this search for a possible current system soon will be reported. We note at this point that the average electron energies within this proton distribution (plasma sheet, near earth plasma sheet, quiescent and storm time ring currents) range from ~0.5 to several keV (Frank, 1967b, 1967c; Bame, 1968; Schield and Frank, 1969) while the corresponding proton average energies are ~3 to 30 keV, factors ~5 to 10 larger than the average electron kinetic energy (and energy densities). Any magnetospheric mechanism or model will necessarily be required to account for this feature of the largest charged particle energy reservoir within the earth's magnetosphere. This relatively high average proton energy, the general homogeneity of the proton spectrums with regard to average energy and broad width from the center of the storm time extraterrestrial ring current at $\sim 4\ R_E$ to the distant plasma sheet at $\gtrsim 35\ R_E$, the presence of magnetosheath proton intensities of the character described herein within the magnetotail, and the overall temporal behavior of the extraterrestrial ring current perhaps suggest that the interplanetary medium should be examined again as a direct source for these proton ($0.5 \lesssim E \lesssim 50$ keV) intensities which possibly gain the magnetosphere via the magnetopause downstream from the earth and are subsequently convected within the magnetosphere by geoelectric potentials and/or fluctuating magnetic fields driven by the solar wind.

4. Interplanetary Medium: 'Invisible Source' for Extraterrestrial Ring Current Protons?

A recent observational survey of the interplanetary medium in the vicinity of the earth has revealed the existence of substantial enhancements of interplanetary proton ($5 \lesssim E \lesssim 50$ keV) intensities, presumably of solar origin, during the developmental phases of two moderate geomagnetic storms (Frank, 1970). Salient features of these two increases of proton intensities were (a) a broad differential energy spectrum of

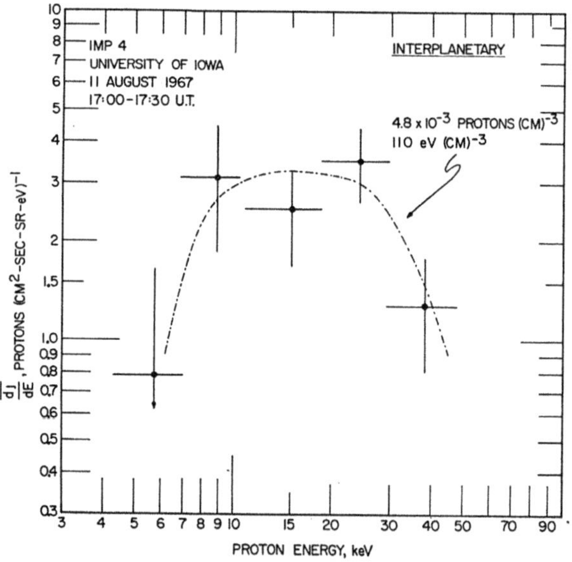

Fig. 7. Differential energy spectrum of *interplanetary* proton intensities averaged over the angular distributions during 17:00-17:30 UT on August 11, 1967 (Frank, 1969b).

intensities over the energy range $5 \lesssim E \lesssim 50$ keV with peak intensities ~ 10 protons $(cm^2\text{-sec-sr-eV})^{-1}$ at ~ 20 keV (Figure 7), (b) peak number densities $\sim 10^{-2}$ protons $(cm)^{-3}$, (c) peak energy densities ~ 300 eV $(cm)^{-3}$, (d) an anisotropy in the angular distribution in the ecliptic plane favoring intensities arriving from near solar directions by factors ~ 2 or 3 (Figure 8), and (e) duration of these events ~ 1 day. A coarse assessment of the relative importance of these proton ($5 \lesssim E \lesssim 50$ keV) distributions to the dynamical character of the interplanetary medium can be gained by noting that the above peak energy densities exceed the 'quiet-sun' ion and electron thermal energy densities by a factor of ~10 and are comparable to or exceed these thermal energy

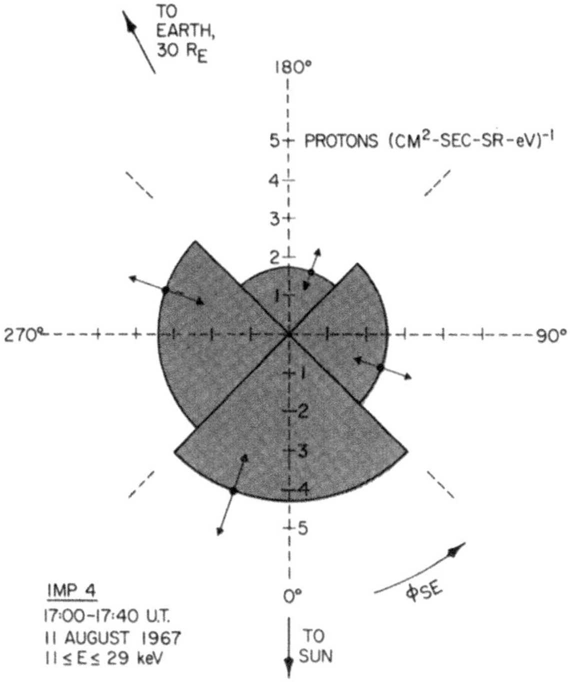

Fig. 8. Angular distribution of *interplanetary* proton ($11 \leqslant E \leqslant 29$ keV) intensities during the event on August 11 (φ_{SE}, satellite centered solar ecliptic longitude) (see also Figure 7) (Frank, 1969b).

densities during solar disturbances. The observations of these increases of interplanetary proton intensities during the developmental phases of two moderate magnetic storms and the hourly Dst(H) values are summarized in Figure 9. Note that these increases commence *hours* before a clear decrease in the ground based magnetometer records is evident and terminate approximately with the occurrence of the maximum main phase decrease. There is a remarkable agreement in chronological sequence among the observational facts that (a) the interplanetary proton ($5 \lesssim E \lesssim 50$ keV) intensities increase hours before the storm, and (b) the early developmental phase of a magnetic storm is characterized by a highly asymmetric injection of protons ($5 \lesssim E \lesssim 50$ keV) into the evening sector of the outer radiation zone, *again*, hours before a decisive decrease is evident in the Dst(H) values (cf., Cahill, 1968; Frank, 1969a). This timely arrival of proton ($5 \lesssim E \lesssim 50$ keV) intensities at the earth just prior to the onset of two small geomagnetic storms and the similarities of the proton differential spectrums within the extraterrestrial ring current (Figure 2) with those in the interplanetary medium provide ample motivation for investigation. They suggest the possibility that the mechanism responsible for supplying a major fraction of the energy within the extraterrestrial ring current is not local acceleration of lower energy protons (approximately several hundreds of eV) but largely convection of

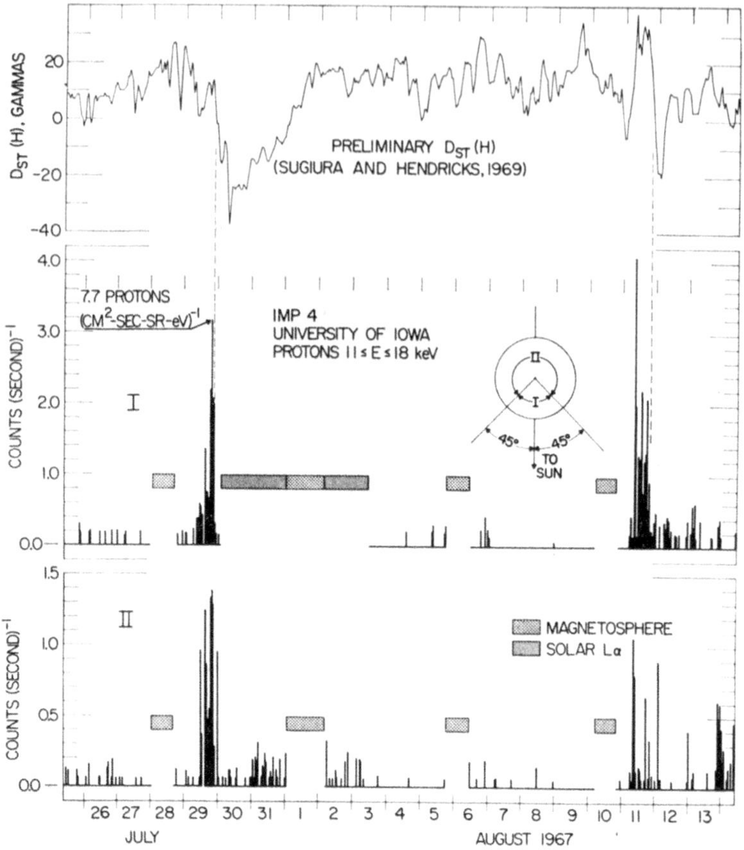

Fig. 9. Hourly average responses of the LEPEDEA instrumentation due to interplanetary proton ($11 \leqslant E \leqslant 18$ keV) intensities arriving at the satellite from directions within a 270° angular sector centered on the anti-solar direction (II, bottom) and within the quadrant centered on the solar direction (I, middle), and hourly Dst(H) values (top) as functions of time for July 25 through August 13, 1967 (Frank, 1969b).

interplanetary protons ($5 \lesssim E \lesssim 50$ keV) to positions deep within the magnetosphere with large enhancements of interplanetary intensities and that, for example, the great variety of magnetic storms (i.e., slow and fast growth phases, multiple storms, etc. (Akasofu, 1966) reflects the temporal variations of solar proton ($5 \lesssim E \lesssim 50$ keV) intensities arriving at the earth (Frank, 1970). It has been shown by Frank (1969b) that it is necessary to convect into the terrestrial magnetosphere only *approximately* 2% of the interplanetary protons ($5 \lesssim E \lesssim 50$ keV) incident on the sunlit magnetopause in order to account for the observed Dst(H) decreases during these two magnetic storms; in other words, it is energetically plausible. A similar result ensues if this calculation is repeated for energy fluxes incident upon the magnetopause at the magnetic tail. The energy fluxes directed *perpendicular* to the magnetopause at ~ 10 to 20 R_E downstream from the earth are $\sim 10^{-3}$ erg(cm^2-sec)$^{-1}$ for magnetosheath protons (Frank, 1970) and $\sim 4 \times 10^{-2}$ erg(cm^2-sec)$^{-1}$ for the above peak interplanetary proton ($5 \lesssim E \lesssim 50$ keV) intensities. As discussed above, magnetosheath protons are observed to gain access to the magnetotail; there is no apparent reason to believe that these more energetic interplanetary protons do not enjoy the same privilege. The reader is reminded at this point that observations of these proton intensities are difficult in the interplanetary medium near the earth and that these intensities are beyond the capabilities of most instruments designed to measure the properties of the solar wind ions (cf., Frank, 1967c, 1970); it is not surprising that this constituent of the interplanetary medium heretofore has been undetected. These are two further series of observational evidences which are directly pertinent to the above suggestion and which are recalled here from previously published literature. First, a similar, and apparently singular, observation of the arrival of relatively high energy densities of solar protons ($E \gtrsim 140$ keV) of $\sim 10^{-9}$ erg(cm)$^{-3}$, *or* 600 eV(cm)$^{-3}$, on September 30, 1961 (Hoffman *et al.*, 1962) was followed several hours later by the growth of a well developed main phase geomagnetic storm. Second, Akasofu (1964) has conducted an extensive correlative study of ground based observations of magnetic storms and solar activity and has concluded that the source for storm time ring current particles was a blast wave of as yet undetected, or 'invisible', corpuscular radiation from the sun; these 'invisible' particles were assumed to be neutral H atoms. Several criticisms of the above suggestion are directed toward the plausibility of relatively large proportions of solar neutral H arriving at the earth (cf., Brandt and Hunten, 1966). Our present interpretation allows us to adopt Akasofu's analyses and conclusions as substantially correct but with the assertion the 'invisible' corpuscular radiations are the *heretofore unobserved solar proton* ($5 \lesssim E \lesssim 50$ keV) *intensities, not solar neutral H.* Throughout the preceding discussions, a strong observational connection among the proton distributions of the plasma sheet and the quiescent and storm time ring currents has been established. Accordingly it should be expected that these proton distributions share a common origin. At present only upper limits for interplanetary proton ($5 \lesssim E \lesssim 50$ keV) intensities during periods of relative magnetic quiescence are available. These upper limits for proton ($5 \lesssim E \lesssim 50$ keV) energy densities are ~ 50 eV(cm)$^{-3}$; this corresponds to an energy flux on the sunlit magnetopause, for example, of $\sim 10^{18}$ erg(sec)$^{-1}$. A rough estimate of the energy input required to maintain these 'quiet-time' magnetospheric proton distributions may be acquired by assuming that the dominant loss mechanism is precipitation into the earth's atmosphere at auroral latitudes. Preliminary estimates (Frank *et al.*, 1969) of proton ($5 \lesssim E \lesssim 50$ keV) energy losses to the earth's upper atmosphere are less than the quiescent energy precipitation due to auroral electrons ($E \gtrsim$ several keV) by factors of ~ 10 to 100. The corresponding rate of energy input into the atmosphere due to electron precipitation is $\sim 2\times 10^{17}$ erg(sec)$^{-1}$ (Craven, 1969) and by adopting an intermediate ratio ~ 30 as noted above the corresponding estimate for proton ($5\lesssim E\lesssim 50$ keV) energy losses into the atmosphere is $\sim 6\times 10^{15}$ erg(sec)$^{-1}$. If this estimate for the power input is reasonable for maintaining the protons within the plasma sheet and extraterrestrial ring current during quiescent conditions, then again it is necessary to transport into the magnetosphere only approximately 1% of the observed upper limit for proton ($5\lesssim E\lesssim 50$ keV) intensities in the interplanetary medium. Clearly there is a need for further measurements of interplanetary and magnetosheath proton intensities over the energy range extending from several keV to tens of keV to further evaluate the interplanetary medium and the magnetosheath as sources for the vast populations of magnetospheric protons with these energies. As a final note, it is stressed that these interplanetary proton ($5\lesssim E\lesssim 50$ keV) intensities recently observed are suggested as a dominant contributor to the proton distributions with similar energies in the plasma sheet and the extraterrestrial ring current, but that the energy required to provide the transport of these protons within the magnetosphere and to provide auroral electron precipitation, energetic charged particle diffusion, etc., must be derived from the primal energy source, the solar wind ions.

Acknowledgments

This research was supported in part by the National Aeronautics and Space Administration under grant NsG-233-62 and contracts NAS5-2054 and NAS5-9074 and by the Office of Naval Research under contract Nonr-1509(06).

References

Akasofu, S.-I.: 1964, *Planetary Space Sci.* **12**, 801.
Akasofu, S.-I.: 1966, *Space Sci. Rev.* **6**, 21.
Bame, S. J.: 1968, in *Earth's Particles and Fields* (ed. by B. M. McCormac), Reinhold Publishing Corporation, New York, p. 373.
Brandt, J. C. and Hunten, D. M.: 1966, *Planetary Space Sci.* **14**, 95.
Burlaga, L. F. and Ogilvie, K. W.: 1969, GSFC Preprint X-616-69-28.
Cahill, L. J.: 1968, U. of New Hampshire Res. Rep. 68-4.
Craven, J. D.: 1969, Ph.D. Thesis, University of Iowa.
Dessler, A. J.: 1968, *J. Geophys. Res.* **73**, 209.
Dungey, J. W.: 1968, in *Earth's Particles and Fields* (ed. by B. M. McCormac), Reinhold Publishing Corporation, New York, p. 385.
Fairfield, D. H. and Cahill Jr., L. J.: 1966, *J. Geophys. Res.* **71**, 155.
Frank, L. A.: 1967a, *J. Geophys. Res.* **72**, 3753.
Frank, L. A.: 1967b, *J. Geophys. Res.* **72**, 1905.
Frank, L. A.: 1967c, *J. Geophys. Res.* **72**, 185.
Frank, L. A.: 1969a, *J. Geophys. Res.*, submitted.
Frank, L. A.: 1969b, *Trans. Amer. Geophys. Union* **50**, 284.
Frank, L. A.: 1970, *J. Geophys. Res.*, **75**, 707.
Frank, L. A., Münch, J. W., and Ackerson, K. L.: 1969, *J. Geophys. Res.*, to be submitted.
Frank, L. A. and Owens, H. D.: 1969, *J. Geophys. Res.*, submitted.
Hoffman, R. A., Davis, L. R., and Williamson, J. M.: 1962, *J. Geophys. Res.* **67**, 5001.
Kanbach, G. W. and Frank, L. A.: 1969, *J. Geophys. Res.*, to be submitted.
Ness, N. F.: 1965, *J. Geophys. Res.* **70**, 2989.

Ness, N. F., Behannon, K. W., Scearce, C. S., and Canterano, S. C.: 1967, *J. Geophys. Res.* **72**, 5769.
Piddington, J. H.: 1968, in *Earth's Particles and Fields* (ed. by B. M. McCormac), Reinhold Publishing Corporation, New York, p. 417.
Schield, M. A.: 1969, *J. Geophys. Res.*, submitted.
Schield, M. A. and Frank, L. A.: 1969, *J. Geophys. Res.*, to be submitted.
Snyder, C. W., Neugebauer, M., and Rao, U. R.: 1963, *J. Geophys. Res.* **68**, 6361.
Speiser, T. W.: 1965, *J. Geophys. Res.* **70**, 4219.
Speiser, T. W.: 1967, *J. Geophys. Res.* **72**, 3919.
Swisher, R. L. and Frank, L. A.: 1968, *J. Geophys. Res.* **73**, 5665.
Taylor, H. E. and Hones, Jr., E. W.: 1965, *J. Geophys. Res.* **70**, 3605.
Verzariu, P., Strong, I. B., and Sugiura, M.: 1969, *Trans. Amer. Geophys. Union* **50**, 279.
Wilcox, J. M., Schatten, K. H., and Ness, N. F.: 1967, *J. Geophys. Res.* **72**, 19.

THE ORIGIN AND DISTRIBUTION OF ENERGETIC ELECTRONS IN THE VAN ALLEN RADIATION BELTS

J. R. WINCKLER
School of Physics and Astronomy, University of Minnesota, Minneapolis, Minn., U.S.A.

1. Introduction

The primary purpose of this paper is to present new results on the injection and distribution of the energetic electron component of the trapped radiation as a result of magnetic storms. The results described herein have been obtained with magnetic deflection electron spectrometers carried in the OGO 1 and OGO 3 satellites in elliptical orbits penetrating both the inner and outer zone regions and in the ATS 1 geostationary orbit satellite in the outer zone. Correlations with ground based magnetic and auroral measurements as well as high altitude balloons have played an important part in the interpretation of these results. The satellite instruments are similar on the various missions and details have been given in several papers and technical reports (Lezniak *et al.*, 1967; Lezniak and Winckler, 1968; Pfitzer, 1968; Pfitzer and Winckler, 1970). The energy range extends from about 50 keV to several MeV and it is possible to study the variation of the directional flux with pitch angle, energy, and position in the trapping regions. The spectrometers on the OGO 1, OGO 3, and ATS 1 satellites now have been directly compared in orbit at appropriate times and positions in the magnetosphere and agree to better than 10% using the laboratory calibration constants.

2. The Long Term Variations of the Inner Zone

The inner radiation belts have been surveyed by the OGO satellites which traverse this region about every 2 days sweeping out along the trajectories a certain combination of L and local pitch angle α. Since the time changes during 1964 and 1965 are relatively slow, data could be accumulated over relatively long intervals to deduce the equatorial pitch angle distribution $F(\alpha_0, L)$ from a number of passes. The details of the reduction of the data to the equatorial pitch angle distribution have been published elsewhere (Pfitzer, 1968; Pfitzer and Winckler, 1968; Pfitzer and Winckler, 1970) and will not be repeated here. Three quiet time pitch angle studies were performed: one during September to December 1964, January to April 1965, and June to August 1966. In each case rather distinctive $\mathrm{d}J/\mathrm{d}E$ vs α_0 curves were obtained for each energy channel (Figure 1). These distributions have been fitted with polynomials in order to deduce the directional flux of particles moving close to the equatorial plane with $\alpha_0 = 90°$, hereafter called 'reduced' flux and presented in the figures that follow. Figure 2 presents in summary form the ratio of the reduced flux in the interval from June to August 1966, compared to September to December, 1964. This figure clearly

B. M. McCormac (ed.), Particles and Fields in the Magnetosphere.

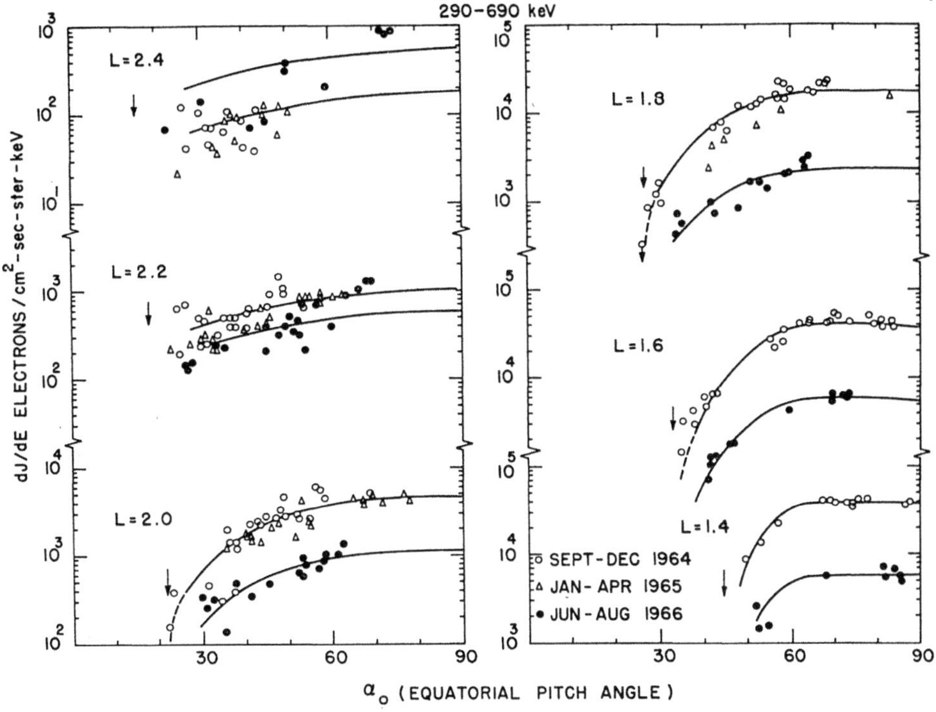

Fig. 1. Equatorial pitch angle distributions in the inner radiation belt for electrons during the solar minimum period. Note the decay of the Starfish radiation and the cutoff of intensity near the loss cone indicated by the vertical arrows.

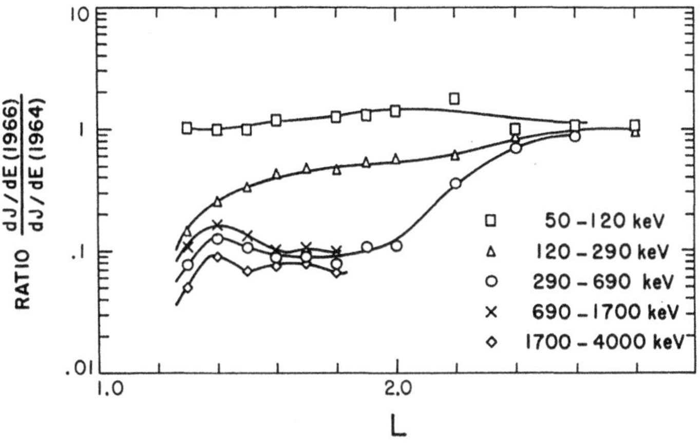

Fig. 2. Long term changes of electron flux of various energy ranges between the epochs 1964 and 1966. The 50 keV electrons remain essentially constant as a function of time and also as a function of L. Below $L = 2.4$ the decay of the Starfish electrons is evident.

shows for energies greater than 290 keV the order of magnitude decrease below $L=2.0$ associated with the continual decay of Starfish injected electrons. For $L>2.0$ in this energy range there is a transition so that after about $L=2.4$ to 2.6 the effects of the Starfish background disappear and a more or less constant intensity ensues. The curve for 120 to 290 keV shows an intermediate behavior while the very low energy channel shows an essentially unchanged intensity during the entire period. One sees thus that the decay characteristics are quite different in different energy ranges and one must suppose that a type of dynamical equilibrium exists. This dynamical equilibrium is able to maintain the 50 keV flux at the constant value and may be a factor in other energy ranges.

The most recent measurements prior to OGO were carried out by Beall *et al.* (1967) with detectors on the polar satellite, 1963-38C measuring the Starfish decay over a 40 mo period beginning about 15 mo after the Starfish explosion and in the energy range greater than 280 keV. (For a general review see Hess (1968) and Freden (1969).) The correlation with the Beall measurements has been reported earlier (Pfitzer and

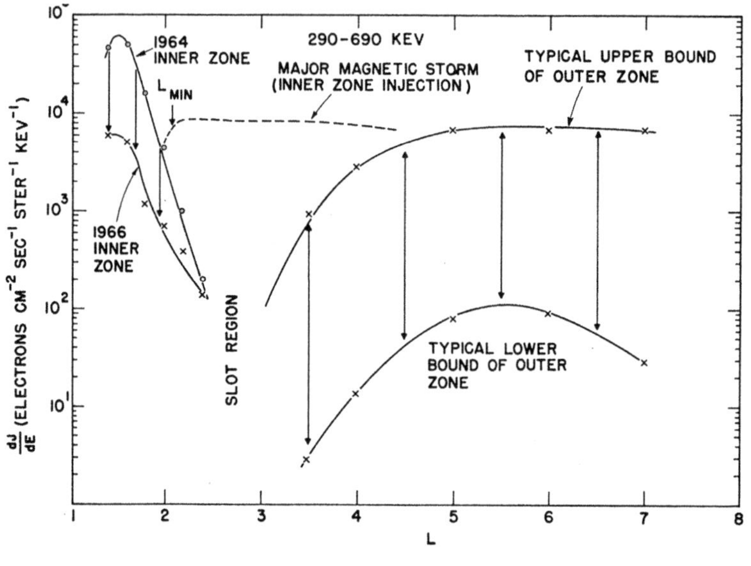

Fig. 3. Schematic summary derived from actual OGO observations of the inner and outer zone and slot region during the early years of present solar cycle. One notes the decay of the deep inner zone Starfish component, but the tendency in the slot region of the electrons to approach an equilibrium value to which the flux returns after magnetic storm injections.

Winckler, 1968) and excellent agreement in the range of pitch angles and energy measured by the Beall group is found. All electron fluxes measured by the OGO satellites above 290 keV in the inner zone for $L<1.8$ before September 2, 1966, possibly can be explained by assuming that the Starfish explosion was the entire source.

3. Storm Time Injection of Inner Zone Electrons

The storm time increases of inner zone electrons to be described in this section follow large fluctuations of the outer zone and slot regions and, therefore, cannot be understood completely without knowledge of the behavior of the entire trapped radiation region. To visualize the time changes, Figure 3 has been constructed from actual OGO data over a wide range of L values, and represents typical reduced flux levels in the inner and outer zone for 290 to 690 keV. The inner zone shows the steady decay of electrons presumably of 'Starfish' origin (see Section 2) from September 1964 to August 1966. The curves near the 'slot' region ($2.4 < L < 3$) represent the quiescent long term profiles. The outer zone ($L > 3$) shows large fluctuations between the upper and lower bonds during small or moderate magnetic storms. A major magnetic storm may fill the slot region and add particles to the inner zone down to L_{min}. The decay times for $L > 2.5$ for fluxes above the quiescent profiles are on the order of a few days and after a major event the slot quickly reappears since particles are not commonly added to this region by the majority of subsequent storm events and a quiescent profile is re-established.

Three major inner zone injection events occurred during the period of observation in April 1965, September 1966, and May 1967. The April event will be discussed elsewhere (Pfitzer and Winckler, 1970). The events on September 2, 1966, and May 25, 1967, are quite similar and the discussion will concentrate on the September event. The 120 to 290 keV channel and the 290 to 690 keV channel behaved almost identically during this event. Figure 4 shows the time history for various L values for the 290 to 690 keV interval. For $L \leqslant 1.8$, the rate after September 2 increased slowly over a period of 20 or more days, and once the new level was reached, no rapid decay was evident and the new level remained stable. At $L = 2.0$ the increase is more rapid reaching a peak in less than 5 days after the event begins. This level is maintained for about 30 days and then an exponential decay drops the flux to a new and apparently stable level. For $L \geqslant 2.2$, the increase is quite large and the exponential decay begins immediately and returns the flux levels to the pre-storm level within 30 to 40 days after the onset of the event. The 50 to 120 keV channel (not shown) behaved similarly to the 120 to 690 keV energy channels but the increases were much smaller ($\simeq 50\%$ for $L < 2.0$).

The 690 to 1700 keV channel (Figure 5) shows no increase for $L < 1.7$ at any time after the event. Instead the fluxes are observed to continue the Starfish decay discussed in the last section. These fluxes are above threshold for $L \leqslant 1.7$.

About 60 days after the onset of the event a new stable inner zone has been created. The fluxes for $L \geqslant 2.2$ have decayed to their pre-storm levels and the fluxes for $L \leqslant 2.2$ have stopped their rapid post-storm decay. This new stable inner zone continues the slow Starfish decay described in Section 2 which continues up to the start of the May 1967 event.

The May event (Figure 4) is more intense than the September event especially at low L values and the initial increases are more rapid with a maximum intensity being

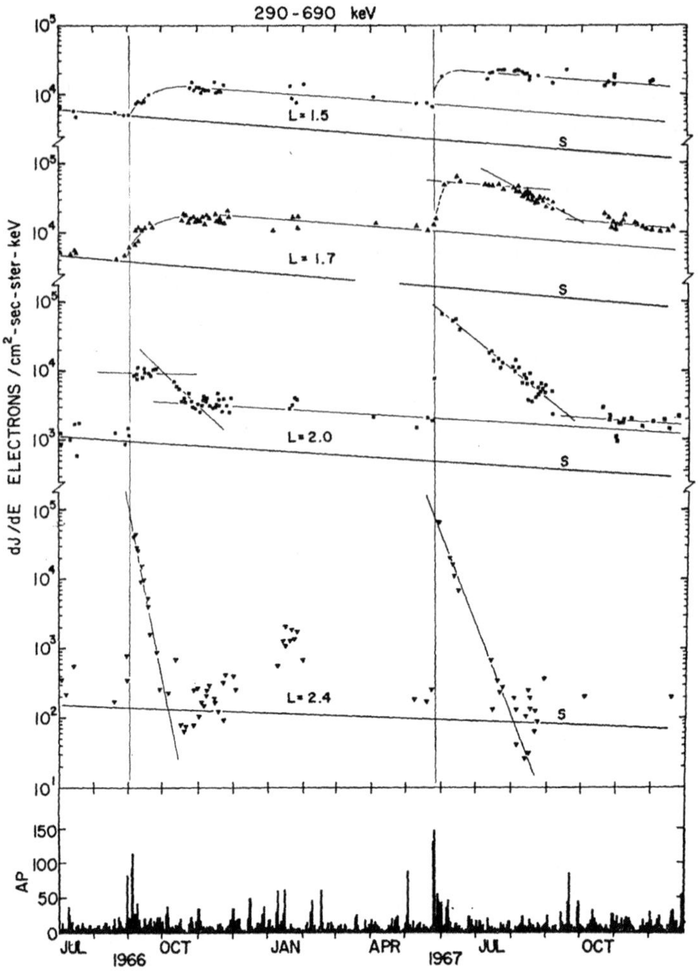

Fig. 4. Time changes of a typical energy range of electrons for various L values in the inner zone. Note the rapid decay after injection for $L = 2.4$ but the slow rise and plateau behavior for lower L's. After a transition period the residual injected flux decays with the same slope as the Starfish line (marked S).

reached within 5 days of the storm instead of the 30 days of the September event. This event also shows the plateau-like behavior followed by an exponential decay to a new apparently stable level. Although the initial increase occurred on May 25, 1966, the inner zone did not return to the equilibrium conditions until October 1966, a time interval of 125 days. The time history of the May event is summarized in Figure 6. Each curve represents the flux versus L profile for the 290 to 690 keV electrons for time intervals separated by 15 days. For $L \geqslant 2.0$ there is in all cases a continuous drop after the initial onset. However, for $L < 2.0$ the decay is delayed until the flux for greater L's drops below a specified value. One surmises that the flux at a given L

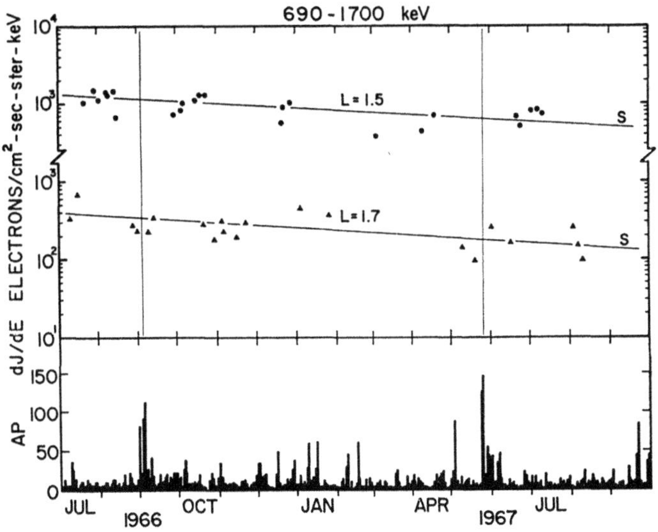

Fig. 5. The steady decay of electrons greater than 690 keV along the Starfish line *S* with no detectable injection from the major magnetic storm events.

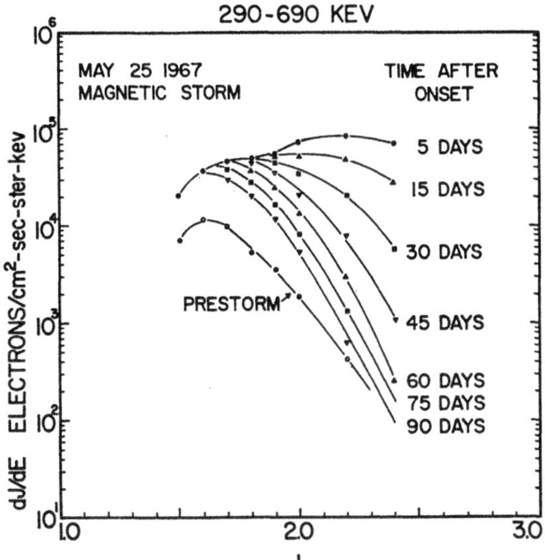

Fig. 6. Summary of the behavior during the May 25 magnetic storm. One notes that the intensity at given *L* value after a quick initial rise drops only after the intensity at larger *L* has fallen to a lower value. These curves may be interpreted on the basis of a diffusive model of the redistribution particles.

value holds constant because of inward diffusion from larger *L*'s as long as the gradient is suitable. Again, as for the September event, the 690 to 1700 keV electron channel shows no detectable increase and the flux levels in this high energy range continue the typical Starfish decay rate for $L \leqslant 1.7$. Figure 7 summarizes the decay characteristics of the three events. Since the decays are approximately exponential, we give the e-fold energy time τ as a function of L for two energy channels. τ is a strong function of L ranging from several months at $L<2$ to a few days in the slot. The 3 storms show very different decay profiles. These curves undoubtedly reflect the

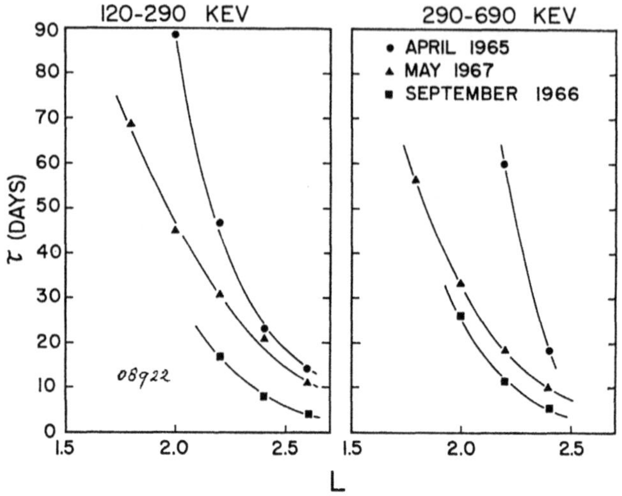

Fig. 7. Lifetime of electrons in several energy ranges showing the very critical dependence on L value on the inner side of the slot region. The time constant from one storm to another varies irregularly and is shortest for the September storm although the May event was larger from the standpoint of geomagnetic indices.

dynamic equilibrium controlled by the loss and gain mechanisms at a given L value. They may depend on the radial distribution of electrons, the state of scattering fluctuations in the magnetic field, and other factors. A complete analysis of the diffusion problem for the entire inner zone may be necessary to give significance to these curves.

4. Summary of Inner Zone Injection Characteristics

(a) Inner zone energetic electron injection events are detected following large magnetic storms for energies below 690 keV. These injected electrons are added to the slowly decaying quiescent inner zone levels. These quiescent levels are determined by the accumulated effects of previous injections, of artificial or natural origin.

(b) The initial increase extends from the outer zone through the 'slot' to some minimum L value which depends on the storm. This initial increase in absolute flux is not strongly dependent on L above the minimum L.

(c) The initial increase occurs within a time interval shorter than our time resolution, which is of the order of 1 day.

(d) A complex readjustment process follows the initial increase which may last up to 4 mo and which finally re-establishes a new quiescent flux profile. Some of the observed characteristics of this readjustment process are:

(i) For L values less than the L_{min} characteristic of the initial increase there may be a slow increase of flux for as much as 40 days near the observable inner limit of the event in L.

(ii) In some cases, a quasi-stable level of elevated flux exists for up to 60 days which at a well defined point in time decays exponentially to the 'final' level.

(iii) This final level is observed to decay very slowly like the Starfish background.

5. Substorm Correlated Electron Increases in the Outer Radiation Belt

One now seeks to establish a link between the initial part of the inner zone injection events and the variations in the outer zone. A significant observation was made by

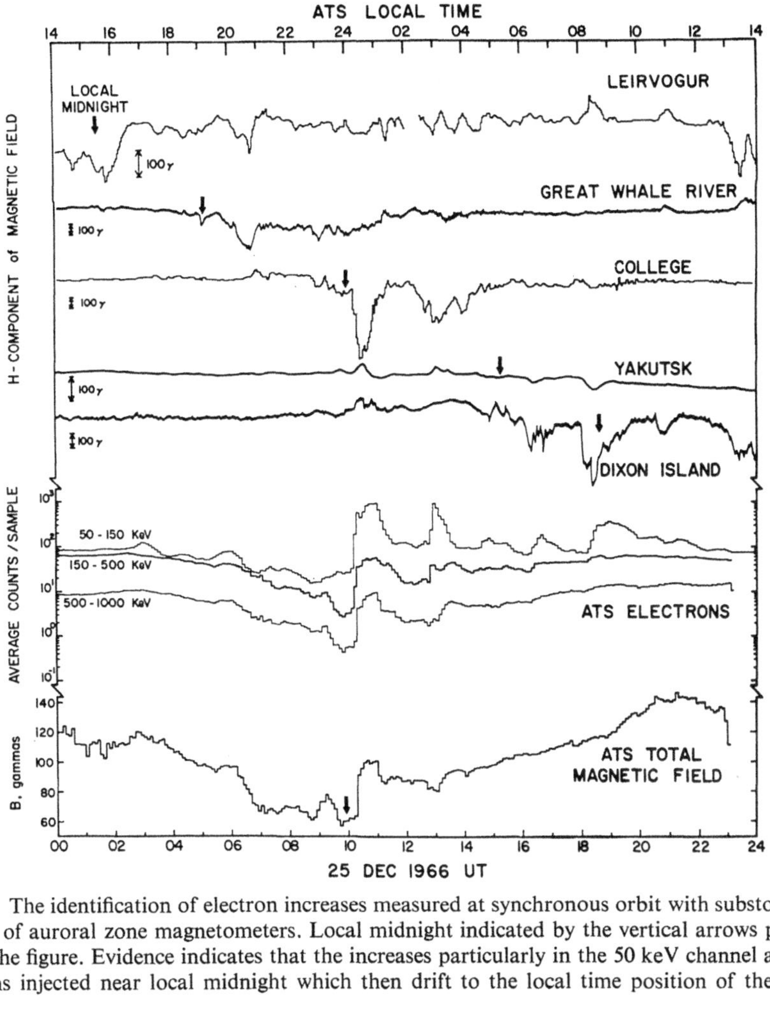

Fig. 8. The identification of electron increases measured at synchronous orbit with substorms from a series of auroral zone magnetometers. Local midnight indicated by the vertical arrows progresses across the figure. Evidence indicates that the increases particularly in the 50 keV channel arise from electrons injected near local midnight which then drift to the local time position of the satellite.

Brown *et al.* (1968) who observed during the April 18, 1968, injection event mentioned earlier that electrons of $E > 300$ keV were seen at intermediate L values ($L = 5.0$) first at the exact time of a substorm near the beginning of this great magnetic storm event. Synchronous orbit observations and correlations between the ATS and OGO satellites have shown that the polar substorm as described by Akasofu (1968) plays a fundamental role in the injection, modulation, and distribution of energetic particles in the radiation belts. The substorm associated variations observed during 24 hr at synchronous orbit are well shown by Figure 8 wherein is plotted appropriate ground based auroral zone magnetic records with local midnight indicated for each. Also shown are the three channels of the ATS spectrometer and the ATS total magnetic field furnished by Cummings, Barfield and Coleman of UCLA. The characteristic feature is the large increase seen in the electron fluxes at local midnight with the largest increase being associated with the low energy channel. The bay association of these increases at or shortly after midnight is well shown by the comparison with the College magnetometer. But even at other local times, one can plainly see by comparison with Leirvogur or Great Whale River records that fluctuations in the electron

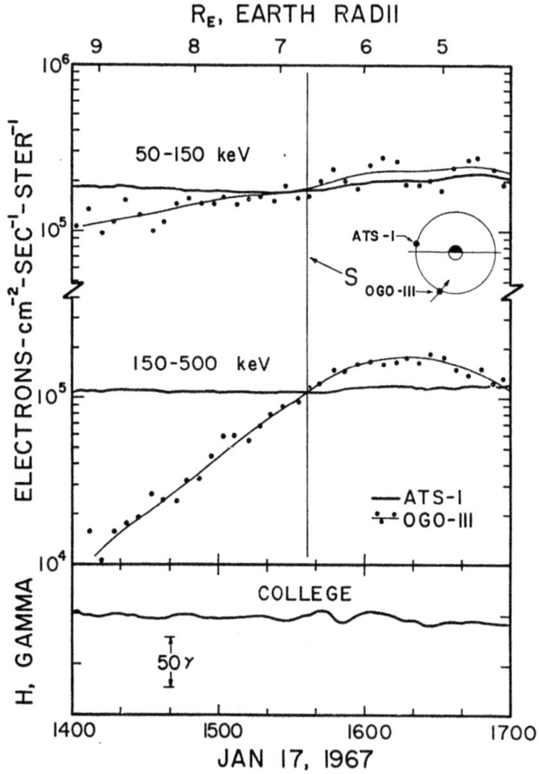

Fig. 9. Quiet day correlation between the OGO 3 and ATS 1 satellites on an absolute flux basis. The vertical line at 1536 UT (S) marks the time when the two satellites were on a common electron drift shell in the distorted magnetosphere.

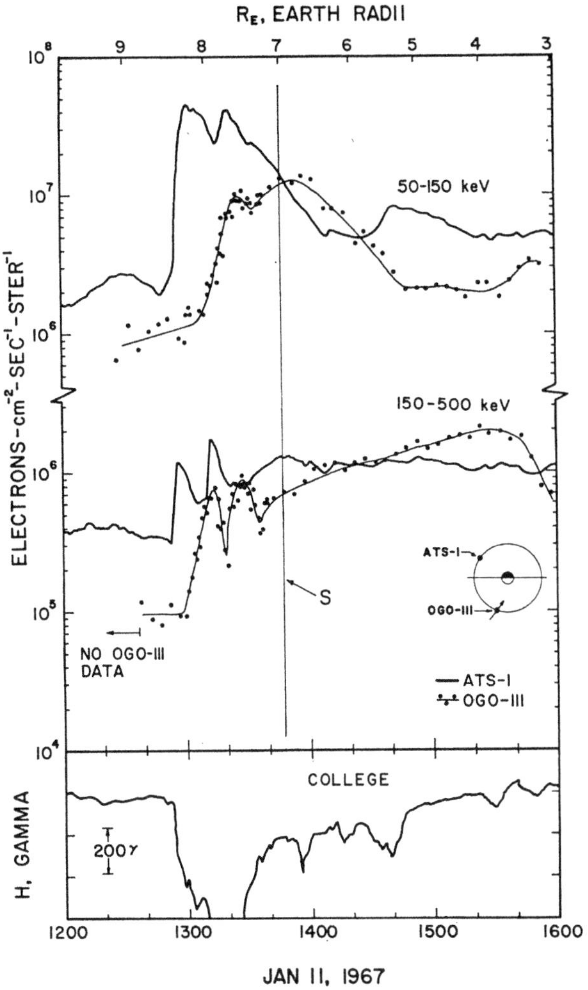

Fig. 10. A substorm associated increase seen by two satellites widely separated in local time. The delay difference in rise and difference in peak intensity is consistent with particles injected near local midnight first seen by ATS 1 and later by OGO 3.

intensity are correlated with the bay event. Likewise, the spacecraft ambient magnetic field responds to these substorms. This correlation has already been discussed by Cummings *et al.* (1968).

Insight into the nature of the outer zone variations has been gained by a close comparison of the electron fluxes measured by OGO 3 and ATS 1. During the spring of 1967, OGO 3 entered the noonside of the magnetosphere close to the equatorial plane and thus was in an ideal position to compare with the ATS 1 satellite which revolved slowly over the 24 hr period at a constant radius of 6.6 R_E. Using the Roederer drift shell program it has been possible to locate the electron drift shells in the distorted magnetosphere and to study the relationship of the quiet day and storm time changes seen on the two satellites (Pfitzer *et al.*, 1969; Pfitzer and Winckler, 1969).

In Figure 9 is shown a quiet day comparison of the two satellites. The vertical line near the center marks the time when according to the computation the two satellites should be on the same drift shell. It is seen that at this time while the College magnetogram shows a quiet period the absolute fluxes of the two spectrometers agree rather well. By contrast, in Figure 10 is shown a substorm increase in the lower energy channels of the two satellites accompanying a bay disturbance at College. The substorm increase observed by the ATS 1 satellite located at about 0200 LT (heavy curve) was observed also by the OGO 3 satellite on the solar side of the magnetosphere but with an onset and general envelope delayed from that seen by ATS 1. The delay in onset is less for the high energy channel than for the low energy, also the rise of the electron increase is less steep as seen by OGO 3 on the solar side than seen by ATS 1 on the nightside. We conclude that this observation shows the presence of a drifting group of electrons which is moving eastward in local time because of the gradient drift process in the magnetosphere. These electrons must have been injected near midnight at the time of the bay disturbance. Since this increase is typical of those seen on the ATS satellite as it remains almost stationary in the magnetic field in the outer zone, we conclude that such a substorm injection on the nightside followed by a drifting and smoothing effect can populate the outer belt with electrons. Other examples of these correlations between the two satellites are shown in Figure 11 in schematic form. One sees that the satellite farthest from local midnight shows the smallest slope and the lowest maximum intensity. Also, in one case the two satellites were interchanged and the OGO 3 sees the event rise before the ATS satellite. A simple computer model has been applied to this situation and confirms quantitatively the delay times, rise times, and energy dependencies of these substorm increases so they may be identified as injection events near local midnight followed by a gradient drift in the nominal magnetosphere.

Similar evidence has been obtained by Arnoldy and Chan (1969) who have compared the time delay between the maximum of the simple bay disturbance seen at midnight with the maximum electron increase seen at synchronous orbit. They find that the delay increases uniformly with the local time as shown in Figure 12. (Midnight is at 10 UT for the ATS 1.) This excellent correlation can be accounted for by drifting bunches of electrons of about 90 keV energy injected into the outer zone near midnight during a polar substorm.

6. Drift Shell Displacement and Magnetosphere Motion During Substorms

The substorm associated changes in the electron flux near midnight and through morning consist of large injection events for the 50 keV electrons accompanied by increases in other channels which may possibly be modulation as well as injection of fresh particles. But the behavior in the afternoon and evening sectors appears to be systematically different. In fact, an examination of various energy particles and their variations during substorms can give apparent clues as to the mode of motion of the

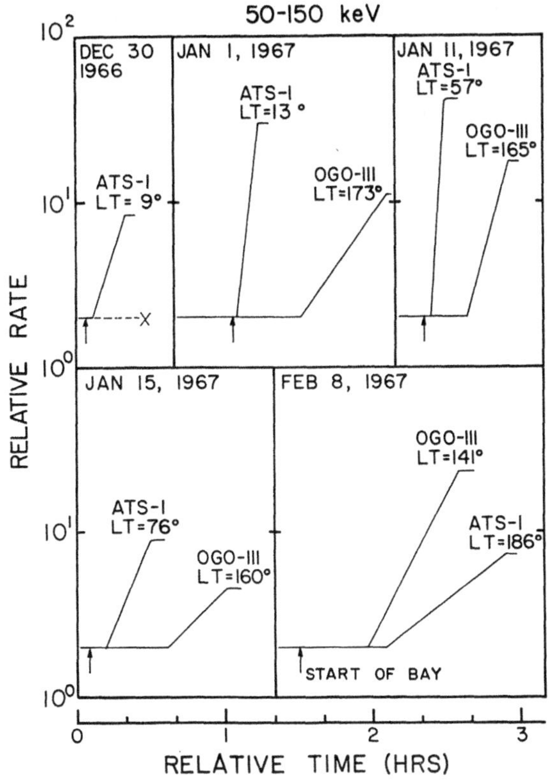

Fig. 11. Schematic summary of five substorm events observed by the two satellites. A simple model shows that these events can be quantitatively explained by electrons injected at local midnight which then disperse and drift eastward reaching first one satellite and then the other.

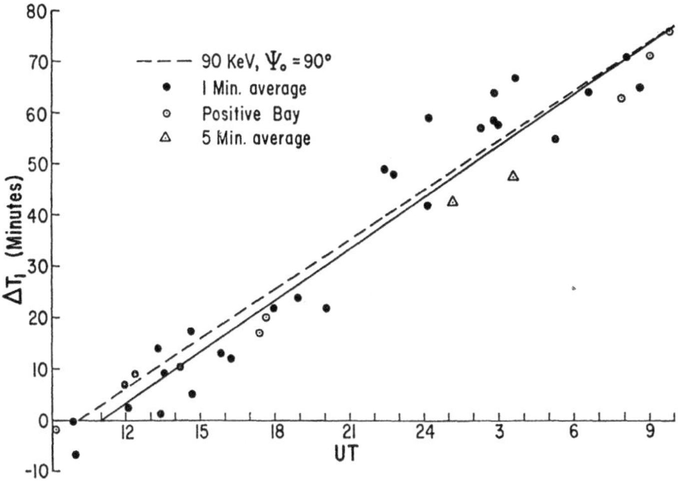

Fig. 12. The delay times measured by Arnoldy and Chan (1969) for 90 keV mean energy electrons observed by the ATS spectrometer showing the systematic delay from the occurrence of a bay at midnight. The delay extrapolates to 0 at 1000 UT which is local midnight for the ATS satellite position.

magnetosphere during the substorm period. For this purpose the most energetic electrons are useful because their energy is affected least by the substorm and their drift period is very short compared to the substorm period of about 2 hr. Figure 13 shows a small substorm in the evening sector observed by the magnetometer at Fort

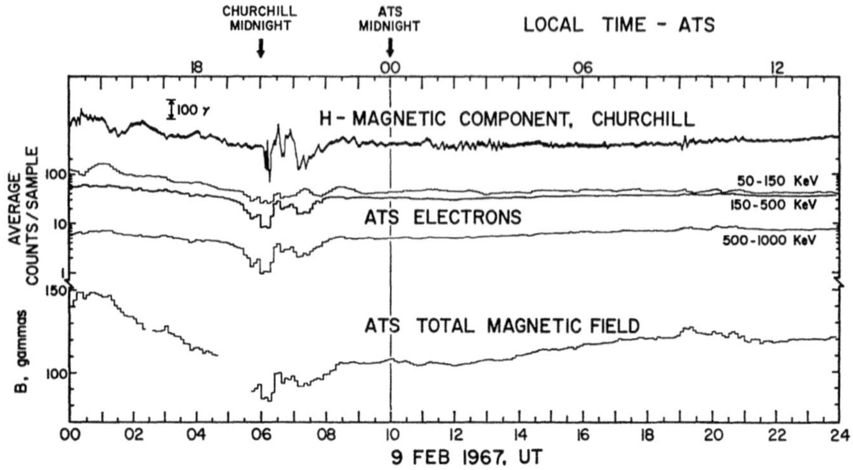

Fig. 13. A small substorm event seen in the evening sector showing the modulation of the electron fluxes believed due to drift shell motion across the ATS orbit. Note the close similarity between the ATS magnetic field and the three electron channels, and also some resemblance to the electrojet current system. An exception are the two smooth increases in the 50 keV channel which are believed due to electrons injected near local midnight which have drifted 20 hr in local time to the satellite position.

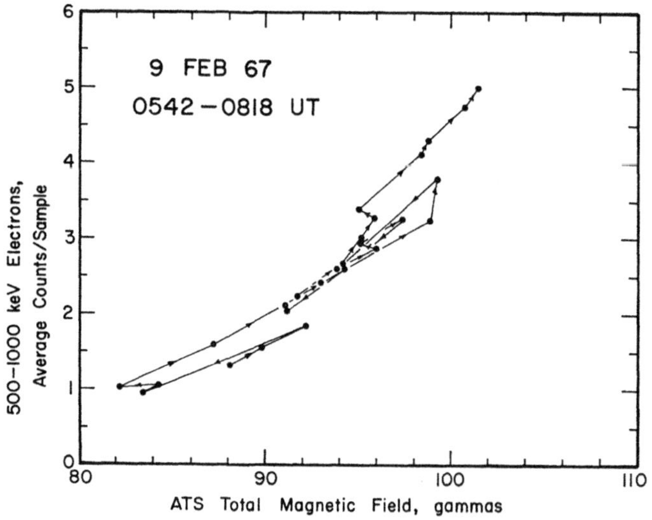

Fig. 14. Correlation curve of the total magnetic field (Cummings, Barfield and Coleman, UCLA) and the high energy electron channel during the small substorm event on February 9, 1967.

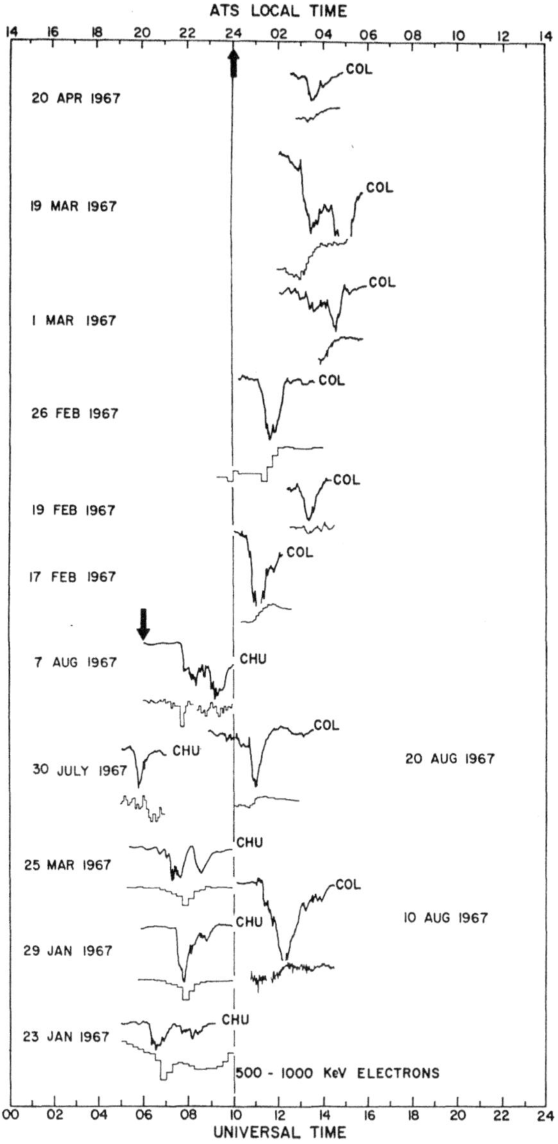

Fig. 15. This figure shows that during bay disturbances, all measured close to local midnight, the high energy electron flux seen by ATS 1 is modulated downward in the evening sector and upward after midnight and in the dawn sector.

Churchill which shows an almost identical modulation of all of the energy channels of the ATS and as well the total magnetic field. In fact, some of the features of the bay electrojet current are reproduced even in the electron variations. A notable exception is the presence in the low energy channel of two smooth increases delayed after the two main parts of the bay. These are almost certainly drifting bunches of electrons which were actually injected near local midnight, and have drifted around almost 24 hr in local time where they are observed in the evening sector. This bay disturbance evidently injected a double peak of low energy electrons. The correlation of the ATS measured magnetic field and the 500 to 1000 keV electron intensity is shown in Figure 14. To first order the particle and field values lie on a correlation curve during the substorm variation shown in the previous figure. We take this to be evidence for a motion of the electron drift shells over the stationary spacecraft due to an effective magnetospheric motion. During a substorm this motion has the sense that the magnetosphere expands in the evening sector with a weakening of the magnetic field and an inward motion of energetic particle drift shells. Investigation shows that at midnight the sense of the motion suddenly reverses and during a substorm there is a collapse of the magnetosphere with an increase in the magnetic field and an outward motion of the drift shells. The evidence for this is presented in Figure 15.

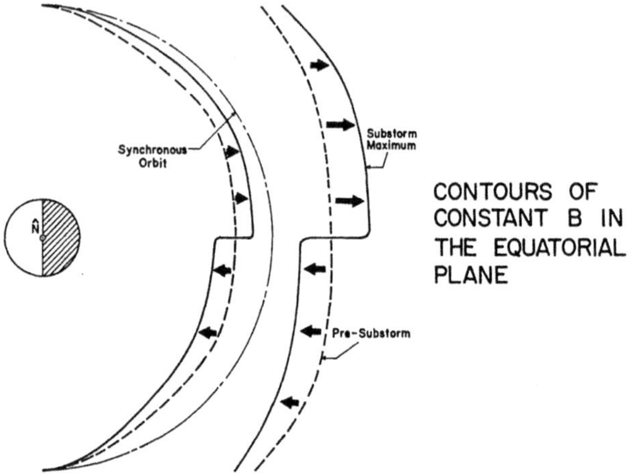

Fig. 16. Inferred contours of constant B showing the displacement during a substorm with a 'fault line' near local midnight. The drift shells would follow such contours if electric field effects are ignored.

Here we show the correlation between the high energy electron channel and magnetic bay disturbances observed for stations near midnight but with the satellite either in the evening or in the post-midnight sector. It is seen that for bay disturbances when the satellite was in the evening sector the trend of the modulation is downward for the energetic electrons, while for bay disturbances when the satellite was in the post-midnight sector the sense of the modulation is upward. One can visualize thus a type of fault line occurring approximately at midnight as shown in Figure 16 with the contours of constant **B** which would be followed by the electrons under ideal conditions shown displaced from their equilibrium position as they move during a substorm. This drift shell motion is consistent with the piling up of the lines of force in the post-midnight sector simultaneously accompanied by an expansion of the magnetic field in the evening. The situation referred to here would be typical of a region near the equatorial plane and near synchronous orbit.

The measurement of substorm effects on the trapped particles is complicated by the continually changing pitch angle viewed by the ATS instruments as the magnetic field tilts and distorts away from its quiet day configuration. There are very large effects on the pitch angle distribution during substorms or even during quiet days because of the drift shell splitting phenomena which occurs in the distorted magnetosphere. It has been possible to quantitatively evaluate the pitch angle distributions at

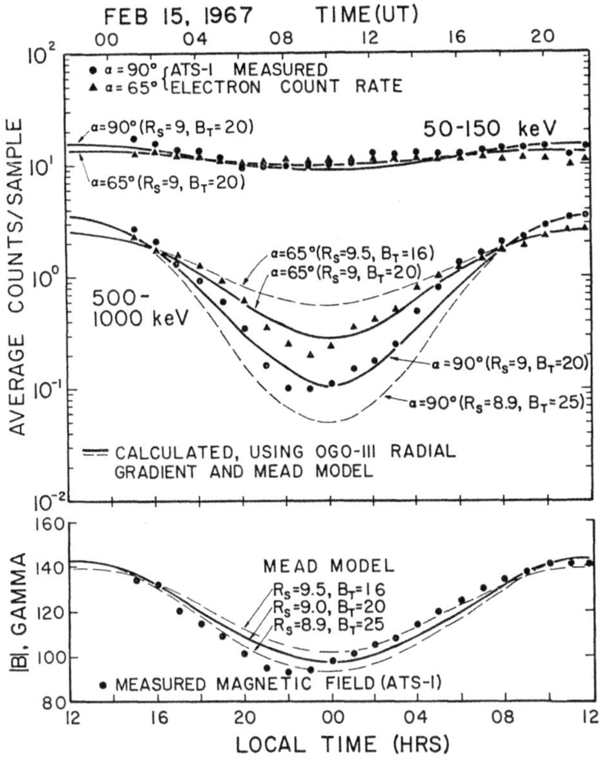

Fig. 17. The quiet day pitch angle variation in ATS orbit showing the greater intensity of field aligned particles at midnight compared to those perpendicular to the magnetic field.

different points in the magnetosphere by comparing the radial distribution measured by OGO 3 with the pitch angle distribution measured at various local times by ATS 1 (Pfitzer *et al.*, 1969). The quiet day example in Figure 17 shows the typical large intensity of particles aligned along the field over the midnight region. By contrast in Figure 18 we see a similar range of pitch angles measured during a time of large substorm activity on February 8, 1967. The striking feature of this figure is that although in the evening and morning sectors, the tendency for the pitch angle distributions is to resemble those on a quiet day, near midnight during the large substorm increase the tendency is completely different and the pitch angle distributions become isotropic. It is believed that this isotropic behavior is consistent with the collapse or compression of the magnetosphere at that time away from the distorted condition so that it resembles more closely a circularly symmetric dipole type configuration in which

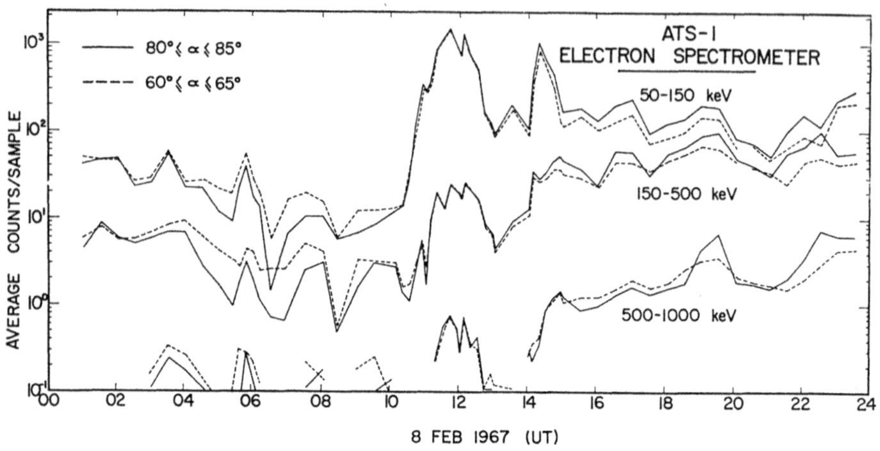

Fig. 18. The time history of two pitch angles during a substorm active period. Note the isotropic behavior during the large midnight substorm increase which contrasts with the previous Figure 17 in the midnight sector.

drift shell splitting is absent. Thus this observation is consistent with the rapid compression during a substorm in the post-midnight region of the magnetic field. Figure 18 also gives indirect evidence for the injection of particles near midnight as one can see that the shell splitting gradually develops towards morning hours beginning with a condition of isotropicity near midnight.

7. The Acceleration of Electrons Near Midnight

Because of the availability of the vector magnetic field for many of the major magnetic storms seen by ATS 1, it is possible to construct a semi-quantitative picture of the midnight acceleration process. In Figure 19 are shown the substorms on February 8 with the total magnetic field and also the angle θ of inclination of the field to the polar axis. Typically during a substorm the magnetic field inflates and tips outward ($\theta > 11°$) in the dusk or pre-midnight local time quadrant. However, the field exhibits a sudden temporal collapse seen by the satellite provided it is within a small local time sector usually located within a few hours of local midnight. This collapse as described previously is accompanied by a magnetic bay disturbance at auroral stations also near local midnight and therefore is a universal time phenomenon and not a stationary pattern through which the satellite moves.

Because of the large values of θ observed in the vicinity of the spacecraft in the evening quadrant, these field lines may extend considerably beyond the synchronous orbit into the geomagnetic tail before returning to the earth. As the collapse begins near midnight these field lines snap inwards very rapidly (the convective velocity is estimated at 0.5 R_E/min) bringing with them and heating in the process magnetotail plasma by means of the betatron and Fermi mechanisms (Figure 20). The observed

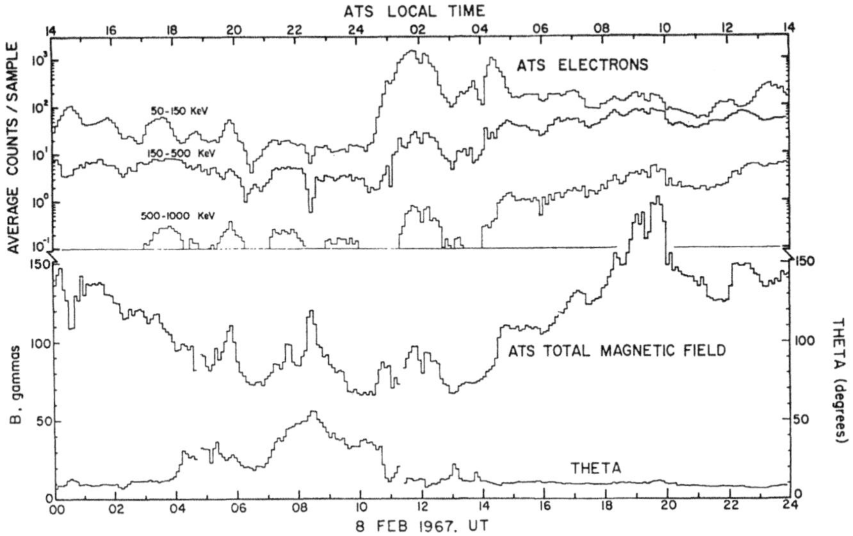

Fig. 19. A day of large substorm activity showing the electron fluxes, the total magnetic field, and the angle of the magnetic field to the polar axis (θ).

snapping of the field lines is sufficiently rapid that the convective velocity component for up to 20 keV particles is larger than the sum of the gradient and curvature drift components and the acceleration may continue for some time before the particles drift easterly out of the convection region. (For 5 keV particles, the sum of the longitudinal drift components is estimated at 0.11 and 0.039 R_E/min at 12 and 7 R_E, respectively;

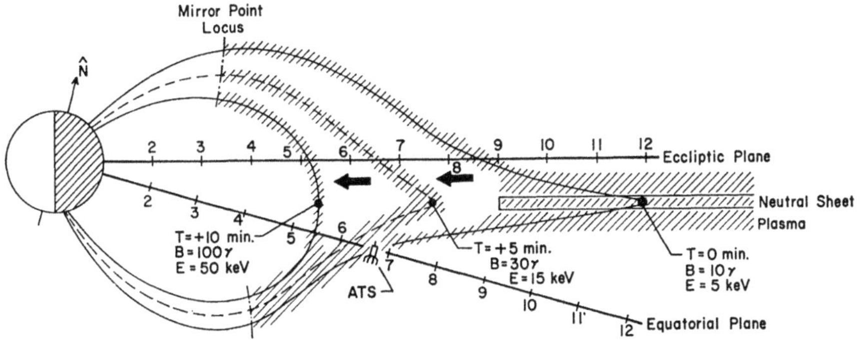

Fig. 20. A plausible geometry derived from actual observations showing the possible Fermi and betatron effects during the post-midnight collapse associated with substorms. The three contours show the possible successive positions of a line of force during this substorm correlated collapse and the resultant field strengths and electron energies at each point.

this is much less than the convective component 0.5 R_E/min.) Since the magnetic field strength on a collapsing line may change tenfold and the length of a field line could decrease by a factor of 3 during the collapse, conservation of the first two adiabatic invariants predicts that particles could have their energies increased tenfold by this process.

One therefore views the injection of energetic electrons into the outer Van Allen belt as an impulsive substorm associated process. As the substorm develops there is an inward convective surge and simultaneous heating of magnetotail plasma as the post-midnight field lines collapse inward. The collapse begins near the synchronous orbit and progresses outward toward the tail as the substorm grows; this region of catastrophic collapse is (at least at synchronous orbit) less than 45° wide and its westward edge is usually located just after local midnight. The particles receive their energy from the snapping in of distended closed field lines – it is not necessary to postulate that field line annihilation is the particle energy source. As the particles convect inward and acquire more and more energy, the gradient and curvature drift velocity components gradually overpower the convective component and the particles drift longitudinally through the electric fields established by the gross magnetospheric motions. This produces further acceleration and deceleration. Computations employing only the betatron mechanism and a typical electron differential energy spectrum at 17 R_E in the magnetotail reported by Bame *et al.* (1967) indicate that this mechanism can account adequately for the peak 50 to 150 keV electron flux observed at synchronous orbit during the midnight collapse.

The observations and ideas on which this picture of electron acceleration is based are well supported by previous experimental and theoretical work. Observations of increases in the outer zone energetic electron flux in correlation with magnetic bays have been observed by the Alouette satellite and reported by McDiarmid *et al.* (1969). Rao (1969) also has described similar increases of energetic electrons closely correlated with bays on the basis of the polar orbiting Injun IV observations. Rao notes that the outer zone flux increases without a change in the trapping boundary latitude as the bay develops but with a poleward motion of the trapping boundary as a bay subsides.

The Alouette evidence is concerned with very large storms and shows that the source may extend well into the morning region. Gradient drift effects, however, may be present in the Alouette data which would localize the source more near the midnight meridian. A connection between electron islands seen in the earth's magnetic tail and substorms has been demonstrated by Rothwell and Wallington (1968), who conclude that the substorm effect on energetic electrons starts inside the magnetosphere on closed field lines and the energetic particles then propagate outwards into the tail region.

One notes that the observations at synchronous orbit are in general agreement with these data but, in addition, because of the slow motion of the ATS 1 satellite through the sun-earth reference frame, time variations may be uniquely seen. The great asymmetry present between the midnight-morning sector and the evening sector in the production of energetic electrons is particularly well shown on the ATS data. Also the successive substorm increases seen everywhere in the ATS orbit which can be attributed to a source near midnight is another unique feature.

Theoretical descriptions particularly applicable to particle acceleration during substorms have been given by Axford (1967) and Atkinson (1966). Axford suggests also the importance of the betatron and Fermi processes in collapsing field lines on the nightside following earlier ideas of Dungey (1961). The discussion by Axford in a recent comprehensive review (Axford, 1969) of convection during substorms based on earlier experiments and theoretical and intuitive ideas describes features of the collapse, inflation, acceleration and propagation of disturbances tailward. Our conclusions support most of these ideas but again the observations at synchronous orbit show that the nightside magnetic field collapse occurs near local midnight in a small local region. It seems possible that viscous friction, discussed by many writers and suggested by Atkinson as a means of inflation of the field in the evening sector coupled with the corotation of the region up to at least 8 R_E on the nightside, might set up the conditions for a substorm collapse of the field. It does not seem necessary to invoke magnetic line annihilation *per se* to energize the particles or even to induce the instability responsible for the substorm. However, observations from one limited region of space such as that covered by the ATS orbit may admittedly give an incomplete and somewhat biased point of view. One is impressed, however, by the fact that on quiet days the corotating magnetosphere on the nightside is somehow able to inflate smoothly and recompress through morning without substorms and without gross instabilities. During the period when substorms are occurring the recompression as one rotates with the earth seems to take place unstably and the extended field lines rush inward near the midnight meridian at a certain instant labeled by the start of a bay disturbance. The corotational energy thus seems to play an essential role in driving the substorm process, coupled with the viscous extension of the tail field lines in a still closed region of the magnetosphere.

Acknowledgments

The efforts of many colleagues at Minnesota have gone into the work reported here, in particular Roger Arnoldy, Karl Pfitzer, Tom Lezniak, and George Parks. As on many other occasions we are indebted to the UCLA group, Professor Paul J. Coleman and David Cummings, for furnishing ATS 1 vector magnetometer data. This research based on the OGO and ATS satellite programs has been supported by NASA under contract NGL-24-005-008.

References

Akasofu, S.-I.: 1968, in *Polar and Magnetospheric Substorms*, D. Reidel Publishing Company, Dordrecht-Holland.

Arnoldy, R. L. and Chan, K.-W.: 1969, *J. Geophys. Res.* to be published.

Atkinson, G.: 1966, *J. Geophys. Res.* **71**, 5157.

Axford, W. I.: 1967, in *Aurora and Airglow* (ed. by B. M. McCormac), Reinhold Publishing Corporation, New York, p. 499.

Axford, W. I.: 1969 in *Magnetospheric Physics* (ed. by D. J. Williams and G. D. Mead), Amer. Geo. Union, Wash., D.C., p. 421 (Feb.-May 1969 issue of *Rev. Geophys.*
Bame, S. J., Asbridge, J. R., Felthauser, H. W., Hones, E. W., and Strong, I. B.: 1967, *J. Geophys. Res.* **72**, 113.
Beall, D. S., Bostrom, C. O., and Williams, D. J.: 1967, *J. Geophys. Res.* **72**, 3403.
Brown, W. L., Cahill, L. J., Davis, L. R., McIlwain, C. E., and Roberts, C. S.: 1968, *J. Geophys. Res.* **73**, 153.
Cummings, W. D., Barfield, J. N., and Coleman, P. J., Jr.: 1968, *J. Geophys. Res.* **73**, 6687.
Dungey, J. W.: 1961, *Phys. Rev. Letters* **6**, 47.
Freden, S. C.: 1969, *Space Sci. Rev.* **9**, 198.
Hess, W. N.: 1968, in *The Radiation Belt and Magnetosphere*, Blaisdell Publishing Company, Waltham, Mass.
Lezniak, T. W., Arnoldy, R. L., Parks, G. K., and Winckler, J. R.: 1967, *Radio Sci.* **3**, 715.
Lezniak, T. W. and Winckler, J. R.: 1968, *J. Geophys. Res.* **73**, 5733.
McDiarmid, I. B., Burrows, J. R., and Wilson, M. D.: 1969, *J. Geophys. Res.* **74**, 1749.
Pfitzer, K. A.: 1968, Ph.D. Thesis, University of Minnesota.
Pfitzer, K. A. and Winckler, J. R.: 1968, *J. Geophys. Res.* **73**, 5792.
Pfitzer, K. A. and Winckler, J. R.: 1969, *J. Geophys. Res.*, **74**, 5005.
Pfitzer, K. A. and Winckler, J. R.: 1970, to be published.
Pfitzer, K. A., Lezniak, T. W., and Winckler, J. R.: 1969, *J. Geophys. Res.*, **74**, 4687.
Rao, C. S. R.: 1969, *J. Geophys. Res.* **74**, 794.
Rothwell, P. and Wallington, V.: 1968, *Planetary Space Sci.* **16**, 1441.

RECENT MEASUREMENTS OF INNER BELT PROTONS

H. ELLIOT and R. J. HYNDS
Imperial College of Science and Technology, London, England

1. Introduction

Two separate but related topics will be discussed in this contribution. They are:

(a) Variations in the intensity of inner belt protons with $E \geqslant 30$ MeV over the solar cycle (Bewick *et al.*, 1970).

(b) A magnetic storm event on October 31, 1968, in which high energy proton precipitation appears to have been observed at the equator at balloon altitudes (Sood *et al.*, 1970).

A. INNER BELT PROTON INTENSITY VARIATIONS OVER THE SOLAR CYCLE

Four years ago at the Advanced Study Institute in Bergen I reported some inner belt proton measurements from the satellite ARIEL 1. These were measurements made with a shielded Anton 302 counter in May, June, and July of 1962 and I compared them with measurements carried out by the Iowa group in 1959 using a similarly shielded counter in Explorer 4 (McIlwain, 1961). The conclusion reached was that the proton flux had certainly not increased between 1959 and 1962 but, if anything, had decreased. This result is hard to reconcile with a CRAND source for these protons coupled with ionization loss and nuclear collisions as the dominating loss process. In particular, the result was shown to be inconsistent with the solar cycle intensity variation predicted by Blanchard and Hess (1964). This, of course, is not the only difficulty to be encountered in ascribing the high energy proton flux to CRAND. Freden (1969) has reviewed some of the other difficulties which include discrepancies in the shape of the spectrum and in the observed flux values which, in some cases, are a factor of 50 greater than the calculated values.

The ESRO satellite IRIS (also known as ESRO II) which was launched in May 1968 from W.T.R. carried an Anton 302 counter similar to that in Explorer 4 and ARIEL 1 for providing a further comparable measurement at a different level of solar activity. IRIS is in a near polar orbit with perigee 330 km and apogee 1091 km and carries a tape recorder which operated satisfactorily for the first 6 mo in orbit. Following failure of the tape recorder, data recovery was limited to real time reception and still continues on this basis. The threshold for protons for the Anton 302 is 30 MeV and 2 MeV for electrons.

Figure 1 shows the times at which the Explorer 4, ARIEL 1' and IRIS observations were made in relation to the solar cycle. The sunspot numbers were 190, 30, and 110, respectively. Figures 2 and 3 show a selection of corresponding equal intensity contours plotted in *B-L* space for Explorer 4 and IRIS. Figure 4 shows the count rates observed on the three spacecraft plotted as a function of *B* for three *L*-values.

B. M. McCormac (ed.), Particles and Fields in the Magnetosphere.

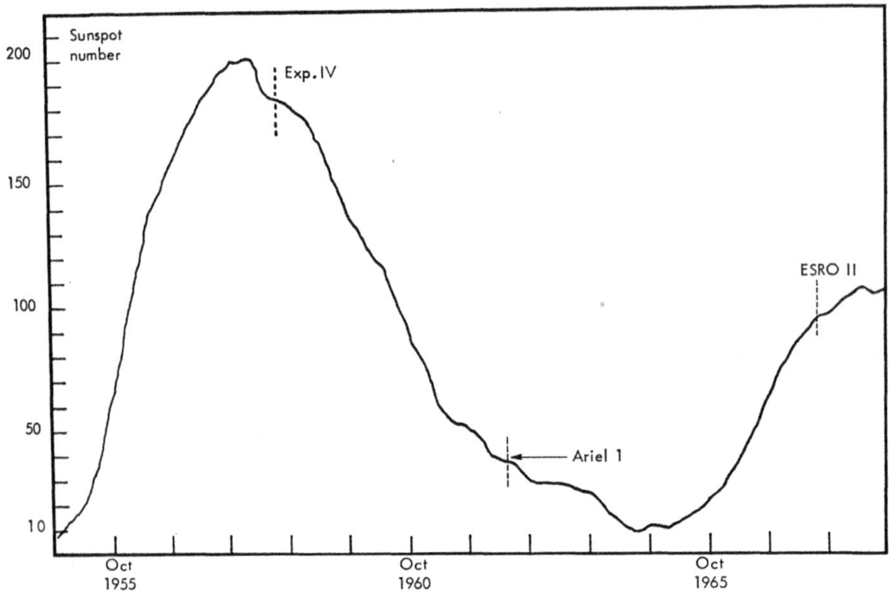

Fig. 1. Times at which measurements were carried out in Explorer 4, ARIEL 1, and ESRO II (IRIS) in relation to the sunspot cycle.

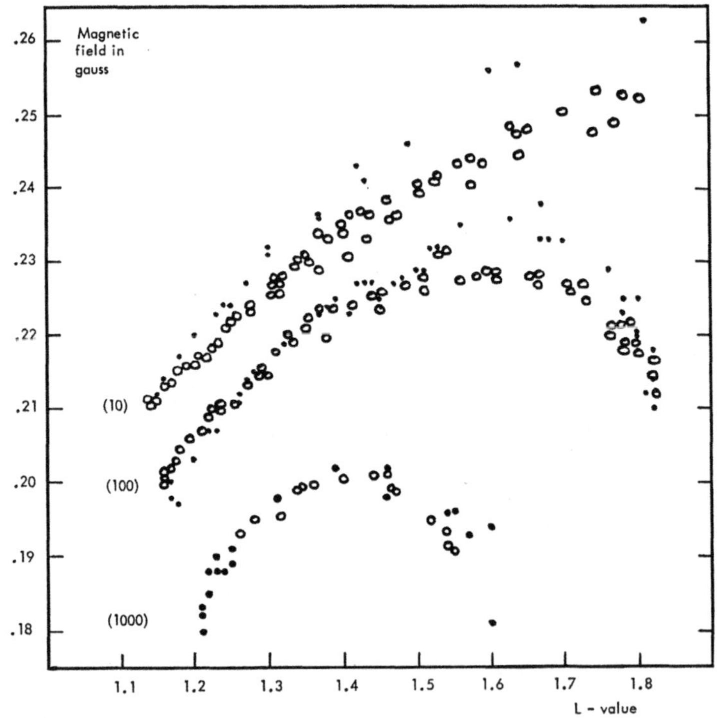

Fig. 2. Equal intensity contours in *B-L* space for Explorer 4 (●) and IRIS (○).

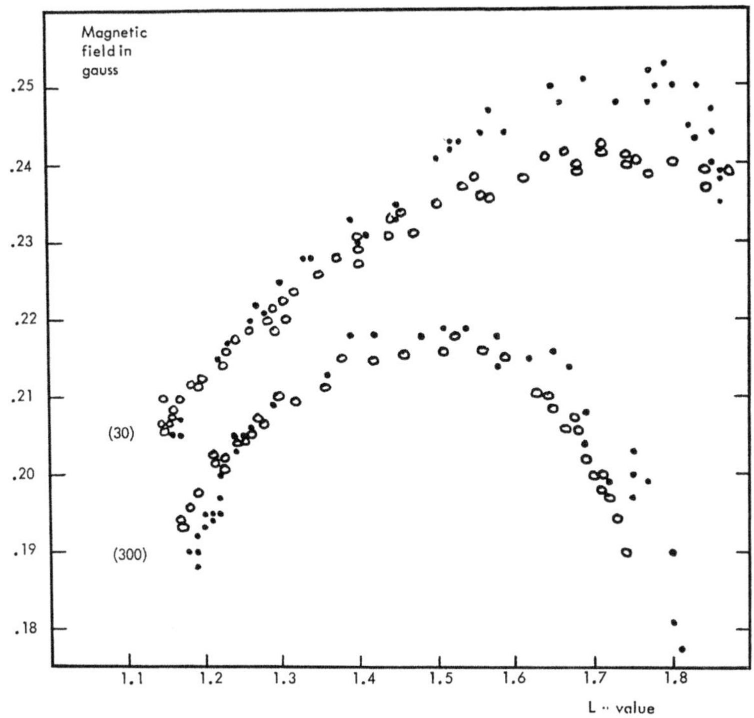

Fig. 3. Equal intensity contours in *B-L* space for Explorer 4 (●) and IRIS (○).

The count rates observed on IRIS agree very well with those recorded on ARIEL 1 and are in general less than those measured on Explorer 4. They confirm the conclusion reached earlier that the inner belt proton flux at energies >30 MeV does not increase as the atmospheric density decreases with diminishing solar activity.

The Explorer 4 and ARIEL I measurements were both made prior to Starfish and the rather limited data on electron fluxes in the inner belt at that time support the view that in each case the G. M. counters were responding primarily to protons. By the time of the IRIS measurements in May and June 1968 the fluxes of β-decay electrons left over from Starfish had decayed to such a level that the G. M. counter onboard that satellite was also counting predominantly protons.

In the range of *B-L* space defined by $1.2 \leqslant L \leqslant 1.4$ and $0.21 \leqslant B \leqslant 0.24$ there is no distinguishable difference between the omnidirectional fluxes measured in 1962 and 1968. The limits of uncertainty are at present in the order of 20 to 25% and we hope to narrow these limits in a more comprehensive analysis in the future. It is apparent from Figure 4 that the flux values measured in both ARIEL I and IRIS are less than the Explorer 4 values in this region of *B-L* space but we have to bear in mind that this difference may be attributable, in part at least, to discrepancies in the *B-L* coordinate system. An error of 1 or 2% in either *B* or *L* would be sufficient. In plotting the Explorer 4 data McIlwain used the Jensen and Whitaker 512 coefficient model where-

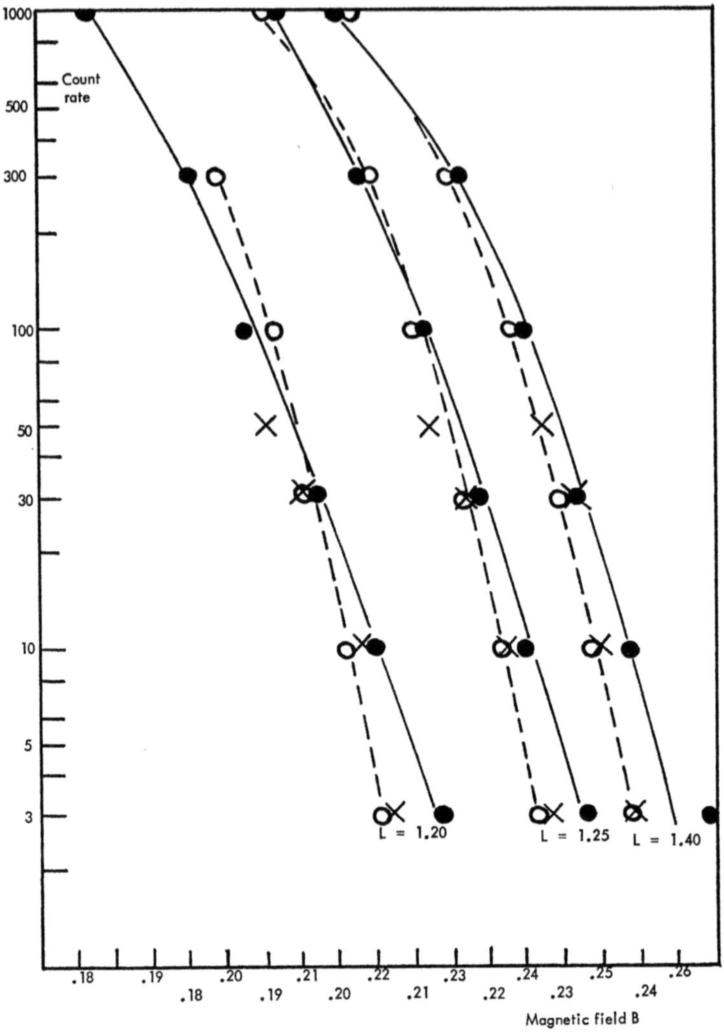

Fig. 4. Count rates as a function of B for a selection of L values. Explorer 4 (●), ARIEL 1 (×), and IRIS (○). Upper scale of B refers to $L = 1.20$ and lower scale to $L = 1.25$ and $L = 1.40$.

as we used the Jensen and Cain 48 coefficient field for the ARIEL I data and the GSFC 64 model for the IRIS data. In the case of ARIEL I and IRIS the counting rates were stored on an onboard tape recorder and are averaged over virtually all longitudes so that they are less sensitive to uncertainties in the coordinate system than was Explorer 4 which relied on real time data transmission only.

Nakano and Heckman (1968) and Heckman and Nakano (1969), from a succession of nuclear emulsion exposures at low altitude in the South Atlantic anomaly, conclude that there have been real variations in the flux of protons with $E \geqslant 63$ MeV. They measured an intensity during the recent sunspot minimum (1963–65) which was about twice that in 1961–62. From 1966–67 the flux decreased by a factor of about 2 bringing it back roughly to the 1961–62 value.

These observations appear to indicate that atmospheric losses play an important role in determining the flux of high energy protons and, therefore, are in disagreement with the conclusion we have reached on the basis of observations spread over an appreciably greater range of solar activity. The situation therefore must still be regarded as unsatisfactory but we expect to be able to make a much more detailed comparison when the analysis of the IRIS data is completed.

I now want to present the results of a rather different kind of observation which appear to have a bearing on this point.

B. A SHORT LIVED INCREASE IN THE CHARGED PARTICLE FLUX AT THE EQUATOR DURING THE MAGNETIC STORM OF OCTOBER 31, 1968

This event was observed in the course of a program of balloon flights carried out from Kampala, Uganda. The primary objective of these flights was to search for π°-decay γ-rays emitted from active regions of the sun. The presence of a weak γ-ray emission from such regions under quiescent (non-flare) conditions would be indicative of the presence of protons with $E \geqslant 200$ MeV trapped in the associated sunspot magnetic fields and this, in turn, has an important bearing on the mechanism of solar flares (Elliot, 1969). Figure 5 shows schematically the type of detector used on most of the flights carried out to date. Half a radiation length of lead converts the γ-rays into electron-positron pairs. The electronics logic requires a coincidence between the elements B, C, and D with an anticoincidence from the plastic scintillator A. B and D are 1 in. thick 'Perspex' Cerenkov radiators and C is $\frac{1}{8}$ in. plastic scintillator. A pulse corresponding to the passage of both the electron and positron through the element is required in B, but only one electron is required in C and D. The Cerenkov elements B and D are both blackened at the top surface, which produces a unidirectional response for the detectors at a confidence level of about 98%. The $\frac{1}{4}$ in. thick scintillator anticoincidence shield A is extended down the side of the 5 in. photomultiplier tube coupled to element B, to eliminate shower particles triggering the γ-ray channel. The fourfold coincidence rate ABCD was monitored continuously as are the individual rates of all four photomultiplier tubes. This gives a thorough check on variations in detector performance during flights. The geometrical factor of the detector is 22 cm^2 sr for an isotropic flux with a geometric half angle of 25°.

The flights were made on the basis of observed and predicted solar activity, the main criterion being the presence of one or more active regions on the visible hemisphere of the sun. On this basis a flight was carried out on October 31, the balloon reaching floating altitude of 4.5 $gm\,cm^{-2}$ at 0630 UT. Telemetred data were received until the balloon drifted out of range at 1530 UT by which time it had descended to an atmospheric depth of 6.3 $gm\,cm^{-2}$.

At 1030 UT the fourfold coincidence rate, which corresponds to the passage of charged particles through all four detecting elements, registered a temporary increase. This increase is shown in Figure 6, together with magnetograms for the magnetic

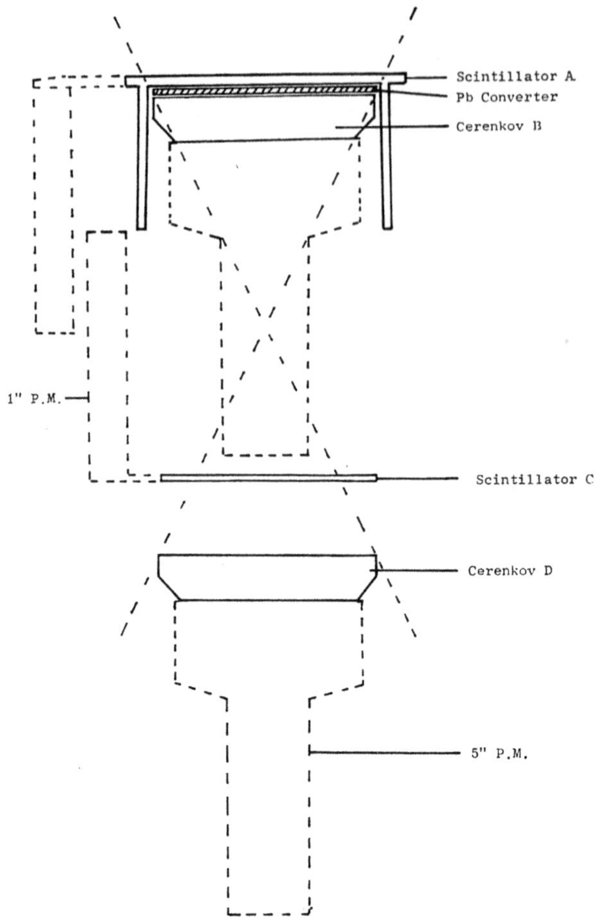

Fig. 5. Schematic layout of the γ-ray detector used in the balloon flights.

storm which was subsequently found to have occurred at the time. A further selection of magnetograms for this storm is shown in Figure 7. For the major portion of the flight the γ-ray flux registered by the payload remained at the level corresponding to normal atmospheric background. We have no reason to suspect any malfunction of the detector and believe that the increased count rate was produced by a real increase in charged particle flux. The minimum energy required to produce a fourfold coincidence is 500 MeV for protons and 20 MeV for electrons, the corresponding geometrical factor being 22 cm^2 sr.

The enhanced particle flux appeared during the bay-like depression in the initial phase of the storm which began with a sudden commencement at 0859 UT. A low energy solar proton event began at 0300 UT on October 31, but no discernible increase was detected on ground level neutron monitors. These monitors, however, did show a Forbush decrease beginning at $\sim$0400 UT on October 30 and on October

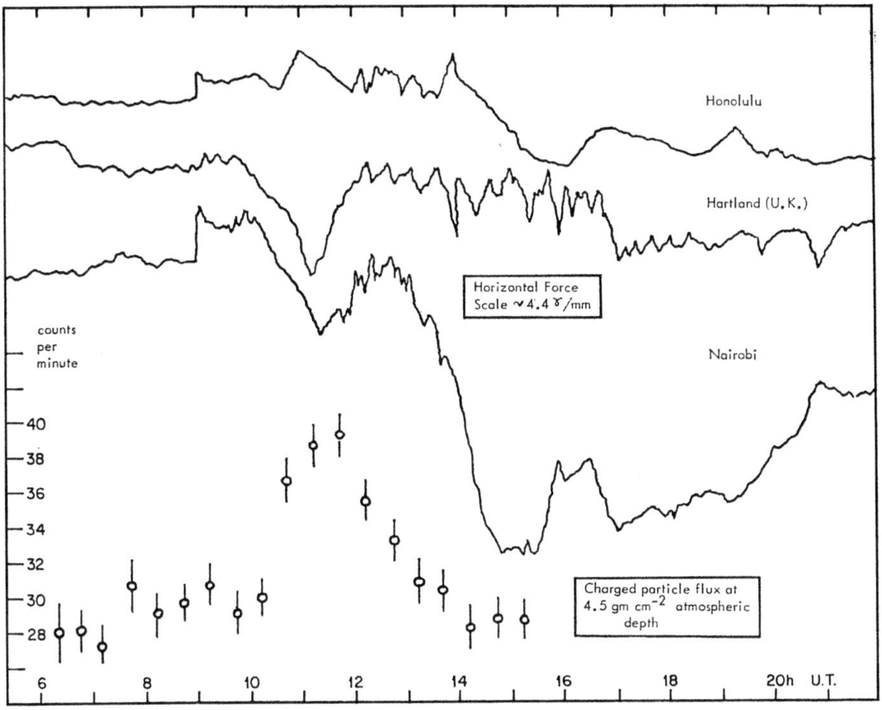

Fig. 6. Increase in ionizing particle count rate plotted as a function of time at balloon altitude during the magnetic storm of October 31, 1968.

31 the Deep River neutron intensity, for example, was $\sim 10\%$ below its pre-decrease value. Figure 8 shows the hourly values recorded by the Climax neutron monitor (provided by Professor J. A. Simpson, University of Chicago) during the day of the flight.

2. Discussion

We have considered a number of possible explanations of the enhanced particle flux and these will now be discussed.

A. SOLAR PARTICLES

The solar particle event in progress at the time of the flight was a predominantly low energy event. It began shortly after 0300 UT on the 31st and continued throughout the whole of that day. Solar particles of up to 60 MeV were observed by IMP F outside the magnetosphere and of up to 90 MeV by ESRO II when over the polar caps. The absence of any ground level neutron event (Figure 8) rules out the possibility that there was any measureable flux of solar protons with energy greater than that corresponding to the geomagnetic threshold at the equator and we do not know of any process whereby the geomagnetic threshold at the equator could be reduced to

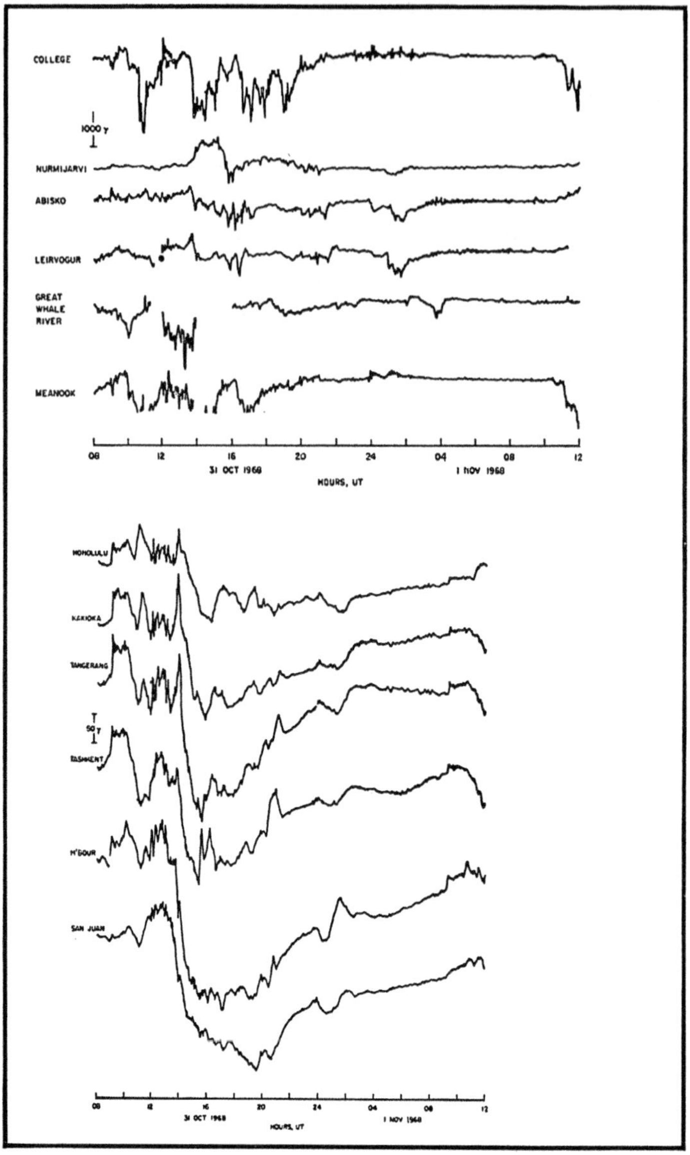

Fig. 7. Selection of magnetograms for the magnetic storm of October 31 taken from the ESSA collection of Solar-Geophysical Data.

a few hundred MeV during the initial phase of a geomagnetic storm. For these reasons it seems unlikely that the increase can be ascribed to solar particles.

B. REDUCTION OF GEOMAGNETIC THRESHOLD FOR COSMIC RAYS

A reduction in geomagnetic threshold below the quiet time value of ~ 15 GV for Kampala would automatically produce an increased cosmic ray flux and indeed,

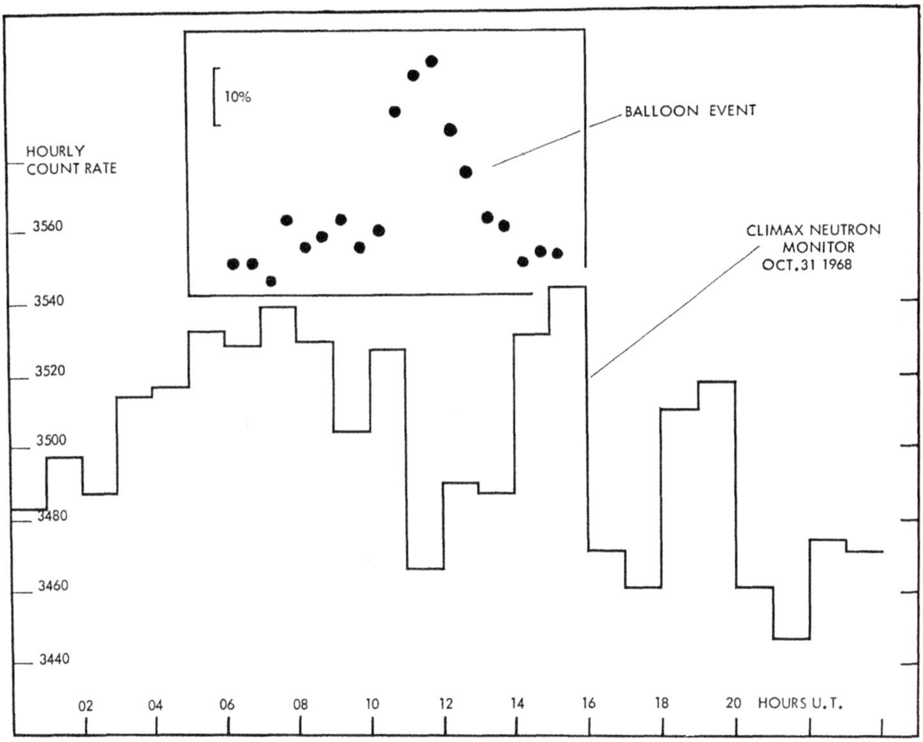

Fig. 8. Climax neutron monitor record for October 31. Charged particle count rate registered on the balloon flight is shown inset on the same time scale.

cosmic ray increases associated with such reductions in threshold rigidity have been described (Kondo *et al.*, 1960). However, this process cannot explain the increase observed on October 31. We know from the slope of the primary spectrum that the threshold rigidity would have had to fall to ~12 GV in order to produce an increase of 30% in the particle flux. The perturbation of the geomagnetic field that this would imply is far in excess of the short lived decrease in H of 100 γ observed at Nairobi at the time of this event. Furthermore, this explanation would have the consequence that there should have been an observable increase in the sea level neutron monitor rates whereas none was observed. In addition, there should have been a much bigger increase in the particle flux at the time of the main phase of the storm when the horizontal component of the field reached minimum value which was well below that recorded at the time of the increased counting rate. In fact, the particle flux had returned to normal by that time.

C. Dumping of Particles from the Inner Radiation Belt into the Atmosphere

The inner radiation belt is the most likely source of the protons and/or electrons required to produce the observed increase. The inner belt proton spectrum extends to sufficiently high energies to allow penetration to balloon altitudes but not high enough to produce a ground level event. The inner edge of the belt lies within a few hundred kilometers of the balloon and comes even lower over the South Atlantic so that a relatively small perturbation of the particle trajectories can produce appreciable atmospheric dumping.

Averaged over the duration of the event, the observed particle density corresponds to a flux of 9×10^{-3} cm^{-2}sec^{-1} sr^{-1} for a period of 6×10^3 sec. The flux of protons with $E \geqslant 500$ MeV in the inner radiation zone depends strongly on position but we may take 3×10^{-4} cm^{-2} sec^{-1} sr^{-1} as a representative value. An order of magnitude calculation shows that if the measured precipitation was uniform over the whole of the bottom surface of the radiation zone this would be sufficient to depopulate the trapping region to a height of ~ 1500 km. If the precipitation was localized the extent of the depopulation, of course, would be reduced accordingly. In view of the low flux values for electrons with $E \geqslant 20$ MeV in the trapping region it seems unlikely that they could provide an appreciable part of the observed precipitation.

The predominant loss mechanism for protons from the inner radiation zone has never been established with certainty, but has been rather generally assumed to be mainly ionization loss at altitudes of several hundred kilometers. The atmospheric scale height at such altitudes varies markedly over the solar cycle, being greatest at solar maximum. This should lead to a systematic variation in trapped proton flux over the solar cycle which has not been observed. The absence of this variation has raised doubts in the past as to whether ionization loss at these altitudes is the predominant loss process. The event described here, if correctly interpreted, provides positive evidence for a catastrophic loss of particles at the time of magnetic storms that may turn out to be more important than ionization loss on the edge of the trapping zone.

Kellogg (1959) was the first to suggest that charged particles can be transported through the geomagnetic field by fluctuation in the field strength with a time scale shorter than the longitudinal drift period for trapped particles. For protons with $E \geqslant 500$ MeV this drift period is a matter of seconds only so that the field fluctuations need to be very rapid if this condition is to be met. It is clear from the selection of magnetograms shown in Figures 6 and 7 that rapid field variations were indeed occurring during the event, but a decision as to whether they are adequate must await a more detailed analysis.

Acknowledgments

The work described here has been carried out in association with J. J. Quenby, A. Bewick, R. K. Sood, and C. A. Green. The IRIS satellite project was carried out under the auspices of the ESRO organization. Much of the work was funded from grants from the Science Research Council.

References

Bewick, A., Elliot, H., Hynds, R. J., and Quenby, J. J.: 1970, to be published.
Blanchard, R. C. and Hess, W. N.: 1964, *J. Geophys. Res.* **69**, 3927.

Elliot, H.: 1969, *Proceedings of the 1968 COSPAR Symposium on Solar Flares*, North-Holland Publishing Company, Amsterdam, Holland, p. 356.
Freden, S. C.: 1969, *Space Sci. Rev.* **9**, 198.
Heckman, H. H. and Nakano, G. H.: 1969, *J. Geophys. Res.* **74**, 3575.
Kellogg, P. J.: 1969, *Nature* **183**, 1295.
Kondo, I., Nagashima, K., Yoshida, S., and Wada, M.: 1960, *Proceedings of IUPAP Moscow Conference* **4**, 210.
McIlwain, C. E.: 1961, *J. Geophys. Res.* **66**, 3681.
Nakano, G. H. and Heckman, H. H.: 1968, *Phys. Rev. Letters* **20**, 806.
Sood, R. K., Green, C. A., and Elliot, H.: 1970, to be published.

ALPHA PARTICLES TRAPPED IN THE EARTH'S MAGNETIC FIELD

S. M. KRIMIGIS
The Johns Hopkins University, Applied Physics Laboratory, Silver Spring, Md., U.S.A.

1. Introduction

The study of geomagnetically trapped α-particles, unlike other charged particle constituents of the radiation belts, has had a comparatively short history. The first comprehensive results were presented two years ago at the preceding Advanced Study Institute (McCormac, 1968) and the emphasis was on the validity of the experimental method as well as the physical significance of the results. Since that time, other experimenters have confirmed and expanded the initial results (Paulikas and Blake, 1968; Fritz and Krimigis, 1969) and now a comprehensive study of the absolute intensities, spatial and angular distributions, energy spectra, and time variations is under way. All of these studies, however, are limited by the difficulty of making reliable experimental measurements with adequate time resolution, so that meaningful experimental results may be obtained.

The main problem posed by the first experimental results (Krimigis and Van Allen, 1967), namely the inadequacy of the *solar wind source-diffusion across lines of force* theory to explain the observed α to proton ratio, has not as yet been resolved in a satisfactory manner. There are two requirements that must be met before an attempt at a solution to the above problem can be made: (a) the accumulation of a large body of experimental data of the nature described above; and (b) the theoretical investigation of loss mechanisms in the magnetosphere for *both* protons and α-particles. Neither of these requirements is satisfied at the present time. In the following discussion a summary of the early results will be given and a modest attempt towards contributing to (a) above will be made by presenting recent data obtained with the University of Iowa satellite Injun 5.

The principal advantages of the Injun 5 data over previous observations are (a) positive particle identification, (b) lower detector energy threshold (to ~1 MeV), (c) determination of spectral form, and (d) faster time resolution.

These preliminary results taken together with the previous data show that the α particle-to-proton ratio j_α / j_p is highly variable at low energies ($E \lesssim 0.5$ MeV/nucleon), but that it appears to be relatively constant at higher energies over a period of approximately 4 yr. Further, the disagreement between the observations and the diffusion from the solar wind model remains unresolved, although it is thought that consideration of all relevant processes may eventually bring agreement between the theory and observations. In addition, the process of direct magnetospheric capture of energetic α-particles from solar flare events is excluded as a possible source of trapped α-particles.

B. M. McCormac (ed.), Particles and Fields in the Magnetosphere.

2. Summary of Previous Results

Attempts to identify α-particles in the radiation zones were made as early as 1959 by use of nuclear emulsions, but no α-particles positively identified as trapped in the geomagnetic field were found. An upper limit was placed on the α-particle to proton ratio (j_α / j_p) of $\sim 1 \times 10^{-3}$ for particles in the range 125 to 185 MeV/nucleon, and

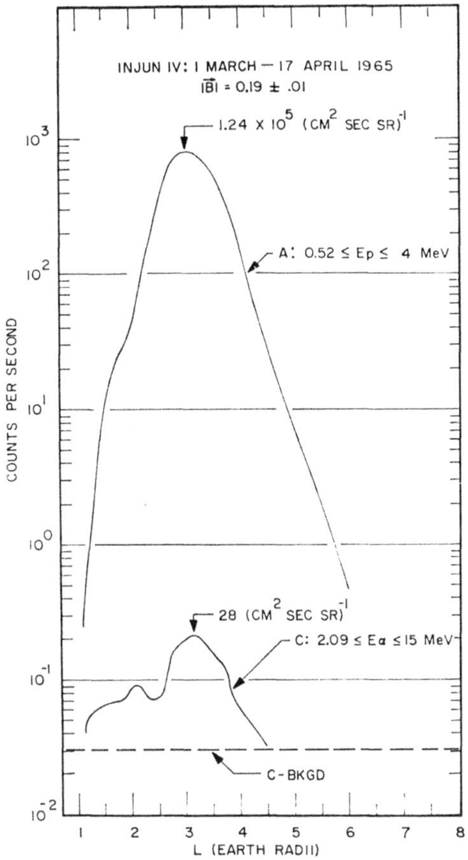

Fig. 1. The distribution in L of protons and α-particles of $E \gtrsim 0.52$ MeV/nucleon, during March–April 1965 as observed by Injun 4.

for L values less than ~ 2. For $L \sim 3$, Fenton (1967) placed a limit of $j_\alpha / j_p \lesssim 6.7 \times 10^{-4}$ for particles in the range 25 to 85 MeV/nucleon. A detailed review of the early work has already been given elsewhere (Krimigis and Van Allen, 1967).

Figure 1 shows the results obtained by Krimigis and Van Allen (1967) using data from the Injun 4 spacecraft at a particle energy of ~ 0.52 MeV/nucleon. Their observations showed that the j_α / j_p ratio at $L \sim 3.1$ and $B \simeq 0.19$ is

$$j_\alpha / j_p \simeq 2.2 \times 10^{-4} .$$

Their results were later confirmed and expanded by Paulikas and Blake (1968), using a substantially different experimental technique, and by Fritz and Krimigis (1969) using data from the OGO 4 spacecraft. The results of Paulikas and Blake (1968) are shown in Figure 2, and it is seen that they are in good agreement with those of

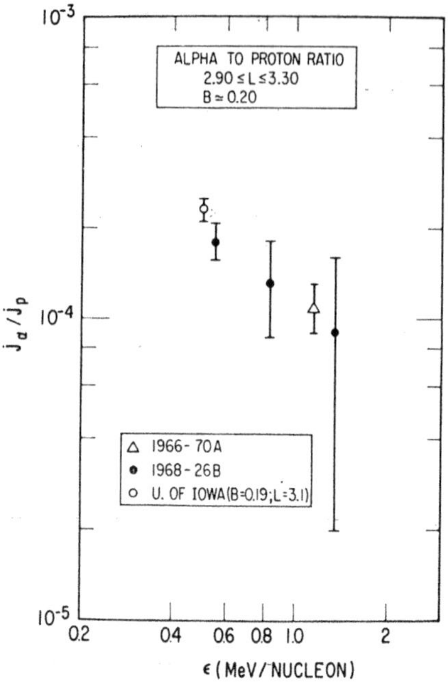

Fig. 2. A summary of the results obtained with satellites 1966–70A, 1968–26B, and of Injun 4 plotted as a function of energy/nucleon (courtesy of Drs. Blake and Paulikas).

Krimigis and Van Allen (1967). Further, their data suggest that the ratio may decrease as the energy/nucleon increases.

The OGO 4 data (Figure 3) extended the measurements down to $\sim$0.26 MeV/nucleon. One observes that the spectrum is probably becoming softer at lower energies when compared to Injun 4 and that the j_α / j_p ratio may increase. The value of the ratio reported by Fritz and Krimigis (1969) at $E \gtrsim 0.35$ and $E \gtrsim 0.46$ MeV/nucleon is $\sim 7 \times 10^{-4}$ and $\sim 2 \times 10^{-4}$, respectively. They also find that the ratio decreases with increasing energy/nucleon, and conclude that solar wind particles are probably injected into the magnetosphere with equal energies, rather than equal velocities.

In regard to time variations, Krimigis and Burns (1968) showed that α-particle fluxes at 0.5 and 0.9 MeV/nucleon have changed by as much as a factor of 5 over a period of $\sim$18 mo. This result is shown in Figure 4, where it is seen that the decay constant for protons and α-particles at $2.7 \leqslant L < 2.8$ following the April 17, 1965, magnetic storm is about 86 days. At other L values, however, the decay modes are different.

The Injun 5 data, presented in the next section will allow the observation of time variations in the α-particle flux following the November 1, 1968, magnetic storm, and also the unique determination of the form of the energy spectrum, which gives information on the source function of such particles.

3. Injun 5 Observations

A. THE SATELLITE AND THE DETECTOR

The University of Iowa Injun 5 satellite was launched on August 8, 1968, into an elliptical orbit with initial apogee of 2525 km, perigee of 644 km, and inclination of

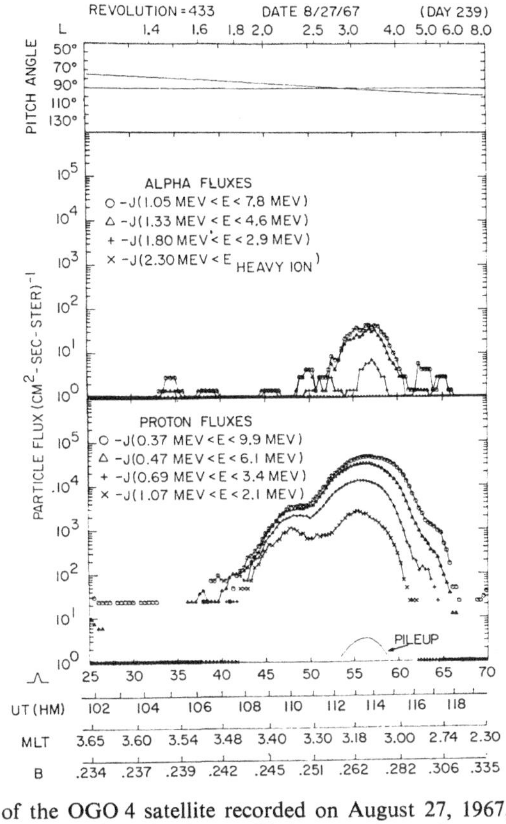

Fig. 3. Typical pass of the OGO 4 satellite recorded on August 27, 1967, showing the counting rates due to protons and α-particles in the specified energy intervals as a function of INL. The values of UT, MLT, and B are shown at the bottom and the pitch angle of a particle entering along the axis of the detector collimators is shown at the top (Fritz and Krimigis, 1969).

80.7°. The solid state detector (SSD) experiment onboard the satellite consists of the proton electron telescope (PET) and the α-particle detector. The latter is the detector of primary interest in the subsequent discussion so that an adequate description is necessary.

The α-particle experiment consists of a surface barrier, totally depleted solid state

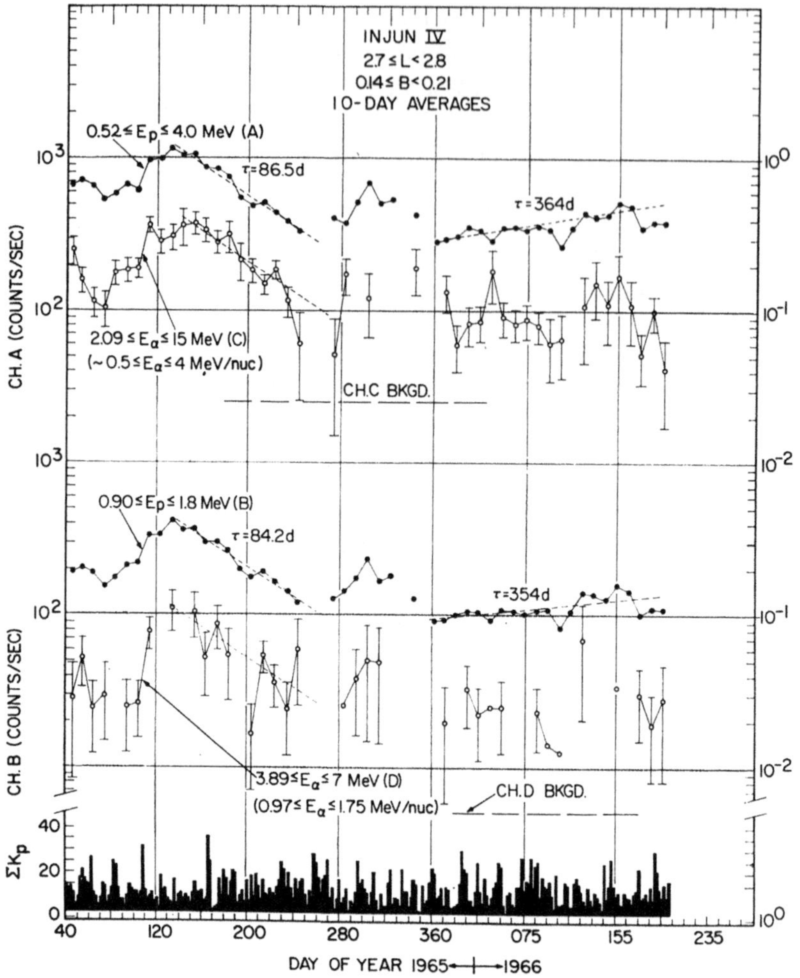

Fig. 4. Time history of protons and α-particles with $E \gtrsim 0.52$ MeV/nucleon and $E \gtrsim 0.94$ MeV/nucleon, as observed with Injun 4, in the indicated L and B intervals. Due to a computer problem the proton counting rates around day 110, 1965 are somewhat inaccurate, but the decay trend is quite clear and appears to be identical for both protons and α-particles of this energy (Krimigis and Burns, 1968).

detector whose thickness is 8.5 μ and whose area is 3.1 ± 0.1 mm^2 (circular). The detector is located inside a conical collimator with a full vertex angle of 26°30′ and is otherwise shielded by a minimum of 11.4 gm/cm^2 of brass corresponding to the range of 95 MeV protons. Four electronic discrimination levels are provided. The first three (channels $\alpha 1$, $\alpha 2$, $\alpha 3$) are sensitive to nuclei with $Z \geqslant 2$ while the last one (channel $\alpha 4$) is sensitive to nuclei with $Z \geqslant 3$ only (Table I). Thus $\alpha 4$ may be used to ascertain (by extrapolation) whether particles observed in $\alpha 1$, $\alpha 2$, and $\alpha 3$ are α-particles or nuclei heavier than α.

The scheme for energy and particle discrimination is similar to the one described earlier for Injun 4 (Krimigis and Van Allen, 1967). With the present detector, the stopping energy for a proton is 0.6 MeV and the first α-particle discrimination level is set at 0.90 MeV. In addition, the electronic pulse width at half maximum is ~100 nsec with a pulse resolution time of 54 nsec. These design features make the instrument insensitive to particles other than those it is designed to measure.

TABLE I

Characteristics of Injun 5 detectors

Detector channel	Unidirectional geometric factor cm² sr	Discrimination level setting (MeV)	Particles to which sensitive
Alpha[a]			
α1	0.005 ± 0.0002	0.90	Alphas: $1.25 \leqslant E_\alpha \leqslant 8.0$ MeV
α2	0.005 ± 0.0002	1.33	Alphas: $1.65 \leqslant E_\alpha \leqslant 4.5$ MeV
α3	0.005 ± 0.0002	1.76	Alphas: $2.03 \leqslant E_\alpha \leqslant 3.35$ MeV
α4	0.005 ± 0.0002	3.43	$Z > 2$, $E > \sim 0.5$ MeV/nucleon
PET[b]			
P1	0.020 ± 0.003	0.21	Protons: $0.306 \leqslant E_p \leqslant 1.4$ MeV
P2	0.020 ± 0.003	0.37	Protons: $0.450 \leqslant E_p \leqslant 1.4$ MeV
P3	0.020 ± 0.003	0.57	Protons: $0.632 \leqslant E_p \leqslant 1.4$ MeV
P4	0.020 ± 0.003	0.75	Protons: $0.804 \leqslant E_p \leqslant 1.4$ MeV
$\overline{\mathrm{P4}}$	0.020 ± 0.003	0.75	Protons: $0.801 \leqslant E_p \leqslant 1.86$ MeV

[a] Shielded from light by 0.19 mg/cm² air equivalent nickel foil.
[b] Shielded from light by 0.17 mg/cm² air equivalent nickel foil.

Below are listed those physical processes that could have caused spurious counts in the α-channels:

(1) Proton pulse pile-up;

(2) Transverse penetration of the detector by high energy protons ($E_p \gtrsim 95$ MeV);

(3) Elastic scattering (and hence greater path length in detector) of protons entering through the collimator;

(4) Inelastic nuclear interactions of protons in the detector and in other material in the satellite; and

(5) Effects of electrons.

The contributions to the counting rate from all of the above processes have been evaluated and found to produce rates of at least two orders of magnitude below those actually observed in flight. A full description of the detailed consideration of each of these processes has been given earlier (Krimigis and Van Allen, 1967), while the details specifically applicable to the Injun 5 experiment have been given by Randall (1969).

In view of the above discussion, we conclude that the response of the alpha experiment is due only to α-particles and heavier nuclei in the energy ranges given in Table I.

To obtain the ratio of the intensities of α-particles to protons, use is made of the data from the PET. This instrument consists of a 24 μ, 10 mm² surface barrier totally depleted silicon detector followed by two detectors whose thicknesses are 862 and 905 μ and area 50 mm^2. The signals from the last two detectors are electronically summed and the combination of three detectors provides several differential and integral proton channels in the range 0.3 to 74 MeV. Those channels that are of interest to the present study are listed in Table I. A full description of the detector system is given elsewhere (Randall, 1969).

B. PRELIMINARY DATA

The physical location of the SSD experiment on the satellite is such that when the spacecraft achieves magnetic orientation, the detector axes become perpendicular to the local magnetic field vector. Satisfactory magnetic alignment was not obtained

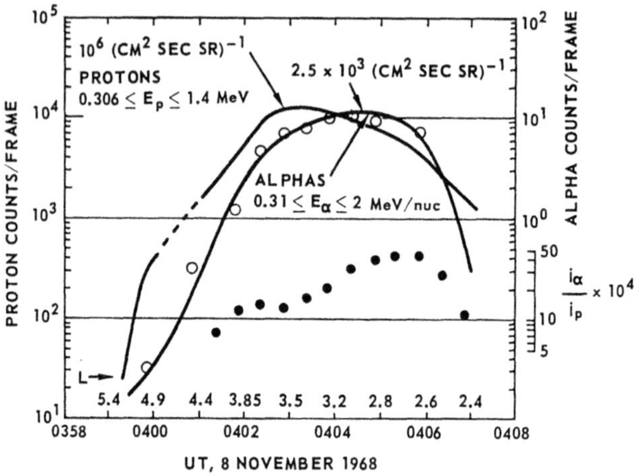

Fig. 5. An Injun 5 pass on November 8, 1968 about 7 days after the geomagnetic storm on Novem ber 1. Note that maximum intensities for protons and α-particles occur at distinctly different values of L. To obtain the value of j_α/j_p ratio, one should multiply the value read on the scale by 10^{-4}.

until the latter part of December 1968, although the satellite did attain orientation for brief periods (~a few minutes) prior to that time. The satellite orientation may be determined at all times by using three mutually orthogonal magnetometers onboard the spacecraft.

Figure 5 shows a satellite pass on November 8, 1968, during which the SSD detector axes were pointing in a direction nearly perpendicular to the local **B** vector. This is the first such pass following the magnetic storms of October 31, and November 1, 1968. The counting rates due to protons of $0.306 \leqslant E_p \leqslant 1.4$ MeV and α-particles of $0.31 \leqslant E_\alpha \leqslant 2$ MeV/nucleon are plotted, along with the intensity ratio of α-particles to protons j_α / j_p. The approximate value of L is indicated and the value of B ranges from 0.21 at $L=4.9$ to 0.17 at $L=2.4$. We observe the following:

(1) The maximum intensity peak for α-particles is located at a different L value from that for protons ($L^p_{max} > L^\alpha_{max}$). This is to be contrasted to previous observations

where the positions of the two peaks were the same (within statistics) for particles of the same energy/nucleon (Krimigis and Van Allen, 1967; Fritz and Krimigis, 1969).

(2) The j_α/j_p ratio has a maximum value of $\sim 4.5 \times 10^{-3}$ at $L \sim 2.8$ and $\sim 3.5 \times 10^{-3}$ at $L \sim 3.1$. The latter value is about a factor of 10 larger than the one reported earlier at 0.52 MeV/nucleon (Krimigis and Van Allen, 1967), at roughly the same point in B–L space.

(3) The j_α/j_p ratio varies by about a factor of 6 in the range $4.3 \lesssim L \lesssim 2.4$. This result differs from that of Krimigis and Van Allen (1967), where the ratio stayed approximately constant in the same range of L.

(4) The peak flux of α-particles at 0.31 MeV/nucleon is $\sim 2.5 \times 10^3$ (cm^2 sec sr)$^{-1}$. Using the α-particle spectrum one may compute the flux at $E \gtrsim 0.52$ MeV/nucleon and compare with the earlier result of Krimigis and Van Allen (1967). One then finds that

$$j_{(E>0.52\ \mathrm{MeV/nucleon})} \simeq 73\,(\mathrm{cm}^2\ \mathrm{sec\ sr})^{-1}$$

compared to ~ 28 (cm^2 sec sr)$^{-1}$ found by Injun 4. The difference is attributed to time variations which may change the flux by factors of 3 to 4 as shown by Krimigis and Burns (1968).

It is probable that the proton and α-particle spatial distribution observed on November 8 does not represent the steady state intensity profile of such particles because of the effects of the November 1 geomagnetic storm. Figure 6 shows a satel-

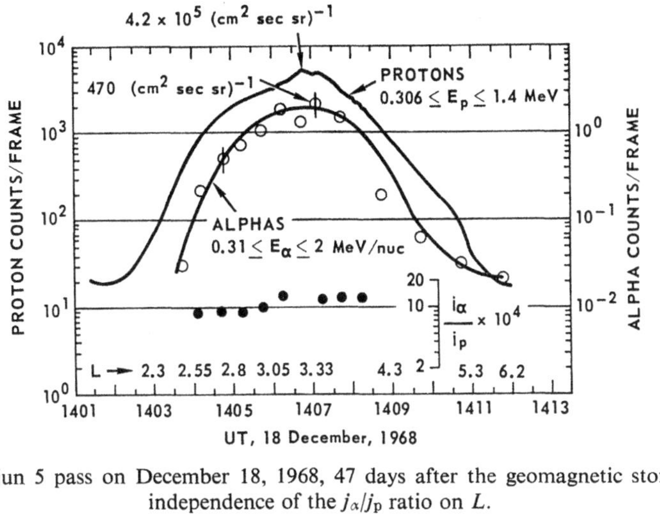

Fig. 6. An Injun 5 pass on December 18, 1968, 47 days after the geomagnetic storm. Note the independence of the j_α/j_p ratio on L.

lite pass on December 18 approximately 47 days after the storm. We observe the following:

(1) The peak intensity for both protons and α-particles occurs at about the same L value. Further, it appears that compared to the November 8 pass the proton peak

has been shifted to a lower L value while the α-particle peak has been shifted to a higher L value.

(2) The peak intensity of α-particles has decreased by about a factor of 5 while that of protons by a factor of 2.

(3) The j_α/j_p ratio has decreased to $\sim 1 \times 10^{-3}$ and is approximately constant as a function of L.

It appears that the particle distribution on December 18 may be representative of the steady state distribution for such particles. To ascertain whether this is indeed the case a pass taken on January 26, 1969, is shown in Figure 7. There were no significant

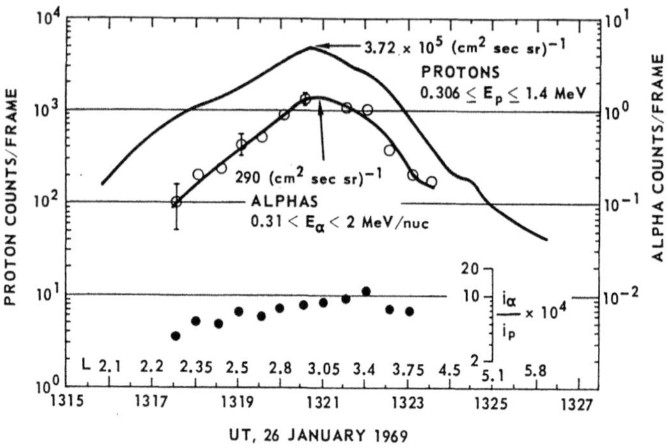

Fig. 7. An Injun 5 pass on January 26, 1969. The proton and α-particle population probably represent a steady-state distribution.

geomagnetic storms between November 1, 1968, and February 2, 1969, so that injection of new particles into the magnetosphere is not expected to have occurred. We observe that there has been further decay in the flux of α-particles and that the j_α/j_p ratio is $\sim 8.5 \times 10^{-4}$. There appcars to be a weak dependence of the ratio on L and the peak fluxes may have shifted to a slightly lower value of L.

It is seen that changes occurring between December 18, 1968, and January 26, 1969 although measurable are not very large. Thus, one may consider the particle intensity profile observed on January 26 as representative of a steady state distribution. Following the February 2, 1969, magnetic storm, a partial redistribution of particles took place so that it is not possible to follow further the time development of the particle distribution formed after the November 1 storm.

Figure 8 shows the j_α/j_p ratio shown in the previous three figures, but plotted in a linear scale. A point obtained prior to the magnetic storm during a period when the detector was pointing perpendicularly to the local **B** vector is also included. It is seen that the major enhancement in the j_α/j_p ratio, and judging from the previous figures in the α-particles, took place in the range $2.5 \lesssim L < 3.5$. The ratio had decreased markedly by December 18 and appears to have returned to its pre-storm value by January 26, 1969, with the rate of decrease being faster at lower values of L. This behavior of the proton to α-particle ratio is roughly consistent with the inward shift in the α-particle peak following the April 17, 1965, magnetic storm (Krimigis and Van Allen, 1967) although in the latter case the protons also moved inward.

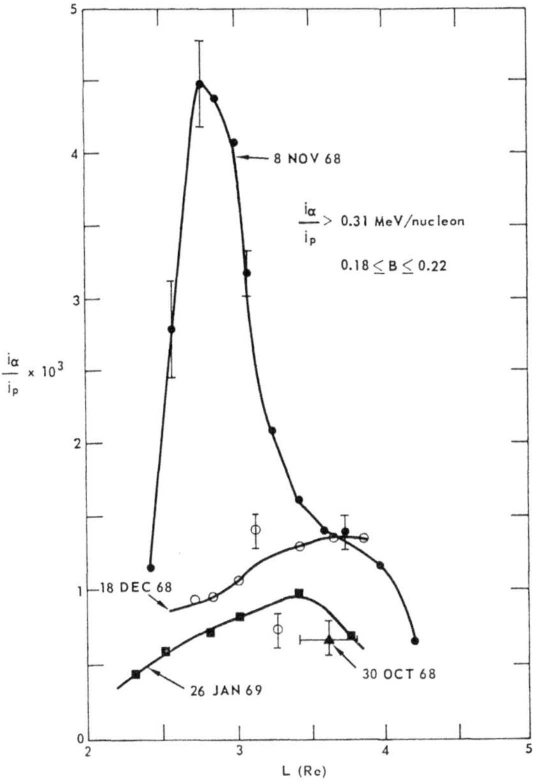

Fig. 8. The j_α/j_p ratio as a function of L from the three previous figures. The point on October 30, prior to the magnetic storm has been included for comparison.

C. ENERGY SPECTRUM

The energy spectrum of protons and α-particles may be used to obtain information on the source function as discussed in Section 4. By using the three ratios $\alpha 1/\alpha 2$, $\alpha 1/\alpha 3$, $\alpha 2/\alpha 3$ one can unambiguously determine the form of the spectrum; further, by using the differences $(\alpha 1-\alpha 2)$ and $(\alpha 2-\alpha 3)$ and assuming that the differential spectrum is a steeply falling function of energy, one obtains two differential energy channels, which should be consistent with the energy spectrum arrived at using the ratios among the three channels. By use of these ratios it was found that the spectrum on November 8 can be easily fitted by an exponential of the form

$$\mathrm{d}j/\mathrm{d}E = Ke^{-E/E_0}$$

where $E_0 \simeq 0.250 \pm 0.005$ MeV. The spectrum on December 18 was consistent with either an exponential of $E_0 \simeq 0.350 \pm 0.02$ MeV, or a power law of the form

$$\mathrm{d}j/\mathrm{d}E = KE^{-\gamma}$$

with $\gamma \simeq 5.5 \pm 0.2$. Finally, the spectrum on January 26 could not be fitted by either a power law or an exponential (Table II). The results on the energy spectra are shown

TABLE II
Energy spectral parameters

Date	Satellite	E_0 [c] (Protons)	E_0 [c] (Alphas)	γ (Alphas)
Nov. 8, 1968	Injun 5	0.14 ± 0.01	0.25 ± 0.005	(6.9, 7.4, 8.2) [a]
Dec. 18, 1968	Injun 5	0.22 ± 0.01	0.35 ± 0.02	(5.3, 5.4, 5.7) [a]
Jan. 26, 1969	Injun 5	0.26 ± 0.01	(0.38, 0.29, 0.23) [a]	(4.7, 6.5, 9.0) [a]
March–April 1965	Injun 4 [b]	0.30 ± 0.01	1.24 ± 0.03	3.0 ± 0.1

[a] The three values are derived from the ratios $\alpha1/\alpha2$, $\alpha1/\alpha3$, and $\alpha2/\alpha3$, respectively.
[b] Injun 4 measurements represent two-point spectral fits.
[c] E_0 is in MeV.

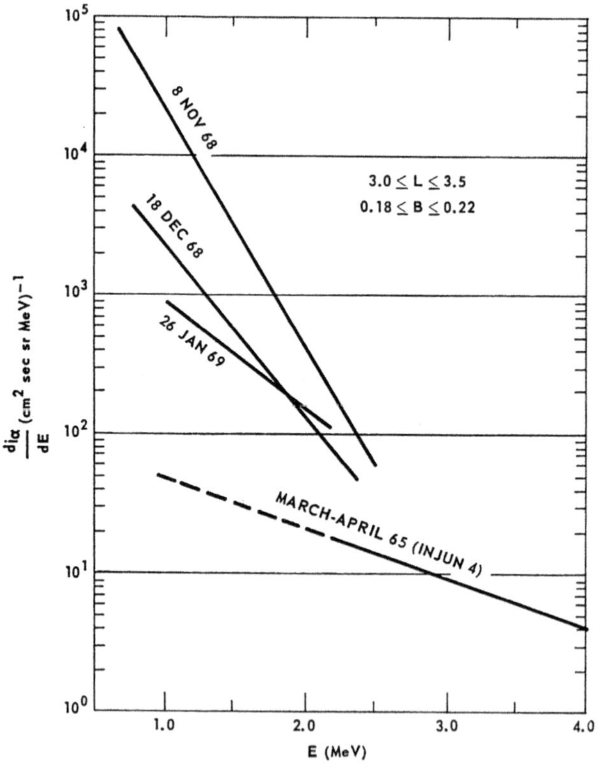

Fig. 9. The differential energy spectrum of α-particles at three different times after the November 1, 1968, magnetic storm. The Injun 4 spectrum has been included for comparison.

in Figure 9, where the January 26 spectrum was obtained using the two differential channels of the detector; the spectrum observed in 1965 by Injun 4 has been included for comparison. We observe the following:

(1) The spectrum becomes progressively harder following the November 1 magnetic storm; and

(2) By comparing the present results to those of Injun 4 it is apparent that the spectrum must become softer at energies below 2.1 MeV (the Injun 4 threshold). In addition, the number of α-particles of $E > 2.1$ MeV is greater now than during the March to April 1965 period.

As was remarked earlier, the steady state spectrum of α-particles on January 26 does not fit an exponential form, although that of protons fits an exponential reasonably well at all times. This point will be examined in more detail in the discussion.

D. j_α/j_p VS. ENERGY/NUCLEON

Figure 10 shows the j_α/j_p ratio in the range $2.5 \lesssim L \lesssim 3.5$ plotted as a function of energy/nucleon for the three Injun 5 passes presented earlier. The ratios were computed by integrating the respective proton and α-particle spectra, so that the energy ranges of protons and α-particles are exactly matched. The data points obtained in

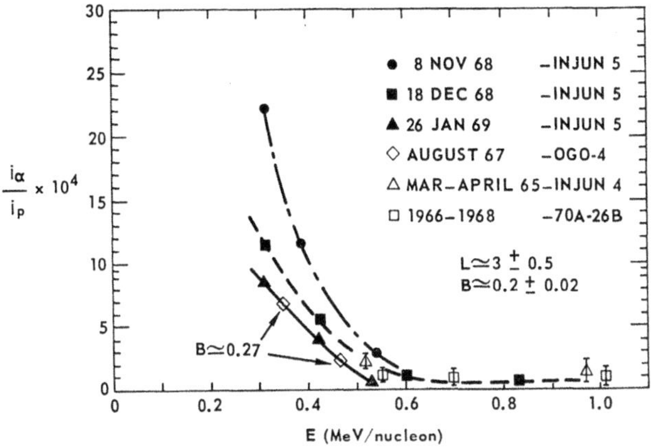

Fig. 10. A summary of j_α/j_p ratio as a function of energy/nucleon, including all the observations reported so far. It appears that in the absence of magnetic storms the steady state value of the ratio would become $\sim 10^{-4}$ at all energies.

previous measurements have been included for comparison. Noteworthy features of this plot include the following:

(1) The j_α/j_p ratio increases sharply below ~ 0.5 MeV/nucleon, reflecting the softening of the α-particle spectrum;

(2) The ratios are highest following the November geomagnetic storm and gradually decay to what may be a quiescent value in about 3 mo; and

(3) At $E \gtrsim 0.5$ MeV/nucleon the ratio appears to be relatively constant and may have had the same value for about 4 yr (March 1965 to January 1969).

One may speculate that if the frequency of occurrence of geomagnetic storms was low enough, the value of the ratio at $E \lesssim 0.5$ MeV/nucleon would approach that at $E \gtrsim 0.5$ MeV/nucleon.

4. Discussion

The study of α-particles in the magnetosphere was undertaken with the objective of providing a set of parameters for one or more 'tracer' elements in the radiation belts, so that various hypotheses on the origin and acceleration of all components of trapped radiation could be evaluated in a manner somewhat analogous to that employed in the study of galactic cosmic rays. It is now believed that the main process by which charged particles populate the radiation belts is radial diffusion across lines of force, with the solar wind being the source of such particles. Although such a process gives some qualitative agreement between the observed proton distribution and the theory (Nakada *et al.*, 1965) there appear to be severe quantitative and qualitative difficulties when the detailed physical mechanisms giving rise to such processes are examined in detail (Fälthammar, 1968; Haerendel, 1968). The model, nevertheless, represents the only one developed in sufficient quantitative detail, so that comparison with experimental data may be made.

Tverskoy (1965) has considered the problem of trapped α-particles in detail and has derived a set of predictions, which may be summarized as follows:

(1) If protons and α-particles are injected into the magnetosphere with the same velocity distribution (as is the case in the solar wind), then

(a) $j_\alpha(4E)/j_p(E) \simeq 0.01$ to 0.1 (the solar wind ratio),
(b) $L^\alpha_{max}(E) = L^p_{max}(E/4)$, and
(c) $S_\alpha(4E, L) = S_p(E, L)$

where E is the particle energy, L_{max} is the value of L where the peak intensity for a particle of the indicated energy occurs and S is the particle distribution function.

As can be seen from Figure 10, the observed ratio is smaller than the predicted value by one to two orders of magnitude. It appears that as the energy/nucleon decreases and immediately after a magnetic storm the ratio *may* approach the solar wind value. But the steady state value of the ratio in the absence of geomagnetic storms would probably approach that at $E \gtrsim 0.5$ MeV/nucleon, i.e., $\sim 2 \times 10^{-4}$. This fact suggests that important loss mechanisms, in addition to coulomb energy loss in the outer atmosphere, must affect α-particle lifetimes relative to those of protons. Possible lines of investigation should include a detailed consideration of charge exchange losses in the ambient plasma and the pitch angle scattering mechanism as it affects both protons and α-particles.

Tverskoy's prediction (b), regarding the value of L at maximum intensity, appears to hold true (to within observational statistics) for the steady state distribution, but does not agree with the observations a few days after the geomagnetic storm of

November 1 as is evident from Figure 5. The prediction, however, was made for the steady state distribution and it agrees with the observations. It may prove useful to investigate those processes which, following a geomagnetic storm, would produce an α-particle distribution different from that of protons of the same energy/nucleon; and those mechanisms which would cause the two distributions to become identical in 30 to 60 days.

The third prediction of Tverskoy states that the α-particle energy spectrum must fit an exponential, as is the case for protons (e.g., Armstrong and Krimigis, 1968), and that $E_{0\alpha} = E_{0p}$. As discussed in 3C above and as seen from Table II, $E_{0\alpha}$ is always different from E_{0p}. Further, the α-particle spectrum fits an exponential on November 8 and December 18, 1968, but does not do so on January 26, 1969, on which date the particle distribution presumably represents a steady state one. Thus, it appears that prediction (c), in general, is not verified by the observations. This conclusion is tentative, pending the outcome from the analysis of additional data.

More recently (Tverskoy, 1969) by examining the theory of turbulent acceleration of charged particles in a plasma (cf., Tverskoy, 1968) has concluded that $E_{0\alpha} = 2E_{0p}$. An examination of Table II shows that this is not the case, although a relation $E_{0\alpha} = 1.5\ E_{0p}$ or $E_{0\alpha} = 5/3\ E_{0p}$ does fit the data.

(2) If protons and α-particles are thermalized in crossing the shock front and transition region, and taking coulomb losses into account, then:

(a) $j_\alpha^{max}(E)/j_p^{max}(E) \sim 10^{-3}$,
(b) $L_{max}^\alpha(E) = 1.3\, L_{max}^p(E)$, and
(c) $S_\alpha(E, L) \sim S_p(E, L)$ for $L > L_l$

where L_l is the value of L where losses for protons and α-particles become important.

The flux of protons of $E > 1.25$, 1.65, and 2.03 MeV is not measured directly in the present experiment, so that it is not as yet possible to determine accurately the L value of maximum intensity and obtain the flux. Using the data of Mihalov and White (1966), one finds that the ratio is in the range 0.1 to 0.01, so that prediction (a) is not verified. Prediction (b) is probably correct [L_{max}^p(1.25 MeV)$\sim$2.8], while prediction (c) is probably not correct (Table II) unless $L_l \gtrsim 4.5$.

In summary, the experimental data show that neither injection of particles with equal velocities from the solar wind, nor thermalization of these particles in the transition region can fully account for the observations. A careful examination of current work on diffusion, however, (Haerendel, 1968; Theodoridis *et al.*, 1969) reveals that there are several loss mechanisms which may be of importance which were not taken into account in Tverskoy's (1965) preliminary considerations. Thus, a detailed reconsideration of the entire problem is clearly indicated before a definite conclusion on a solar wind source is reached.

It has been suggested (Krimigis and Van Allen, 1967) that direct magnetospheric capture of energetic α-particles emitted in solar particle events may account for the observed α-particles trapped in the geomagnetic field. There are several arguments against this hypothesis:

(1) Lanzerotti (1968) has made simultaneous observations of such particles inside (at 6.6 R_E) and outside the magnetosphere. He finds that although particles have immediate access to such L values, they disappear as soon as the outside source is removed, i.e., no permanent trapping.

(2) Blake *et al.* (1968) observed quasi-trapped α-particles over the polar caps during the August to September 1966 solar particle events. They find that such particles do not persist 'significantly longer (on a time scale of hours) than the incoming solar flux.' Further, the regions in which trapped α-particles are observed are not accessible to particles of this energy from outside the magnetosphere, even if they arrive via the magnetosphere tail (Gall *et al.*, 1968).

(3) The most decisive argument against direct magnetospheric capture is obtained by inspection of Figure 4. The α-particle flux was enhanced following the April 17, 1965, magnetic storm in spite of the fact that there were no solar energetic protons or α-particles associated with this storm.

Thus, current evidence argues strongly against direct magnetospheric capture of energetic solar α-particles.

The possibility of an *in situ* acceleration of ambient α-particles from the earth's exosphere has not been, as yet, quantitatively assessed, so that a comparison with experimental data cannot be made.

Acknowledgments

The author wishes to thank Drs. T. P. Armstrong and T. A. Fritz and Mr. B. A. Randall for permitting the use of the Injun 5 data presented jointly at the 1969 American Geophysical Union Meeting prior to the publication of the full paper. Thanks are also due to Dr. J. A. Van Allen for support and useful discussions and to Mr. P. Verzarin for assistance with the Injun 5 data.

The Injun 5 spacecraft was developed and flown under contract NAS1-2973 with the National Aeronautics and Space Administration/Langley Research Center. Analysis and publication have been performed in part under contract L-17659 between NASA/LRC and The Johns Hopkins University/Applied Physics Laboratory.

References

Armstrong, T. P. and Krimigis, S. M.: 1968, *J. Geophys. Res.* **73**, 143.
Blake, J. B., Paulikas, G. A., and Freden, S. C.: 1968, *J. Geophys. Res.* **73**, 4927.
Fälthammar, C.-G.: 1968, in *Earth's Particles and Fields* (ed. by B. M. McCormac), Reinhold Book Corporation, New York, p. 157.
Fenton, K. B.: 1967, *J. Geophys. Res.* **72**, 3889.
Fritz, T. A. and Krimigis, S. M.: 1969, *J. Geophys. Res.* **74**, 5132.
Gall, R., Jimenez, J., and Camacho, L.: 1968, *J. Geophys. Res.* **73**, 1593.
Haerendel, G.: 1968, in *Earth's Particles and Fields* (ed. by B. M. McCormac), Reinhold Book Corporation, New York, p. 171.
Krimigis, S. M. and Burns, A. L.: 1968, International Symposium on the Physics of the Magnetosphere, Washington, D.C.
Krimigis, S. M. and Van Allen, J. A.: 1967, *J. Geophys. Res.* **72**, 5779.

Lanzerotti, L. J.: 1968, *Phys. Rev. Letters* **21**, 1929.
McCormac, B. M., Ed.: 1968, in *Earth's Particles and Fields*, Reinhold Book Corporation, New York.
Mihalov, J. D. and White, R. S.: 1966, *J. Geophys. Res.* **71**, 2207.
Nakada, M. P., Dungey, J. W., and Hess, W. N.: 1965, *J. Geophys. Res.* **70**, 3529.
Paulikas, G. A. and Blake, J. B.: 1968, International Symposium on the Physics of the Magnetosphere, Washington, D.C.
Randall, B. A.: 1969, Master's Thesis, University of Iowa, unpublished.
Theodoridis, G. C., Paolini, F. R., and Frankenthal, S.: 1969, *J. Geophys. Res.* **74**, 1238.
Tverskoy, B. A.: 1965, *Geomagnetizm i Aeronomiya* **5**, 617.
Tverskoy, B. A.: 1968, *Soviet Phys. JETP* **26**, 821.
Tverskoy, B. A.: 1969, private communication.

MEASUREMENTS OF TRAPPED α-PARTICLES: $2 \leqslant L \leqslant 4.5$

J. B. BLAKE and G. A. PAULIKAS

Aerospace Corporation, Los Angeles, Calif., U.S.A.

We have continued a program of measuring the fluxes and energy spectra of trapped α-particles because such measurements should give valuable clues as to the source and loss mechanisms of the much more abundant protons. A four element counter telescope was flown aboard satellite 1968-26B. This instrument was sensitive to both protons and α-particles with energies from 0.5 to 15 MeV/nucleon and resolved the spectrum into seven channels per particle species. The data presented here were taken from April 4 to 7, 1968.

Figure 1 presents proton and α-fluxes measured around the intensity peak ($L=3$) of the lowest energy α-channel ($E_\alpha=2.5$ MeV). The agreement with the earlier

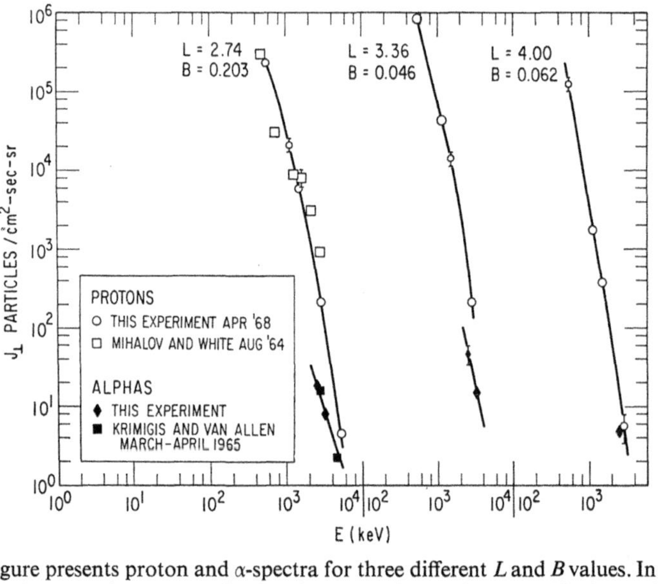

Fig. 1. This figure presents proton and α-spectra for three different L and B values. In addition some data from two other experiments are displayed: proton fluxes given by Mihalov and White (1966) and α-fluxes given by Krimigis and Van Allen (1967). The latter measurements were plotted by integrating their given channel widths over the experimentally determined spectral shape to find the mean energy. Representative error bars are shown.

measurements shown in Figure 1 is quite good. However, temporal changes are known to occur in these particles (Krimigis and Van Allen, 1967; Krimigis, 1968); thus the agreement must be partially fortuitous although it does indicate a relative stability over a substantial fraction of a solar cycle.

B. M. McCormac (ed.), Particles and Fields in the Magnetosphere.

Two features to be noted are:

(a) the proton spectra soften appreciably between 1.5 and 3 MeV; and

(b) the α-spectra are harder for the same total energy.

The data in Figure 2 show:

(a) In terms of energy/nucleon the maximum α/p ratio is $(2 \pm 0.2) \times 10^{-4}$. Over the range $2.25 \leqslant L \leqslant 4.25$ the ratio is constant to within a factor of approximately 4.

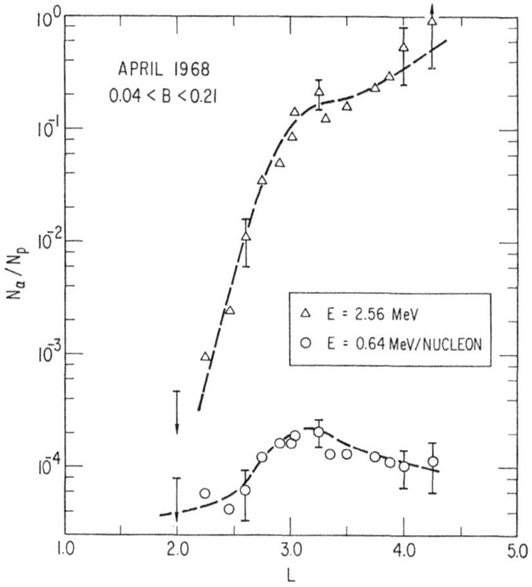

Fig. 2. This figure presents the relative abundance of α-particles to protons for $E = 0.64$ MeV/nucleon and for $E = 2.56$ MeV as a function of L.

(b) In terms of total energy the ratio is relatively constant for $L \gtrsim 3$. The ratio increases gradually with L and around $L \simeq 4.25$ the ratio reaches unity; at $L \simeq 4.5$ proton and α-intensities reach background. For $L < 3$ the ratio decreases precipitously as the α-intensities decline rapidly without a corresponding decrease in protons.

The intensity at constant B vs L profiles in Figure 3 for 2.5 MeV α, 550 keV protons, and 2.9 MeV protons displays the causes of the L dependence of the abundance ratios given in Figure 2. (The abundance ratios plotted in Figure 2 used the proton spectra; they are not simply the ratio of channels shown in Figure 3.)

If the peak intensity of 2.5 MeV α's (at $L = 3.0$) is compared to the peak intensity of protons of 2.5 MeV ($L \simeq 2.3$) one obtains

$$\frac{N_\alpha(E)}{N_p(E)} = (1.9 \pm 0.2) \times 10^{-2}.$$

This ratio is the same as the results of Fritz and Krimigis (1969) for $E_{p,\alpha} > 1.06$ MeV.

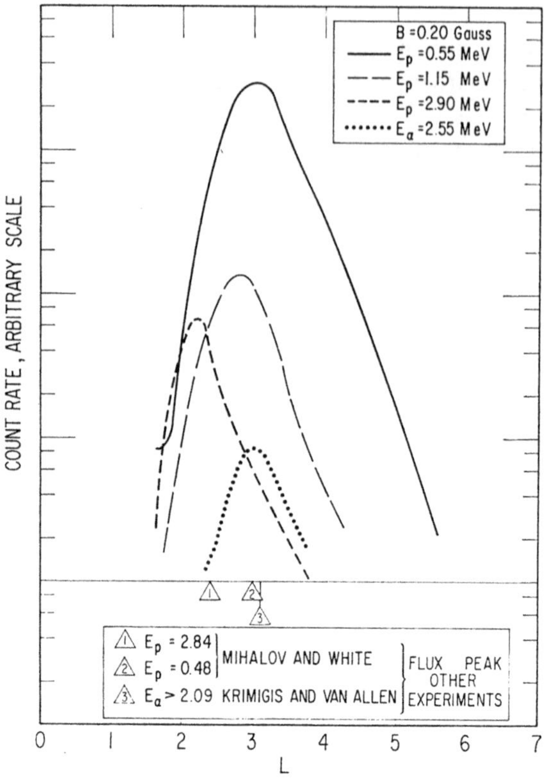

Fig. 3. This figure presents the particle intensity vs. L profiles at $B \simeq 0.20$ G for 0.55 MeV protons 1.15 MeV protons, 2.9 MeV protons, and 2.5 MeV α-particles. Also shown is the location of peak intensities of protons and α-particles from the work of Mihalov and White (1966) and Krimigis and Van Allen (1967).

The predictions of Tverskoy (1965), which include only Coulomb loss, are compared in the following table with the results of this experiment.

Due to the limited α spectral data available no more can be said about the particle distribution function predictions except that the same velocity prediction is closer to the experimental results.

This comparison does not provide a convincing answer as to whether particles are injected into the magnetosphere with the same velocity, the same energy, or neither. In terms of the same total energy at injection, it is not clear why the experimental abundance ratio at 2.3 MeV is essentially the solar wind value if indeed enhanced Coulomb losses of α-particles relative to protons are responsible for the experimental result that $1.3\, L_p(E) \simeq L_\alpha(E)$ (Tverskoy's analysis). It is known, however, that diffusion processes conserving the first and second invariants become much less effective inside an $L \simeq 3$ (Mihalov and White, 1966; Hess, 1968). Possibly added losses as the protons move inside $L = 3$ compensate for the larger Coulomb loss rate for α's.

The work of Tverskoy does not include charge exchange losses; the relevant cross

TABLE I

Relative energies at magnetospheric injection	N_α/N_P		Relative location intensity peak		Particle distribution function	
	Predicted	Experimental	Predicted	Experimental	Predicted	Experimental
$E_p = E_{\alpha/4}$ Same velocity	5×10^{-2}	2×10^{-4}	$L_P(E/4) = L_\alpha(E)$	Agrees	$S_\alpha(E, L) = S_P(E/4, L)$	Better agreement
$E_p = E_\alpha$ Same total energy	10^{-3}	2×10^{-2}	$1.3\ L_P(E) = L_\alpha(E)$	Agrees	$S_\alpha(E, L) = S_P(E, L)$	Worse agreement

sections are such that charge exchange should affect the relative abundances during the period when the particles have energies of a few tens of keV or less. This could explain why, if injection is at equal velocities, the relative abundance is so low. Examination of pitch angle scattering also is needed.

We conclude that the experimental data indicate the need for more realistic theoretical studies.

References

Fritz, T. A. and Krimigis, S. M.: 1969, *J. Geophys. Res.* **74**, 5132.
Hess, W. N.: 1968, in *The Radiation Belt and Magnetosphere*, Blaisdell Publishing Company, Mass.
Krimigis, S. M. and Van Allen, J. A.: 1967, *J. Geophys. Res.* **72**, 5779.
Krimigis, S. M.: 1968, in *Earth's Particles and Fields* (ed. by B. M. McCormac), Reinhold Publishing Corporation, New York, p. 89.
Mihalov, J. D. and White, R. S.: 1966, *J. Geophys. Res.* **71**, 2207.
Tverskoy, B. A.: 1965, *Geomagnetizm i Aeronomiya* **5**, 617.

PART VII

ACCELERATION AND MOTION OF PARTICLES

INTRODUCTORY SURVEY OF RADIATION BELT DIFFUSION

CARL-GUNNE FÄLTHAMMAR

Division of Plasma Physics, Royal Institute of Technology, Stockholm, Sweden

1. Observational Evidence of Diffusion

At the previous Advanced Study Institute held at Freising in 1967, a summary of observational evidence up to that time was given by Roederer (1968a). Since then further observations have been reported, which confirm the conclusion that radial diffusion does play a role in radiation belt dynamics.

The most direct indications of diffusion come from observations of time variations of flux profiles. There are cases where readily identifiable features show an inward motion and/or radial spreading. A typical example of this has recently been given by Vernov *et al.* (1969) and is reproduced in Figure 1. From observations of this kind

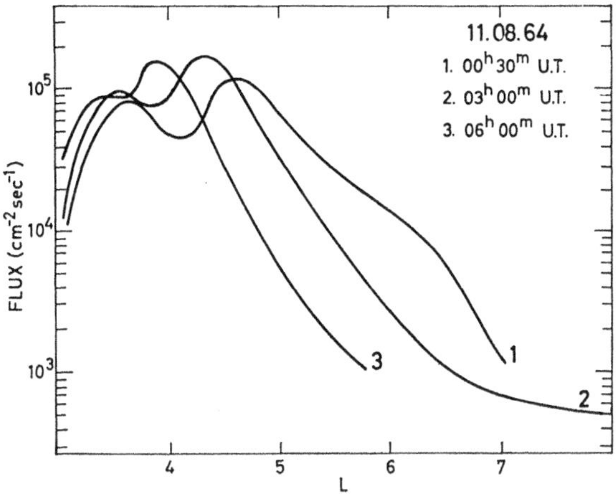

Fig. 1. Inward motion of a peak in the flux distribution of electrons above 400 keV according to data from the satellite Electron 3 (Vernov *et al.*, 1969).

several authors (e.g., Frank, 1965; Craven, 1966; McIlwain, 1966; Vernov *et al.*, 1968; Williams *et al.*, 1968) have deduced a 'diffusion velocity'. A compilation of results from such analyses is shown in Figure 2. As the flux distribution is deformed during its motion the velocity is not well defined and different parts of it (maxima, inflection points, etc.) appear to move at different velocities.

A more adequate approach is to analyze these time variations by fitting solutions of an appropriate diffusion equation to the observational data and so obtain a value for the diffusion coefficient. This has been done by Newkirk and Walt (1968b) for a couple of cases. From the spreading of the electron ($E_e > 1.9$ MeV) shell from the

November 1, 1966, Russian nuclear explosion, they found a diffusion coefficient of $D_2 = 1.3 \times 10^{-5}\ R_E^2$/day at $L = 1.765$ in good agreement with an early estimate by Brown (1966). (D_2 is the second Fokker-Planck coefficient and equal to twice the ordinary diffusion coefficient.) This is shown in Figure 3, where for comparison are

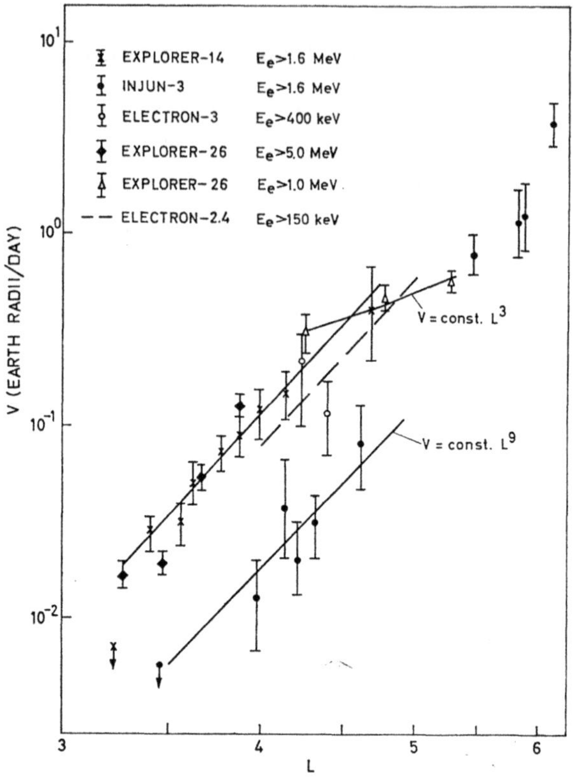

Fig. 2. Compilation of observed velocities of inward motions reported by various authors.

also given the corresponding values from the three U.S. Argus experiments. The D_2 values were derived by Walt (1970) on the basis of a report by Manson *et al.* (1969). Newkirk and Walt (1968b) also analyzed the observations (Frank *et al.*, 1964) that had been discussed previously by Frank (1965) in terms of diffusion velocity (Figure 2). The best fit to the observations was found for $D_2 = 10^{-8} L^{10}\ R_E^2$/day for electrons above 1.6 MeV in the radial interval $2.6 < L < 5$. This is shown in Figure 3.

Another way in which the presence of radial diffusion has been inferred is from analysis of observed equilibrium states. Such considerations were first applied to the outer zone soft protons, whose stable fluxes were suggestive of diffusive equilibrium. More recently, equilibrium analyses have been found fruitful to use also for electrons. In the inner belt, where the atmospheric losses are well known (Walt, 1964; Walt and Newkirk, 1966), it was found by Imhof *et al.* (1967) that the electrons produced by the U.S. nuclear detonation on July 2, 1963 ('Starfish'), had a long term decay rate twenty times smaller than would be expected from atmospheric losses. Interpreting the slow apparent decay as being due to replenishment by radial diffusion from above, Newkirk and Walt (1968a) derived diffusion coefficient values in the range $1.15 < L < 1.21$ for electrons above 1.6 MeV (cf., also Walt and Newkirk, 1968; Farley, 1969a). Their result is shown in Figure 3.

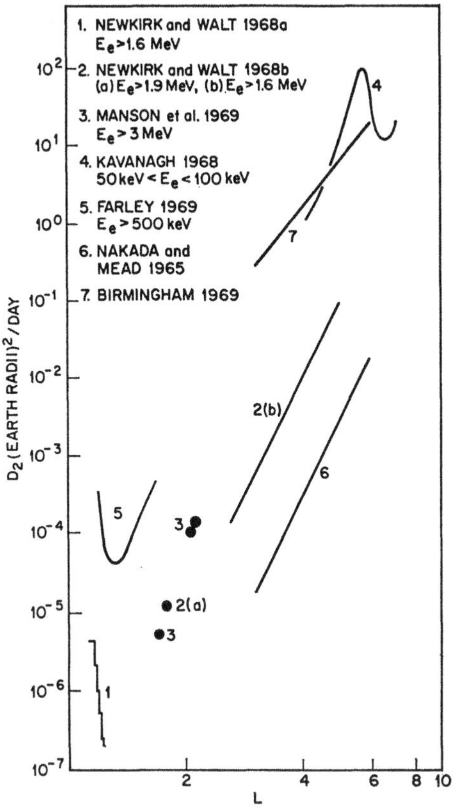

Fig. 3. Summary of empirically derived and theoretically predicted diffusion coefficients.

At somewhat higher L values, around $L = 1.4$, the Starfish electrons were found to decay *faster* than predicted from atmospheric loss only. Recently, Farley (1969b) interpreted this observation in terms of radial diffusion acting as a *loss* mechanism in this region by allowing particles to move away to both higher and lower L shells. On the assumption that this was the only additional loss process, the diffusion coefficient was calculated and found to be as shown in Figure 3. The result is of course an upper limit because if loss due to pitch angle scattering was present simultaneously, which may well be the case, the amount of radial diffusion needed to explain the observations would be correspondingly smaller.

Computing diffusion coefficients for outer belt electrons from equilibrium considerations meets with greater difficulties. For one thing the outerbelt electron fluxes are highly variable even on short time scales, so that it is a debatable question what distribution to use for the equilibrium distribution. Furthermore, the losses, which in this region are believed to be due mainly to pitch angle scattering, are not well known and in an analysis of this kind (using the equilibrium assumption) the result depends critically on the loss mechanism. In both these respects the method of calculating the diffusion coefficient from the equilibrium distribution is at a disadvantage compared with the analysis of time dependent features in the distribution. The uncertainty of evaluating diffusion coefficients from equilibrium states of the radiation belt population has been emphasized by Haerendel (1968), who concluded that for protons in the range $1.5 < L < 4$ the results were seriously distorted due to unknown loss processes. However, an attempt has been made by Kavanagh (1968) who for the equilibrium distribution of electrons in the energy range 50 to 100 keV used an average over quiet day distributions and for the loss rate an estimate for 40 keV electrons given by Kennel and Petschek (1966). As seen in Figure 3 the resulting diffusion coefficient is large and varies strongly with L. However, the result must be seen against the above mentioned uncertainties involved in its derivation. Detailed discussion of some of the diffusion coefficients compiled in Figure 3 will be given by Walt (1970).

Whereas many determinations of radial diffusion coefficients have been made, as seen from Figure 3, determinations of the diffusion coefficient along magnetic field lines are more scarce. However, Imhof (1968) has made one on the basis of the longitude variation of fluxes of 0.4 MeV electrons with low mirror points. When drifting past the Brazilian magnetic anomaly such particles are lost in the atmosphere, but during the continued eastward drift of the remaining particles (with higher mirror points) the orbits with low mirror points are repopulated by pitch angle scattering (mirror point diffusion). From the flux distribution at different longitudes east of the anomaly one can calculate the rate at which this repopulation occurs, and assuming the diffusion to obey a Fokker-Planck equation find the diffusion coefficient. Imhof's result is shown in Figure 4.

Some of the recently observed cases of apparent radial diffusion are remarkable in that they show electrons making their first appearance well inside the trapping region and performing a diffusion from there on inwards or both outwards and inwards. For example, Williams *et al.* (1968) have reported several cases where in connection with magnetic storms electrons above 300 keV initially appear as low as at $L = 3$ or 4 and subsequently diffuse to higher and lower L values.

An example of particles starting their diffusion even further in is shown in Figure 5. It was reported by Pfitzer (1968) and Pfitzer and Winckler (1968). It shows the appearance at $L \approx 2.2$ and subsequent inward spreading of electrons. Four energy ranges from 50 to 1700 keV are shown.

Observations of energetic electrons appearing well inside the trapping boundary also have been reported by Rao (1969) and by Rothwell *et al.* (1969). In the latter

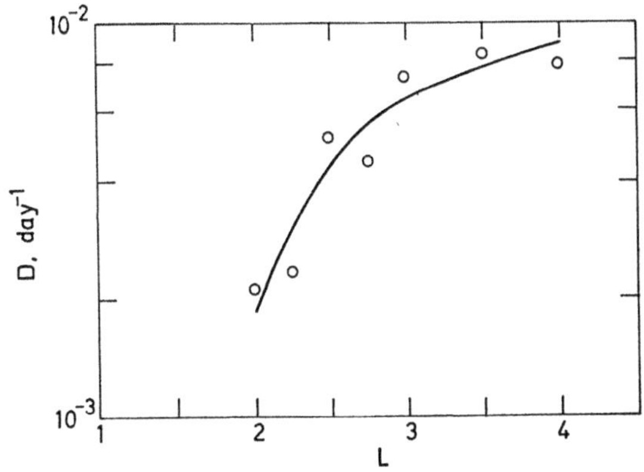

Fig. 4. Diffusion coefficient for pitch angle diffusion of 0.4 MeV electrons according to Imhof (1968).

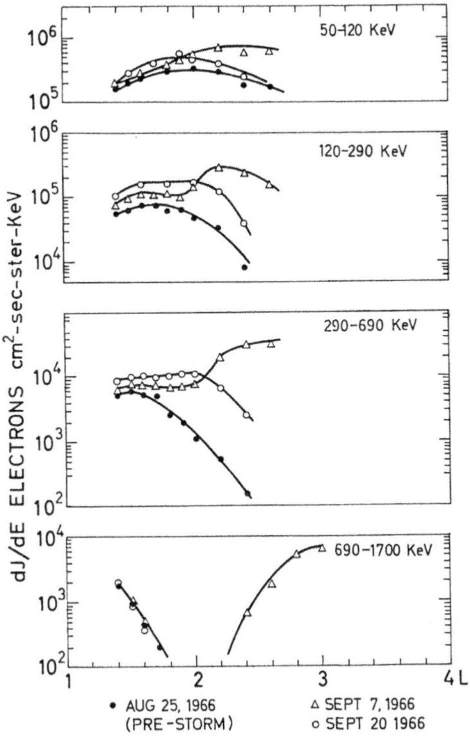

Fig. 5. Time variation in the radial distribution of electrons in four energy ranges between 50 and 1700 keV. The enhancement of fluxes followed the September 2, 1966, solar flare event (Pfitzer and Winckler, 1968).

case, which refers to the June 9, 1968, magnetic storm the electrons were reported to show a progressive spectral hardening at $L=4$ which is suggestive of local acceleration.

The complicated and varied behavior of the outer zone electrons may be further exemplified by the observations of Lanzerotti (1968b), which have been conjectured to be manifestations of internal acceleration mechanisms in the magnetosphere.

On the other hand, the protons observed by Rothwell *et al.* (1969) showed little change during the magnetic storm and exhibited an e-fold energy varying with L in the range $2<L<5$ in a manner consistent with radial diffusion conserving the first two adiabatic invariants. Such radial variation of e-fold energy has been reported by several authors and is another indication of radial diffusion. However, it must be remembered that this is only an indication and no proof because alternative explanations may be possible.

Among other observations of interest is the rapid access of solar protons to the interior of the magnetosphere observed by Lanzerotti (1968a) and interpreted as being due to rapid diffusion across magnetic field lines.

Still another observation relating to the diffusion concept is that of McDiarmid and Burrows (1967) who observe a solar cycle variation of the position of energetic (above 250 keV) electrons in qualitative agreement with diffusion from an external source. However, they also notice that the higher energy part of this population (1 to 3 MeV) is present at so large L values that they could not have gained their energy by inward motion conserving the first two adiabatic invariants.

2. Theoretical Considerations

At the Advanced Study Institute two years ago a fairly complete understanding had been reached of what radial diffusion will result from any *given* electric and magnetic fields varying on time scales much larger than the bounce frequency (Fälthammar, 1968). The diffusion coefficient for particles of any given category depends on the spectral power densities of the spatial Fourier components of the fields at the drift frequency and its harmonics for these particles. Furthermore, the diffusion coefficient can be expressed in a simple explicit way in terms of these spectral densities.

As disturbances on the time scales mentioned will conserve the first two adiabatic invariants, a well known relation (Nakada *et al.*, 1965) prevails between energy and position. For exponential distributions of nonrelativistic particles this means an e-fold energy that increases when the L value decreases (cf., above).

If, in addition to the slow variations of the fields, there are higher frequency fluctuations present, causing violation of the second and/or the first adiabatic invariants, the position energy relation of Nakada *et al.* (1965) breaks down, but the contribution of the slow fields to the radial diffusion is still well represented by the formulas referred to above. (The position energy relation of course may also be changed by other causes such as energy dependent loss.)

Violation of the second invariant by bounce resonance can contribute to radial diffusion or to pitch angle diffusion or to both depending on the details of the interaction. The contribution to radial diffusion recently has been studied by Southwood *et al.* (1969). An extensive analysis of bounce resonance has been published by Roberts and Schultz (1968) and is also covered in the recent review by Roberts (1969). Finally, gyroresonance leading to violation of the first adiabatic invariant will primarily be a source of pitchangle diffusion (see Roberts, 1969). It does not, in itself, cause any appreciable motion across field lines, but combined with azimuthal drift in a distorted magnetic field it will cause a certain amount of radial diffusion of the kind proposed by Roederer (1968b) (see below).

An outstanding difficulty in applying the diffusion theory to the magnetosphere is the lack of knowledge of the electric and magnetic perturbation fields that are responsible for the diffusion. In the case of the large scale fields that are of interest in connection with the radial diffusion by drift resonance it is largely a lack of observational information. In the case of the small scale fields the difficulty is more serious because the fields are produced largely by the particles themselves. What is required to solve this problem is a treatment of the *mutual* interaction between particles and fields, carried all the way into the nonlinear regime where instability growth ceases and equilibrium is reached. This is a formidable task. Important initial progress was achieved by Cornwall (1966), and Kennel and Petschek (1966) and recent advances will be reported later by Haerendel (1970).

Most of what has been published in radiation belt theory since the Advanced Study Institute at Freising has been concerned with radial diffusion. An exception to this trend is an extensive study by Gendrin (1968) of pitch angle diffusion by gyroresonant interaction with Alfvén waves. This work is intimately related with the work to be presented in Haerendel's (1970) presentation and is assumed to be covered there.

A basic question concerning the spatial character of electric induction fields due to magnetic disturbances causing radial diffusion which had been raised by Kavanagh (1967) was recently settled by Birmingham and Jones (1968).

The general expressions for diffusion coefficients given at the Freising Institute (Fälthammar, 1968) applied to diffusion in a pure dipole field, but the static deformation of the real geomagnetic field can readily be accounted for by simply adjusting the magnetic field strength values in the final formulas. At $L=6$ the correction is of the order of 20%. Schultz and Eviatar (1969) have generalized the calculations so as to take full account of the higher harmonics of the magnetic field which were found to contribute about 5% at $L=7$.

It has been known for some time that time dependent electric potential fields are likely to be more important for radial diffusion than magnetic fluctuations (and their associated electric induction fields). This view has been further advocated by Birmingham (1969), who, applying a previously developed formalism (Birmingham *et al.*, 1967) and assuming an rms electric field strength of 0.2 mV/m in the outer magnetosphere found a diffusion coefficient given by $D_L = (\frac{1}{2}) D_2 = 1.3 \times 10^{-4} L^6 \text{ day}^{-1}$. As seen in Figure 3, this exceeds by several powers of 10 the value obtained by Nakada and Mead (1965) from magnetic impulses. However, as there still exists very little observational knowledge of the electric fields and in particular of their time and space variation, the estimate must be taken with caution.

Tverskoy (1969) claimed to have found a much larger diffusion coefficient than that of Nakada and Mead in spite of the fact that he used the same physical processes, namely magnetic impulses, and even the same data as these authors. A somewhat different result is expected because Tverskoy allowed for additional field asymmetry due to tail currents, but it is unclear how this could produce the large increase of the diffusion coefficient (by about a factor of 50) that is reported.

Another result published by Tverskoy (1969) has to do with the mirror latitude dependence of the diffusion coefficient caused by magnetic impulses. The variation he estimates is much smaller than previously reported in the literature, and he ascribes the difference to the fixed phase relation between the Fourier components of his model disturbance. However, more detailed calculation shows that for Tverskoy's model, too, the same result is found as previously reported (Conrath, 1967; Fälthammar, 1968), i.e., the diffusion coefficient due to magnetic disturbances is reduced by about a power of 10 for particles with high mirror latitudes. On the other hand, diffusion coefficient due to time dependent electric potential fields is independent of mirror latitude (Fälthammar, 1968).

The so-called shell splitting, whose role for radial diffusion has been emphasized by Roederer (1968b) has recently been verified observationally by Pfitzer *et al.* (1969). Quantitative studies of the diffusion have been made by Fälthammar and Walt (1969) and by Roederer and Schultz (1969) and will be discussed later in this session Walt, 1970). The possible role of this process for populating the high energy tail of the trapped particle distribution has been pointed out by Roederer (1968b) and analyzed in detail by Theodoridis *et al.* (1969).

As the diffusion in L, pitch angle, and energy are mutually coupled a simultaneous account of all of these factors would be desirable, but such a synoptic treatment is still far out of reach. However, Cornwall (1968) has shown that a nonlinear formalism where pitch angle diffusion is treated as a loss term has considerable merit. Walt (1970) has developed a method of eigenfunction expansion to analyze simultaneous diffusion in L and in pitch angle.

References

Birmingham, T. J.: 1969, *J. Geophys. Res.* **74**, 2169.
Birmingham, T. J. and Jones, F. C.: 1968, *J. Geophys. Res.* **73**, 5505.
Birmingham, T. J., Northrop, T. G., and Fälthammar, C.-G.: 1967, *Phys. Fluids* **10**, 2389.
Brown, W. L.: 1966, in *Radiation Trapped in the Earth's Magnetic Field* (ed. by B. M. McCormac) D. Reidel Publishing Company, Dordrecht-Holland, p. 757.
Conrath, B. J.: 1967, *J. Geophys. Res.* **72**, 6069.
Cornwall, J. M.: 1966, *J. Geophys. Res.* **71**, 2185.
Cornwall, J. M.: 1968, *Radio Sci.* **3**, 740.
Craven, J. D.: 1966, *J. Geophys. Res.* **71**, 5643.
Fälthammar, C.-G.: 1968, in *Earth's Particles and Fields* (ed. by B. M. McCormac), Reinhold Book Corporation, New York, p. 157.
Fälthammar, C.-G. and Walt, M.: 1969, *J. Geophys. Res.* **74**, 4184.
Farley, T. A.: 1969a, *J. Geophys. Res.* **74**, 377.

Farley, T. A.: 1969b, *J. Geophys. Res.* **74**, 3591.
Frank, L. A.: 1965, *J. Geophys. Res.* **70**, 3533.
Frank, L. A., Van Allen, J. A., and Hills, H. K.: 1964, *J. Geophys. Res.* **69**, 2171.
Gendrin, R.: 1968, *J. Atmos. Terr. Phys.* **30**, 1313.
Haerendel, G.: 1968, in *Earth's Particles and Fields* (ed. by B. M. McCormac), Reinhold Book Corporation, New York, p. 171.
Haerendel, G.: 1970, this volume, p. 416.
Imhof, W. L.: 1968, *J. Geophys. Res.* **73**, 4167.
Imhof, W. L., Reagan, J. B., and Smith, R. V.: 1967, *J. Geophys. Res.* **72**, 2371.
Kavanagh, L. D., Jr.: 1967, *J. Geophys. Res.* **72**, 6120.
Kavanagh, L. D., Jr.: 1968, *J. Geophys. Res.* **73**, 2959.
Kennel, C. F. and Petschek, M. E.: 1966, *J. Geophys. Res.* **71**, 1.
Lanzerotti, L. J.: 1968a, *Phys. Rev. Letters* **21**, 929.
Lanzerotti, L. J.: 1968b, *Ann. Geophys.* **24**, 847.
Manson, D. J., Fennell, J. F., George, J. A., Hickeson, J. L., Maldonado, G. V., and Weber, A. H.: 1969, DASA Rept. No. 2309.
McDiarmid, I. B. and Burrows, J. R.: 1967, *Can. J. Phys.* **45**, 2873.
McIlwain, C. E.: 1966, *Proc. IUCSTP Symposium on Solar Terrestrial Relations, Belgrade 1966* (also University of California, San Diego, Report No. SP-66-5).
Nakada, M. P. and Mead, G. D.: 1965, *J. Geophys. Res.* **70**, 4777.
Nakada, M. P., Dungey, J. W., and Hess, W. N.: 1965, *J. Geophys. Res.* **70**, 3529.
Newkirk, L. L. and Walt, M.: 1968a, *J. Geophys. Res.* **73**, 1013.
Newkirk, L. L. and Walt, M.: 1968b, *J. Geophys. Res.* **73**, 7231.
Pfitzer, K. A.: 1968, Ph.D. Thesis, University of Minnesota Cosmic Ray Group Technical Report CR-123.
Pfitzer, K. A., Lezniak, J. W., and Winckler, J. R.: 1969, University of Minnesota Cosmic Ray Group Technical Report CR-133.
Pfitzer, K. A. and Winckler, J. R.: 1968, *J. Geophys. Res.* **73**, 5792.
Rao, C. S. R.: 1969, *J. Geophys. Res.* **74**, 794.
Roberts, C. S.: 1969, *Rev. Geophys.* **7**, 305.
Roberts, C. S. and Schultz, M.: 1968, *J. Geophys. Res.* **73**, 7361.
Roederer, J. G.: 1968a, in *Earth's Particles and Fields* (ed. by B. M. McCormac), Reinhold Book Corporation, New York, p. 143.
Roederer, J. G.: 1968b, in *Earth's Particles and Fields* (ed. by B. M. McCormac), Reinhold Book Corporation, New York, p. 193.
Roederer, J. G. and Schultz, M.: 1969, *J. Geophys. Res.* **74**, 4117.
Rothwell, P. L., Webb, J. H., and Katz, L.: 1969, Trapped and Polar Particles During the June 9, 1968 Magnetic Storm, preprint.
Schultz, M. and Eviatar, A.: 1969, *J. Geophys. Res.* **74**, 2182.
Southwood, D. J., Dungey, J. W., and Etherington, R. J.: 1969, *Planetary Space Sci.* **12**, 349.
Theodoridis, G. C., Paolini, F. R., and Frankenthal, S.: 1969, *J. Geophys. Res.* **74**, 1238.
Tverskoy, B. A.: 1969, *Rev. Geophys.* **7**, 219.
Vernov, S. N., Kuznetzov, S. N., Logachev, Yu. J., Lopatina, G. B., Sosnovets, E. N., and Stolpovsky, V. G.: 1968, *Space Res.* **8**, 120.
Vernov, S. N., Gorchakov, E. V., Kuznetsov, S. N., Logachev, Yu. J., Sosnovets, E. N., and Stolpovsky, V. G.: 1969, *Rev. Geophys.* **7**, 257.
Walt, M.: 1964, *J. Geophys. Res.* **69**, 3947.
Walt, M.: 1970, this volume, p. 410.
Walt, M. and Newkirk, L. L.: 1966, *J. Geophys. Res.* **71**, 3265.
Walt, M. and Newkirk, L. L.: 1968, in *Earth's Particles and Fields* (ed. by B. M. McCormac), Reinhold Book Corporation, New York, p. 209.
Williams, D. J., Arens, J. F., and Lanzerotti, J. J.: 1968, *J. Geophys. Res.* **73**, 5673.

TRAPPED PROTONS $\geqslant$100 keV AND POSSIBLE SOURCES

D. J. WILLIAMS

NASA Goddard Space Flight Center, Greenbelt, Md., U.S.A.

1. Introduction

In this brief review we are able only to consider the general characteristics of the trapped $\geqslant$100 keV proton population, the principal time variations observed, and the status of major source and transport mechanisms with primary emphasis on diffusion by violation of the third adiabatic invariant.

2. Survey

The initial indication of the existence of a lower energy proton population in the trapping regions was obtained from a rocket borne emulsion experiment conducted by Naugle and Kniffen (1961). These early results showed a steep rise in the 8 to 20 MeV proton intensities in the high latitude portion ($L \geqslant 1.64$) of the region of observation, $1.5 \leqslant L \leqslant 1.8$ and 1400 to 2000 km. Soon after this observation, the first definitive measurements of trapped protons down to energies of $\sim$1 MeV were reported by Bame *et al.* (1962) using a 1960 rocket observation at $L \sim 2.6$.

In 1962 the results of the first survey of low energy protons throughout the outer trapping regions and the first observations of protons down to energies of $\sim$100 keV were reported by Davis and Williamson (1963) using data from the Explorer 12 satellite. Figure 1, from Davis and Williamson, shows three typical spectra obtained at various altitudes in the magnetosphere. They found in general that

(a) protons of energies $\geqslant$100 keV were trapped throughout the region $L=2$ to the magnetopause;

(b) the proton spectrum softened with increasing radial distance (Figure 1) and could be well represented as an exponential;

(c) during a $1\frac{1}{2}$ mo geomagnetically quiet period the proton intensities were stable to $\pm 30\%$; and

(d) a proton intensity enhancement of a factor of $\sim$3 was observed in the $L=3$ to 4.5 region during a magnetic storm.

Extensive additional observations from Explorer 14, 15, and 26 (Davis, 1965; Davis and Williamson, 1966; Soraas and Davis, 1968), from Mariner 4 (Krimigis and Armstrong, 1966), from satellite 1964–45A (Mihalov and White, 1966), from Explorer 33 (Armstrong and Krimigis, 1968), from Injun 4 (Krimigis, 1968), and from OGO 4 (Fritz and Krimigis, 1968) have generally verified the earlier results and have added further significant details to the behavior of low energy ($\gtrsim$100 keV) outer zone protons. Extensions of and additions to the above list of general behavioral characteristics have been

B. M. McCormac (ed.), Particles and Fields in the Magnetosphere.

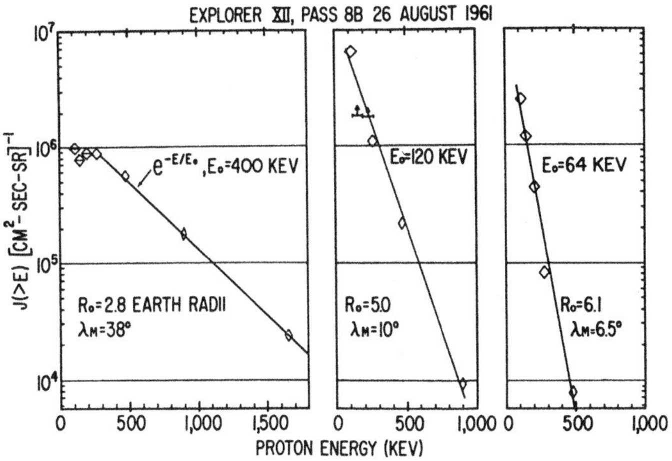

Fig. 1. Integral proton spectra measured on three field lines on August 26, 1961. Note exponential representation and spectral softening with increasing radial distance (Davis and Williamson, 1963).

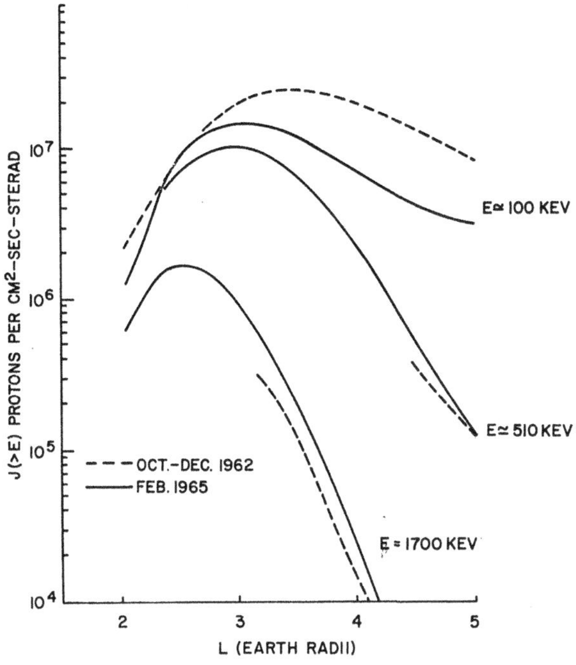

Fig. 2. Proton intensity profiles from 1962 and 1965 for equatorial pitch angles of 60°. Note low energy enhancement and high energy depletion which occurred between the indicated epochs (Davis and Williamson, 1966).

(a) the low energy proton population is quite stable during geomagnetically quiet times;

(b) the characteristic e-fold energy E_0 of the proton spectrum varies, with minor perturbations, as L^{-3} out to $L \sim 5$ during quiet periods;

(c) large magnetic storms produce nonadiabatic changes down to L values of 3 while even small magnetic disturbances produce similar changes at $L \gtrsim 5$; and

(d) the adiabatic redistribution of the low energy proton population due to ring current effects is very important and must be compensated for in order to investigate sources, losses, and invariant violating transport mechanisms.

The much greater stability of the $\geqslant 100$ keV proton population relative to the low energy (10 to 100 keV) trapped electron population has been demonstrated by Davis (1965). However, long term variations do occur in the proton population and are illustrated in Figure 2 (Davis and Williamson, 1966). Note that between the time periods shown proton intensities have been enhanced at low energies and depleted at high energies. This result has been since verified and shown to be an effect which occurs during geomagnetic storms (Soraas and Davis, 1968).

In addition, the occurrence of proton intensity enhancements in coincidence with magnetic substorms has been reported by Davis and Williamson (1966) and Konradi (1967). An event discussed by Davis and Williamson has been studied in detail by Brown *et al.* (1968) using all available electron, proton, and magnetic field data from the Explorer 26 satellite. Although the data did not allow an unambiguous interpretation, they did clearly indicate the importance of the magnetic substorm in particle acceleration processes. Konradi has indicated further that substorm associated protons appeared to have drifted to the satellite location following injection in the nightside hemisphere.

To end this brief survey of outer zone proton characteristics, we show in Figure 3 an R-λ plot of a model $\geqslant 400$ keV trapped proton population as compiled by the National Space Science Data Center (King, 1967). Data for Figure 3 were obtained

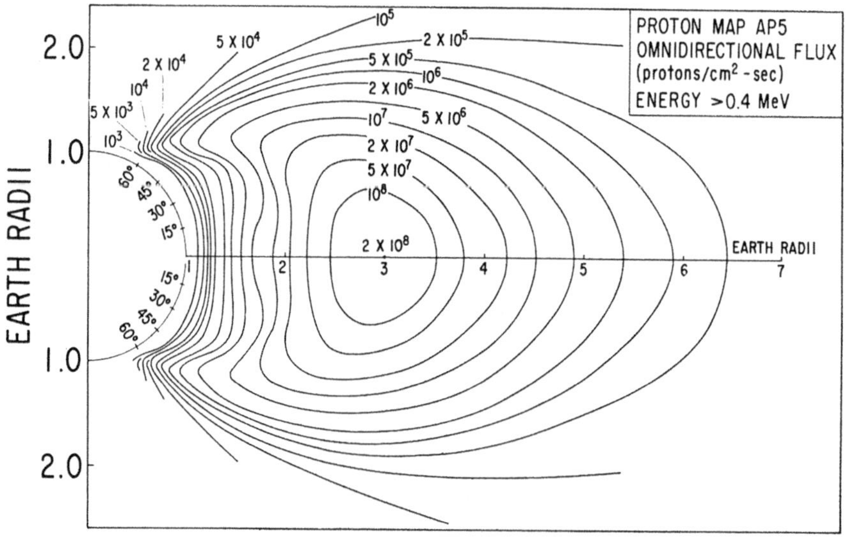

Fig. 3. R-λ flux map of model (AP5) proton environment. Compensation not made for time variations (King, 1967).

from 7 satellites, 6 research groups, and span the period August 1961 to April 1965. Because of time variations Figure 3 is not necessarily representative of the actual situation at any given time. However, intensity levels at a given point in space are generally stable to factors of 2 to 3 and it is to within this accuracy that the model of Figure 2 applies.

3. Time Variations

We mentioned earlier that redistributions of the proton population due to ring current effects are significant and must be accounted for to discuss sources, losses, and other nonadiabatic processes. Such changes, resulting from a particle's response to slow variations in the earth's magnetic field while conserving the three adiabatic invariants, have been reported for energetic protons (40 to 110 MeV) at $L \leqslant 2.4$ by McIlwain (1966).

A recent study by Soraas and Davis (1968) clearly shows similar adiabatic (invariant conserving) variations occurring in the low energy proton population. They have used 5 mo of data (January to June 1965) to study the temporal variations of the trapped 100 to 1700 keV proton population. Adiabatic effects are removed by transforming the data from the time dependent field to a reference field where the ring current field is effectively zero. The transformation is obtained for protons mirroring at the equator using a disturbance field

$$\Delta B(t, R) = CD_{ST}(t) f(R)$$

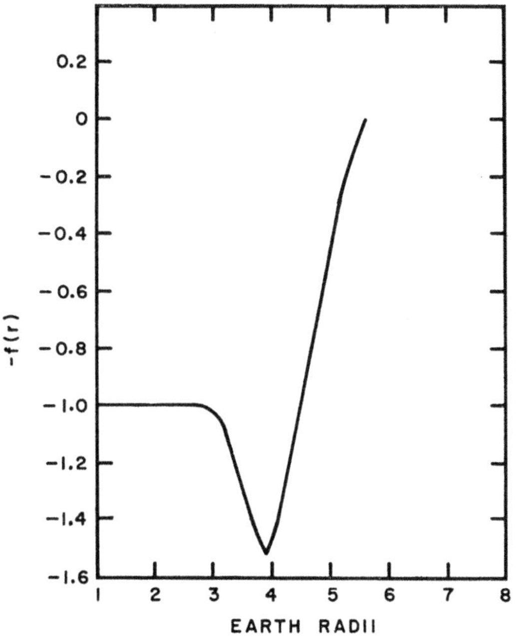

Fig. 4a. Radial dependence of ring current magnetic field used in model calculations of adiabatic effects on trapped protons (Soraas and Davis, 1968).

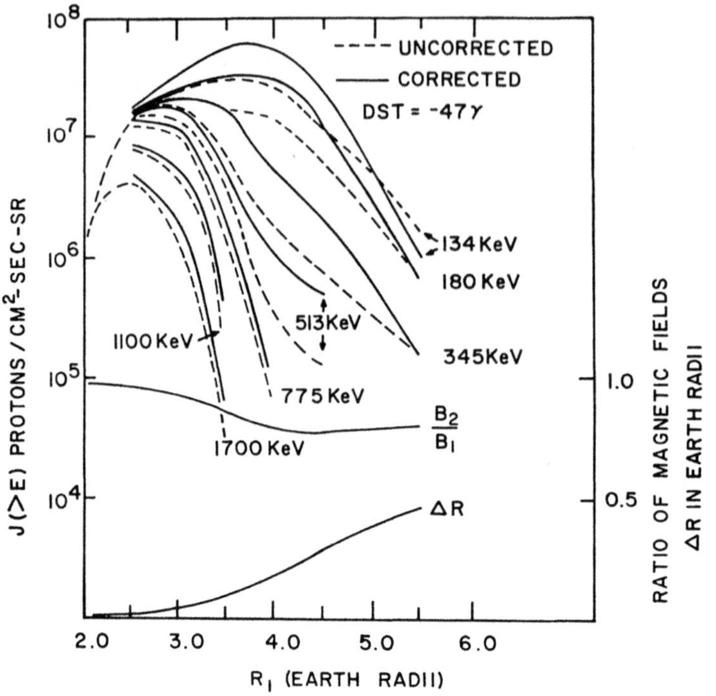

Fig. 4b. Radial proton integral intensity profile as measured on day 109, 1965 when Dst = $-47\ \gamma$, along with transformed profile corresponding to Dst = 0 assuming conservation of the three adiabatic invariants. Radial dependence of the ratio of the magnetic field after and before buildup of ring current and radial movement ΔR of the particles also is shown. Note adiabatic effects can produce both increases and/or decreases at any given position in trapping region (Soraas and Davis, 1968).

where $C = 0.7$ and corrects for the induction field of the earth, t = time of observation and $f(R)$ = radial dependence of a model ring current field. The function $f(R)$ and an example of the effects of the transformation are shown in Figure 4. It should be noted that whether such adiabatic redistributions produce intensity increases or decreases at a given point in space depends strongly on the existing proton energy spectra and spatial gradients as well as the specific field perturbation in effect at the time. Remaining variations in the time histories of the corrected fluxes are now inter-

---

Fig. 5. The time behavior of protons mirroring at the equator for the eight integral energies measured at $L = 3$. The curves are displaced in order to avoid overlap and the values read from the curves A to H must be multplied by 10 raised to the following exponents -1, -0.75, -0.50, -0.25, 0.0, 0.25, 0.25, and 0.25 in order to get the integral proton intensity above a certain energy in protons/cm²-sec-sr. Below the proton data are plotted the hourly average Dst values (Soraas and Davis, 1968). (a) Uncorrected fluxes. The different curves are marked with letters running from A to H corresponding to the energies 98, 134, 180, 345, 513, 775, 1140, and 1700 keV. (b) Corrected fluxes after the adiabatic effects are removed. The different curves are marked with letters running from A to H corresponding to the energies 134, 180, 220, 345, 513, 775, 1140, and 1700 keV. Note absence of variations at approximately day 169 indicating perturbations in Figure 5a due entirely to adiabatic redistribution at this time.

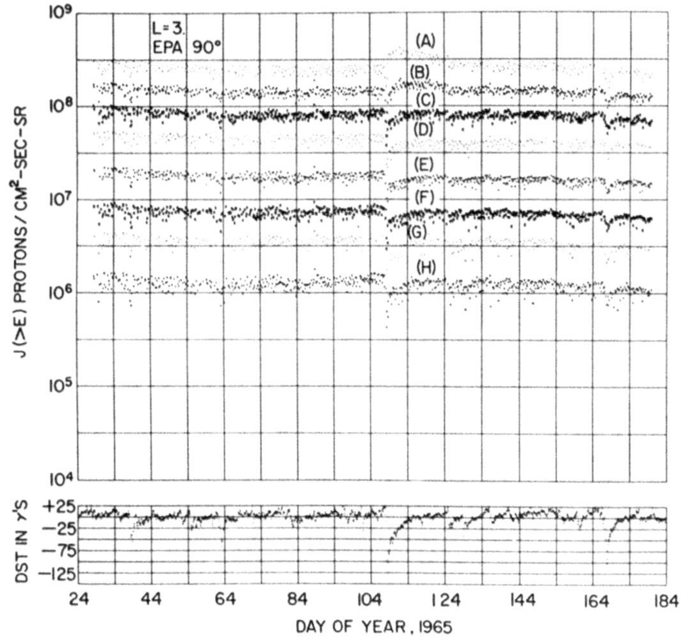

a

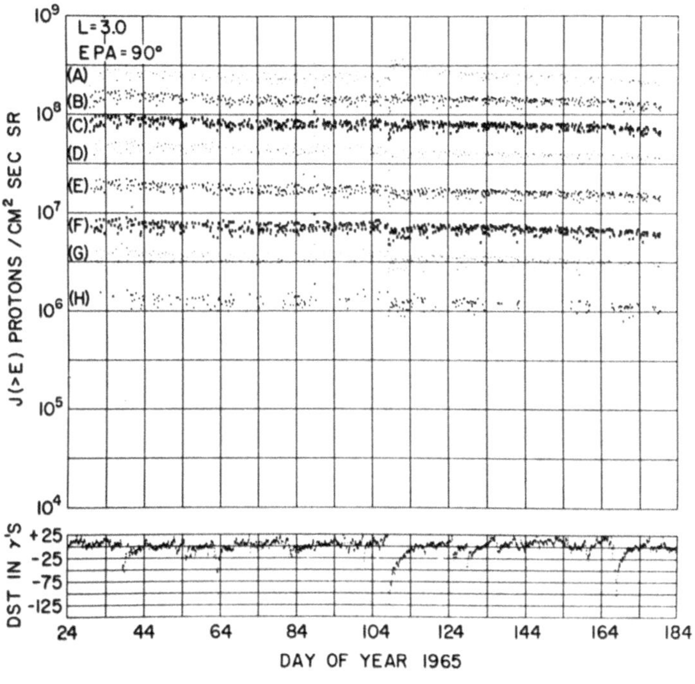

b

Fig. 5a and b.

preted as being indicative of the occurrence of nonadiabatic (invariant violating) processes.

Figures 5 and 6 present the uncorrected and corrected time histories of the equatorially mirroring 100 to 1700 keV trapped proton population at $L=3$ and 4 respectively. The corrected data are seen to be clearly more ordered than the uncorrected data. In fact Soraas and Davis report that the corrected data yield a regression coefficient of intensity on Dst which is a factor of 5 smaller than that obtained with the uncorrected data. This removal of the intensity Dst correlation is to be expected if the transformation used has properly accounted for the adiabatic variations in the data. Note that the intensity variation at $L=3$ on day 168 seems to have been purely an adiabatic redistribution during the small magnetic disturbance shown in the Dst values.

Figures 5 and 6 indicate and Soraas and Davis (1968) show that three basic large scale time variations in 100 to 1700 keV proton intensities were in evidence during the period under study:

(a) Adiabatic redistributions of the proton population which vary directly with the magnetic field variations and during which the three particle adiabatic invariants are conserved;

(b) Rapid nonadiabatic variations occurring in the main phase of geomagnetic storms during which the low energy proton intensities are enhanced and high energy intensities depleted; and

(c) A slow nonadiabatic post-storm variation during which both low and high energy intensities recover toward the pre-storm levels; low energy intensities decay and high energy intensities increase during this period.

The above results apply directly to equatorially mirroring protons. However, Soraas and Davis (1968) report that variations in the proton pitch angle distributions which are both correlated and noncorrelated with magnetic variations are observed. Compensation therefore must be made for adiabatic variations in proton intensities at all pitch angles before quantitative comparisons can be made with invariant violating processes such as cross L diffusion.

A next step in the above approach is the extension of the calculations to all pitch angles in order to obtain the pitch angle dependence of the nonadiabatic processes. In addition, it may be necessary to obtain the function $f(R)$ separately for each storm and to extend its functional form to higher altitudes.

Fig. 6. The time behavior of protons mirroring at the equator for the eight integral energies measured at $L=4$. The curves are displaced in order to avoid overlap and the values read from the curves A to H must be multiplied by 10 raised to the following exponents -1, -0.75, -0.50, -0.25, 0.0, 0.25, 0.25, and 0.25 in order to get the integral proton intensity above a certain energy in protons/cm^2-sec-sr. Below the proton data are plotted the hourly average Dst values (Soraas and Davis, 1968). (a) Uncorrected fluxes. The different curves are marked with letters running from A to H corresponding to the energies 98, 134, 180, 345, 513, 775, 1140, and 1700 keV. (b) Corrected fluxes after the adiabatic effects are removed. The different curves are marked with letters running from A to H corresponding to the energies 134, 180, 220, 345, 513, 775, 1140, and 1700 keV.

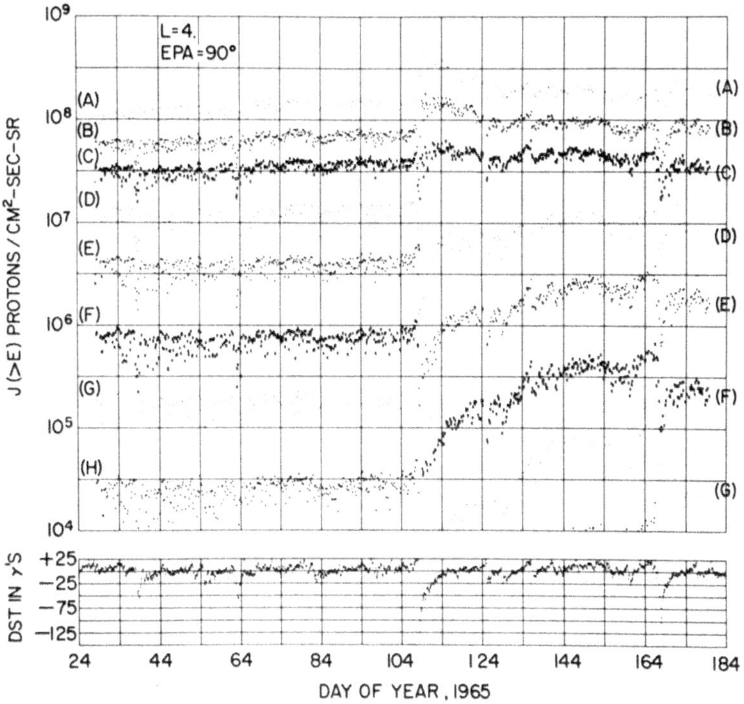

a

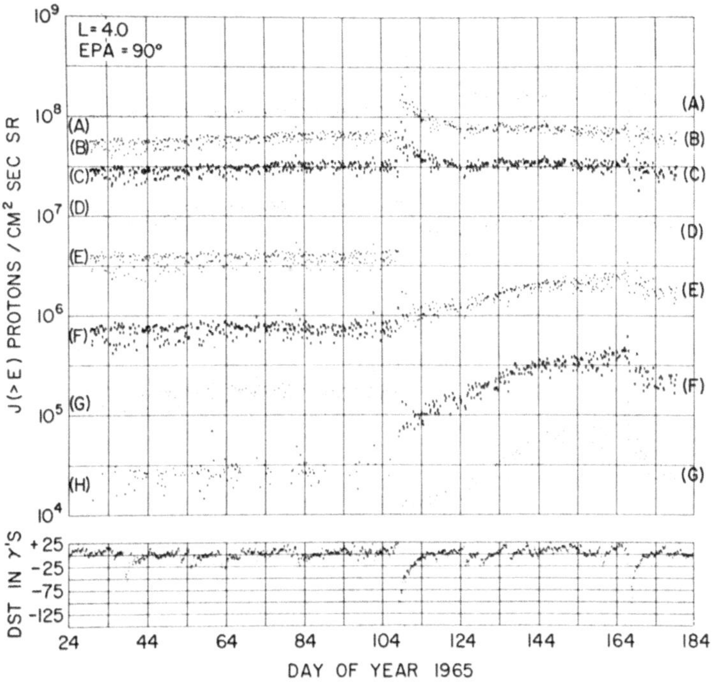

b

Fig. 6a and b.

4. Transport and Sources

A. STEADY STATE

The initial work of Kellogg (1959), Herlofson (1960) and Parker (1960), showed that the transport of charged particles across L shell under violation of the third adiabatic invariant while conserving the first and second invariants was a potentially effective way of populating the trapping regions with energetic particles. The mechanism envisioned as causing this diffusion was the occurrence of magnetic perturbations on time scales causing a violation of the third invariant. With a source of particles at the magnetopause the net result of this process would be a group of particles diffusing inward across field lines and becoming energized due to the conservation of the first and second invariants. Following this initial work several excellent theoretical studies have been reported concerning charged particle diffusion in the magnetosphere (Davis and Chang, 1962; Tverskoy, 1964, 1965; Dungey *et al.*, 1965a, 1965b; Fälthammar, 1965, 1966, 1968; Nakada and Mead, 1965; Haerendel, 1968).

An early comparison of proton observations in the outer zone with diffusion theory by Dungey *et al.* (1965a, 1965b) showed reasonable agreement at several pitch angles for the radial dependence of the proton's characteristic e-fold E_0. The L^{-3} dependence expected for E_0 is simply a consequence of the diffusion of an exponential energy spectrum under conservation of the first and second invariants. As the proton spectrum is reasonably approximated by an exponential the results of Dungey *et al.* provide an argument in favor of diffusion. For other initial spectral forms, the evolution of the proton energy spectrum as the protons diffuse across field lines has to be calculated. Thus observation of characteristic energy dependencies other than L^{-3} does not by itself disprove the existence of a diffusive process.

More quantitative studies have been completed using a Fokker-Planck formalism:

$$\frac{\partial n}{\partial t} = -\frac{\partial}{\partial r} D_1 n + \frac{1}{2}\frac{\partial^2}{\partial r^2} D_2 n - \frac{\partial}{\partial \mu}\left\langle\frac{\Delta\mu}{\Delta t}\right\rangle n - \frac{n}{\tau}.$$

Here n = number of particles in dr, dμ, and dJ, μ = first adiabatic invariant, the magnetic moment, J = second adiabatic invariant, $D_1 = \langle \Delta r/\Delta t\rangle$ = average radial displacement per unit time, $D_2 = \langle (\Delta r)^2/\Delta t\rangle$ = average radial displacement squared per unit time, and τ = e-fold lifetime for charge exchange.

Using the relation between D_1 and D_2 obtained by Fälthammar (1966) for a dipole field

$$D_1 = \frac{r^2}{2}\frac{\partial}{\partial r}\frac{D_2}{r^2} = r^2 \frac{\partial}{\partial r}\frac{D}{r^2}; \quad D \equiv \frac{D_2}{2}.$$

The above may be written in a 'diffusion-like' form

$$\frac{\partial n}{\partial t} = \left\{\frac{\partial}{\partial r}\frac{D}{r^2}\frac{\partial}{\partial r}(r^2 m)\right\} - \frac{\partial}{\partial \mu}\left\langle\frac{\Delta\mu}{\Delta t}\right\rangle n - \frac{n}{\tau}.$$

The last two terms on the right above represent losses due to coulomb collisions and charge exchange, respectively.

This equation has been solved by Davis and Chang (1962), Tverskoy (1964), and Nakada and Mead (1965). The latter authors evaluated the size of the diffusion coefficient D by obtaining the rate and size of sudden impulses occurring in the magnetosphere and by using a distorted magnetosphere without a tail configuration (Mead, 1964). They obtained a value of

$$D = 0.3(10)^{-9} r^{10} R_E^2/\text{day}.$$

Using this value for D, Nakada and Mead (1965) compared the predicted and observed proton steady state ($\partial n/\partial t = 0$) distribution functions. These are shown in Figure 7.

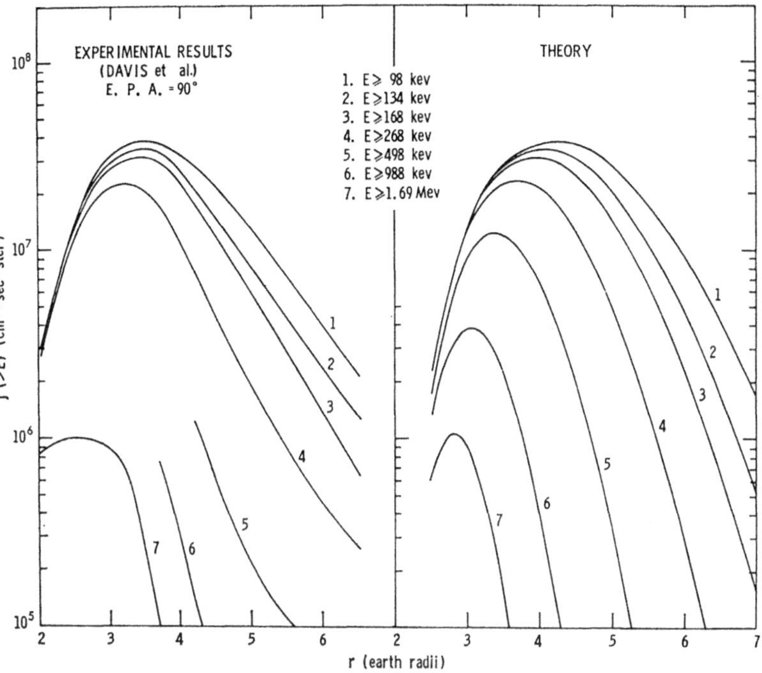

Fig. 7. Comparison of integral fluxes as measured by Davis *et al.* (1964) and calculated integral fluxes. The calculated curves are normalized to the same peak flux for the lowest energy threshold (Nakada and Mead, 1965).

The basic difference between the observed and calculated curves in Figure 7 is that the peaks of the observed distributions are at a lower altitude than the theoretical curves. Nakada and Mead (1965) found that an increase of a factor of 8 in D would bring the observed and calculated peaks into coincidence. They also argued that a larger value of D should be expected since

(1) inclusion of tail field effects would increase D by about a factor of 4, and

(2) perturbations other than sudden impulses probably contribute to the diffusion process. The importance of other variations (including electric fields) may be seen from the fact that D can be expressed in terms of the power spectrum of the disturbance (Fälthammar, 1965, 1968).

In addition, Tverskoy (1965) after an earlier low estimate, has reevaluated D and obtained

$$D = (4\text{–}13)\ 10^{-9} r^{10} R_E^2/\text{day},$$

a value which would produce agreement between the observed and calculated proton intensity distributions.

Therefore, with apparently reasonable values for the diffusion coefficient the steady state equatorially mirroring proton distributions can be described by diffusion theory. Absolute flux comparisons remain to be done, i.e., an observed steady source of low energy protons at high altitudes has still to be used to predict observed fluxes of higher energy protons at low altitudes.

B. NONSTEADY STATE

An important advance in further estimating the importance of diffusion via third invariant violation as a transport process in the outer zone has been reported by Soraas (1969). He has considered the nonsteady state ($\partial n/\partial t \neq 0$) condition by studying the post-storm long term nonadiabatic recovery of proton intensities toward pre-storm values (the third basic time variation discussed in the previous section). Using the same formalism as Nakada and Mead (1965), Soraas (1969) obtains a value for the diffusion coefficient by best fitting the time evolution of the measured distribution function.

The measured proton distribution function representing the initial conditions for the solution of the Fokker-Planck equation is shown in Figure 8 for the post-storm period following the April 18, 1965, storm. The measured and computed time evolution of this distribution is shown in Figure 9. Calculated curves are shown for the best fit value of D

$$D = 2.4(10)^{-9} r^{10} R_E^2/\text{day}$$

as well as for the earlier estimate of Nakada and Mead.

The experimental and theoretical results show good agreement at a value required by Nakada and Mead to fit the equilibrium distribution and in good agreement with the estimate of Tverskoy (1965).

The calculated curves for both values of D in Figure 9 are identical for the three lowest energies shown, indicating that the proton time behavior is dominated by losses at these energies. Also the low energy protons decay faster than the predicted curves. Therefore in the loss dominated region, either the coulomb and charge exchange loss terms are not properly accounted for or additional loss mechanisms are operating. Soraas suggests that pitch angle scattering by ion cyclotron noise (Kennel and Petschek, 1966) may be forcing the enhanced low energy intensities to decay rapidly back to their stable upper limit.

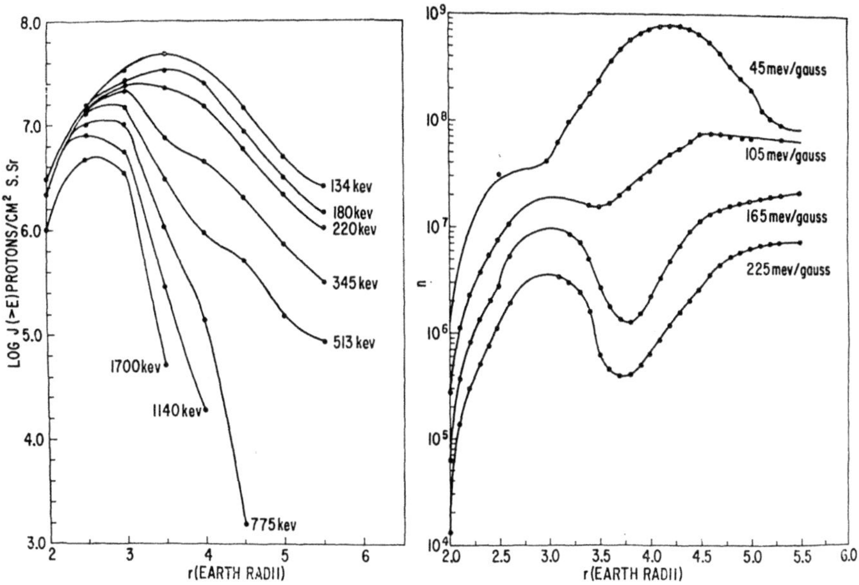

Fig. 8. The initial distribution. The left hand side of the figure shows the integral proton fluxes above various energies vs radial distance as measured after the April 18 storm on day 111 of 1965. The right hand side of the figure shows the distribution function n for different values of the magnetic moment plotted vs radial distance (Soraas, 1969).

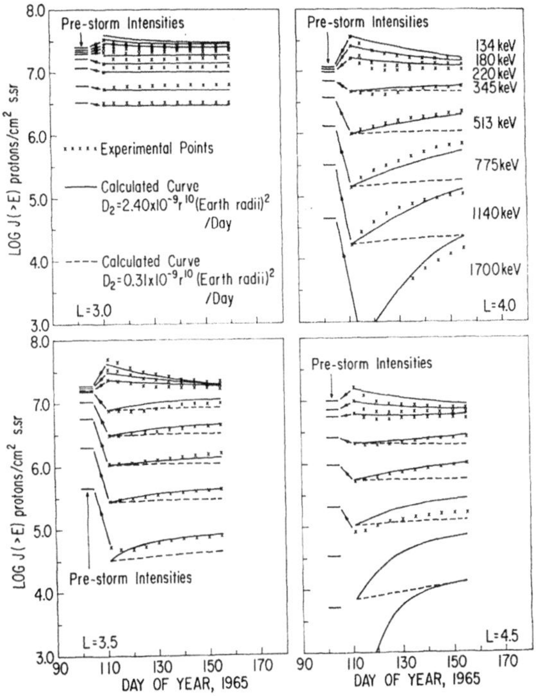

Fig. 9. The time behavior of the integral proton intensities of different L values computed from the transport equation with two values of the diffusion coefficient is compared with the experimentally measured values at $L = 3.0$, 3.5, 4.0, and 4.5 (Soraas, 1969).

The major role played by loss mechanisms shows the importance of including all appropriate loss terms in the Fokker-Planck equation prior to quantitative comparisons with theory, especially if proton and α-particle distributions are to be compared as a test of diffusion processes (Tverskoy, 1965; Fritz and Krimigis, 1968; Krimigis, 1970).

It appears thus that diffusion theory employing violation of the third adiabatic invariant and conservation of the first and second invariants predicts distribution functions which are in good agreement with

(1) the observed steady state ($\partial n/\partial t=0$) proton distribution and

(2) the observed time evolution ($\partial n/\partial t\neq 0$) of the proton distribution during the slow nonadiabatic recovery toward pre-storm intensities following a magnetic storm.

The value of the diffusion coefficient in both instances is $2.4(10)^{-9}\ r^{10}\ R_E^2$/day.

The cause of the rapid nonadiabatic variations occurring during magnetic storms is as yet unknown. Present evidence indicates they are intimately connected with the occurrence of magnetic substorms. The importance of these substorm associated variations as a source of outer zone protons as compared to the steady inward diffusion of a high altitude low energy population remains to be determined.

Note added in proof. The conversion from flux to density given by Nakada and Mead (1965) is $n \propto rj$ where n = number of particles in dr, dμ, dJ, r = radial distance and j = differential directional flux. However, the proportionality term contains a factor of μ^{-1} which should be included in the conversion if μ is not conserved. The omission of this term by Nakada and Mead (1965) and by Soraas (1969) leads to errors in the Coulomb loss term on page 404 which can be as large as a factor of 2 for the cases considered. However uncertainties in the ambient atmosphere will overshadow a discrepancy of this size and the results reviewed herein are in the main unaffected. I am indebted to T. Birmingham, L. Davis, G. Mead, and M. Walt for discussions on this point.

Acknowledgments

I wish to thank Mr. L. R. Davis and Drs. T. Birmingham, S. M. Krimigis, and G. D. Mead for many very helpful discussions. I further thank Messers. L. Davis and F. Soraas for allowing me to discuss some of their results prior to final publications.

References

Armstrong, T. P. and Krimigis, S. M.: 1968, *J. Geophys. Res.* **73**, 143.

Bame, S. J., Conner, J. P., Hill, H. H., and Holly, F. E.: 1962, *J. Geophys. Res.* **67**, 1628.

Brown, W. L., Cahill, L. F., Davis, L. R., McIlwain, C. E., and Roberts, C. S.: 1968, *J. Geophys Res.* **73**, 153.

Davis, L. R.: 1965, in *Proceedings of the Plasma Space Science Symposium*, D. Reidel Publishing Company, Dordrecht-Holland, p. 212.

Davis, L. R., Jr. and Chang, D. C.: 1962, *J. Geophys. Res.* **67**, 2169.

Davis, L. R. and Williamson, J. M.: 1963, *Space Res.* **3**, 365.

Davis, L. R., Hoffman, R. A., and Williamson, J. M.: 1964, *Trans. Amer. Geophys. Union* **45**, 84.
Davis, L. R. and Williamson, J. M.: 1966, in *Radiation Trapped in the Earth's Magnetic Field* (ed. by B. M. McCormac), D. Reidel Publishing Company, Dordrecht-Holland, p. 215.
Dungey, J. W., Hess, W. N., and Nakada, M. P.: 1965a, *Space Res.* **5**, 399.
Dungey, J. W., Hess, W. N., and Nakada, M. P.: 1965b, *J. Geophys. Res.* **70**, 3529.
Fälthammar, C.-G.: 1965, *J. Geophys. Res.* **70**, 2503.
Fälthammar, C.-G.: 1966, *J. Geophys. Res.* **71**, 1487.
Fälthammar, C.-G.: 1968, in *Earth's Particles and Fields* (ed. by B. M. McCormac), D. Reidel Publishing Company, Dordrecht-Holland, p. 157.
Fritz, T. A. and Krimigis, S. M.: 1968, Univ. of Iowa Rep. 68-65.
Haerendel, G.: 1968, in *Earth's Particles and Fields* (ed. by B. M. McCormac), D. Reidel Publishing Company, Dordrecht-Holland, p. 171.
Herlofson, N.: 1960, *Phys. Rev. Letters* **5**, 414.
Kellogg, P. J.: 1959, *Nature* **183**, 1295.
Kennel, C. F. and Petschek, H. E.: 1966, *J. Geophys. Res.* **71**, 1.
King, J. H.: 1967, NASA Rept. SP-3024.
Konradi, A.: 1967, *J. Geophys. Res.* **72**, 3829.
Krimigis, S. M.: 1968, in *Earth's Particles and Fields* (ed. by B. M. McCormac), D. Reidel Publishing Company, Dordrecht, Holland, p. 89.
Krimigis, S. M.: 1970, this volume, p. 364.
Krimigis, S. M. and Armstrong, T. P.: 1966, *J. Geophys. Res.* **71**, 4641.
McIlwain, C. E.: 1966, *J. Geophys. Res.* **71**, 3623.
Mead, G. D.: 1964, *J. Geophys. Res.* **69**, 1181.
Mihalov, J. D. and White, R. S.: 1966, *J. Geophys. Res.* **71**, 2207.
Nakada, M. P. and Mead, G. D.: 1965, *J. Geophys. Res.* **70**, 4777.
Naugle, J. E. and Kniffen, D. A.: 1961, *Phys. Rev. Letters* **7**, 3.
Parker, E. N.: 1960, *J. Geophys. Res.* **65**, 2117.
Soraas, F.: 1969, GSFC Rept. X-612-69-241.
Soraas, F. and Davis, L. R.: 1968, GSFC Rept. X-612-68-328.
Tverskoy, B. A.: 1964, *Geomag. Aeron.* **3**, 351.
Tverskoy, B. A.: 1965, *Geomag. Aeron.* **5**, 617.

RADIAL DIFFUSION OF TRAPPED PARTICLES

MARTIN WALT

Lockheed Palo Alto Research Laboratory, Palo Alto, Calif., U.S.A.

1. Introduction

Evidence for the radial motion of trapped particles was reviewed by Fälthammar (1970). Many experimental results of a varied nature support the contention that particles diffuse across L shells, even in magnetically quiet periods. During magnetically active periods more violent changes in the trapped radiation population take place, indicating that strong acceleration and/or injection is occurring. These impulsive events will be ignored in this paper, and only the gradual diffusive motion will be considered. In the diffusive motion it is generally assumed that the radial velocity caused by the perturbing electric and magnetic fields is small compared to the normal azimuthal drift velocity. Thus, the diffusive motion can be described by a Fokker-Planck equation such as derived by Haerendel (1968). The role of this slow type of diffusion in the overall formation and decay of the radiation belts is uncertain at the present time. However, if the magnitude and characteristics of the process can be established, its importance can be evaluated.

The theoretical and experimental values of the diffusion coefficient describing the radial motion have been collected by Fälthammar (1970; see Figure 3). In spite of the meager data available, there is a substantial scatter in the experimental values, reflecting that either the diffusion coefficient varies with time or particle energy or that the assumptions used in deducing the coefficients from the experimental data are not valid in all cases.

This paper will consist first of a brief review of the theoretical formulation underlying the analysis of radial diffusion data. Section 3 will outline a method of treating radiation belt data which circumvents some of the more serious approximations employed to date.

2. Diffusion Equations

At the previous Advanced Study Institute Haerendel (1968) presented the general diffusion equation for trapped particles in the form

$$\frac{\partial f}{\partial t} = \frac{\partial}{\partial x_j}\left[\mathscr{J} D_{jk} \frac{\partial}{\partial x_k}\left(\frac{f}{\mathscr{J}}\right)\right], \tag{1}$$

where $f(\bar{x})$ is the distribution function of the trapped particles and $\mathscr{J}$ is the Jacobian relating the adiabatic invariants μ, J, Ψ to the independent variables x_1, x_2, x_3 being used for a particular case. The tensor D_{jk} is the diffusion coefficient and the repeated indices in Equation (1) are summed. In most cases of radial diffusion it has been assumed that the first two adiabatic invariants μ and J are conserved. Hence, if one

B. M. McCormac (ed.), Particles and Fields in the Magnetosphere.

uses μ, J, and L as coordinates and $n(\mu, J, L, t)$ is the distribution function in these variables, the appropriate diffusion equation in the absence of sources and losses is

$$\frac{\partial n}{\partial t} = \frac{\partial}{\partial L}\left[\frac{D_{LL}}{L^2}\frac{\partial}{\partial L}(nL^2)\right]. \tag{2}$$

The radial flow of electrons depends on $\partial/\partial L(nL^2)$ as is apparent from Equation (2) above. Since $n \propto (L\Phi/\mu)$, where Φ is the differential directional flux of particles $(\mathrm{cm}^2 \text{ sec MeV sr})^{-1}$, the radial transport depends on gradients in ΦL^3 where $\Phi(L)$ is evaluated at constant μ and J for all L. Nakada *et al.* (1965) showed that for protons $L^3\Phi|_{\mu, J=\text{const.}}$ increased monotonically with increasing L and cited this fact as evidence that trapped protons were diffusing inward from the outer boundary of the radiation belts. Similar results have been found for electrons in the outer belt and are illustrated in Figure 1 which gives curves of $L^3\Phi|_{\mu, J=\text{const.}}$ for equatorially trapped electrons. The

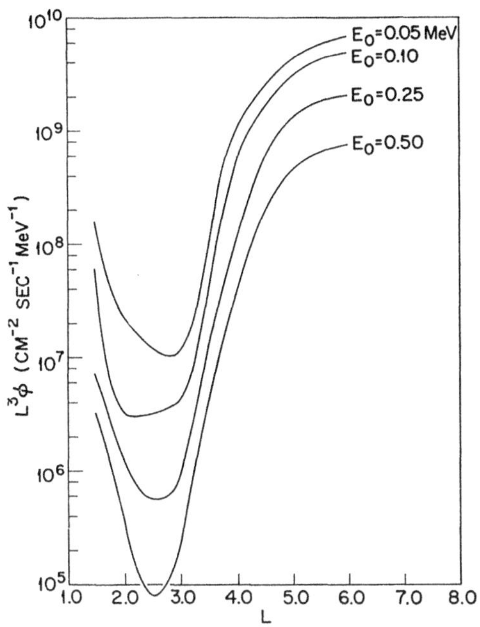

Fig. 1. Variation of $L^3\Phi$ with L, where Φ is the differential directional flux of equatorially trapped electrons evaluated at constant μ and J. E_0 denotes the electron energy at $L=6$.

electron fluxes were taken from the AE2 environment of Vette *et al.* (1966), and the values of E_0 denote the energy of each electron group at $L=6$. The positive slope of $L^3\Phi$ is consistent with a net inward drift of electrons into the radiation belts. Below $L=2.5$ the distributions are dominated by the presence of Starfish electrons and no conclusion regarding the diffusion of natural electrons is possible there.

If the processes causing the radial motion do not preserve μ and J, an equation different from Equation (2) results. For example, in the case of radial diffusion caused by pitch angle scattering in an asymmetric geomagnetic field (the L shell splitting process described by Roederer (1968)) the appropriate diffusion equation is (Fälthammar and Walt, 1969)

$$\frac{\partial f}{\partial t} = \frac{\partial}{\partial L}\left[L^2 D_{LL} \frac{\partial}{\partial L}\left(\frac{f}{L^2}\right)\right] + \frac{\partial}{\partial x}\left[x N_2 D_{xx} \frac{\partial}{\partial x}\left(\frac{f}{x N_2}\right)\right]. \tag{3}$$

The distribution function $f(x, L, E, t)$ is the number of particles in $dL\, dE\, dx$ where E is the particle energy and x is the cosine of the equatorial pitch angle. N_2 is a function of x only and is equal to $1/L$ times the spiral distance between mirror points. In this case $f \propto \Phi L^2 x N_2$ so that the radial flow of particles depends on the derivative of Φ with respect to L. For $L \lesssim 4.5$ this derivative is negative and indicates that the L shell splitting process will cause a net flow of electrons out of the radiation belts.

All attempts thus far to derive empirical values of the diffusion coefficient from the observed behavior of electron fluxes have used Equation (2) with some approximation to simulate losses. Usually the losses are represented by an added term n/τ where τ is the appropriate particle lifetime either from collisional losses or from pitch angle diffusion. Thus,

$$\frac{\partial n}{\partial t} = \frac{\partial}{\partial L}\left[\frac{D_{LL}}{L^2} \frac{\partial}{\partial L}(nL^2)\right] - \frac{n}{\tau}. \tag{4}$$

Most of the experimental data available have not included pitch angle distributions or energy spectra so additional assumptions are needed to convert the measured omnidirectional fluxes (usually above some energy threshold) to the distribution function n needed in Equation (4).

For protons the application of Equation (4) may be adequate although energy loss by ionizing collisions must be included. For electrons, however, there is ample evidence that pitch angle scattering occurs at least as rapidly as radial diffusion. Hence, the assumption that particles diffuse at constant μ and J is undoubtedly incorrect, and a more general treatment is needed. In particular one must take account of the pitch angle distribution as this factor will influence both the loss rates and the radial transport.

3. Simultaneous Radial and Pitch Angle Diffusion

In any process involving simultaneous radial and pitch angle diffusion all three of the adiabatic invariants are likely to change values. Hence, Equation (1) requires a diffusion term for each of the three independent variables as well as for the three possible cross terms. As such an approach is beyond convenient calculational techniques, some approximations must be made. It will be assumed therefore that the radial motion and pitch angle scattering can occur simultaneously, but the mechanisms are independent so that the time average of $\Delta L \Delta x$ vanishes. In the pitch angle scattering process it will be assumed that energy is conserved, and in the L shell diffusion process it will be assumed that μ and x are constant. The assumption that x remains constant rather than the integral invariant J introduces no error for equatorial particles and little error for off-equatorial particles as can be seen from the traces of mirror points of particles diffusing at constant μ and J (Haerendel, 1968; see Figure 1).

With these approximations, the natural coordinates for the diffusion equation are x, L, and z, where

$$z = \frac{(E + 2m_0c^2)\, EL^3}{m_0^2 c^4} . \tag{5}$$

Note that z remains constant for both the pitch angle scattering and the L shell motion so that the overall diffusion involves only two independent variables.

Using the formalism of Haerendel (1968) the equation for the distribution function $f(x, z, L, t)$ is given by

$$\frac{\partial f}{\partial t} = \frac{\partial}{\partial x}\left[xN_2 D_{xx} \frac{\partial}{\partial x}\left(\frac{f}{xN_2}\right)\right] + \frac{\partial}{\partial L}\left[\frac{D_{LL}}{L^{5/2}} \frac{\partial}{\partial L}(L^{5/2} f)\right]. \tag{6}$$

The solution of Equation (6) is obtained by first expanding the distribution function in eigenfunctions of the pitch angle diffusion term

$$f(x, z, L; t) = \sum_n a_n(t, L, z)\, g_n(x, L) \tag{7}$$

where

$$\frac{\partial}{\partial x}\left[xN_2 D_{xx} \frac{\partial}{\partial x}\left(\frac{g_n}{xN_2}\right)\right] = -\lambda_n(L)\, g_n \tag{8}$$

with the boundary conditions

$$g_n(x_c) = 0$$

$$\frac{\partial}{\partial x}\left(\frac{g_n}{xN_2}\right)_{x=0} = 0 .$$

The eigenfunctions g_n are orthogonal with the weighting function $1/xN_2$ and are normalized such that

$$\int_0^{x_c} \frac{g_m g_n}{xN_2}\, dx = \delta_{mn} . \tag{9}$$

The L dependence of g_n enters weakly through the cutoff pitch angle cosine $x_c(L)$ and through $D_{xx}(L, x)$. Substituting Equation (7) into (6), multiplying the result by g_m/xN_2 and integrating on x from 0 to x_c gives for each m

$$\frac{\partial a_m}{\partial t} = \int_0^{x_c} \left(\frac{g_m}{xN_2}\right) \frac{\partial}{\partial L}\left[\frac{D_{LL}}{L^{5/2}} \frac{\partial}{\partial L}\left(L^{5/2} \sum_n a_n g_n\right)\right] dx - \lambda_m a_m . \tag{10}$$

Equation (10) cannot be treated easily without additional approximations. Therefore, let x_c be independent of L, which is a good approximation in the outer zone.

Also assume D_{xx} is independent of L so that the g_n are functions of only x. Little data on D_{xx} are available although Imhof (1968) has found D_{xx} for electrons to vary only slightly with L in the range $3<L<5$. The relatively constant electron lifetime in the outer belt also supports this result. With these approximations the equation for a_m is

$$\frac{\partial a_m}{\partial t} = \frac{\partial}{\partial L}\left[\frac{1}{L^{5/2}} \sum_n D_{LL}^{mn} \frac{\partial}{\partial L} (L^{5/2} a_n)\right] - \lambda_m a_m \tag{11}$$

where the matrix

$$D_{LL}^{mn} = \int_0^{x_c} D_{LL}(x) \frac{g_m g_n}{x N_2} \, dx . \tag{12}$$

Equation (11) gives the diffusion equation for the amplitude of the mth pitch angle mode. If one neglects all modes but $m=1$, Equation (11) reduces to Equation (4) (after appropriate changes of variables) with a loss term given by $\lambda_1 a_1$. Thus, it is apparent that the use of Equation (4) is correct only when μ and J are conserved or when the pitch angle distribution at all L is in the lowest diffusive mode $g_1(x)$ and D_{LL}^{mn} is diagonal. The matrix D_{LL}^{nm} will be diagonal for D_{LL} independent of x as is approximately true for diffusion driven by electric potential fields. However, for magnetic perturbations $D_{LL}(x)$ has a maximum at $x=0$ and falls rapidly with increasing x (Fälthammar, 1968) so that the off-diagonal elements are appreciable. For magnetic perturbations therefore the radial diffusion leads to a mixing of the pitch angle eigenfunctions; the preferential radial transport of equatorially trapped particles will populate higher modes even if the initial distribution is in the lowest pitch angle mode at all L.

The application of this formalism to trapped radiation should result in more realistic values of the diffusion coefficients both for pitch angle scattering and for radial diffusion. Consider first the diffusion caused by electric potential fields since in that case the matrix D_{LL}^{nm} is diagonal and Equation (11) contains terms only in a_m. Therefore,

$$\frac{\partial a_m}{\partial t} = \frac{\partial}{\partial L}\left[\frac{1}{L^{5/2}} D_{LL}^{mm} \frac{\partial}{\partial L} (L^{5/2} a_m)\right] - \lambda_m a_m . \tag{13}$$

For assumed values of D_{LL}^{mm} the equilibrium distribution for the time independent case can be calculated for specified boundary conditions, usually $a_m=0$ at an inner boundary $L\approx 1$ and $a_m=A_m$ at some outer boundary. The values of D_{LL}^{mm} and D_{xx} (and therefore λ_m) could be adjusted then to give agreement with the observed equilibrium distribution. For arbitrary D_{LL}^{mm} it would be necessary to numerically integrate Equation (13), and it is also necessary to find the g_m and λ_m by numerical integration of Equation (8). For $D_{LL}^{nn} \propto L^k$ the solution to Equation (13) can be expressed in terms of modified Bessel functions of nonintegral order.

For the time dependent case, again assuming diagonal D_{LL}^{mn}, the time and radial variables can be separated and a_m expressed as a sum of terms each of which is the product of a time dependent function and a space dependent function. The radial equation again must be integrated numerically except for the special case of $D_{LL}^{mm} \propto L^k$ where a Bessel function solution is possible.

If the diffusion is driven by magnetic perturbations, the off-diagonal elements of the matrix do not vanish, and the solution is no longer simple. Fortunately, the off-diagonal terms are about a factor of 5 smaller than the diagonal terms, and the diagonal terms decrease with increasing m. Completely neglecting the off-diagonal terms in Equation (11) would still lead to a better approximation than the use of Equation (4) since (11) allows for the more rapid decay of the higher pitch angle modes and for the x dependence of D_{LL}. It should be possible to improve this result by iteration, first calculating $a_m^{(0)}$ for all m by neglecting the off-diagonal elements and then calculating $a_m^{(1)}$ by evaluating the off-diagonal contribution using a_n^0. This iteration could be repeated until convergence is reached.

To apply this method in the most general way requires rather complete information on the trapped particles. The theory provides for including the energy spectrum, pitch angle distribution, and radial variation of the flux, all as a function of time. If only part of this information is available, for example if only omnidirectional fluxes above a given threshold are measured, arbitrary assumptions must be made for the energy and angular distributions, and some of the definiteness of the calculation is lost. Complete data of the type needed are not available as yet. However, recent measurements with directional spectrometers on OGO 5 may provide data for an analysis of the type proposed here.

References

Fälthammar, C.-G.: 1968, in *Earth's Particles and Fields* (ed. by B. M. McCormac), Reinhold Publishing Corporation, New York, p. 157.

Fälthammar, C.-G.: 1970, this volume, p. 387.

Fälthammar, C.-G. and Walt, M.: 1969, *J. Geophys. Res.* **74**, 4184.

Haerendel, G.: 1968, in *Earth's Particles and Fields* (ed. by B. M. McCormac), Reinhold Publishing Corporation, New York, p. 171.

Imhof, W. L.: 1968, *J. Geophys. Res.* **73**, 4167.

Nakada, M. P., Dungey, J. W., and Hess, W. N.: 1965, *J. Geophys. Res.* **70**, 3529.

Roederer, J. G.: 1968, in *Earth's Particles and Fields* (ed. by B. M. McCormac), Reinhold Publishing Corporation, New York, p. 193.

Vette, J. I., Lucero, A. B., and Wright, J. A.: 1966, *Models of the Trapped Radiation Environment* **II**; *Inner and Outer Zone Electrons*, NASA SP-3024.

ON THE BALANCE BETWEEN RADIAL AND PITCH ANGLE DIFFUSION

G. HAERENDEL

Max-Planck-Institut für Physik und Astrophysik, Institut für extraterrestrische Physik, Garching/Munich, Germany

1. General Characteristics of Diffusion and Quasi-Linear Theories

The various attempts to mathematically describe the observed populations of trapped particles can be summarized by the following set of equations expressing conservation of particle number f and energy:

$$\frac{\partial}{\partial t} f(x^\mu, t) = \frac{\partial}{\partial x^\mu}\left(D^{\mu\nu} \frac{\partial f}{\partial x^\nu}\right) - L(x^\mu) + S(x^\mu) \tag{1}$$

and conservation of electromagnetic wave energy ε:

$$\frac{\partial}{\partial t} \varepsilon(\omega, t) = -\operatorname{div}(\varepsilon \mathbf{v}_g) + 2(\gamma - l)\, \varepsilon . \tag{2}$$

The variables x^μ represent the adiabatic invariants which undergo small fluctuations by either interaction with the electromagnetic wave field or collisions. $D^{\mu\nu}$ is the diffusion tensor which so far has been considered mostly for L and pitch angle diffusion. S represents any internal particle source like the neutron albedo decay and L losses due to collisions.

From a given volume in the magnetosphere wave energy may escape by propagation with group velocity $\mathbf{v}_g$ or by absorption (coefficient l) as it should occur in the ionosphere, or by Landau or resonance damping of any other kind. γ is the growth rate of the wave amplitude.

Equations (1) and (2) or similar equations are the basis of the quasi-linear theory of trapped ratiation. The coupling between the two equations arises physically from instabilities of the particle distribution f leading to a positive growth rate and a subsequent growth of at least one of the diffusion coefficients $D^{\mu\nu}$, which is closely related to the power in the wave field.

Mathematically the coupling is provided by the growth rate being a functional of the particle distribution:

$$\gamma = \gamma[f] \tag{3}$$

and by

$$D^{\mu\nu} = D^{\mu\nu}(\varepsilon) . \tag{4}$$

The relation between the particle variables x^μ and the wave frequency ω is established by the resonance condition for the interaction and the dispersion relation of the wave.

In the case of a pure diffusion theory the coefficients $D^{\mu\nu}$ are assumed to be predetermined; thus, $\varepsilon(\omega, t)$ is treated as a known function. The corresponding case on

B. M. McCormac (ed.), Particles and Fields in the Magnetosphere.

the wave side is represented by calculations of the amplification of waves by the energetic particle population. Here $f(x^\mu)$ is assumed to be predetermined and unaffected by the wave growth.

In order to account for some of the features of trapped particles Andronov and Trakhtengerts (1964), Cornwall (1966), and Kennel and Petschek (1966) have chosen the quasi-linear approach in the stationary case for dealing with pitch angle scattering. In the stationary case, where one deals only with a limited volume element, a flux tube or L shell, and restricts the propagation of wave energy to the direction parallel to **B**, Equation (2) simplifies to either:

$$\gamma[f] - l = 0 \tag{5}$$

or an equation, to be given below (Equation (12)), expressing the balance between the loss of wave energy through the 'walls' of the system (for instance the ionospheric ends of a flux tube) and internal production. In this equation the wave energy no longer appears. Instead, we obtain a certain limitation on the particle fluxes in so far as they should arrange themselves in a way as to generate wave energy as fast as it is lost from the system. By such reasoning and simple but obvious assumptions on the loss terms, limits to the stably trapped fluxes were derived by Kennel and Petschek (1966).

The level at which the wave field is maintained appears now as a consequence of Equation (1) in the stationary limit. Since the particle flux is already determined by the conservation of wave energy (Equation (5)), a specification of the source of particles readily determines the diffusion coefficients expressing the interaction with the generated wave field. If a strong source is acting, the wave field has to be maintained at a high level in order to keep the particle flux at the level which is just sufficient to generate wave energy with the same rate at which it disappears from the system. Strong variations in the source can normally be taken care of by slight variations in the growth rate γ because the wave energy grows exponentially with γ. Therefore, the particle flux obviously can be rather stable in the presence of strong variations in the particle source and diffusion current.

Such behavior was observed for the low energy trapped protons by Davis and Williamson (1966). This finding is our main justification for the following treatment.

Before going into details a few preliminary words shall be said about the limitations and possible extensions of such a theory. In many cases the above argumentation should apply only for part of the particle population, namely, the part which constitutes most of the energy content. This means that at energies which are well separated from the peak of the spectral distribution other types of interactions may as well dominate any tendency towards instability. It may for instance be that these particles play the role of test particles moving in a wave field that is determined by the dominant component. These particles, of course, should show effects of variations in the source or wave field.

Extensions of this work should include also the explicit treatment of the wave field. In this case time variations most likely have to be considered. If applied to some of the striking types of emissions wave-wave coupling and nonlinear effects should not be negligible. An attempt in this direction was made by Cocke and Cornwall (1967).

2. Observational Evidence

A large body of data showing the presence of either pitch angle scattering or transport across L shells has been accumulated. The situation of the protons in many respects seems to be less complex than that of the electrons. Our following considerations therefore shall be restricted to protons.

As the main hints towards the joint action of L diffusion and pitch angle scattering we regard the following observations: (a) the well known variation of the characteristic proton energy expressing the conservation of μ for transport across L shells, (b) the fact that the direction of the gradient of the proper distribution function is directed outward, (c) the occurrence of H precipitation even at low latitudes, and (d) the striking temporal stability of the dominant part of the proton flux, even under severe geomagnetic perturbations. Williams (1970) presented data from Soraas and Davis (1968) which convincingly demonstrate this point and also indicate the limitations in as much as at higher energies variations become more and more pronounced.

If only radial diffusion without any losses were acting $B^{-1}j_{\perp}(\mu, L)$ for equatorial particles should be nearly constant over most of the radiation belt because of the expected strong increase of the radial diffusion coefficient with L (Haerendel, 1968). The same should apply to $B^{-2}j_{\perp}(>\mu, L)$. Data collected by Haerendel show marked gradients of this quantity directed outward. This indicates, on the one hand, that at μ values below 300 MeV/Gauss inward diffusion is the dominant source, and on the other hand that the diffusion is accompanied by significant losses. Since the distribution in L depends very sensitively on the ratio of loss to diffusion rate, the remarkable temporal stability suggests an intimate relation between the source and the loss processes. The pitch angle instability process naturally provides such an interdependence. The theory presented subsequently so far can only be successfully applied for L values larger than about 3.5 to 4.0, where the slope of the distribution function in L space is considerably flatter. Whether a theory like that of Nakada and Mead (1965) which considers only L diffusion and collisions can account for the protons on the inner L shells as well as for the higher energies as suggested by Williams (1970) is still uncertain in the light of results on empirical diffusion coefficients derived by Haerendel (1968). Below we shall present some evidence for the continuation of pitch angle scattering for $L<4$. However, it may well be that the responsible wavefield is no longer generated by an instability of the protons themselves, but rather externally determined.

3. Brief Summary of Wave Particle Interaction

The following three sections contain a summary of results to be published elsewhere (Haerendel, 1970).

At first we shall consider a situation with homogeneous magnetic field and cold plasma background density. Therefore, the gyrofrequency ω_g and the Alfvén velocity v_A are constants. For an energetic particle to be in resonance with a circularly polarized wave propagating along field lines the resonance condition

$$\omega(1 - v_{\|}/v_{ph}(\omega)) = \omega_g \tag{6}$$

has to be fulfilled. Positive ω stands for left handed (L) waves, negative for right handed (R). Oppositely directed $v_{\|}$ and v_{ph} have opposite signs. Such a situation applies to (p, L) and (e, R) interactions. In (p, R) and (e, L) interactions $v_{\|}/v_{ph} > 1$. In the following treatment we shall regard only the (p, L) interaction.

The phase velocity $v_{ph}(\omega)$ is determined by the dispersion relation of ion cyclotron waves:

$$v_{ph} = v_A \sqrt{1 - \omega/\omega_g}. \tag{7}$$

In a frame of reference traveling with the wave's phase velocity, the electric vector vanishes and therefore the kinetic energy is conserved in the interaction. If we assume the changes of the particle velocity to be small, we get

$$v_{\perp}\, \mathrm{d}v_{\perp} + (v_{\|} - v_{ph})\, \mathrm{d}v_{\|} = 0. \tag{8}$$

With constant v_A and ω_g, Equation (8) can be integrated by using Equations (6) and (7) according to Gendrin (1968):

$$v'^2 = v_{\perp}^2 + (v_{\|} - v_{ph})^2 + v_A^2 \ln(1 - v_{ph}^2/v_A^2) = \text{const}. \tag{9}$$

Although in the course of many (random) scatterings $v_{\|}$ is changed and with $v_{\|}$ also the resonant frequency ω and $v_{ph}(\omega)$, the quantity v' remains at its initial value. This makes v' a convenient variable.

If we consider the particle distribution function instead of depending on $v_{\|}$ and $v_{\perp}$, as usual, as relying on $v_{\|}$ and v' and call it $F(v', v_{\|}, t)$, the expression for the growth of a wave interacting with the particle population F can be written as:

$$k_i(\omega) = \frac{4\pi^3 e^2}{m_e} \frac{k}{3c^2k^2 - \omega^2} \int \mathrm{d}v' v' v_{\perp}^2 (v'_{\perp} v_{\|}) \left.\frac{\partial F}{\partial v_{\|}}\right|_{v_{\|} = v_{\mathrm{res}}(\omega)}. \tag{10}$$

k_i is the imaginary part of the wave vector (real part k) which we have chosen instead of the growth rate γ. $v_{\perp}^2$ appears as a function of v' and $v_{\|}$ by use of Equation (9).

The nature of the instability is quite apparent from the last equations. We derive from Equation (8) the change of kinetic energy E:

$$\mathrm{d}E = m v_{ph}\, \mathrm{d}v_{\|}. \tag{11}$$

According to Equation (7) only waves with $\omega < \omega_g$ can propagate; thus, $v_{\|} \cdot v_{ph}$ must be negative. Equation (11) shows that a scattering leading to $\mathrm{d}v_{\|} > 0$ necessarily results in a loss of energy and a decrease in pitch angle. If the velocity distribution has negative gradients with respect to $v_{\|}$ at constant v', a rearrangement towards a smoother distribution results in a loss of particle and gain of wave energy. Evaluated for a certain resonance velocity $v_{\text{res}}(\omega)$ and integrated over all possible values of v', the wave can grow if $k_i<0$.

4. The Model

We shall now apply the above relations to trapped protons with the following assumptions:

(a) Pitch angle scattering dominates all other loss processes,

(b) Interactions occur only close to the equator in an interval of latitudes $\Delta\lambda$,

(c) Waves propagate only parallel to field lines, and

(d) Wave-wave interaction is negligible, i.e., particles with different $v_{\parallel}$ cannot interact with each other.

We follow Kennel and Petschek (1966) and consider the reflection of a wave train between the ionospheres along a flux tube. When traveling through the region of interaction of length Δs the wave amplitude A will grow by a factor $\exp[-k_i\,\Delta s]$. When reflected only the fraction R of the wave returns to the region of amplification. We have steady state if the losses in the ionosphere equal the gain at the equator (Figure 1):

$$R\exp[-k_i\,\Delta s]=1\,. \tag{12}$$

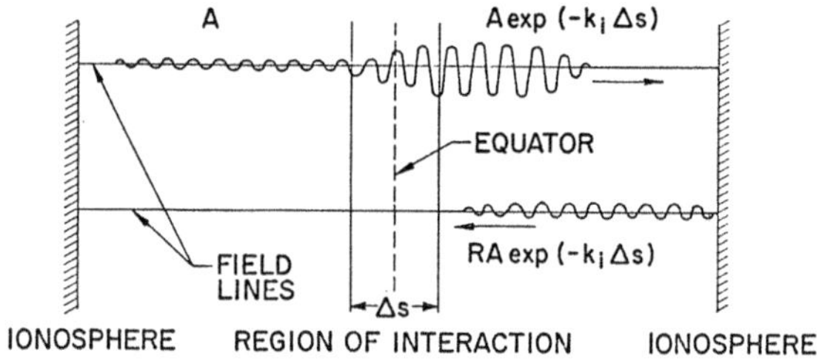

Fig. 1. Schematic representation of the changes of wave amplitude A by amplification close to the equator and partial reflection at the ionospheres.

For a certain value of ω Equations (10) and (12) result in an integro-differential equation as formerly derived by Kennel and Petschek (1966). If considered for all possible values of ω the whole distribution will be covered. However, this would imply that the whole wave field was directly generated in the instability process and that the entire particle distribution was determined by the particle wave energy transfer at the equator. Such an assumption does not appear to be likely. We can only hope that it applies in a limited band of frequencies which is in resonance with the dominant part of the differential particle spectrum.

Instead of attempting to find a general solution for Equation (12) one can start with a certain parametric ansatz for the distribution function. v' is more or less a measure of the particle's kinetic energy. As it is observed that the differential proton spectrum inside $L=6$–7 is rather narrow (Mihalov and White, 1966), we choose the simplest possible way, a delta function in v':

$$F(v'v_{\parallel}) = \frac{1}{2\pi v_0}\,\delta(v' - v_0)\,G(v_{\parallel}). \tag{13}$$

This means that $\int G(v_{\parallel})\,dv_{\parallel}$ is the number density of energetic protons. The choice of a delta function for the v' dependence is consistent with the pitch angle scattering process, since v' is conserved when $v_{\parallel}$ changes. With this ansatz we are not careful with the special shape of the differential energy spectrum; we care only about total fluxes and the average energy.

Equations (10), (12), and (13) result in a simple differential equation in $G(v_{\parallel})$, which is readily integrated if the reflection coefficient R and the length of the interaction region Δs are known. As a boundary condition we have a vanishing flux at the loss cone.

Instead of using $v_{\parallel}$ as a variable we switch to the resonance frequency ω by means of Equations (6) and (7):

$$v_{\parallel}/v_A = (1 - y)^{3/2}/y\,, \quad y = \omega/\omega_g\,. \tag{14}$$

Since v_A and ω_g depend only on L this choice is very convenient. The total velocity parameter v_0 will be replaced by

$$U = (v_0/v_A)^2\,. \tag{15}$$

The solutions of Equation (12) can be conveniently written for the directional flux j (Haerendel, 1970):

$$j(L, E, \alpha) = K(L)\,g(U, y, L) \tag{16}$$

with

$$\begin{aligned} K(L) &= \frac{3cB_0}{4\pi^3 eR_e}\left(-\frac{\ln R}{\Delta\lambda}\right)L^{-4} \\ &= 1.45 \times 10^6\left(-\frac{\ln R}{\Delta\lambda}\right)\left(\frac{4}{L}\right)^4 \quad [\mathrm{cm}^{-2}\,\mathrm{sec}^{-1}\,\mathrm{sr}^{-1}]\,. \end{aligned} \tag{17}$$

Neither the particle energy nor the ambient plasma density appear in the constant $K(L)$, which is entirely determined by the assumed dipole geometry, the reflection coefficient R, and the half width $\Delta\lambda$ of the latitude interval inside which particles of a given pitch angle would experience a doppler shifted wave frequency sufficiently close to the equatorial gyrofrequency. In subsequent numerical estimates we shall allow for a deviation of both frequencies of 10%. For simplicity the reflection coefficient is assumed to be independent of frequency for a given L.

The function $g(U, y, L)$ is almost independent of L at pitch angles not too close to the loss cone since L enters only via the boundary condition. The parameter U contains the choice of the invariant proton velocity v_0 and of the Alfvén velocity i.e., of the exospheric cold plasma density n_{H}:

$$U = \frac{(m/2)\,v_0^2}{B^2/8\pi}\cdot n_{\mathrm{H}}\,. \tag{18}$$

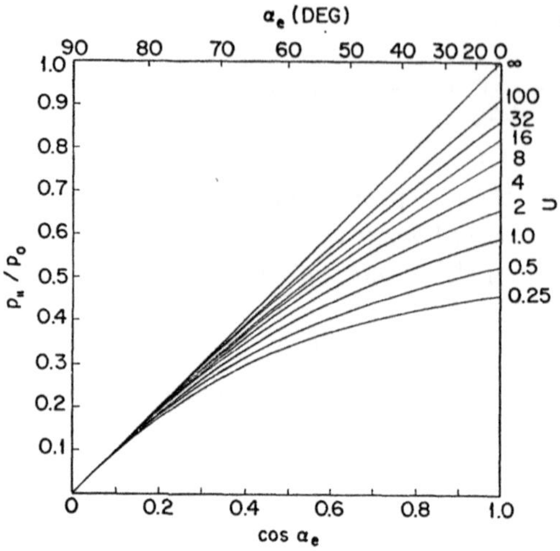

Fig. 2. Relative variation of the parallel component of the proton momentum with pitch angle under the scattering process for various values of U.

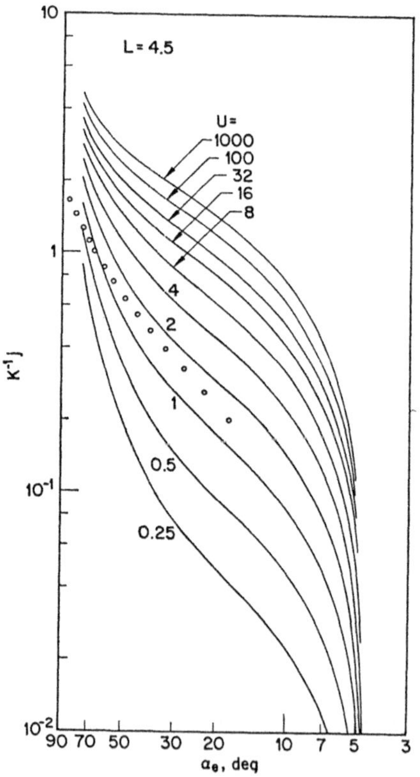

Fig. 3. Theoretical stationary proton distribution at $L = 4.5$ for different values of parameter U (18) in arbitrary units. Data after Davis and Williamson (1966) represent j (> 100 keV) normalized to fit the theoretical curve with $U = 1.5$.

If particle and wave velocities are comparable, the relative change in energy after the scattering is considerable as seen from Equation (11). For large particle velocities the kinetic energy stays almost constant. As the growth rate is a function of the gradient in $v_{\|}$ rather than in pitch angle (Equation (10)) and as in the case of significant energy loss $v_{\|}$ varies very little with pitch angle, steep gradients in $v_{\|}$ correspond to rather soft gradients in α (Figure 2). Consequently, lower and less anisotropic fluxes j satisfy the equilibrium condition in Equation (12) at low values of U as seen in Figure 3, in which the solutions $g(U, y, L)$ for $L=4.5$ were plotted. With increasing U the solutions converge. In former treatments of this problem only the limit of large U was considered explicitly.

The divergence of the solutions for near equatorial pitch angles is a consequence of the decreasing opportunity for particles and waves to be in resonance. For equatorial pitch angles of more than 65° or mirror latitudes $\lambda_m < 12°$, other instabilities or interactions with waves emitted at off-equatorial latitudes will prevent an excessive growth of the flux close to the equator.

5. Comparison with Observations

The removal of the degeneracy of the solutions by taking into account the energy variation in principle could help in selecting the most realistic solutions. If the flux is plotted as a function of $\cos^2 \alpha_e$, it turns out to behave very nearly exponentially in the pitch angle interval $30° < \alpha_e < 65°$. If we define a measure for the anisotropy by

$$A_2 = \frac{\log[j(65°)/j(30°)]}{\cos^2 65° - \cos^2 30°} \tag{19}$$

we find the following values listed in Table I which allow a direct comparison with the data of Soraas and Davis (1968).

TABLE I

U	A_2
1000	−0.45
100	−0.49
32	−0.54
16	−0.59
8	−0.65
4	−0.73
2	−0.84
1	−0.96
1/2	−1.11
1/4	−1.26

During undisturbed periods the data between $L=4$ and 5 lie in the interval $-1.1 < A_2 < -0.9$, corresponding to $1/2 < U < 1.5$. So, we are dealing with a range of U

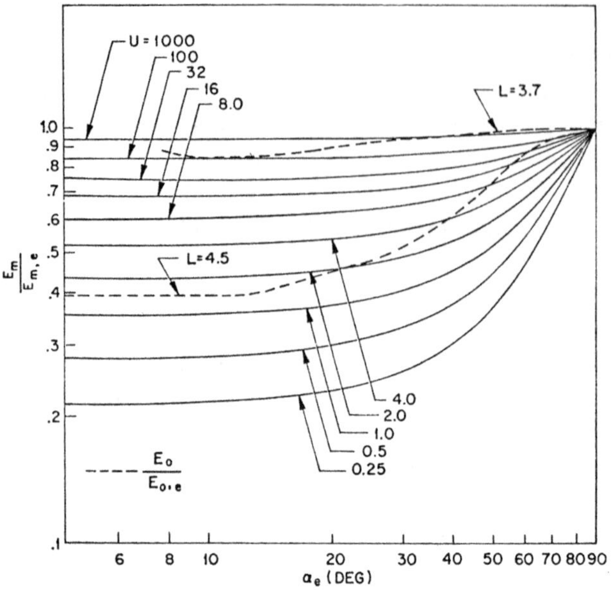

Fig. 4. Relative variation of the proton energy with pitch angle for various values of U. Index "e" refers to equatorial values. The dashed lines represent data on the characteristic energy E_0 in an exponential fit of the integral energy spectrum after King (1967).

values where the energy variation is considerable. This is shown in Figure 4 where the kinetic energy normalized by $(m/2)\,v_0^2$ is plotted as a function of pitch angle. These curves are compared with the variation of the characteristic energy E_0 used in experimental fits of the integral spectrum as taken from the trapped radiation environment (King, 1967). For $L=4.5$ there is reasonable agreement.

If we take the peak of the differential energy spectrum at

$$E_{m,e} = E_1/L^3 \tag{20}$$

by assuming that the so far unspecified source of particles is given by L diffusion in a narrow range of pitch angles, and if we choose $E_1 = 16$ MeV (Mihalov and White, 1966) we expect from Equation (18) that U should vary with L approximately as

$$U = 6.7 \times 10^{-3}\, L^3 n_{\mathrm{H}}\,. \tag{21}$$

For $L=4.5$ and $1/2 < U < 1.5$ we get

$$3\ \mathrm{cm}^{-3} < n_{\mathrm{H}} < 9\ \mathrm{cm}^{-3}\,.$$

These numbers are quite consistent with the densities measured outside the plasmasphere.

In Figure 3 we have chosen $U = 1.5$ at $L = 4.5$ as an example. In order to match the absolute value of the particle fluxes we need a value of $K(L=4.5) = 1.2 \times 10^7\ \mathrm{cm}^{-2}\ \mathrm{sec}^{-1}\ \mathrm{sr}^{-1}$. With $\Delta\lambda = 8\overset{\circ}{.}5$ we obtain this value from Equation (17) by choosing a reflection coefficient of 14%. It turns out that the emitted frequencies lie in the range

$0.48 < \omega/\omega_g < 0.62$, corresponding to 5 to 6.5 Hz. Theories of various authors give greatly varying answers for the reflection coefficient R. However, all agree in that the absorption of wave energy is rather strong at frequencies >1 Hz. The number here obtained lies well within the range of theoretical values.

Comparison of the variation of E_0 as a function of pitch angle or B for different L shells shows a sudden change close to $L=3.9$. For smaller L values the variation along B is nearly negligible (see $L=3.7$ trace in Figure 3) whereas at higher L, E_0 is a clearly decreasing function of B. This change is attributed to the discontinuity in plasma density at the plasmapause. During magnetically quiet periods particles at L shells <3.9 are mostly entirely inside the plasmasphere. The parameter U should be on the order of 100. Scattering would occur without significant changes in kinetic energy. At high L values ($L \gtrsim 5$) the peak of the energy spectrum is below 100 keV. Since E_0 was determined from particle fluxes with higher energies, these curves can no longer be used for comparison.

Although the absence of a significant variation of the mean energy along an L shell inside the plasmasphere is in agreement with the expectations of this theory, we find quite a difference between theoretical and observed pitch angle distributions. The observed proton fluxes turn out to be much more anisotropic than predicted. In addition the absolute fluxes are lower and vary with L quite differently than predicted

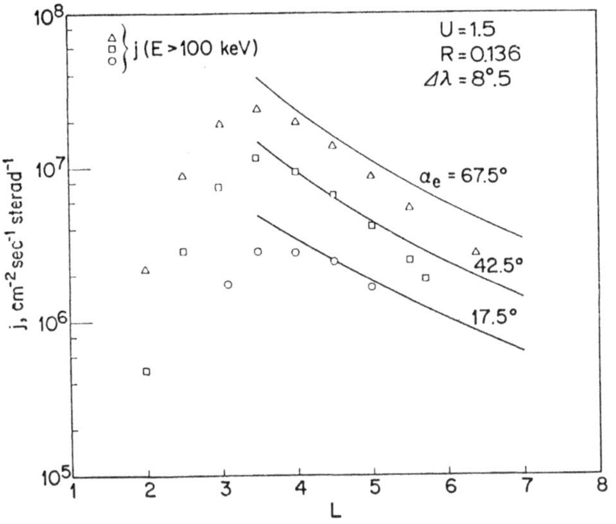

Fig. 5. Theoretical and observed proton distributions for three values of equatorial pitch angle as a function of L. Data after Davis and Williamson (1966).

by this theory. Figure 5 shows the radial distributions as calculated with constant R and constant values of U inside and outside the plasmasphere for three different pitch angles, demonstrating the lack of agreement at smaller L values. The faster decrease at $L>5$, however, must be attributed to the fact that the measured fluxes do not include a significant part of the protons, since the lower energy limit is above the peak of the spectrum. It is expected that their inclusion by future measurements will lead to closer agreement.

For $4.0<L<5.5$ the pitch angle distributions seem to fit the observed ones well except for near equatorial values (Figure 6).

6. Discussion

A simple application of the theory of pitch angle instability can be used to derive model distributions of proton fluxes in the outer radiation belt which are in reasonable agreement with the data. The pitch angle anisotropy and the softening of the energy spectrum with increasing B for constant L depend on the ratio of proton vs. Alfvén

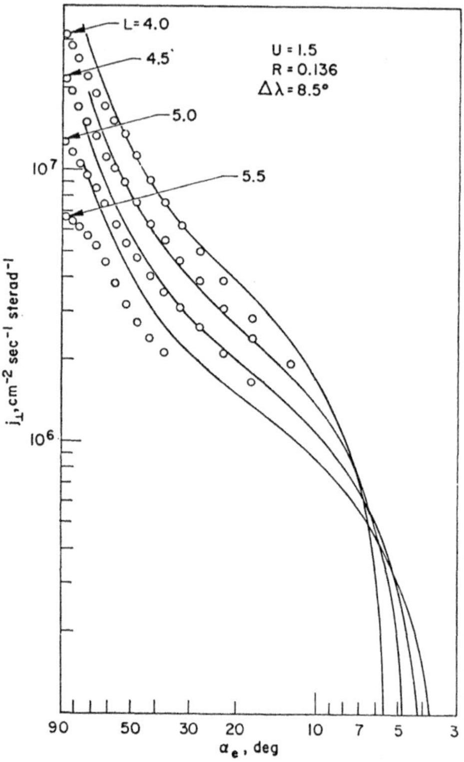

Fig. 6. Comparison of theoretical proton fluxes vs. pitch angle at different L values with data on j (> 100 keV) from Davis and Williamson (1966).

velocity, which outside $L=4$ turns out to be close to unity. Estimates of the cold plasma density n_H from best fits to the data lead to a few particles per cm^3 in accordance with measurements outside the plasmasphere. The magnitude of the proton fluxes is greatly determined by the absorption of the emitted ion cyclotron waves in the ionosphere.

Although L diffusion under conservation of μ is assumed to be the source of particles, this assumption does not appear explicitly in the solutions in Equation (16). Only in determining the energy parameter U in Equation (21) was use made of the constancy of μ. However, L diffusion at all possible pitch angles is not consistent with maintaining a δ function like differential energy spectrum (Equation (13)) for a wide range of L, which was used here for mathematical convenience. In spite of this shortcoming the author believes that the observed variation of E_0 along a field line is an expression of the pitch angle scattering dominating the proton distribution, whereby the adiabatic invariance of μ is violated. This is contrary to the suggestion of Nakada *et al.* (1965) that the variation of the spectral distribution with L and B reflects pure L diffusion under conservation of μ and the second invariant J. Furthermore, the radial distribution of the absolute particle flux external to the plasmasphere is not understood as a consequence of a pure diffusion problem with additional loss terms as in the work of Nakada and Mead (1965).

Inside the plasmasphere there is no good agreement of theory and data. The proton flux drops faster than expected towards the earth and the anisotropy is greater. In this region the approach by a diffusion equation as used by Nakada and Mead (1965) may be valid. However, it is still uncertain whether collisions dominate the losses by pitch angle scattering. External wave sources (modes other than the ion cyclotron waves discussed here) may lead to an efficient scattering. In fact, the rather strong change of the E_0 vs. B variation at $L=3.9$ suggests pitch angle scattering still to occur inside the plasmasphere. An abrupt decrease of the radial diffusion coefficient at the plasmapause in combination with an externally impressed loss term may prevent the distribution from reaching the stably trapped limit for proton-ion cyclotron wave interactions.

In the approach presented here time variations can arise from variations in the external parameters, i.e., U and R. Increases of the integrated flux during magnetic storms and subsequent decreases as observed by Soraas and Davis (1968) may be in accordance with this theory. The increase observed in April 1965 was accompanied by a dccrease of the pitch angle anisotropy. The experimental values could be accounted for by a temporary increase of the parameter U caused by an enhancement of the cold plasma density n_H (see Equation (18)). However, so far there is no observational evidence available to support such a conclusion.

The differential energy spectrum was described by a delta function; therefore, particles at the high energy part of the spectrum are not treated. Their distribution could be affected by the waves emitted by the dominant particles. To get interaction absorption has to occur at slightly off-equatorial latitudes where the doppler shift has to be larger for interaction with the same wave frequency. This was excluded in the present model. Such a theory amounting to an exchange of particle energy between different parts of the spectrum by emission and reabsorption of waves and a resulting local acceleration of the higher energetic particles should be further investigated. Thereby, it may also be possible to remove the divergence of the theoretical particle distribution at mirror points close to the equator.

References

Andronov, A. A. and Trakhtengerts, V. Y.: 1964, *Geomag. Aeron.* **4**, 181.
Cocke, W. J. and Cornwall, J. M.: 1967, *J Geophys Res.* **72**, 2843.
Cornwall, J. M.: 1966, *J. Geophys. Res.* **71**, 2185.
Davis, L. R. and Williamson, J. M.: 1966, in *Radiation Trapped in the Earth's Magnetic Field* (ed. by B. M. McCormac), D. Reidel Publishing Company, Dordrecht-Holland, p. 215; also, private communication.
Gendrin, R.: 1968, *J. Atmos. Terr. Phys.* **30**, 1313.
Haerendel, G.: 1968, in *Earth's Particles and Fields* (ed. by B. M. McCormac), Reinhold Publishing Company, New York, p. 171.
Haerendel, G.: 1970, *J. Geophys. Res.*, to be submitted.
Kennel, C. F. and Petschek, H. E.: 1966, *J. Geophys. Res.* **71**, 1.
King, J. H.: 1967, NASA Document NASA SP-3024.
Mihalov, J. D. and White, R. S.: 1966, *J. Geophys. Res.* **71**, 39.
Nakada, M. P. and Mead, G. D.: 1965, *J. Geophys. Res.* **70**, 4777.
Nakada, M. P., Dungey, F. W., and Hess, W. N.: 1965, *J. Geophys. Res.* **70**, 3529.
Soraas, F. and Davis, L. R.: 1968, GSFC Report X-612-68-328.
Williams, D. J.: 1970, this volume, p. 396.

PART VIII

SUMMARY

PANEL REPORT

MARTIN WALT
Lockheed Palo Alto Research Laboratory, Palo Alto, Calif., U.S.A.

Following two weeks of presentations and discussion, the final period of the conference was devoted to a review in which selected speakers summarized the most noteworthy points raised in the previous two weeks. The purpose of this session was to remind the Institute members of the content of the previous talks and also to indicate areas of principal uncertainty where future experiments would be most fruitful. This type of review is quite subjective and perhaps a different panel might have stressed different points. Nevertheless, an overall summary is desirable in that it can emphasize areas of agreement and disagreement and can place the remarks made throughout the Institute in their proper perspective.

In reviewing the material presented at this Institute it is instructive to compare the emphasis of the present meeting with that of its predecessors, the 1967 meeting at Freising and the one in 1965 at Bergen. The trend in space research toward lower energy particles, already evident at Freising, has continued and a larger fraction of particle observation and interpretation is directed toward electrons and protons in the kilovolt and lower energy range. Another noticeable trend in experimental work is the move toward measurements at larger radial distances. Thus, while at Bergen approximately half the papers were on the stably trapped particles, a major field of current interest is the particles and fields in the distant magnetosphere, the sheath region, and the bow shock. This shift of interest reflects both the maturity of the near earth trapped particle knowledge and the increased availability and sophistication of satellites traversing the more remote regions of the geomagnetic field.

The realization that electric fields, both static and dc, play a pivotal role in the dynamics of the earth's particles was recognized at Freising. At Santa Barbara almost one fifth of the material dealt with observations of these fields and the theoretical interpretation of their consequences. Interest in plasma instabilities has continued also and these experimental data are approaching the quality and coverage required to differentiate between the multitude of theoretically predicted instabilities.

In the following sections different aspects of the conference are summarized. Dr. Vette describes the current status of our knowledge of the stably trapped particles. Dr. Fredricks summarizes the results on electromagnetic waves in the magnetosphere and the structure of the collisionless bow shock. Dr. Vasyliunas deals with large scale magnetospheric structure and the character of the low energy particles which have so much influence on the dynamics of the outer magnetosphere. Finally, Dr. Haerendel reviews the experimental and theoretical work on plasma instabilities in the magnetosphere.

B. M. McCormac (ed.), Particles and Fields in the Magnetosphere.

1. Trapped Particles*

Several new results in trapped radiation were reported at the Institute. The measurements on trapped α-particles have been extended down to 0.26 MeV per nucleon, which is about a factor of 2 lower in energy than previous measurements. About 18 mo worth of continuous data are available, so that one can begin to study the time behavior of the α-particle fluxes. It is quite clear now that α-particles are trapped, and with present techniques they are amenable to measurement. It is also clear that although α-particles make a convenient tracer for understanding sources and losses, one has to be very careful about treating the loss mechanisms. Since sources and losses are pretty much tied together in the diffusion equations, one must treat the loss processes carefully and consider how the uncertainties in loss processes affect the derived results on source characteristics. Measurements of α-particles will undoubtedly continue and hopefully at the next Institute there will be more definite conclusions as to their origin.

The fluxes of inner zone protons seem to be quite stable. This feature has been known for some time but was confirmed by the information from Dr. Elliot. At the very low altitudes there have been measurable changes since 1962, and these changes were reported earlier. However, the point stressed at this Institute is that if one uses the best numbers available on atmospheric densities and the albedo neutron source strength, one cannot inject enough protons to give agreement with experiment. There are uncertainties in the calculation, of course, and the point is well taken that one needs better values for the neutron flux as well as the loss rates. The calculations presented by Dr. White indicate that some processes other than the atmospheric losses are needed to explain the time variations of high energy protons as measured by nuclear emulsions.

Another significant result concerning the inner zone trapped radiation was the clear observation of changes in the natural electron population at energies as high as 1 MeV. It appears that once these electrons are injected, at least in the very stable region below $L = 1.7$, the lifetimes seem to be consistent with the lifetimes of Starfish electrons. At higher L values, as Dr. Winckler showed, the lifetimes of the freshly injected electrons depend on other conditions. Each one of the events seems to have a different lifetime, so there obviously must be different transport mechanisms operating at different times, possibly resulting from different power spectral densities in the fluctuating magnetic and electric fields that might be present. In the future we will see more of these injection events and because the disturbances in the inner zone relax more slowly than in the outer zone, one may be able to better understand the loss mechanisms in the inner zone. It is apparent that one can begin to consider the inner zone more as a natural phenomenon now that the Starfish electrons are disappearing.

Winckler and others presented evidence on the injection of particles in the midnight

* Presented by J. Vette.

side of the magnetosphere. Measurements of this type have been reported before, but one is now getting closer to seeing the actual dynamics of the injection, i.e., the slingshot effect. Another result reported by Winckler was the experimental verification of L shell splitting. These observations clearly established the value of multiple satellites with comparable instruments. One of the more firmly established theories in trapped radiation is the adiabatic motion of particles and the L shell splitting is another demonstration of that motion.

The measurements of Dr. Frank, because they represent a new energy range that has not been amenable to measurement before, bring new information and new ideas into the study of trapped radiation. For example, from these data one is beginning to understand empirically the behavior of the ring current, the asymmetric injection, the dynamics, and the decay. More cases are needed, but clearly the measurements that Frank already has will bring new understanding to this subject.

From many of the papers presented here it is clear that the distribution of particles in the magnetosphere varies greatly with the energy threshold and also with the intensity threshold of the detector. At present we do not know completely the physical boundaries that are important. Many boundaries such as the plasmapause, the stable trapping boundary, and the pseudo-trapping boundary have been identified, but there is still some confusion as to where various classes of particles are really bounded. For example, the measurements that Frank presented showed the plasma sheet electrons existing right into the plasmapause where the higher energy electrons decrease in intensity to form some kind of a distinguishable boundary. The lower energy electrons showed a sharp intensity peak right at the plasmapause. Earlier measurements indicated that the plasma sheet boundary was perhaps more closely identified with the pseudo-trapping boundary as defined by 40 keV electrons. Perhaps Frank's measurements will give some clarification of the distribution of 40 keV particles in the tail since measurements with different intensity threshold levels show different average spatial distributions. I think the low energy particle distributions are very complicated and depend very much on the parameters that one is measuring.

Dr. Williams, in giving the important results of Davis and Soraas, pointed out what can be done, at least in some cases, to remove the adiabatic effects produced by the ring current. To study the dynamics of trapped particles and investigate the features one doesn't understand, it is best to remove first those effects that one does understand. This approach seems possible in some cases for the protons; on the other hand, electrons in the outer zone still seem to be too dynamic to be treated in the same way.

A few comments on the trapped particle environment models are probably needed, as these environments were used frequently to compare with theoretical results. Dr. Haerendel used the proton environment to infer a possible inner zone source and Dr. Walt used the electron environment in some of his calculations. These environments are based on empirical data and not on theory since present theories do not seem to fit the data too well. In some areas where there are no data we still try to extend the model to make it a complete one for certain purposes. One needs to be aware of this fact, as Haerendel pointed out in discussing the E_0 variations of low energy protons as a function of L at $5<L<6$. In that region the model curves are simply guesses because there are no data. I want to make sure that people understand this point; where data are available we try to represent it as best we can, but in the absence of data the models are extrapolations.

The ability to construct these environments depends on reducing these data to various compatible forms. One of the factors we must come to terms with in the future is the coordinate system with which people organize their data. Until recently, the Jensen and Cain 1960 magnetic field model has been used. Now, there are a number of more sophisticated field models that one can use depending on his specific interest. Perhaps some discussion and decision on what field should be used for the general organization of trapped radiation data is again needed.

At the present time work is progressing on a new high energy proton environment. New data are now available and it is clear from these data that previous estimates are off by a factor of 3 at the equator. In the outer zone the large time variations in the electron flux cause considerable difficulties in compiling models. The best that we can do under these conditions is to give a time averaged environment and some quantitative measure of the variability. This concept was presented in some detail at the beginning of the conference.

2. Waves and Particles in the Magnetosphere and Bow Shock*

This summary on the new measurements of particles, waves, and the bow shock is not complete, and other members of the Institute will have their own prejudices. The points stressed here are those which I feel will lead to greater advances in understanding the magnetosphere and the interactions of the solar wind with the geomagnetic field.

The mapping of the wave activity in the magnetosphere was begun by Helliwell's group at Stanford with ground based measurements. This work was extended by the OGO 1 and OGO 3 experiments, and more recently by OGO 5. OGO 5 is perfectly stabilized, has a long apparent lifetime, and has not only magnetic search coil experiments but also an experiment which senses electric fields in the frequency range 200 Hz to about 70 kHz. The results show the occurrence of electromagnetic and electrostatic waves between about 5 and 10 R_E in the tail and in a rather thin layer centered around the geomagnetic equator. The wave activities increase during magnetic bay activity, and considerable noise is seen on nearly every pass of the satellite through the thin equatorial region. The electromagnetic waves, generally chorus emission, are both continuous and structured. These emissions have been seen before by Helliwell and others, but the previous satellites did not map the equatorial region with the same volume of mapping that OGO 5 is able to do. The waves, which were described by Dr. Russell, have frequencies which vary between about a quarter to one-half of the equatorial gyrofrequency.

* Presented by R. Fredricks.

During these measurements of the magnetic field the electric field sensor quite often detects the electric component of these electromagnetic waves. When the conditions are favorable for comparison, one can measure the index of refraction of the medium by measuring E and B, taking the ratio, and assuming that the waves are whistler mode noise. In the region between the 5 and 10 R_E around the equator the indices of refraction are on the order of 5, corresponding to electron and ion densities on the order of $\frac{1}{10}$ per cm^3, in agreement with other direct and indirect evidence. These chorus waves may be the turbulence that Kennel and Petschek need for the whistler induced particle precipitation to limit the flux of belt electrons. With an inferred refractive index of about 5, the waves would resonate with electrons somewhere between 40 and 100 keV, so they should have something to do with the control of fluxes of energetic electrons in this range.

In addition to the electromagnetic waves, the group at TRW has observed very strong electrostatic waves which are in narrow bands at various frequencies in the vicinity of the half gyrofrequency, three halves the gyrofrequency, just above the gyrofrequency, and at the first harmonic of the gyrofrequency. There are also waves tentatively identified as broadband emisssions, and these could be near either the upper hybrid or the plasma frequency. These waves are quite often observed at the same time as chorus, but they are measured also when chorus is not there. Sometimes chorus is present when the electrostatic waves are absent. With amplitudes up to tens of millivolts per meter these electrostatic waves, since they are slow waves, could resonate either with soft electrons or with 100 keV protons. Drs. Kennel, Scarf, and myself are trying to determine the origin of these waves and what they will do to the particles.

It is unfortunate that there is no reliable electric field measurement below 200 Hz in the deep tail region beyond 5 R_E. This is one experiment which should be done because electrostatic drift waves may be there. These waves would have frequencies of a few Hertz or less and would be predominantly electric. They should have something to do with the proton distributions, but as yet they have not been measured.

In regard to particle measurements, Dr. Frank's LEPEDEA sensor is giving some very badly needed data on the particle distributions and dynamics. When we can digest this, we will have a much better picture of what is going on. If we could add dc or ELF electric field measurements to the particle data, we would have an even better picture. However, we will not know much more about the tail region until new space probes with better instrumentation are available.

In the inner magnetosphere some of the most impressive work was the Injun 5 experiment reported by Dr. Gurnett. In particular, his measurements of the Poynting vector direction are some of the most interesting new results. The evidence that chorus is propagating downward is consistent with the OGO 5 result that chorus appears to be generated in the equatorial regions. It is not necessarily generated everywhere, but does propagate from the equatorial regions as Helliwell's rather heurestic model predicts. Also ELF and dc measurements from Injun 5 near the plasmapause boundary and the higher auroral region are very important. Dr. Frank has a LEPEDEA experiment aboard Injun 5 and has obtained a great deal of very interesting data over the whole orbit but in particular in the high auroral regions.

Dr. Wolfe gave an historical sketch of the bow shock observations, describing the improvements in measurement techniques beginning with Pioneer 1 and ending with OGO 5. It appears that descriptions of the small scale structures of the collisionless shock are now becoming available because of the high data rate satellite information. The results presented by Dr. Scarf from OGO 5 and by Dr. Montgomery from Vela 4 are particularly interesting. The high data rates of OGO 5 allow at least the field experiments to sample both magnetic and electric fields with up to 18 msec time resolution. The particle probes have to accumulate enough counts or enough flux to give statistical accuracy and generally have lower sampling rates. However, some measurements on the spacecraft, the total ion flux for instance, are available at 600 msec resolution. In the Vela 4 satellites the electrostatic analyzer is capable under some circumstances of forming velocity distribution functions on time scales of a few seconds. These data are what is needed to understand the structure and the physics of the collisionless shock.

The hydromagnetic signature of the shock and the electrostatic wave turbulence that accompanies this hydromagnetic structure have been measured by the OGO 5 instruments, and along with these field measurements there are time correlated particle data. By combining the available OGO 5 shock data with the Vela 4 particle measurements, one can obtain a picture of the shock structure which is consistent and makes physical sense. The electrostatic wave turbulence develops in the hydromagnetic structure whenever there are large enough current densities in the magnetic gradients. These currents are diamagnetic currents without any doubt, and if the drift velocity corresponding to these diamagnetic currents becomes on the order of either the electron or the proton thermal speed, a variety of streaming instabilities can develop. Under these conditions large proton scattering occurs. The largest proton scattering occurs when the electrostatic turbulence levels reach a saturation value. In the satellite reference frame this can appear either deep inside the shock or fairly close to the front of the structure, depending on the state of motion of the structure. In the rest frame, the proton scattering probably occurs continuously at the same position.

Electrons are probably energized at the very front edge of the shock as indicated by the Vela 4 measurements of Montgomery. The electrons are heated; they are not randomized, but their distribution is changed drastically in the front of the shock. The evolution of the electron and proton distributions in the shock from Vela 4 would indicate that once the plateau in electron velocity develops the protons continue to flow. They may slow down, but they are not thermalized. The main body of the distribution is broadened significantly.

At present there is no self-consistent theory for high Mach number, high β shocks, such as the bow shock. Any theory must explain the appearance of the nonlinear wave trains which in general lead the ramp or jump in the magnetic field, and this nonlinear wave train must be constructed in such a way that the protons flow through it without being randomized. The theory must account for the nearly reversible wave structure, but it must have some electrostatic plasma wave turbulence associated with it in a self-consistent way. It also has to explain the fact that the scale length in the hydromagnetic structures can only be related to the electron inertial length (velocity of light divided by the plasma frequency) and the theory must explain why at some point fairly deep in the structure there is a very abrupt and sudden proton heating. The rest of the shock structure downstream has to be explained also but is probably less important to the topic than is the shock front itself.

There is one theory that I think might be modified to describe the shock, and this is a two fluid theory of oblique solitary pulses by Paul Kellogg published in *Physics of Fluids* in October 1964. The paper gives some basic ideas, and it generates solitary pulse solutions or dissipation pulse solutions which have the right characteristic length and are oblique waves. Perhaps this theory could be modified for large β and finite Larmor radius without losing the principal features.

3. Low Energy Particles*

This section will review the Institute highlights relating to the topics of low energy particles and magnetospheric models. For low energy particles (meaning particles with energies from ~10 keV to ~100 eV and below) it becomes rather hard to draw the line between trapped radiation and the particles associated with the outer magnetosphere and the magnetotail. The low energy particles appear to form a single complicated structure encompassing the entire magnetosphere and the tail, so that any division into trapped and untrapped categories is bound to be more or less arbitrary. The studies of trapped radiation at energies of ~40 keV and above in many respects, have been an extension of cosmic ray physics into lower energies: in both cases we observe only the "high energy tail", a small fraction of the total particle population, and we discuss the observations in terms of particle intensity at a given energy and angle. By contrast, in the study of low energy particles we are now able to observe the major part of the particle population; we can describe the observations in plasma physics terms and discuss collective properties of the entire particle population, such as density or pressure.

The present status of the observations of low energy particles is similar to that of higher energy particles at the time of the Bergen meeting. Certain regions have been very well explored and in others a detailed examination of particle distributions is just beginning. In particular, the overall structure of electron fluxes in the magnetotail at ~18 R_E and in the evening sector of the outer magnetosphere is now fairly well known, thanks to the Vela and OGO satellite experiments. We are beginning to look at finer details such as the relation of particle fluxes to magnetic bay activity, pressure balance between plasma sheet and solar wind particles (both discussed at this meeting by Dr. Hones), and the connection between plasma sheet electrons and the auroral

* Presented by V. Vasyliunas.

oval (discussed by Vasyliunas). Our knowledge of electrons in the magnetosphere at other local times is still very much in a survey stage, and low energy particles at close distances (within the plasmasphere) have been hardly explored yet.

From the proton measurements of Dr. Frank we now know the gross properties of the ring current during both quiet and storm times. In particular, the existence of a permanent quiet-time ring current has been established and the asymmetric formation of the storm-time ring current, long suspected from magnetic data, has been verified by direct observation. Frank also reported the detection (for the first time, to my knowledge) of protons in the high latitude magnetotail.

Observational data on particle structures throughout much of the magnetosphere and tail already exist and measurements from numerous satellites are being analyzed at various institutions. Therefore, I think we can confidently expect that by the next meeting of this series we should have a fairly complete general picture of the distribution of low energy particles in the magnetosphere. One important point to be studied is the detailed correlation of low energy particle fluxes and geomagnetic activity. In great contrast to, e.g., the inner zone protons, these low energy particle fluxes are highly variable. They form the main part of the energy content of the magnetosphere and are evidently very strongly coupled to the whole range of auroral and substorm phenomena. Plasma physics processes such as instabilities and particle wave interactions appear to play an important role that deserves much further study.

The suprathermal protons (~ 10 keV) detected in the solar wind by Frank represent a previously unsuspected particle population. They are undoubtedly a very interesting phenomenon as far as the solar wind is concerned, but what role (if any) they play in magnetospheric processes is not yet clear.

Several new results in the seemingly well studied field of 40 keV electron observations were presented. Meng reported a uniform layer of 40 keV electrons at the magnetopause. Dr. Singer studied the angular distribution of 40 keV electrons in the magnetotail and observed significant anisotropies in the morning side; no obvious interpretation of this result is apparent, but it should tell us something about processes going on in the tail.

A major gap still remains in our knowledge of the particle spectrum at extremely low energies, in the few eV range. Attempts to observe these particles directly by means of instruments on satellites have been pretty unsuccessful; it is very difficult to cope with photoelectrons and other interference from the satellite. There have been many very important observations using whistlers, but these have necessarily been limited to the inner magnetosphere, out to L values of 5 to 6. It would be very useful to find a technique (perhaps also based on wave propagation) that would give some information on the extremely low energy particles at larger distances. These particles play no role in the energy balance of the outer magnetosphere and perhaps little in the particle density balance. However, they may be quite important in determining the behavior of plasma waves, as was pointed out by Drs. Kennel and Hasegawa. Also, the field aligned currents that are being postulated for various reasons will be carried, at least in part, by low energy particles coming up from the ionosphere.

Turning now to magnetospheric models, our present knowledge of the structure of the magnetosphere and its particle populations is largely derived from observations alone, and theoretical understanding in terms of first principles is still very sketchy. Beginning with the convection theory of Axford and Hines and the reconnection theory of Dungey eight years ago, a number of ideas have been discussed in qualitative fashion, suggesting how particles can be brought from the magnetotail into the inner magnetosphere and accelerated and thus predicting some of the gross features such as an asymmetric ring current and the DS current system. Only recently have there been attempts to construct simple mathematical models to test the ideas and see what the predictions really are. Such models of electric fields and currents were discussed at this meeting by Drs. Dessler and Vasyliunas (relating to convection) and by Dr. Akasofu (relating to substorms). I think it is fair to say that this work is still at a somewhat primitive stage. There does seem to be general agreement that electric fields in the ionosphere and the magnetosphere, together with field aligned currents, play a crucial role although the relation between electric fields in the two regions is still a controversial matter.

In the past, the ionospheric electric field could only be estimated indirectly from ground level magnetic disturbances. Now, with the Ba cloud technique the electric field can be measured directly and its relation to magnetic disturbances studied. Eugene Wescott presented results of Ba cloud releases in the auroral zone which, confirming the earlier work of Haerendel, Lüst and their group, seem to establish rather convincingly the existence of a real eastward electrojet and an associated electric field; thus, the basic substorm ionospheric current appears to consist of two cells rather than one as had been recently proposed. Dr. Wescott also reported a Ba cloud release over the polar cap. His results show that the simple picture of a magnetic disturbance produced by an overhead current related to the local electric field is not valid in the polar cap; field aligned currents certainly play an important role in determining the total observed ground level magnetic disturbance. Similar conclusions were reached by Haerendel from a Ba cloud release deep in the magnetosphere.

The question whether the magnetosphere is open or closed has been debated for years; ready access (or otherwise) of solar flare particles to the magnetotail or the polar cap has generally been considered as the crucial observational test. At this Institute no new theoretical arguments were brought forth for either point of view. A fair amount of new observational material was presented. As in the past, the new observations do not seem to fit either theory very well or else can be reconciled with both. Dr. Van Allen presented some very clean-cut data demonstrating the ready access of solar electrons to the magnetotail, obtained with simultaneous measurements from two satellites, one inside and one outside the magnetosphere. (He also brought out one point not emphasized before: in going from interplanetary space to the magnetotail these particles have not crossed any significant electric potential differences.) Other data on this point were much less clear. The solar proton observations, reported by Drs. Paulikas, Stone, Burrows and others do not indicate such a ready access, and there are a number of features such as structure in the polar cap and enhancements at the lower latitude edge of the polar plateau that are not readily explained by either theory. The observations do indicate that a close coupling between the magnetosphere and interplanetary space exists; the simple old picture of the magnetospheric boundary as an impermeable membrane is not valid. The exact nature of the coupling, whether it is direct field line connection, a viscous drag layer, or perhaps something quite different is not known. Understanding the physics of the magnetospheric boundary appears to be essential for the interpretation of solar particle observations, as well as for magnetospheric physics in general; processes at the magnetospheric boundary are believed to be ultimately responsible for magnetospheric convection and may play a major role in the formation of the plasma sheet.

4. Plasma Instabilities*

During the conference there were many remarks on the observational evidence for plasma instabilities and several papers were entirely devoted to such instabilities. In the latter classification were contributions by Drs. Kennel, Cornwall, and Hasegawa. Kennel gave a survey of what he thinks are the dominant processes for instabilities, beginning with the whistler mode instabilities for which he is well known. Although there is much evidence from hiss and chorus events that these instabilities are operating in the magnetosphere, it is quite apparent that the whistlers are not the full story. For example we heard from Dr. Fredricks about the electrostatic waves which must also be created in instabilities. One remark that was quite interesting, I think, is that in the present state of the art it is not possible to account for the chorus events. One really has to consider nonlinear effects, and this point I think is really a challenge to all the theorists working with these events.

For those who are dealing with electric fields, a factor of utmost interest is the anomalous resistivity effect. In the discussion of field aligned currents and strong precipitation Dr. Kennel presented the work of one of his students on the limiting precipitating fluxes as a function of altitude. This result was quite interesting because one could see that the limiting fluxes were comparable to those seen in auroras. There appears to be a strong tendency for the ionosphere to create an increased resistivity. Thus, we can have parallel electric fields and can expect electrostatic acceleration of particles. Hopefully, we will be able to find the link between the particle energy spectra in aurora and the spectra in the magnetotail. In the tail or in the plasma sheet there is no clear indication of rather narrow peaks in the energy distribution, but in the aurora this is a common phenomenon. It may well be that these energy peaks are generated somewhere between the plasma sheet and the aurora and perhaps only at altitudes of a few thousand, or perhaps tens of thousand, kilometers.

The presentation of Dr. Cornwall was mainly a discussion of the philosophy of the coupling of two instabilities. The instabilities considered are the drift instability, which arises from gradients in either the particle density or temperature, and the

* Presented by G. Haerendel.

instability involved in pitch angle scattering. The very interesting suggestion was made that the drift instabilities, which lead to rather sharp radial diffusion, may couple with the pitch angle scattering process. In this coupling the pitch angle scattering acts rapidly and by destroying the large scale features has a stabilizing effect on the drift instabilities. At present the theory is not developed sufficiently to compare with experiment, but Cornwall showed how the pitch angle diffusion coefficient entered as a stabilizing factor in the equation describing the growth of the drift wave. If this coupling really is operating, it should cause rather stable radial diffusion if it is applied as Cornwall suggested to the inward diffusion of trapped particles. If it applies at all, this stabilizing effect of the pitch angle distribution would also stabilize the rate at which particles were thrust inward toward the earth and would lead to a rather stable situation overall. I don't know how much observational evidence we have, but it will be an interesting question to test the observations in this respect.

One of the possible candidates for the promotion of these drift instabilities may be the plasmapause, since we saw in the presentation of Dr. Chappell that the density gradients can be tremendous. I think this boundary is quite likely to be unstable, and it would be valuable to see if there is evidence of instabilities in that region.

Finally, Dr. Hasegawa, who restricted his presentation to hydromagnetic waves, described characteristics of the mirror instability. He related this instability to some observations obtained by Brown with Explorer 26. The observed wave in both the particle flux and magnetic field intensity is the so-called nonoscillatory wave. Really, it is just a local instability and does not propagate although it appeared in the satellite record as a wave. In this talk Dr. Hasegawa made it quite clear that this appearance is due to the gradient B drift of those particles that carry the inhomogeneities past the satellite. The process is somewhat similar to the striations that are seen in the Ba clouds. These striations, which are clearly due to instabilities, do not seem to propagate. If they move at all it is just due to that part of the electric field that perhaps does not create the instability but carries the Ba ions in some horizontal direction. I need not go into further details of this mirror instability as we all know from plasma physics that particles enhance the mirror configuration by concentrating in the low B field, and this condition can be unstable and subject to further growth.

In the discussion following Hasegawa's paper we had a very general definition of an instability. I do not recall the words exactly, but in spirit the definition was that an instability is "something that grows." Similarly I would like to have a definition of a wave. In the light of the nonoscillatory wave studied by Dr. Hasegawa, perhaps we need a more general definition of a wave that will include this case.

CONCLUSIONS

BILLY M. McCORMAC
Lockheed Palo Alto Research Laboratory, Palo Alto, Calif., U.S.A.

The purpose of the Institute was to ensure that the experimental and theoretical results were presented and adequately discussed and no attempt was made to reach agreements or determine popular opinion. Each individual must evaluate the importance and credibility of this information for himself. These personal conclusions about the Institute are presented below from the point of view of future research requirements. The Institute did much to identify important areas of agreement and disagreement and focused attention on multidisciplinary research and key problems of the future. Most of the participants concluded that another two week Institute on this subject should be held in 1971. The persons mentioned below attended the Institute and either have chapters in this book or made their contributions during the discussion. An important point emphasized a number of times during the Institute was that multiple satellites are essential in order to separate spatial and temporal variations in all but a few aspects of the inner radiation belt.

1. Magnetospheric Models

The presentations and discussions of the various aspects of the magnetospheric models emphasized the physical phenomena in a very excellent manner. Although models change with time, they help to identify observational and theoretical problems which may advance our state of knowledge. The various models seem to be slowly converging as a result of an increase in data.

A. MAGNETOTAIL AND ENTRY OF THE SOLAR WIND PARTICLES

The question of open vs. closed magnetosphere is of fundamental importance. Particle observations will probably provide the answer and the moon as an absorbing shield will help. The data seem to favor an extended magnetotail resulting from connection of interplanetary field lines, the magnetotail being drawn back into the plasma. The large scale plasma motion energy is obtained from the solar wind via inward motion of annihilated magnetic field lines in the neutral sheet. There are two surfaces of field line reversal: the neutral sheet and magnetopause. There are three different plasma regions: trapped radiation, ring current, and plasma sheet.

It is now accepted that some merging of the magnetic field occurs at the magnetopause and neutral sheet. The question is, "How does merging behave and how important is it?". According to Dessler, the region of tail merging has not been experimentally identified. Also, the merging mechanism may be different in the neutral sheet from that at the magnetopause.

B. M. McCormac (ed.), Particles and Fields in the Magnetosphere.

The importance of Birkeland (field aligned) currents also is raised. Birkeland currents are difficult to observe from the earth's surface, especially in the vicinity of the auroral electrojet. There seemed to be general agreement that there were Birkeland currents.

The length of the magnetotail will depend on the sensing technique, but the entry of solar flare particles will define the effective length. There seems to be no concern about field line mixing, but the ability of electrons and protons to make the turn where the interplanetary and magnetospheric field lines are connected is important. Electrons seem to have an efficiency near to 100% while protons are much less than this. Alpha particles may even be less efficient, thus helping explain the low α-particle to proton ratio in the magnetosphere compared to the solar wind.

According to Anderson, 50 keV solar flare electrons appear in the tail with little delay and enter the tail within 5000 R_E. Protons of 1 to 10 MeV show some delay and enter within a few 100 R_E. The electron and proton data differ on tail length. The ability to turn the corner may affect the length for each. It apparently still has not been proven that protons have access by direct magnetic field connection. Measurements show that the plasma sheet field lines are closed. At the magnetopause there is no evidence of chaotic mixing which might allow ready access of solar wind particles. There does not seem to be an explanation of how to supply electrons for islands, although Vette suggests that the energization occurs within the plasma sheet. Anderson also believes that the particle observations eliminate the importance of ac and dc electric fields in particle access processes.

A detailed study of the energetic particles in the tail was made by Van Allen. An investigation of the shadowing by the moon shows that the same electrons are in the magnetotail and the solar wind from impulsive events with very short time delay. These data are interpreted to show that there is an open magnetosphere, no closed electric potential surfaces within 64 R_E, solar electrons enter within 64 and 900 R_E, and that the motion of the interplanetary field past the earth drives plasma convection and polar electrical currents. Diffusion across a closed tail with a ΔE across the tail would have changed the kinetic energy of the electrons in an observable fashion. Electron shadowing by the moon shows that there is no significant diffusion perpendicular to the magnetic field at $R_E < 64$. Diffusion or not beyond 64 R_E is a critical question. Better particle measurements are needed to define the flow of particles and means of entering the magnetotail. PCA observations, pitch angle distributions, and vector measurements of the magnetic field show little or no promise.

Both solar protons and electrons are observed over the polar cap and neither are believed to be accelerated by an electric field (Rothwell). Solar electrons seem to have a very high efficiency for access to the polar cap, perhaps 100% (Paulikas).

B. PLASMA SHEET BEHAVIOR IN A M SUBSTORM

In a *M* substorm the plasma sheet first thins and then expands. Electron precipitation occurs on the auroral oval as electrons disappear at 17 R_E (Hones). Then as electron precipitation decreases, the electrons reappear at 17 R_E. An interpretation is that the plasma is squeezed earthward and the sheet thins, then thickens as plasma is forced back into the tail. Some of the plasma precipitates and some reinflates the tail. The tubes of force where energization occurs do not extend out to 17 R_E during the substorm.

In the plasma sheet the energy density is greatest near to the neutral sheet. Most of the kinetic energy is in protons. At a substorm most of the change in energy density is due to an increase in particle density. The plasma sheet pressure responds within minutes to changes in dynamic pressure of the solar wind. The tail is kept inflated to 50% of the pressure of the solar wind.

Principal auroral precipitation occurs on dipole-like field lines that carry Birkeland currents. These currents may originate at the inner edge of the plasma sheet.

C. M SUBSTORM CURRENT SYSTEM

A single current system model during a *M* substorm has been proposed by Akasofu. This current system consists of an auroral electrojet, field aligned currents, and an asymmetric ring current in the day-evening sector. The field aligned currents are outward in the evening and inward in the morning; thus, electrons are accelerated downward in the evening and in the morning protons are accelerated upward.

D. MAGNETOSPHERIC CONVECTION

Convection with a region of enhanced pressure in a narrow longitudinal sector leads to a field aligned current flowing into the ionosphere at its western edge and out of the magnetosphere at its eastern edge.

The origin of the driving *E* field and of the plasma sheet must be known but are not. Properties of plasma sheet particles lead to an interpretation that the plasma sheet is located within the closed magnetosphere. Electric fields associated with open field lines depend on the details of the merging processes and are unknown.

A two cell equivalent current system which agrees with observations in the plasmasphere and shows the main features of the *DS* current system has been proposed by Vasyliunas. There is an *E* field across the polar cap. There are field aligned currents confined to the polar cap boundary toward earth on the dawn side and away from the dusk side.

E. SOURCE OF TRAPPED RADIATION AND AURORAL IONS

Although the solar wind is generally assumed to be the primary source of trapped radiation and auroral particles, Axford believes that it has not been experimentally proven and that the ionosphere may be the source. We still can not follow the morphology of auroral particles. If the field lines close on the polar wind particles, they could become a part of the plasma sheet. Even though most of the polar wind particles do escape the magnetosphere, the remainder could provide a source of belt and auroral particles. Good ion specie observations are needed to resolve the source of radiation belt and primary auroral ions, and experimentally it will not be easy.

F. SUMMARY

As a result of detailed discussions after presentation of the various models, the following can be concluded:

(1) There is disagreement on the termination of the inner edge of the plasma sheet. Vasyliunas believes that it is terminated by pitch angle scattering while Dessler suggests gradient drift.

(2) It is agreed that the inner edge of the plasma sheet is on closed field lines and that all auroras except polar cap aurora occur on closed field lines. This seems to demolish the Dungey-Speiser model.

(3) It was agreed that the location of merging in the neutral sheet is very uncertain. There is no agreement as to whether merging is important or not. No one could agree on a good test for merging.

(4) There was no agreement as to how aurora are produced. While merging may help furnish energy, acceleration of electrons may occur at much closer distances than where merging could occur.

(5) Particle measurements are needed to resolve the source of auroral particles, the openness of the magnetosphere, and the effective length of the magnetotail.

(6) Measurements need to resolve whether there are one or two current cells during *M* substorms and convection. Field aligned currents require determination but are difficult to measure.

(7) Whether diffusion beyond 60 R_E is important or not needs to be determined.

2. The Bow Shock

The collisionless shock produced by the interaction between the solar wind and the earth's magnetosphere is the bow shock.

A. THE SOLAR WIND

As K_p increases, the solar wind velocity, density, and proton temperature all increase; but the parameter that correlates best is the proton temperature (Bame).

The solar wind also contains flare associated shocks which have a transit time of 50 to 70 hr (Hundhausen). These solar wind shocks have a velocity of about 90 km sec^{-1} for an observer moving with the solar wind and seem to be 'driven'-type shocks. The bow shock is stronger than the solar wind shocks. There is a problem relating these solar wind shocks to geomagnetic phenomena. These transit times are longer than the times derived from flare-SC correlation studies. There are a number of hydromagnetic shock phenomena observed in the solar wind (Ogilvie) which are yet to be explained.

In the solar wind upstream transverse shock waves are a permanent feature (Fairfield). They have a velocity of 2.5 times the solar wind velocity and a period of 20 to 60 sec. They are believed to be waves generated by protons going back upstream.

B. BOW SHOCK STRUCTURE

Electrostatic ion wave turbulence is believed to be the important dissipation mechanism in the bow shock. This requires that electron preheating occurs in the leading edge so that $Te \gg Ti$. Observations indicate that Te increases in 300 m (1 sec) while the Ti increases over 5 to 10 sec (Montgomery). At the bow shock there is a rapid rise in the plasma thermal energy. The bow shock location seems to correlate with the dynamic solar wind pressure. Aggson's observations show that there are no large E fields near to the shock.

3. Magnetospheric Particles

Much more data on the pitch angle distributions, energy spectrum, and flux of particles are required throughout the magnetosphere, especially the lower energy particles. It is difficult at this time to emphasize key points or make any conclusions about these particles.

4. Electric and Magnetic Fields

Since many theoretical predictions and interpretations involve electric fields, it is essential that they be measured throughout the magnetosphere.

A. DC MAGNETIC FIELDS

The magnetotail field lines are slightly divergent with distance from the earth and deviate from the earth-sun line due to field spread and aberration. The neutral line is generally beyond the moon. The magnetotail seems to be structured at $\sim 500\ R_E$ and thus between 100 and 500 R_E ceases to be a well ordered entity (Behannon).

The vertical component of the magnetic field in the tail has not been measured. It is estimated to be only a few per cent of the total field so that its measurement will be difficult.

The changes in magnetic field strength observed during a substorm are difficult to explain. There are spatial discontinuities wherein traversals of $\beta < 1$ to $\beta > 1$ boundaries are observed. The high β regions are attributed to high proton kinetic energy.

B. AC MAGNETIC FIELDS

Magnetic field fluctuations are difficult to explain. Pulsating auroras may be related. A good pattern does not exist between magnetic field effects at the surface and at synchronous altitudes. There may be a variation in the relative influence of ring currents and magnetotail currents (Coleman).

Magnetic waves are observed in the magnetosphere with periods from hundreds of sec to frequencies of kHz (Russell). The lowest frequency waves appear to be field line resonances, possibly driven by the motions of the magnetospheric boundary. The highest frequency waves include auroral or VLF hiss, whistlers, and chorus. The intermediate frequencies are difficult to interpret.

C. DC ELECTRIC FIELDS

There is no simple proportionality between $E_\perp$ and ΔH, thus the ionospheric conductivity must play an important role during the substorm. Surface magnetospheric disturbances can be used to predict the direction of $E_\perp$ but not the magnitude. Much more information about the magnetospheric current systems is needed.

Observations of Ba releases in the auroral electrojet region are in agreement with the two cell current configuration (Lüst; Wescott). The electric field is not uniform in space and time, and is of low value inside auroral forms.

Barium releases in the polar cap region show that the drift vectors were generally aligned toward the sun which are similar to polar cap auroras. An appreciable component of the polar cap disturbance is not due to horizontal ionospheric currents.

The high latitude E field is irregular. The lower latitude boundary of the irregularity is highly variable. No variation in the E field is seen at the plasmapause boundary (Gurnett).

D. AC ELECTRIC FIELDS

Occasionally weak E fields are seen at the plasmapause. There are strong ac E fields at latitudes poleward of the inner edge of the auroral oval. It has not been determined if there are electromagnetic or electrostatic waves and no correlation has been found yet with particles (Maynard).

Some long period (15 to 20 sec) E field oscillations have been found to be conjugate at 70 to 80° INL suggesting convective processes on closed field lines (Gurnett).

5. Wave-Particle Interactions

Much more data concerning wave-particle interactions are required before much progress is to be made in understanding the phenomena.

Both electrostatic and electromagnetic ac waves play a role in wave-particle interactions. Electrostatic waves seem to be important in providing collisionless shock dissipation. In the solar wind, magnetosheath, and bow shock the most intense signals are generally the low frequency noise bursts detected in close association with steep magnetic field compressions. It is believed that the oscillations are electrostatic ion sound waves generated by two stream instabilities. Electromagnetic waves are presumably radiated by superthermal electron streams in the solar wind (Scarf).

Both electrostatic and electromagnetic emissions are found in the magnetosphere, especially at the equator on auroral oval magnetic field lines.

High frequency electrostatic waves are observed in deep space but are difficult to observe on the earth's surface. Several instabilities can result in the generation of electrostatic waves by electrons, but much better observations are required to select which are important (Kennel). In fact, two instabilities may be competing for the plasma free energy making the interpretation of results difficult (Cornwall). Electrostatic waves are believed to be very important in precipitating particles.

Hydromagnetic instability waves have been observed. The most feasible types are the mirror instability and the current driven transverse instability. The mirror instability has been used to explain hydromagnetic waves observed during a storm. The current driven transverse, which is driven by a current parallel to B requires additional study (Hasegawa).

6. Radiation Belt Observations

Most of the particle population data describe the proton and electron distributions. Certainly much more α-particle and low energy proton data are needed. A very excellent summary of the data to-date is provided by Vette. The identification of regions and boundaries often depends on the magnetic activities, particle type and energy. For example, during some days the plasma sheet boundary moves into the stable trapping region. For both protons and electrons, each storm exhibits different characteristics and is very complex.

A. PROTONS IN THE RING CURRENTS

A major area of interest is the 0.5 to 50 keV proton population, which is the largest single charged particle energy reservoir in the magnetosphere. The source of these protons is not understood.

There is a quiet time ring current at $L \simeq 6.5$ with average proton energies of 5 to 20 keV (Frank). There are no evidences of local acceleration and no good correlations with any solar wind parameters. During a storm the ring current penetrates deep into the outer radiation belt ($L \sim 3$ to 4) producing the world-wide decrease in H. The motion of this ring current is not understood but seems quite important.

Frank finds keV protons increase in interplanetary space hours before storms. It is suggested that the early development phase of a storm results from high asymmetric injection into the evening sector of the outer belt. Convection of only 2% of the energetic protons incident on the sunlit magnetosphere can produce the storm time ring current. The data indicate that there is no acceleration and diffusion of these keV protons in the magnetotail. More detailed measurements of the spectra of several to tens of keV protons are needed in interplanetary space magnetosheath and magnetosphere. It should be determined if there is a difference in access of solar protons to the auroral oval vs the polar cap. The loss of the ring current protons in the outer belt is attributed to charge exchange.

B. ELECTRONS

Substorm electron injection occurs near local midnight and is followed by gradient drift (Winckler). The delay at locations from midnight can be accounted for by drifting bunches of 90 keV electrons injected in the outer zone near midnight during a polar substorm. The injection into the Van Allen belt could be an impulsive substorm process accelerating keV electrons.

The injection at midnight seems to occur simultaneously over a large ΔL. The drift shells are modulated. The magnetosphere expands in the evening sector with a weakening of the magnetic field and an inward motion of energetic particle drift shells. At midnight there is a collapse at the magnetosphere with an increase in the magnetic field and an outward motion of drift shells.

C. INNER ZONE PROTONS

The inner belt proton flux at energies >30 MeV does not increase as atmospheric density decreases with diminishing solar activity and is not compatible with a CRAND source (Elliot).

Efforts to explain the source of the Starfish high energy protons have still been unsuccessful (White).

Elliot presented evidence for a catastrophic loss of protons at the inner edge of the trapping region which may be more important than ionization loss at the edge of the ionosphere.

D. ALPHA PARTICLES

Injection of α-particles from the solar wind with a velocity equal to the protons followed by thermalization in the magnetosheath can not account for the observations (Krimigis). Much more data are needed before the data on α-particle to proton ratio, etc., can be fit to a solar wind source, explain diffusion across field lines, and explain loss mechanisms. Diffusion does not seem to fit the observations. Investigations considering that the α-particles have equal energy to the protons look even less attractive than the equal velocity case (Blake). The suitability of using α-particles to trace magnetospheric phenomena remains to be proven.

7. Acceleration and Motion of Particles

Radial diffusion plays a complex role in radiation belt dynamics. Diffusion velocity is difficult to define and measure. Diffusion coefficient computations are difficult because of a lack of knowledge of loss mechanisms. There is also the problem of the sudden appearance of particles deep into the belt. The complicated temporal and spatial behavior of the electron population does not allow the radiation belt dynamics to be described in terms of a simple balance between radial diffusion from a source at the boundary competing with internal losses. Diffusion has not been directly verified experimentally.

There may be simultaneous radial and pitch angle diffusion (Walt). There is much scatter in the experimentally determined diffusion coefficients. The experimental data are inadequate as the energy spectrum, pitch angle distribution, and radial variation of flux must be detailed as a function of time.

During storms the trapped protons are redistributed; small storms perturb in to $L \sim 5$ and large storms to $L \sim 3$. The redistribution seems to be produced by ring currents (Williams). First there is an adiabatic redistribution which is directly proportional to ΔH. Then, during the main phase of the storm there is a rapid nonadiabatic change where the low energy proton intensities are enhanced and high energies depleted. The cause of these rapid nonadiabatic variations as a source of outer zone protons as compared to inward diffusion of a high latitude low energy population remains to be determined. Finally, there is a slow nonadiabatic recovery as both low and high energy protons recover toward prestorm levels.

APPENDIX I: GLOSSARY

Most of the abbreviations utilized in this book are obvious. Several whose meaning may not be apparent are listed below:

AA	Auroral Absorption
AU	Astronomical Unit
CNA	Cosmic Noise Absorption
FD	Forbush Decrease
FP	Fabry-Pérot
FWHM	Full Width Half Maximum
IN	Invariant
INL	Invariant Latitude
INLT	Invariant Local Time
IR	Infrared
LT	Local Time
MHD	Magnetohydrodynamics
MLT	Magnetic Local Time
PCA	Polar Cap Absorption
PEJ	Polar Electrojet
SC	Sudden Commencement
SI	Sudden Impulse
UT	Universal Time
UV	Ultraviolet

INDEX OF SUBJECTS

Alpha particles 53–56, 88–91, 99, 141, 317, 432, 449
– trapped 364–378, 380–384
Auroral electrojet 35–42, 229–235, 439
Auroral particles 22, 43–57
– electrons 24–33, 42–47, 309–310
– protons 42–55, 309–310
Auroral zone
– electric field 214, 229–238
– magnetic field 26–31

Barium ion clouds 40
– ionospheric 213–219, 229–238
– magnetospheric 219–228
Bow shock 5, 80, 83, 95–100, 102–104, 176, 436–437, 445

Charged particles. *See* Alpha particles, electrons, and protons
Chorus 195–196, 206–211, 279
Collisionless shock 102–107, 275
Conductivity, ionospheric 64–70, 213–234
Convection, magnetospheric 18, 24, 50, 60–70, 213, 228, 239, 444
Currents
– auroral 34–45
– Birkeland 18, 22
– field aligned 18, 22, 41, 61–70, 193, 227
– ionosphere 61–70, 213, 216, 238, 439
– magnetospheric 61–70
– neutral sheet 166, 170
– ring 35–45, 47, 165–167, 180–182, 192, 319–330, 448

Diffusion 6–9, 13, 16, 113, 118–121, 387–394, 410–415
– pitch angle 6–7, 13, 412–427
– radial 387–394, 404–408, 410–427, 449

Electric field
– auroral 214, 229–238
– ac 9, 16, 102–107, 247–253, 275–283, 447
– conjugate 244–245
– dc 9, 239–245, 252–253, 446–447
– electromagnetic 276–282
– electrostatic 102, 276, 282–283
– ionospheric 61–70, 213–228, 239–245, 247–253
– magnetosheath 275–281
– magnetospheric 62–70, 111–121, 213, 228, 247, 275, 282
– polar cap 229–238, 242–243, 248–249
– solar wind 275–278
Electrojet 35–42, 229–235, 439
Electromagnetic waves 434–435
Electrons
– auroral 24–33, 42–47, 309–310
– bow shock 95–100
– diffusion 387–394, 410–415
– magnetopause 122–127
– magnetosheath 122–131
– magnetosphere 6–16, 22, 27–28, 62, 111–121, 437–438
– magnetotail 6–16, 29, 111–131, 307
– – diffusion 113, 118–121
– neutral sheet 122, 129
– plasma sheet 24–33, 122–131, 305–309
– polar cap 132–147
– solar wind 6–16, 24, 29–33, 47, 76, 98–100, 111–121, 141–147
– trapped 132–140, 306–317, 332–351, 432–433, 448
Electrostatic instabilities 266–274, 440–441
Electrostatic turbulence 95, 100, 102, 436
Electrostatic waves 257–264, 434–435

Helium ions 48–57, 148–153
Hiss 195–196, 207–211
Hydromagnetic
– instabilities 266–274, 284–290, 440–441
– shock 83–93
– wave 19, 198, 284–290

Interplanetary magnetic field 4–16, 19–21, 30, 77
Interplanetary shock 31–32, 75–77, 103
Instabilities
– electrostatic 266–279, 440–441
– hydromagnetic 266–274, 284–290, 440–441
Ionosphere
– conductivity 64–70, 213–234
– currents 61–70, 213, 216, 238, 439
– electric fields 61–70, 213–228, 239–245, 247–253
– ions 46–49
– magnetic field 213–228

Magnetic bay 24–28, 31–32

Magnetic field
- auroral 26–31
- fluctuations 102–107, 171–193, 195–211, 446
- interplanetary 4–16, 19–21, 30
- ionosphere 213–228
- magnetosheath 165–170
- magnetosphere 165–170, 213–228
- magnetotail 157–164
- merging 4, 18–22, 24, 64
Magnetic shock 31–32
Magnetic substorms 24–35, 48, 171–193
Magnetopause 4–16, 18–22, 24, 49–51, 102, 114–115
- electrons 122–127
Magnetosheath 10, 30, 75, 83, 96, 103–105
- electric field 275–281
- electrons 122–131
- magnetic field 165–170
- protons 324–326
Magnetosphere 3–16, 18–22
- alpha particles 53–55
- convection 18, 24, 50, 60–70, 213, 228, 239, 444
- currents 61–70
- electric fields 62–70, 111–121, 213, 228, 247, 275, 282
- electrons 6–16, 22, 27–28, 62, 111–121, 437–438
- electrostatic waves 257–264
- hydromagnetic waves 284–290
- magnetic field 165–170, 213–228
- magnetotail 3–16, 18–22, 24–33, 48–49, 61–62, 111–121
- - electrons 6–16, 29, 111–131, 307
- - magnetic field 157–164
- - protons 11–16, 307, 323–325
- neutral sheet 3–5, 18–22, 25–29, 122, 129, 166, 170
- oxygen ions 48, 148–153
- pitch angle diffusion 6–7, 13, 412–427
- plasmapause 148–153, 243–244
- plasma 18–19, 24–33
- plasma sheet 10–11, 22, 24–33, 48–49, 122–131, 157, 305–309, 319–323, 443–444
- protons 7–16, 22, 62, 437–438

Plasma sheet 10–11, 22, 24–33, 48–49, 157, 443–444
- electrons 24–33, 122–131, 305–309
- protons 28–33, 49, 306, 319–323
Plasmasphere 206–209
- ions 148–153
Polar cap
- electric field 229–238, 242–243, 248–249
- electrons 132–147
- protons 132–147
Polar wind 46–51

Protons
- auroral 42–55, 309–310
- diffusion 404–408, 420–427
- magnetosheath 324–326
- magnetosphere 7–16, 22, 62, 437–438
- magnetotail 11–16, 307, 323–325
- neutron albedo decay 46
- plasma sheet 28–33, 49, 306, 319–323
- plasmasphere 148–153
- polar cap 132–147
- precipitation 47–49, 55, 357–362
- solar wind 7–16, 22, 24, 28–33, 47–53, 75–80, 82–92, 96–100, 105–106, 141–147, 326–330
- trapped 47, 54, 132–140, 309–317, 319–323, 353–357, 364–378, 380–384, 396–408, 432–434, 449

Ring current 35–45, 47, 165–167, 180–182, 192, 319–330, 448

Satellites
- Ariel 1: 353–357
- Alouette: 308–309, 350
- ATS 1: 40, 55, 171–193, 198–202, 311–313, 332, 340–351
- ERS 13: 314–315
- ERS 17: 308, 312–315
- ESRO 2: 353–359
- Explorer 4: 353–357
- Explorer 6: 198, 313–315
- Explorer 12: 397, 313–315
- Explorer 14: 157, 312–315, 388, 390
- Explorer 26: 312, 388, 396–398
- Explorer 33: 111–121, 157–159, 164, 396
- Explorer 34: 83, 88–89, 161
- Explorer 35: 111–121, 157–161
- HEOS 1: 219–224
- IMP 1: 18, 157, 312–314
- Injun 3: 305, 388
- Injun 4: 365–369, 374–375, 396
- Injun 5: 205, 239–245, 364–376
- OGO 1: 102, 197–198, 205, 280, 301, 305, 312
- OGO 2: 198, 216
- OGO 3: 102, 197–198, 203–205, 208–209, 305, 309, 322, 332, 340–343, 347
- OGO 4: 198, 204, 366–367, 375, 396
- OGO 5: 103–107, 198, 202–206, 257–264, 275–280
- OGO 6: 247–253
- OV1-10: 239, 247–253
- OV3-3: 276–277
- VELA 3: 24–33, 76–78, 176
- VELA 4: 24–33, 75–76, 79, 95–100, 122–131
Solar wind
- access to magnetosphere 111–121, 442–443
- alpha particles 54–56, 88–91, 99, 141

- bulk velocity 75–80, 82–92, 96–100, 102–105, 162
- direction 75–78
- electric field 275–278
- electrons 6–16, 24, 29–33, 47, 76, 98–100, 111–121, 141–147
- filaments 13–15
- ion 124–127
- ion temperature 32, 75–80, 82–92, 96–100
- magnetospheric interactions 3–16, 75–78, 176–177
- pressure 3, 19–21, 24, 31–32, 75–78
- protons 7–16, 22, 24, 28–33, 47–53, 75–80, 82–92, 96–100, 105–106, 141–147, 326–330
- shock waves 79–81

Sudden commencement 32, 81, 84–86
Sudden impulse 84–86
Spacecraft
- Mariner 5: 90–92, 201
- Mariner 7: 157, 162–164
- Mariner 8: 103–104, 157, 163–164, 275–276, 280
- Mariner 9: 104, 275

Trapped radiation. *See* alpha particles, electrons, and protons

VLF emissions 105–106, 195–196, 205–211, 292–301, 435

Wave-particle interactions 275–284, 296–301, 418–420, 447
Whistlers 102, 107, 195–196, 205–206, 210, 292–293

X-rays 27–28

ASTROPHYSICS AND SPACE SCIENCE LIBRARY

Edited by

J. E. Blamont, R. L. F. Boyd, L. Goldberg, C. de Jager, Z. Kopal, G. H. Ludwig, R. Lüst, B. M. McCormac, H. E. Newell, L. I. Sedov, Z. Švestka, and W. de Graaff

1. C. de Jager (ed.), *The Solar Spectrum. Proceedings of the Symposium held at the University of Utrecht, 26–31 August, 1963.* 1965, XIV + 417 pp. Dfl. 50.—
2. J. Ortner and H. Maseland (eds.), *Introduction to Solar Terrestrial Relations. Proceedings of the Summer School in Space Physics held in Alpbach, Austria, July 15–August 10, 1963 and Organized by the European Preparatory Commission for Space Research.* 1965, IX + 506 pp. Dfl. 65.—
3. C. C. Chang and S. S. Huang (eds.), *Proceedings of the Plasma Space Science Symposium, Held at the Catholic University of America, Washington, D.C., June 11–14, 1963.* 1965, IX + 377 pp. Dfl. 68.—
4. Zdeněk Kopal, *An Introduction to the Study of the Moon.* 1966, XII + 464 pp. Dfl. 72.—
5. Billy M. McCormac (ed.), *Radiation Trapped in the Earth's Magnetic Field. Proceedings of the Advanced Study Institute, Held at the Chr. Michelsen Institute, Bergen, Norway, August 16–September 3, 1965.* 1966, XII + 901 pp. Dfl. 130.—
6. A. B. Underhill, *The Early Type Stars.* 1966, XIII + 282 pp. Dfl. 56.—
7. Jean Kovalevsky, *Introduction to Celestial Mechanics.* 1967, VIII + 427 pp. Dfl. 30.—
8. Zdeněk Kopal and Constantine L. Goudas (eds.), *Measure of the Moon. Proceedings of the Second International Conference on Selenodesy and Lunar Topography held in the University of Manchester, England, May 30–June 4, 1966.* 1967, XVIII + 479 pp. Dfl. 90.—
9. J. G. Emming (ed.), *Electromagnetic Radiation in Space. Proceedings of the Third ESRO Summer School in Space Physics, held in Alpbach, Austria, from 19 July to 13 August, 1965.* 1968, VIII + 307 pp. Dfl. 58.—
10. R. L. Carovillano, John F. McClay, and Henry R. Radoski (eds.), *Physics of the Magnetosphere. Based upon the Proceedings of the Conference held at Boston College, June 19–28, 1967.* 1968, X + 686 pp. Dfl. 130.—
11. Syun-Ichi Akasofu, *Polar and Magnetospheric Substorms.* 1968, XVIII + 280 pp. Dfl. 55.—
12. Peter M. Millman (ed.), *Meteorite Research. Proceedings of a Symposium on Meteorite Research held in Vienna, Austria, 7–13 August, 1968.* 1969, XV + 941 pp. Dfl. 160.—
13. Margherita Hack (ed.), *Mass Loss from Stars. Proceedings of the Second Trieste Colloquium on Astrophysics, 12–17 September, 1968.* 1969, XII + 345 pp. Dfl. 65.—
14. N. D'Angelo (ed.), *Low-Frequency Waves and Irregularities in the Ionosphere. Proceedings of the 2nd ESRIN-ESLAB Symposium, held in Frascati, Italy, 23–27 September, 1968.* 1969, VII + 218 pp. Dfl. 43.—
15. G. A. Partel (ed.), *Space Engineering. Proceedings of the Second International Conference on Space Engineering, held at the Fondazione Giorgio Cini, Isola di San Giorgio, Venice, Italy, May 7–10, 1969.* 1970, XI + 728 pp. Dfl. 140.—
16. S. Fred Singer, *Manned Laboratories in Space. Second International Orbital Laboratory Symposium.* 1969, XIII + 133 pp. Dfl. 30.—

SOLE DISTRIBUTORS FOR U.S.A. AND CANADA:

Vols. 2–6, and 8: Gordon & Breach Inc., 150 Fifth Ave., New York, N.Y. 10011
Vols. 7 and 9ff.: Springer Verlag New York, Inc., 175 Fifth Ave., New York, N.Y. 10011